SEMICONDUCTOR MICRODEVICES AND MATERIALS

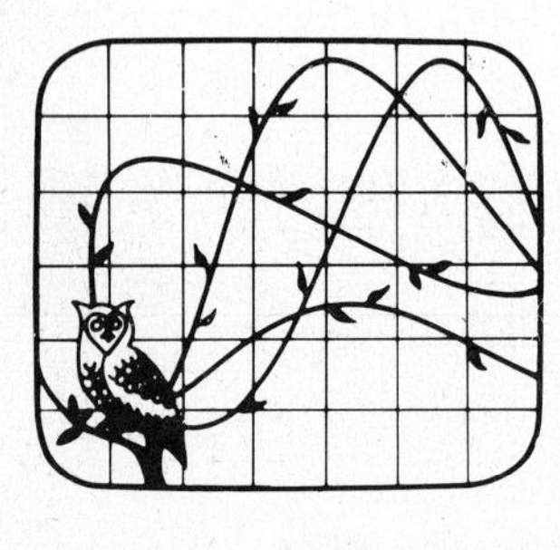

HRW
Series in
Electrical and
Computer Engineering

M. E. Van Valkenburg, Series Editor Electrical Engineering
Michael R. Lightner, Series Editor Computer Engineering

P. R. Bélanger, E. L. Adler, and N. C. Rumin INTRODUCTION TO CIRCUITS WITH ELECTRONICS: AN INTEGRATED APPROACH
L. S. Bobrow ELEMENTARY LINEAR CIRCUIT ANALYSIS
L. S. Bobrow FUNDAMENTALS OF ELECTRICAL ENGINEERING
C. T. Chen LINEAR SYSTEM THEORY AND DESIGN
D. J. Comer DIGITAL LOGIC AND STATE MACHINE DESIGN
D. J. Comer MICROPROCESSOR BASED SYSTEM DESIGN
C. H. Durney, L. D. Harris, and C. L. Alley ELECTRIC CIRCUITS: THEORY AND ENGINEERING APPLICATIONS
M. S. Ghausi ELECTRONIC DEVICES AND CIRCUITS: DISCRETE AND INTEGRATED
G. H. Hostetter, C. J. Savant, Jr., and R. T. Stefani DESIGN OF FEEDBACK CONTROL SYSTEMS
S. Karni and W. J. Byatt MATHEMATICAL METHODS IN CONTINUOUS AND DISCRETE SYSTEMS
B. C. Kuo DIGITAL CONTROL SYSTEMS
B. P. Lathi MODERN DIGITAL AND ANALOG COMMUNICATION SYSTEMS
C. D. McGillem and G. R. Cooper CONTINUOUS AND DISCRETE SIGNAL AND SYSTEM ANALYSIS Second Edition
D. H. Navon SEMICONDUCTOR MICRODEVICES AND MATERIALS
A. Papoulis CIRCUITS AND SYSTEMS: A MODERN APPROACH
S. E. Schwarz and W. G. Oldham ELECTRICAL ENGINEERING: AN INTRODUCTION
A. S. Sedra and K. C. Smith MICROELECTRONIC CIRCUITS
N. K. Sinha CONTROL SYSTEMS
M. E. Van Valkenburg ANALOG FILTER DESIGN
W. A. Wolovich ROBOTICS: BASIC ANALYSIS AND DESIGN

SEMICONDUCTOR MICRODEVICES AND MATERIALS

David H. Navon
UNIVERSITY OF MASSACHUSETTS
AMHERST, MASSACHUSETTS

HOLT, RINEHART AND WINSTON
New York Chicago San Francisco Philadelphia
Montreal Toronto London Sydney Tokyo
Mexico City Rio de Janeiro Madrid

To
my wife, Bobbe,
my daughter, Beth,
and my son, Marc,
and to the memory
of my mother, Anna,
who made all this possible

Acquisitions Editor: Deborah L. Moore
Production Manager: Paul Nardi
Project Editor: Bob Hilbert
Interior Design: Rita Naughton
Illustrations: Scientific Illustrators

Address correspondence to:
383 Madison Avenue, New York, NY 10017

Library of Congress Cataloging-in-Publication Data

Navon, David H.
Semiconductor microdevices and materials.

Includes index.
1. Semiconductors. I. Title.
TK7871.85.N394 1985 621.3815′2 85-16445

ISBN 0-03-063983-2

Printed in the United States of America
Published simultaneously in Canada

6 7 8 038 9 8 7 6 5 4 3 2 1

CBS College Publishing
Holt, Rinehart and Winston
The Dryden Press
Saunders College Publishing

Contents

Preface

Goals and Methods

The objective of this book is to provide the reader with a reasonably comprehensive introduction to the wonderland of microchip integrated circuits. To achieve this goal it is necessary to provide some understanding of the semiconductor materials used to fabricate these chips, the physics of the microdevices employed, the integrated circuits utilized, and the technology used in monolithic microcircuit manufacture. The intent is not only to provide an understanding of present-day microchip integrated circuits, but also to present sufficient device and materials physics for the reader to be able to comprehend any future developments in this rapidly moving field.

Audience Addressed

The prerequisites for a good comprehension of the subject matter presented in this text are basic first-year undergraduate courses in college mathematics and physics. A prior study of classical physics and the mathematics of calculus is essential; an introduction to modern physics is useful, as well as some knowledge of differential equations. The practicing engineer new to the field of monolithic integrated circuits should also find this book helpful.

Contents

The treatise begins at the beginning. After a short description of the history of the development of the microchip (Chapter 1), an introduction to the crystallography of the solid state is presented, including the structure of semiconductor crystal materials (Chapter 2). Then the energy band theory of crystals is developed (Chapter 3), followed by an introduction to the quantum theory of semiconductor materials (Chapter 4). Next, the electrical carrier transport mechanisms in uniform semiconductor materials are described (Chapter 5). This is followed by a development of the theory of the electrical properties of the basic *p-n* junction (Chapter 6). Next, some applications of *p-n* junction diodes are described as well as the high frequency and fast switching behavior of these devices (Chapter 7). The physics of the bipolar junction transistor is then developed (Chapter 8), followed by a discussion of the operation of these devices in circuits and at high frequencies (Chapter 9). The physics of operation of unipolar field-effect transistors is next described, including the high frequency and the fast switching performance of these devices (Chapter 10). Junction FETs, MESFETs, MOSFETs, and HEMT devices are also considered.

After completing this introdution to semiconductor materials and the basic devices utilized in the design of integrated circuits, the process physics and methods of fabrication of microchips are described. Modifications of the form of these semiconductor devices needed to adapt them for the monolithic technology are discussed (Chapter 11). Next, the versions of digital and analog circuits that are popular in their integrated form are presented. Limitations on microminiaturization are described, as well as the computer-aided design of microchips. Finally, the physics of other semiconductor electronic devices used primarily in high frequency and/or high power applications is developed (Chapter 12). The physical operation of Gunn-effect devices, IMPATT diodes, the semiconductor laser, the charge-coupled device (CCD), and the family of power thyristors is described.

A summary of the major subjects treated is provided at the end of each chapter. Also there, is a set of problems, each of which is designed to illustrate the major points made in the chapter.

Course Format

The organization of this book provides several possibilities for course presentations. The first ten chapters will provide material for a one-semester undergraduate course on fundamental semiconductor devices and materials. The sections marked with asterisks can be omitted without affecting the continuity of later material. This course should follow the standard two-semester sequence in electronics.

The entire text can be covered in two quarters for those schools operating on the quarter system. For those schools which provide their students with a strong introduction to semiconductor devices as a part of their electronics sequence, Chapters 8 through 12 will provide a detailed introduction to the design of microdevices and microcircuits.

The practicing engineer with a prior knowledge of semiconductor devices will find Chapters 11 and 12 useful in understanding recent developments in integrated-circuit design and the operation of the newer semiconductor devices. The first-year graduate student may want to read the entire text in preparation for advanced study of semiconductor devices or computer engineering.

Acknowledgments

The author would like to thank the Motorola Corporation, the Digital Equipment Corporation, the Monsanto Corporation, Texas Instruments, Inc., and Analog Devices Semiconductor, Inc., for supplying some of the photographs and device specification sheets used in this book. He would also like to acknowledge the assistance of the following professors in reviewing the original manuscript and for making many recommendations: Peter Blakey of the University of Michigan, George Prans of Manhattan College, Murray Gersherzon of the Unviersity of Southern California, Richard Williams of the University of Akron, Lester Eastman of Cornell University, Anab Kulkarni of the Michigan Technological Institute, Carlos Araujo of the University of Colorado, Steve Schwarz of the University of California at Berkeley, Archie McCurdy of the Worcester Polytechnic Institute, and M. E. Van Valkenburg of the University of Illinois. In addition, the expert help provided by Mrs. Marge Olanyk in preparing this manuscript is gratefully acknowledged.

David H. Navon
Amherst, Massachusetts

The first elementary integrated circuit fabricated by Jack Kilby of Texas Instruments Incorporated which produced a complete circuit function.

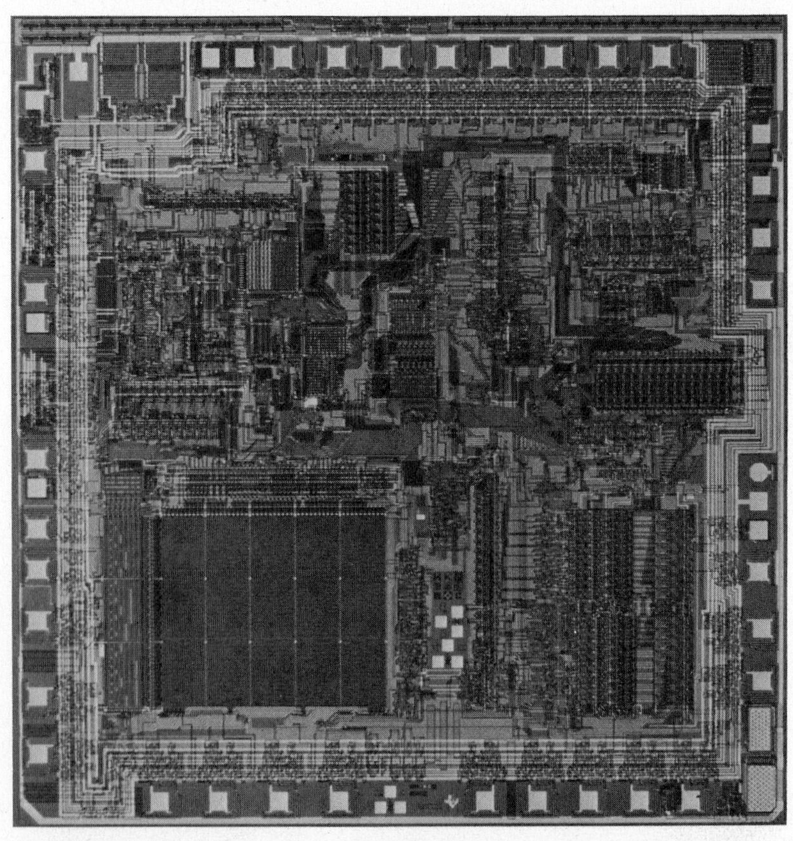

Photograph of a microprocessor or ''computer on a chip'' produced in 1984. (Courtesy of Texas Instruments Incorporated, Dallas, TX.)

CHAPTER 1

Introduction to Semiconductor Devices

1.1 INTRODUCTION

The invention of the transistor in 1948 served as a beginning for the revolution in electronics which was to take place in this second half of the twentieth century. This device uses a semiconductor crystal as its basic starting material. The microchip, the fundamental building block of modern-day computers, contains a large number of transistor devices and employs typically the semiconductors silicon or gallium arsenide as substrate materials. It is the purpose of this book to supply sufficient understanding of the physics of semiconductor crystal materials that an adequate description of the operation of semiconductor devices can be presented. The physics of the internal behavior of the semiconductor devices that are employed in the design and fabrication of microchip integrated circuits as well as the construction of optoelectronic devices and circuits is the primary subject of this book.

The material presented herein is intended to provide the microchip designer with a fundamental understanding of the operation of the variety of semiconductor devices which are used in the design of microchip integrated circuits. It is possible to design these electronic circuits by knowing only the terminal characteristics of devices such as transistors. That is, a circuit model can be constructed for the device using only the electrical characteristics measured at the three device terminals, at a fixed electrical bias point. However, measurements would have to be made at a variety of different bias points and frequencies to provide for a generally applicable circuit model. It is useful to understand the physics of the device in order to extrapolate circuit behavior to another bias point or frequency. In fact it is necessary to have a knowledge of device physics so that a circuit model can be constructed for the device which will be useful over a broad range of operating conditions, and which can be utilized in the computer-aided circuit analysis of newly designed microchip circuits. In addition, the circuit designer should have some insight into the physics of device operation in order to be better qualified to take full advantage of the device's multifold uses.

The microchip device designer of course should have an even more profound knowledge of semiconductor material physics as well as device physics, not only for the design of the devices of today, but to anticipate and invent the devices of tomorrow. A vast majority of today's solid-state devices such as switching transistors for digital circuits, communication lasers, light-emitting display diodes, optical detectors, and Gunn microwave oscillators use semiconductor crystals as the basic starting material. Hence an introduction to the physics of semiconductor materials and devices is essential for an understanding of modern electronics.

1.2 TRANSISTOR DEVICES AND MICROCHIP INTEGRATED CIRCUITS

The invention of the transistor in 1948 by Shockley, Brattain, and Bardeen caused a revolution in electronics which was to take place in less than ten years. Up to that time the vacuum tube diode and triode were the most used electronic devices. Although the transistor inventors were all physicists, their unusual insight into the requirements for a good amplifying device was essential in exploiting this physical discovery. At the time of the discovery the point-contact (whisker) semiconductor diode had already been developed. Now an active device[1] was sought using a surface field-effect principle. In fact, it was the basic study of semiconductor surfaces which was being pursued by these physicists at the time of their epoch-making invention. This dramatic development is well documented[2] and pinpoints how an understanding of semiconductor material behavior can lead to new device development. In fact, the later invention of the bipolar junction transistor of Shockley, which superseded the

[1]An active electronic device is one which provides signal amplification.

[2]For example, see "Special Report: The Transistor," *Electronics*, **41**, 77 (1968).

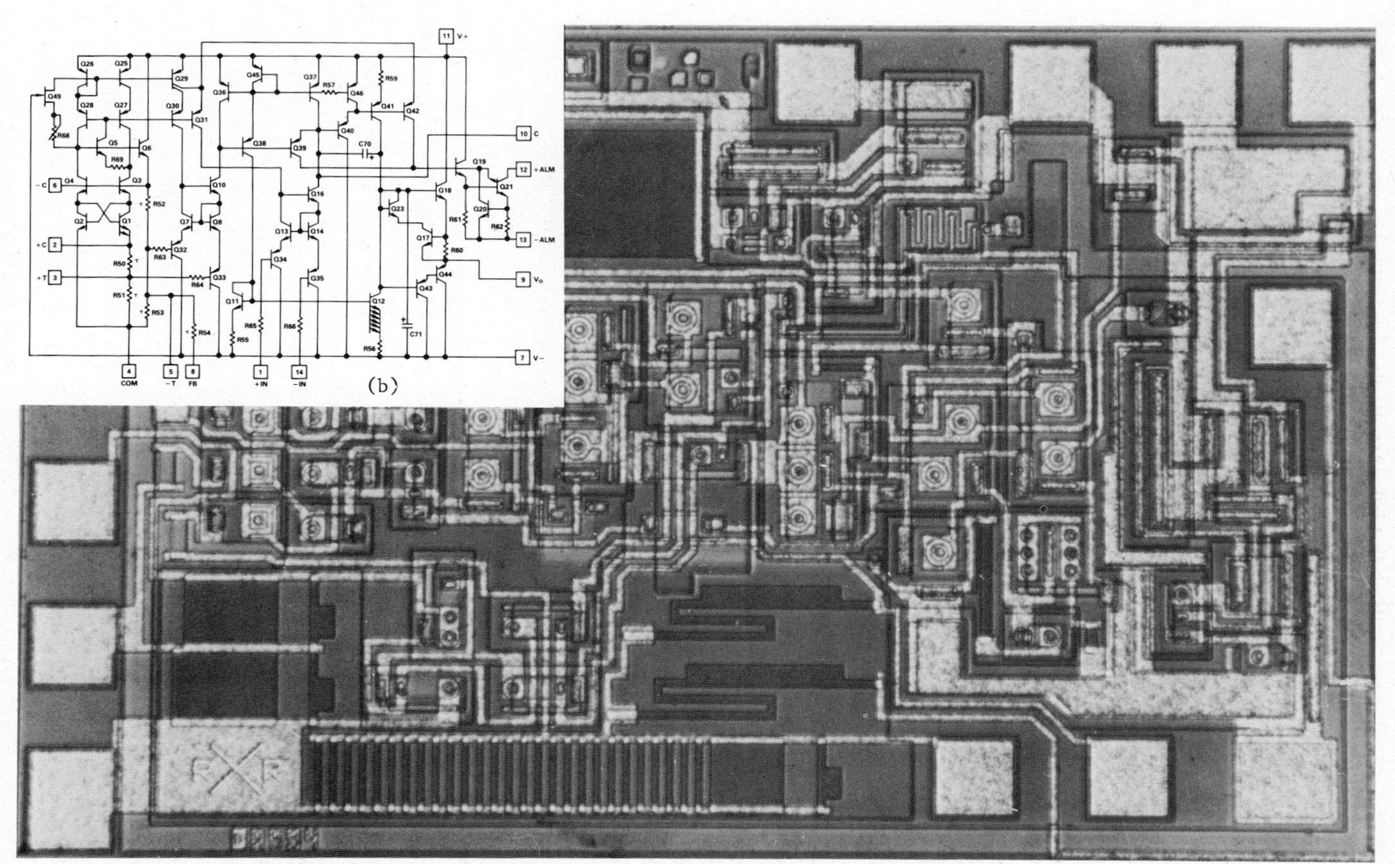

Figure 1.1 (a) Actual photograph of a silicon digital integrated circuit showing the various planar transistors, diodes, and resistors. The large white islands on the periphery are metallized contacts for external connection. (b) A schematic circuit diagram of the circuit in (a).

point-contact device, was another triumph of keen physical insight into the properties of semiconductor materials. This discovery subsequently gave rise to the development of the most important electronic circuit building block of today—the *microchip integrated circuit*—which derives from transistor devices and performs a complete electronic circuit or computer function.

1.3 THE MICROCHIP INTEGRATED CIRCUIT

Although the bipolar or field-effect transistor is indeed the basic building block of computers, other circuit elements such as diodes, resistors, and capacitors are also needed. In fact, a typical information storage circuit commonly used, the "flip-flop"[3] can contain perhaps 20 or 30 circuit elements such as transistors, diodes, resistors, and capacitors. In addition a variety of digital circuit elements can be combined on a single chip of semiconductor in order to perform complex computational functions. A device which performs a *complete* circuit function is known as a *monolithic*[4] integrated circuit or microchip. The concept of this integrated circuit as "one piece of solid into which have been included several components, passive as well as active, without external connection between these devices" was proposed by both J. S. Kilby of the Texas Instruments Corporation and R. H. Noyce of the Fairchild Semiconductor Corporation in 1959.[5] Figure 1.1 shows a schematic diagram and a microphotograph of such a circuit. Some of these devices have as many as 120 terminals or leads compared to only 3 for a transistor. Recently silicon chips less than 1 cm^2 in area containing 500,000 or more devices have been fabricated. The technology permits the combining of arithmetic, logical, and even memory functions on a single-crystal semiconductor block constituting a complete "microcomputer on a chip." One such chip contains a microprocessor.

The introduction of integrated circuits about the year 1959 was a natural outgrowth of a development which had only begun ten years earlier. This element has now become the basic building block of electronic computers, communication modems, etc. The circuit designer of the future must be able to choose appropriate microchip integrated-circuit blocks and interconnect them so as to create a complete electronic system. Perhaps he will require the fabrication of integrated circuits which are not normally available. He must then communicate to the device fabricator his circuit needs. This communication link will be much more efficient if the circuit designer has some detailed knowledge of the physics and technology of integrated circuits. Of course the device

[3]This element is maintained in a particular state indefinitely even after removal of input signal. Hence it exhibits a "memory" function.

[4]Monolithic refers to the fact that all of the electrical action takes place on a block of semiconductor material. silicon, gallium arsenide, or another substrate material.

[5]A court decision on a patent interference case between these corporations awarded a priority on four of the claims to Kilby (Texas Instruments) and two claims to Noyce (Fairchild Semiconductor). See footnote 2 and also J. S. Kilby, "Semiconductor Solid Circuits," *Electronics,* **32,** 110 (1959).

designer now must have knowledge of electronic circuit design as well as the physics and technology of integrated circuits. The many aspects of this subject will be outlined in the chapters of this book.

□ SUMMARY

A brief introduction was presented on the events which led to the development of the modern-day integrated-circuit microchip. The extent of the knowledge of semiconductor device physics required by microchip device designers and by microcircuit designers was discussed.

CHAPTER 2

Semiconductor Crystal Structure

2.1 CRYSTAL STRUCTURE OF SOLIDS

The vast majority of electronic devices today are fabricated from single-crystal solid-state materials. Some common starting materials used in the manufacture of microchip and optoelectronic devices are single crystals of silicon, gallium arsenide, aluminum gallium arsenide (a mixture of gallium arsenide and aluminum arsenide), and gallium arsenide indium phosphide (a mixture of indium phosphide and gallium arsenide). All of these materials are classified as electronic semiconductors and have similarities in the regular arrangement of atoms in their solid, single-crystal structures. Silicon of course is one of the basic elements, while the other materials are termed **compound semiconductors.** This chapter defines the form of the solid known as the single crystal and explains how the valence form of the atoms and their geometric arrangement in the crystal give rise to certain electrical properties of the solid of interest in connection with electronic devices.

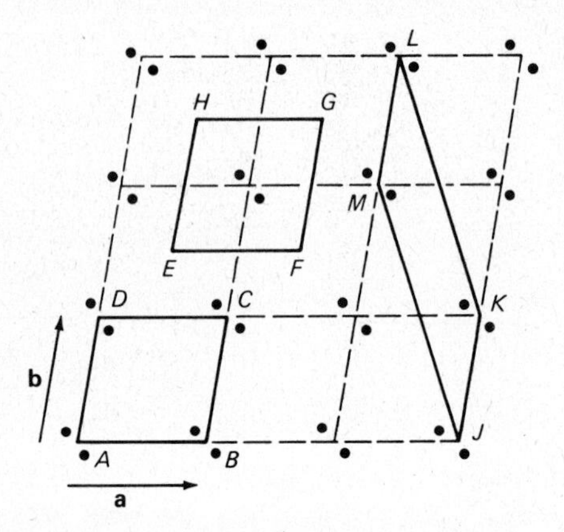

Figure 2.1 Two-dimensional representation of a single crystal. The lattice structure is indicated by the dotted lines. There are two atoms (black dots) indicated per lattice site. Three possible unit cells are indicated by the parallelograms *ABCD, EFGH,* and *JKLM.* Each cell contains two atoms. Any of these basic blocks placed side-by-side will reconstruct the total crystal. The basic vectors for defining the lattice points are denoted by **a** and **b**. Every point in the lattice may be determined by translating from point to point by a vector **a** or **b**.

2.1.1 Crystallinity

Solids occur in the **crystalline** or **amorphous** state. The crystalline solid refers to the organization of about 10^{22} atoms per cubic centimeter arranged in three dimensions in a regular manner. This structure may be obtained by repeating in three dimensions an elementary arrangement of some atoms or building blocks called **unit cells.** This is shown in two dimensions (for simplicity) in Fig. 2.1. If the periodic arrangement occurs throughout the volume of a solid sample, this constitutes a **single crystal.** However, if the regular structure occurs only in portions of a solid and the different portions are aligned arbitrarily with respect to each other, the material is said to be **polycrystalline;** the individual regular portions are referred to as **crystallites** or **grains** and are separated from each other by **grain boundaries.** If the individual crystallites are reduced in extent to the point where they approach the size of a unit cell, periodicity is lost and the material is called **amorphous.**

2.1.2 The Single Crystal

A system of atoms in thermal equilibrium at a particular temperature tends toward a specific crystalline state if this is the lowest free energy configuration of the solid.[1] Of course, different types of arrangements represent different energies and hence a certain atomic configuration is favored over others for a

[1]The fact that a system in equilibrium tends toward a minimum free energy is one of the most important laws of physics. Symmetry arguments dictate that a regular array of atoms will have a lower total energy than a corresponding group of randomly arranged atoms. The free energy of a system represents the energy available to do work.

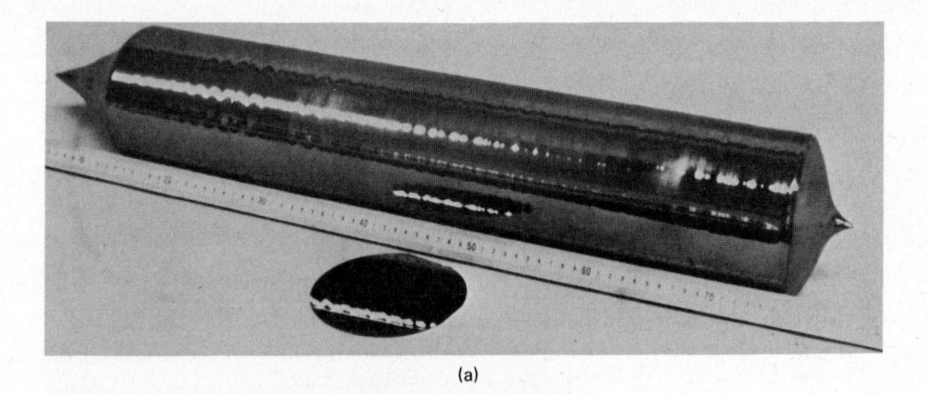

(a)

(b)

Figure 2.2 (a) A large single-crystal of silicon. (Photo supplied by the Monsanto Chemical Co.)
(b) A 6-in. diameter single-crystal of silicon just withdrawn from the melt. The single-crystal "seed" used to trigger single crystal formation is still shown attached to the top of the grown crystal. (Photo supplied by the Dynamit Nobel Silicon SpA, Merano, Italy.)

given material at a particular temperature. The most stable arrangement at any temperature depends in a complex manner on the forces between atoms, the space between them, the size of the atoms, etc. In common usage the words "crystal" and "gem" are sometimes used interchangeably. This has led to the idea that all single crystals must exhibit the natural geometric shape typical of gems. If all crystals were grown under truly equilibrium conditions without outside forces acting, this would tend to be true. However, good single crystals do not always have a completely regular outward appearance when grown under normal laboratory conditions. Figures 2.2a and 2.2b show photographs of a few single crystals of a semiconductor material prepared for use in device

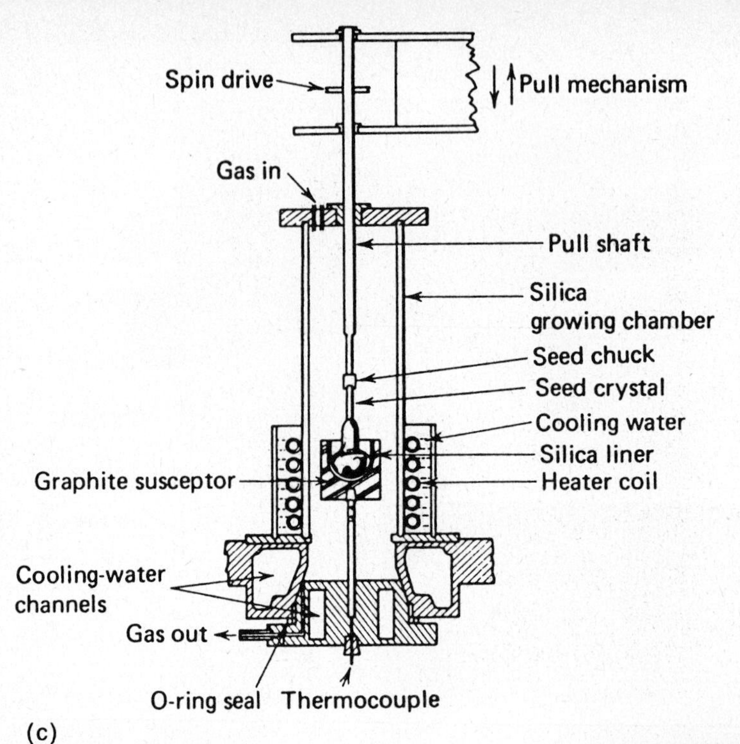

(c)

Figure 2.2 (Continued) (c) Typical crystal pulling apparatus. A crystal is grown by dipping a seed (small piece of single crystal) into the molten material and slowly withdrawing it so that the remainder freezes out on the seed. [From W. R. Runyon, *Silicon Semiconductor Technology* (McGraw-Hill, New York, 1965), Figs. 4-2, 4-21b, 4-18.]
(d) A photo of a commercial crystal-growing system. (Courtesy of Cybeq Systems, Menlo Park, CA.)

(d)

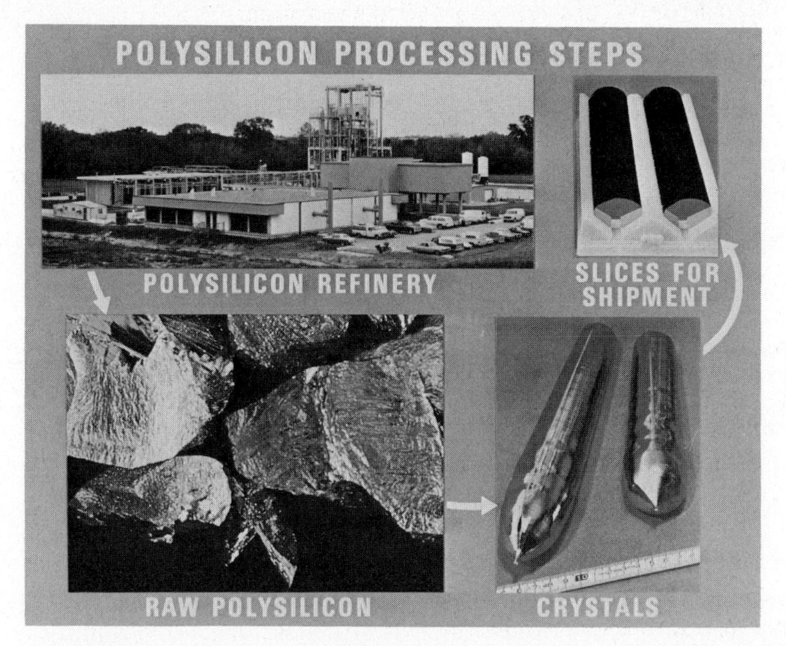

A manufacturing plant for producing pure silicon using $SiCl_4$ as a starting material. First, raw polycrystalline silicon is produced and then melted down to grow silicon single crystals. These are then sliced for use as substrates to produce semiconductor devices and integrated circuit chips. (Courtesy of the Texas Instruments Incorporated, Dallas, TX).

fabrication. Some tendency toward regular gemlike facets is observed, but perhaps not quite that expected of natural quartz or rochelle salt crystals. A sketch of a typical crystal pulling apparatus for "growing" these crystals vertically is shown in Fig. 2.2c. This gives rise to crystals which are grown from the liquid phase by dipping a seed (small piece of crystal) into the molten material and slowly withdrawing it so that the remainder freezes out on the seed. A photo of a commercially manufactured crystal-growing system is shown in Fig. 2.2d. This technique may also be used to grow crystals horizontally. Another method of single-crystal preparation extensively used in preparing the substrate material for fabricating microchips is called **epitaxial growth.** This is a technique for growing thin-film semiconductor layers from the vapor phase onto single-crystal, liquid-melt-grown substrates and will be discussed in Chapter 11.

Most solid-state devices employ single-crystalline substances as starting materials. The reason for this is that it is easier to control the properties and electrical behavior of these crystals whose properties are everywhere the same, compared to polycrystalline materials whose grains are of random size and in which foreign substances tend to collect along grain boundaries. This uniformity of the electrical properties of crystals is particularly important when microchips are fabricated from pieces of single-crystal semiconductors. Since *all* of the thousands of devices on the chip must uniformly operate as designed, the accurate control of the properties of the entire chip is essential. Also, the crystalline form makes analysis and characterization of the material easier. For

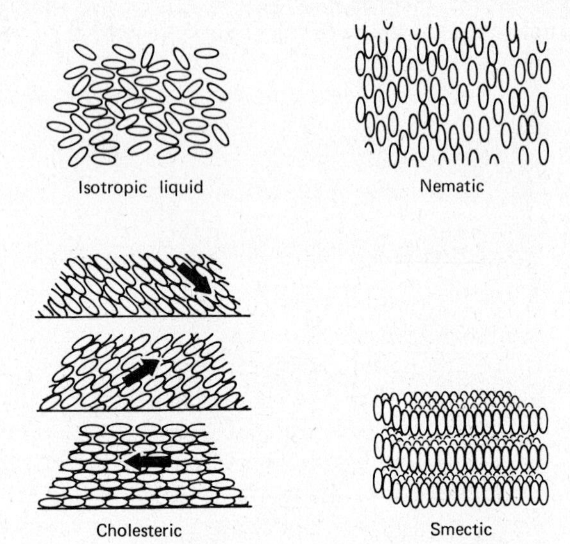

Figure 2.3 Schematic drawing indicating three types of liquid crystals having different molecular alignments. Also indicated is the random molecular orientation in these liquid crystals at high temperature when they become isotropic liquids. The nematic form is used in constructing electronically activated numeric readout display devices which however become inoperative at higher temperatures when the material becomes isotropic.

example, extraneous impurities are more easily identified and detected by using the electrical properties of crystals. Nevertheless, single crystallinity is not essential for solid-state devices and, for example, amorphous semiconductor materials are in limited-use device applications.[2]

2.1.3 Liquid Crystals

Although the topic of crystalline solids is of primary interest here, there is a class of **liquid crystals** which has been known for some time but which has only recently come into use for device applications. These comprise a class of materials that have only two- or one-dimensional regularity. The term "liquid crystal" refers to materials which pour like a liquid and assume the shape of their container, yet possess a degree of regularity in their molecular ordering. Their optical reflection and transmission properties indicate regularities unlike an ordinary isotropic[3] liquid. They occur generally as organic molecules, cigarlike in shape and organized or packed in "layered structures" as shown in Fig. 2.3. The significant interest in **nematic** liquid crystals generates from their use as a working material in the numeric readout displays such as are employed in hand calculators. This display's importance stems from the fact that it consumes very little battery energy.[4] Nematic liquid-crystal device operation depends on the material's property of transparency or opacity, depending on whether or not an electric field is applied to it. The cholesteric liquid crystal

[2]S. R. Ovshinsky, "Reversible Electrical Switching Phenomena in Disordered Structures," *Phys. Rev. Lett.*, **21**, 1450 (1968).

[3]One having no particular directional properties.

[4]For a survey of numeric readout devices, including the gallium arsenide phosphide light-emitting diodes, see "Special Report: Numeric Readout Displays," *Electronics*, **44**, 65, May 24, 1971.

changes the color of reflected light with heating, and hence can be used to sense temperature.

Another device which is extensively used for readout displays is the light-emitting diode (LED).[4] This is constructed of single-crystal semiconductor materials such as gallium arsenide phosphide. This device will be discussed in detail in a subsequent chapter.

2.2 SPACE LATTICES AND CRYSTAL STRUCTURE

A single crystal has already been defined as a regular repetition of a basic building block of atoms throughout the volume of a solid. The geometric shape of the building block is a three-dimensional parallelepiped and contains one atom in simple crystals like copper, silver, and sodium but may contain many thousands of atoms in complex organic protein crystals. The enormous variety of crystal structures can be defined by arranging atoms systematically about a regular or periodic arrangement of points in space called a **space lattice.** The lattice is defined by three fundamental or basic translation vectors **a**, **b**, and **c** so that arrangement of atoms in a crystal infinite in extent looks *identical* when observed from any point, displaced a distance r from an origin, as when viewed from the point r', where

$$\mathbf{r}' = \mathbf{r} + n_1\mathbf{a} + n_2\mathbf{b} + n_3\mathbf{c} \tag{2.1}$$

and n_1, n_2, and n_3 are integers. This is shown in a two-dimensional representation in Fig. 2.4. The set of points defined by the vector $\mathbf{r}'$ for all integer values of n_1, n_2, and n_3 is periodic in space and constitutes the space lattice.

2.2.1 The Primitive and Unit Cell

The volume defined by the vectors **a**, **b**, and **c** constitutes the **primitive cell** of the lattice if it is the *smallest* cell which when periodically stacked fills all of the crystal space. Since the angles of this volume are oblique in general, it is sometimes more convenient, but not always possible, to work with a unit cell which is defined by three axes at right angles to one another (orthogonal axes).

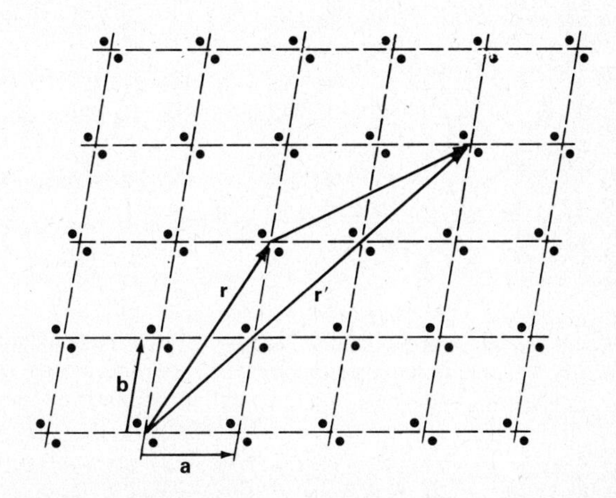

Figure 2.4 A two-dimensional representation of a space lattice is shown. The lattice points are at the intersection of the parallel lines. A cluster of two atoms is shown at each lattice point. The basic translation vectors **a** and **b** are indicated. The lattice looks identical when viewed from the point denoted by the vector **r** as from **r**′.

Periodic stacking of the unit cell will also map out the whole crystal lattice, but it is not always the smallest volume with which this task can be accomplished. The length of the edge of the unit cell is called the **lattice constant.**

2.2.2 Four Simple Crystal Structures

A. SIMPLE CUBIC (sc)

If single atoms or groups of atoms are placed at the corners of a simple cubic lattice, this constitutes a **simple cubic** crystal. The unit cell also happens to be a "primitive" cell because it contains only *one* lattice point and hence is the smallest volume which when repeated describes the whole crystal. A simple cubic structure has a single atom per corner site and hence contains one atom per primitive cell. The one atom per cell derives from the fact that each of the eight corner atoms contributes only one-eighth of an atom to each of the eight cells surrounding each lattice point. CsCl, TlBr, and NH_4Cl form crystals of the simple cubic type. The only element to crystallize in this form is Po.

B. BODY-CENTERED CUBIC (bcc)

The cubic cell of the body-centered crystal lattice not only contains a site in each corner of the cube but also one at the center of the volume, as shown in Fig. 2.5; hence the bcc cell contains two lattice points and is not primitive. Also shown in Fig. 2.5 is the bcc primitive cell which contains only one lattice site. Note that the angles of this structure are oblique and not identical. This accounts for the conventional usage of the cubic unit cell for describing the bcc crystal structure. Typical materials which crystallize in bcc form are Na, Rb, Cr, Ta, and W.

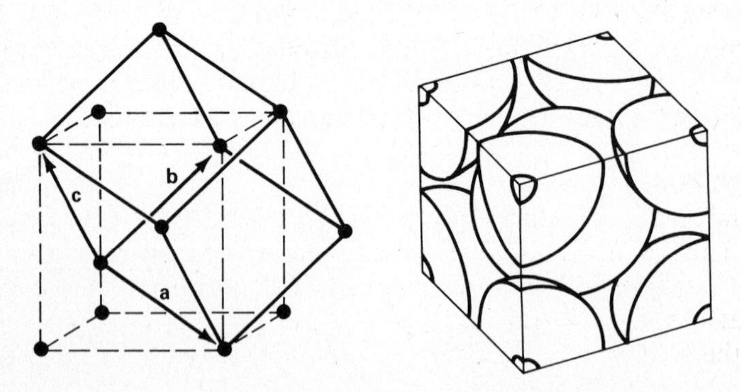

Figure 2.5 On the left the primitive (solid lines) and conventional (dashed lines) unit cells of the bcc lattice are sketched. Vectors **a**, **b**, and **c** constitute the basic vectors for this lattice. On the right the atoms in the bcc unit cell are represented. (Adapted from Fig. 4-6, p. 65, "Material Science for Engineers," by L. H. Van Vlack, Addison-Wesley Publishing Co., Reading, MA, 1970.)

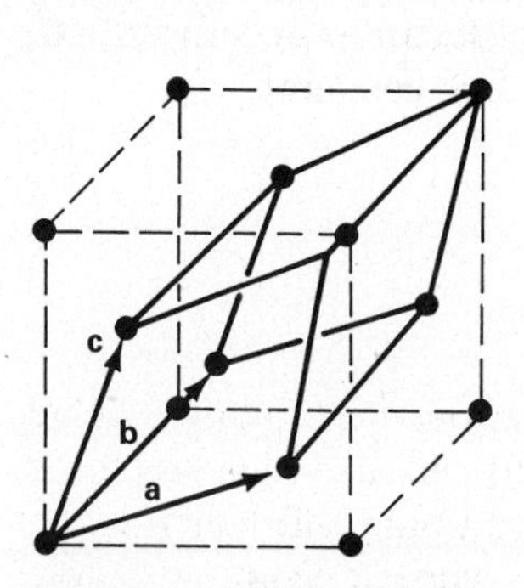

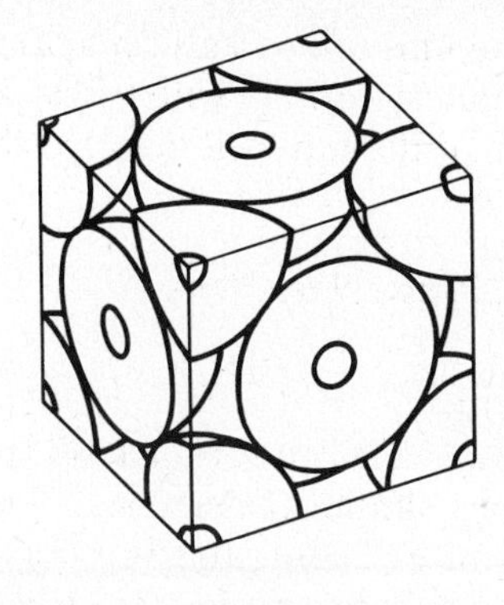

Figure 2.6 On the left the primitive (solid lines) and conventional (dashed lines) unit cells of the fcc lattice. Vectors **a**, **b**, and **c** constitute the basic vectors for this lattice. On the right the atoms in the fcc lattice are represented. (Adapted from Fig. 4-5, p. 63, "Material Science for Engineers," by L. H. Van Vlack, Addison-Wesley Publishing Co., Reading, MA, 1970.)

C. FACE-CENTERED CUBIC (fcc)

The unit cell of the face-centered lattice contains eight corner positions plus one site in each of the six cube faces as shown in Fig. 2.6. Since the face-centered sites are shared between two adjacent cells and the corner sites are shared by eight cells, there are four lattice points in the conventional face-centered cubic unit cell. Also, as in the body-centered case, one can define a primitive fcc cell, and this is also shown in Fig. 2.6. Again this primitive cell contains one site but has oblique angles and is more difficult to visualize than the conventional unit cell. Some materials which crystallize in the fcc form are Al, Ca, Ni, Cu, Pb, Ag, NaCl, LiH, PbS, and MgO.

Any atom in the fcc structure can be shown to be surrounded by 12 adjoining atoms or **nearest neighbors.** This is the maximum possible in an arrangement of solid spheres. In contrast, the bcc structure has only eight nearest neighbors and hence is more loosely packed. This is because the bcc atoms touch one another along the cube diagonal rather than along a cube edge.

D. DIAMOND STRUCTURE

The conventional unit cell of the diamond structure is basically cubic as pictured in Fig. 2.7a. This configuration is even more loosely packed than the previous structures discussed. Each atom has only four nearest neighbors forming a **tetrahedral** bond. This basic configuration may be constructed by placing an atom at the body center of a cube, two atoms at opposite corners of the top face of this cube, and two atoms at opposite corners of the bottom cube face, but twisted 90° with respect to the top-face atoms. This configuration is shown in Fig. 2.7b but it is *not* the unit cell. The tetrahedral bonds of the four corner atoms to the central atom are very strong and highly directional, occurring at angles of about 109.5°. The total diamond structure can be visualized

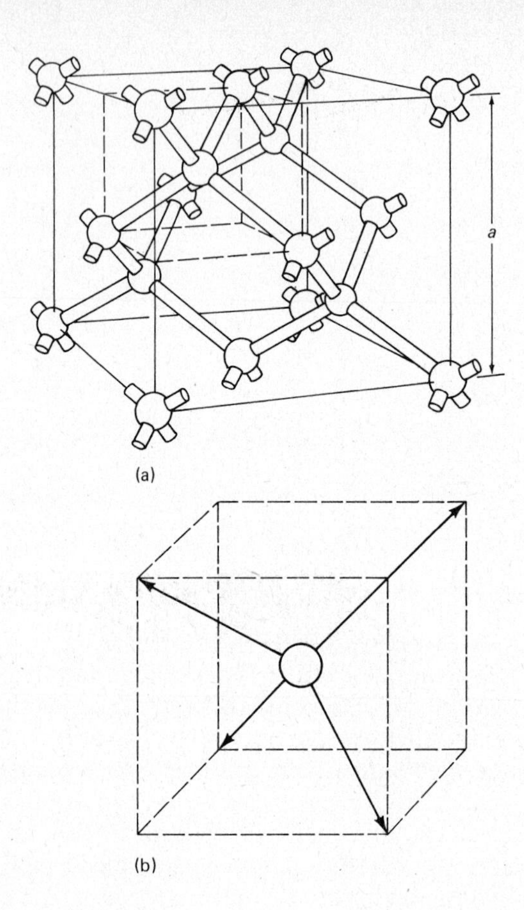

Figure 2.7 (a) Diamond crystal structure. For C, Si, and Ge: a = 3.56, 5.43, and 5.65 Å, respectively. Nearest-neighbor spacing is $a\sqrt{3}/4$, and occurs with tetragonal symmetry (dotted cube). The large cube (volume a^3) effectively contains eight atoms. [After W. Shockley, *Electrons and Holes in Semiconductors* (Van Nostrand, New York, 1950).]
(b) The tetrahedral bond joins the center atom to four nearest-neighbor atoms along the lines indicated by the arrows.

as two interpenetrating fcc lattices, one displaced from the other by one-fourth the length and along a cube diagonal.

This type of crystal structure is particularly of interest in the study of electronic materials since many important properties of the elemental semiconductors silicon and germanium derive from the basic tetrahedral structure. A closely related structure is the so-called **zincblende** crystalline configuration of the intermetallic semiconductor gallium arsenide, gallium arsenide phosphide, and cadmium sulfide. Here the center atom in the tetrahedral configuration is of atomic species A, while the four corner atoms are of species B (e.g., A is Ga, B is As). Silicon forms the tetrahedral covalent bond by joining together atoms, each with electron valence four; one atom shares an electron with each of the four corner atoms. Gallium arsenide consists of a compound containing the atom gallium, with three outer electrons, and arsenic, with five outer electrons. Here again they join together to form, on the average, a stable shell of eight electrons. This also applies to cadmium with two electrons and sulfur with six electrons to yield cadmium sulfide.

EXAMPLE 2.1

Silicon crystallizes in the diamond structure as shown in Fig. 2.7a. The dimension of the unit cell of this basically cubic structure is 5.43 Å (1 Å $= 10^{-8}$ cm). The atomic weight of silicon is 28.1. Find

(a) The nearest-neighbor distance between atoms (the bond length).
(b) The atomic radius of a silicon atom in this structure.
(c) The density of silicon using the above data.

Solution

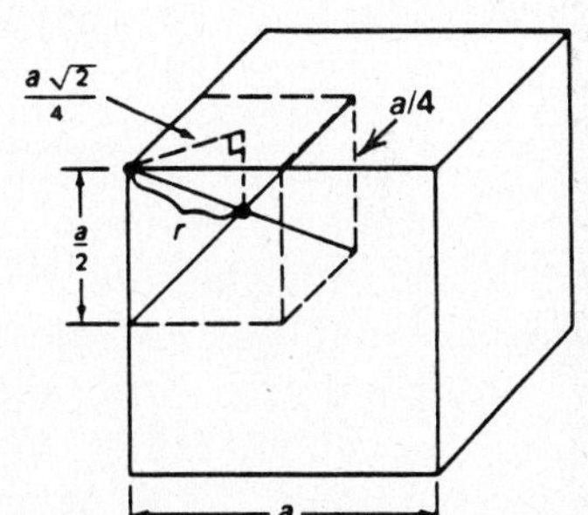

(a) From Fig. 2.7a:

$$r^2 = \left(\frac{a}{4}\right)^2 + \left(\frac{a\sqrt{2}}{4}\right)^2$$

$$r = \frac{a\sqrt{3}}{4} = \frac{(5.43)(1.73)}{4} = \underline{2.35 \text{ Å}}.$$

This is the closest approach of any two atoms in the lattice to each other.

(b) For the calculation of atomic radius, it is assumed that the atoms are stacked as rigid spheres. The atoms in closest proximity to each other are treated as spheres in contact. Hence the atomic radius $R = \frac{1}{2}r = \underline{1.17 \text{ Å}}$.

(c) Per unit cell, there are
$\frac{6}{2}$ atoms on the six cube faces (each atom is shared by two adjoining unit cells);
$\frac{8}{8}$ atoms on the eight corners (each atom is shared by eight adjoining unit cells);
four atoms contained entirely within the body of the unit cell. (only four of the eight small cubes have central atoms)

Hence the total atoms/unit cell $= \frac{6}{2} + \frac{8}{8} + 4 = 8$,

$$\frac{\text{mass}}{\text{unit cell}} = \frac{\text{no. atoms/unit cell}}{\text{no. atoms/gram-atom}} \frac{\text{grams}}{\text{gram-atom (or mole)}}$$

$$= \frac{8}{6.02 \times 10^{23}} \times 28.1 = 3.73 \times 10^{-22} \text{ g/unit cell}$$

so

$$\text{Density} = \frac{\text{mass/unit cell}}{\text{volume/unit cell}} = \frac{3.73 \times 10^{-22} \text{ g/unit cell}}{(5.43 \times 10^{-8})^3 \text{ cm}^3\text{/unit cell}}$$

$$= \underline{2.33 \text{ g/cm}^3} \text{ (experimental value} = 2.33 \text{ g/cm}^3\text{)}.$$

2.2.3 Miller Indices and Crystal Plane and Direction Identification

The geometric configuration of the atoms in the surface plane of a microchip of a single-crystal semiconductor material is characteristic of the crystal structure of that material. This arrangement of surface atoms can influence the electrical behavior of the semiconductor devices fabricated on these chips. Hence it is important to have a method to identify each specific planar atomic arrangement.

Parallel sets of planes passing through specific lattice points in a space lattice can be identified by a set of numbers called **Miller indices.** These numbers can be obtained by determining the intercepts of one of these planes on the three coordinate axes. For example, choose the origin of the coordinate system to coincide with a lattice site in one of these planes. Assume that the next parallel plane in this set has intercepts on the x-, y-, and z-axes equal to x_1, y_1, and z_1, correspondingly. Now take the *reciprocals* of these numbers and multiply each by a common integer so as to obtain three new numbers which are integers. Commonly the three smallest such integers are sought and these are referred to as Miller indices h, k, and l; they are normally written surrounded by parentheses $(h\ k\ l)$.

An example of this procedure of plane identification is shown in Fig. 2.8. The problem is to determine the Miller indices of a set of parallel planes, one of which passes through the origin of the coordinate system shown; the next parallel plane intersects the x-, y-, and z-axes at 6, 3, and 2 unit distances, respectively. The reciprocals of these numbers are $\frac{1}{6}$, $\frac{1}{3}$, and $\frac{1}{2}$, and multiplying

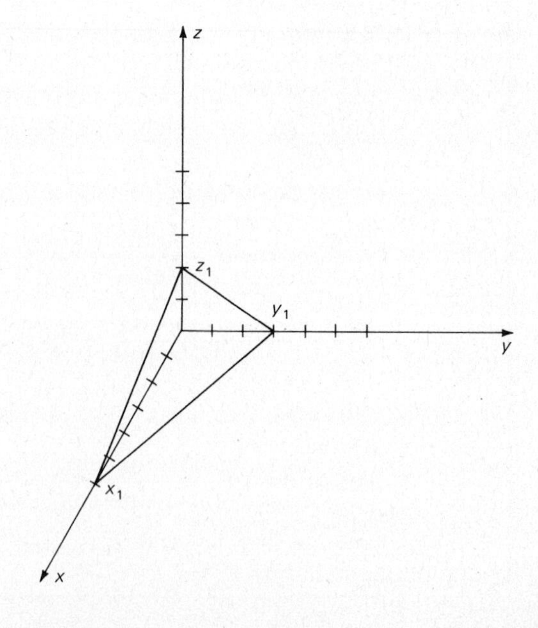

Figure 2.8 The plane shown is identified by the Miller indices (1 2 3). This plane intersects the x-axis at 6 unit distances, the y-axis at 3 units, and the z-axis at 2 units. The axes are orthogonal to one another and hence the lattice indicated is basically cubic. However, the Miller scheme is more general and may be applied to oblique axes as well.

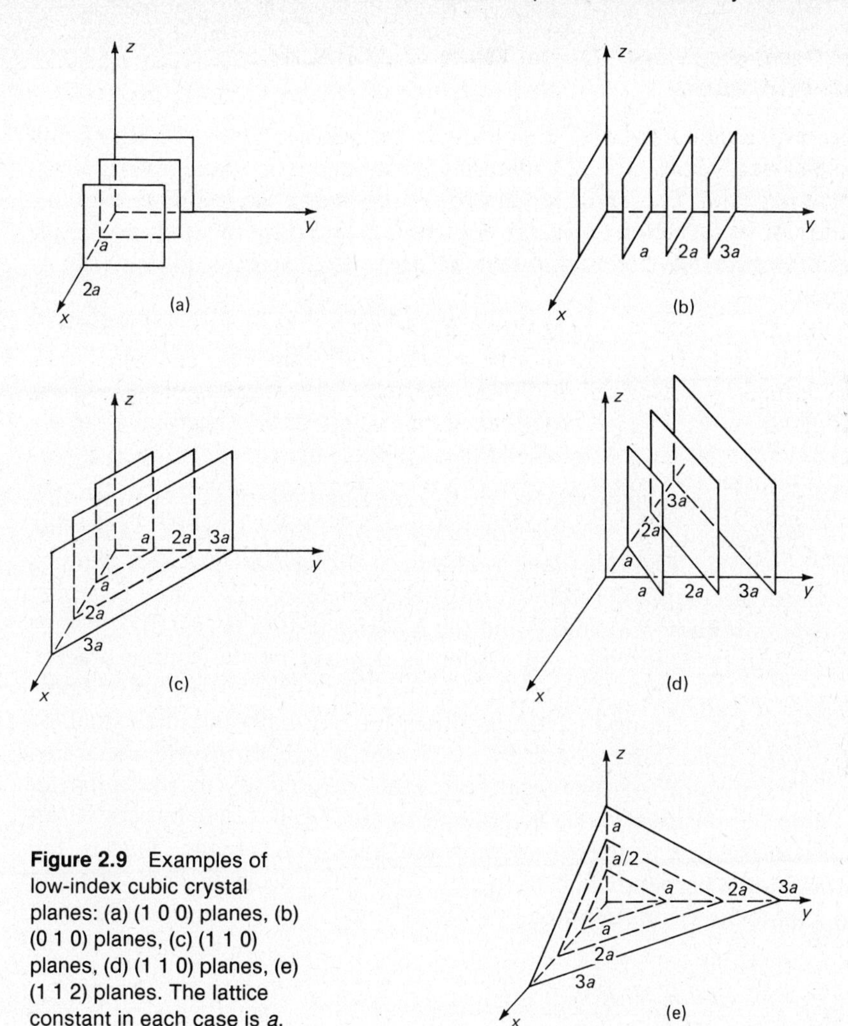

Figure 2.9 Examples of low-index cubic crystal planes: (a) (1 0 0) planes, (b) (0 1 0) planes, (c) (1 1 0) planes, (d) (1 1 0) planes, (e) (1 1 2) planes. The lattice constant in each case is *a*.

each of these by 6 yields the Miller indices (1 2 3). If the x-intercept were negative then the three reciprocal intercept numbers would be $-\frac{1}{6}$, $\frac{1}{3}$, and $\frac{1}{2}$ and the Miller notation would be $(\bar{1}\ 2\ 3)$. Here the overbar denotes a negative intercept. It can be seen that the family of planes (1 2 3), $(1\ 2\ \bar{3})$, $(1\ \bar{2}\ 3)$, $(\bar{1}\ 2\ 3)$, $(1\ \bar{2}\ \bar{3})$, $(\bar{1}\ 2\ \bar{3})$, $(\bar{1}\ \bar{2}\ 3)$, and $(\bar{1}\ \bar{2}\ \bar{3})$ are all equivalent for a *cubic* lattice in the sense that the number of atoms per square centimeter plus the interplanar spacing are the same. Furthermore, $(\bar{1}\ 2\ 3)$ and $(1\ \bar{2}\ \bar{3})$ denote the same set of parallel planes; likewise, $(\bar{1}\ \bar{2}\ \bar{3})$ and (1 2 3), $(\bar{1}\ \bar{2}\ 3)$, and $(1\ 2\ \bar{3})$, etc. Similarly for a cubic lattice permuting the numbers 1, 2, and 3 in the Miller notation results in planes equivalent in the above sense. The equivalent set of crystallographic planes is then written as {1 2 3}. Examples of several sets of cubic crystal planes are given in Fig. 2.9.

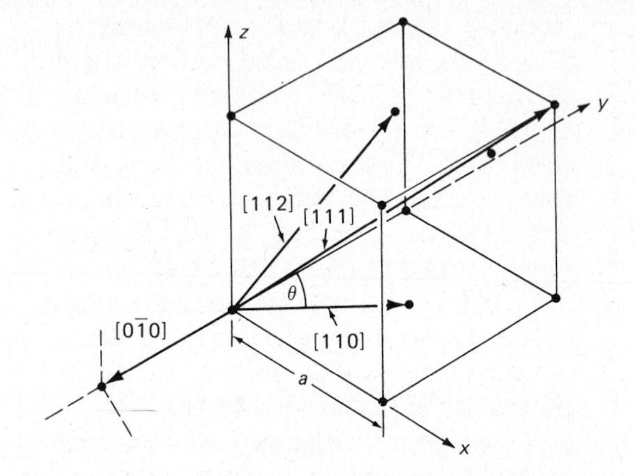

Figure 2.10 The crystal directions [1 1 2], [1 1 1], [1 1 0], and [0 $\bar{1}$ 0] are indicated by vectors; θ is the angle between the [1 1 1] and [1 1 0] directions. Equation (2.2) gives $\cos\theta = \frac{1}{3}\sqrt{6}$.

A plane perpendicular to the x-axis intersecting that axis at two unit distances from the origin, for example, intersects neither the y- nor the z-axis. For purposes of identification, this plane is assumed to intersect these axes at infinity. The coordinate-axis intersections for this plane are at 2, ∞, and ∞ units; the reciprocals are then ½, 0, and 0 and the Miller notation is (1 0 0).

A crystallographic direction in a solid is denoted by the square bracket notation $[h\,k\,l]$. Here the numbers h, k, and l refer to the x-, y-, and z-components of a vector defining a particular direction. Normally the three smallest numbers whose ratio corresponds to the vector length ratios are used. For example, the direction of the negative x-axis corresponds to the notation [$\bar{1}$ 0 0], while the positive y-axis is denoted by [0 1 0]. In *cubic* lattices it can be shown that the $[h\,k\,l]$ direction is normal to the $(h\,k\,l)$ plane. That is, the [1 2 3] direction is perpendicular to the (1 2 3) plane. The angle θ between two crystallographic directions denoted by $[h_1\,k_1\,l_1]$ and $[h_2\,k_2\,l_2]$ is given for a *cubic* lattice by

$$\cos\theta = \frac{h_1h_2 + k_1k_2 + l_1l_2}{(h_1^2 + k_1^2 + l_1^2)^{1/2}(h_2^2 + k_2^2 + l_2^2)^{1/2}} \,. \tag{2.2}$$

This is illustrated in Fig. 2.10.

EXAMPLE 2.2

Prove that for a cubic lattice with lattice constant $a = 5.43$ Å the $[h\,k\,l]$ direction is normal to the $(h\,k\,l)$ plane for the (0 $\bar{1}$ 0) and (1 1 0) planes. Then determine the angle between the [0 $\bar{1}$ 0] and [1 1 0] directions.

Solution

The direction of the negative y-axis may be represented by a unit vector whose y-component is -1 and x- and z-components are 0. In Miller notation this direction is indicated by the $[0\,\bar{1}\,0]$ direction. By Miller notation a $(0\,\bar{1}\,0)$ plane intercepts the negative y-axis at one lattice distance and the x- and z-axes at ∞, and so is perpendicular to the y-axis. Therefore the $[0\,\bar{1}\,0]$ direction is normal to the $(0\,\bar{1}\,0)$ plane.

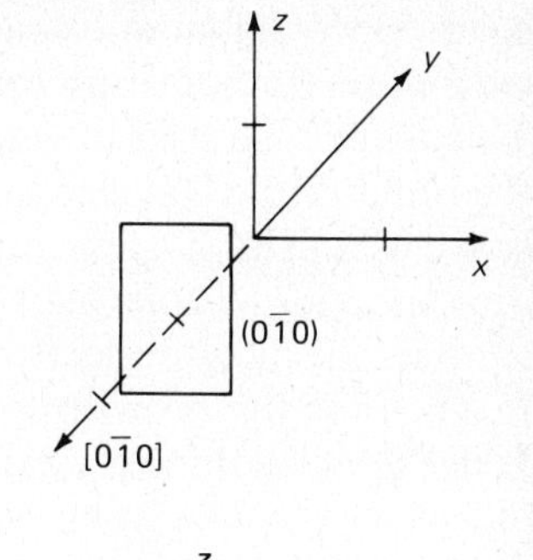

The [1 1 0] direction is that of a vector whose x- and y-components are 1 and z-component is 0. It is a vector in the x-y plane at an angle of 45° with the x- and y-axes. By Miller notation a (1 1 0) plane intercepts the x- and y-axes at one lattice distance and intercepts the z-axis at ∞; hence it is a plane which intersects the x-y plane in a line which makes an angle of 45° with both the x- and y-axes. Therefore the [1 1 0] direction is normal to the (1 1 0) plane.

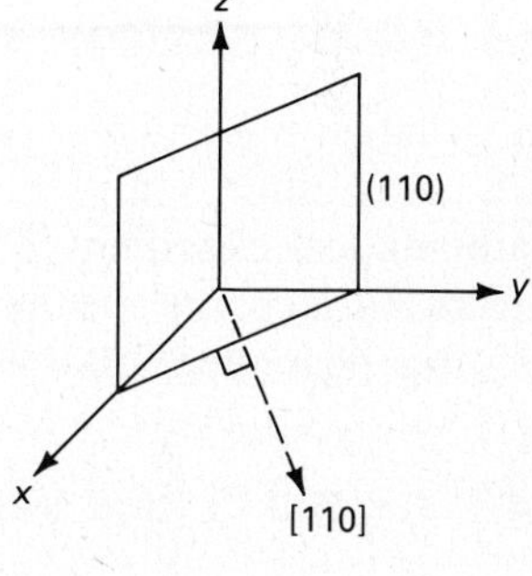

The $[0\,\bar{1}\,0]$ and [1 1 0] directions are specified by

$$h_1 = 0, \qquad h_2 = 1,$$

$$k_1 = -1, \qquad k_2 = 1,$$

$$l_1 = 0, \qquad l_2 = 0.$$

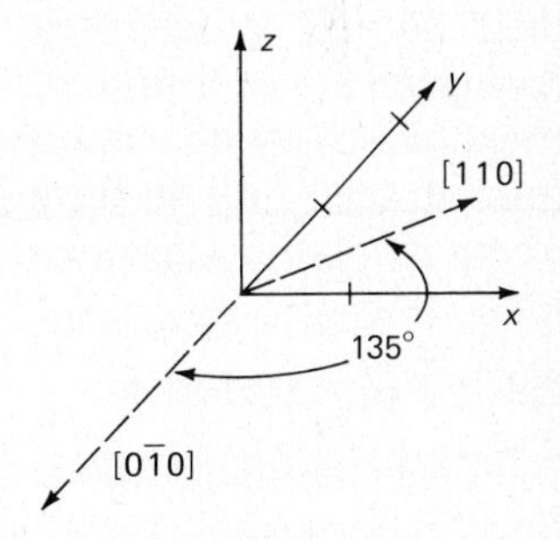

The angle θ between these two directions is obtained from Eq. (2.2),

$$\cos\theta = \frac{(0)(1) + (-1)(1) + (0)(0)}{[0^2 + (-1)^2 + 0^2]^{1/2}(1^2 + 1^2 + 0^2)^{1/2}} = -\frac{1}{\sqrt{2}}$$

$$\underline{\theta = 135°.}$$

2.3 CRYSTAL STRUCTURE ANALYSIS

In order to understand the basis of some of the physical, electrical, and optical properties of crystals it is necessary to know the atomic arrangement in the solid. Since the atomic spacings are of the order of a few angstrom units (10^{-8} cm), normal optical microscope techniques do not have sufficient resolution. This is the case since it can be shown that the resolution of a microscope is of the order of the wavelength of light used for viewing. Since the optical microscope uses light of a few thousand angstroms wavelength, atoms cannot be seen directly with visible light. Instead they must be "viewed" with the aid of shorter-wavelength electromagnetic radiation. X-rays are a form of electromagnetic radiation with wavelength of the order of a few angstrom units and hence are useful for crystal structure determination.

*2.4 IMPERFECTIONS IN SOLIDS

This chapter has thus far been devoted to the description of crystal structures with every atom in a precisely defined periodic position. This is, of course, a mathematical expedient. Real crystals have several types of **imperfections** or **defects.** In fact, the mechanical, electrical, and magnetic properties of solids are often dependent on these crystal defects in a significant way.

2.4.1 Crystal Lattice Vibrations

At any finite temperature the atoms in a crystal lattice have thermal energy imparted to them and may be pictured as vibrating in a random fashion about their normal lattice sites. This departure from strict periodicity of atom arrangement is a form of crystal imperfection and accounts partly for the electrical resistivity of a material, for these **lattice vibrations** tend to scatter electrons in their motion through a crystal under the influence of an electric field. These oscillations also account for much of the heat capacity of a solid.

2.4.2 Point Defects

Although thermal lattice vibrations cause the atoms to deviate temporarily from their normal crystal site, on the average they still maintain their lattice positions. When there are atoms that are more permanently displaced from their normal lattice sites at isolated locations in the crystal, these are referred to as **point defects.** Lattice sites where atoms are missing are called **vacancies.** Extra atoms which are located in the crystal between normally occupied lattice sites are called **interstitials.** Foreign atoms which occur in lattice sites normally occupied by the host crystal atoms are called **substitutional impurities.** Foreign atoms which are present in a host crystal between occupied lattice sites are known as **interstitial impurities.** These imperfections have dimensions of the order of the lattice spacing and yet can be detected in semiconductor lattices by careful measurements of the electrical properties of the crystal. Figure 2.11 illustrates these different types of point defects.

Atoms in a crystal can be knocked out of place in a radiation environment consisting of high speed electrons, neutrons, or protons. When an atom is dis-

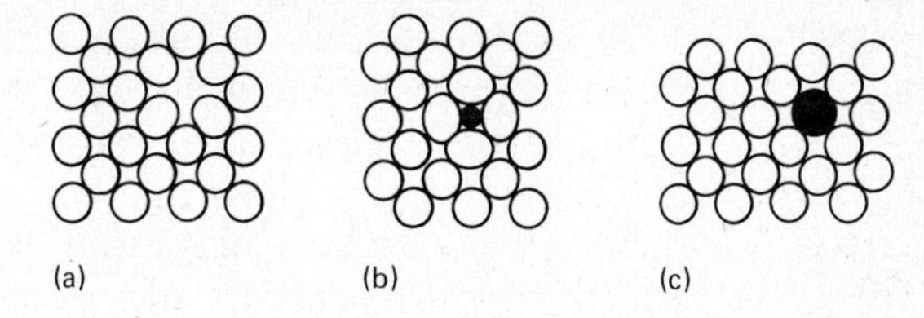

Figure 2.11 Three common point defects: (a) vacancy, (b) interstitial, (c) substitutional impurity.

placed from its normal lattice site to a nearby interstitial position, forming a **vacancy–interstitial pair,** this is known as a **Frenkel defect.** Should a displaced atom find its way to the surface, this is known as a **Schottky defect.** These defects account for the sensitivity of the electrical properties of semiconductor materials and devices to atomic radiation.

The energy necessary to create a vacancy is of the order of an electron-volt. Hence atoms can be removed from their normal lattice by thermal energy if the crystal is raised to an elevated temperature. The number of vacancies N_v in a crystal in thermal equilibrium at an absolute temperature T containing N atomic sites is given essentially by

$$N_v = \gamma N e^{-E_v/kT}, \tag{2.3}$$

where E_v is the energy required to produce a vacancy, k is the Boltzmann gas constant, and γ is a structure constant. Hence a crystal at any temperature will contain a certain number of defects in thermal equilibrium with the crystal lattice.

2.4.3 Line Defects: Dislocations

The **line defect** refers to the displacement of whole rows of atoms from their regular lattice positions. This more complicated type of gross crystal imperfection is known as a **dislocation.** The **edge dislocation** refers to the insertion of an extra half-plane of atoms into a regular crystal. This is pictured in Fig. 2.12 and causes the crystal to distort. Most of the stress is concentrated at the furthest edge of the extra half-plane penetrating the crystal. This line of maximum stress is the dislocation line.

A **screw dislocation** is formed by the relative motion of one part of a crystal with respect to another. To illustrate its formation, consider making a fine line cut part way into a regular crystal. Now displace the crystal on one side of the cut relative to the other side by one atomic distance. This is illustrated in Fig. 2.13. Maximum stress is present along the edge of the cut; this constitutes the dislocation line. If a plane of atoms in the crystal perpendicular to the screw dislocation line is followed through the formation of this type of dislocation, it will now tend to form a surface which spirals around the dislocation line. Hence the designation "screw dislocation."

Line defects in solids tend, in general, to be made up of combinations of edge and screw dislocations. Note that in the forming of the defect, the crystal displacement is parallel to the screw dislocation line whereas the crystal dis-

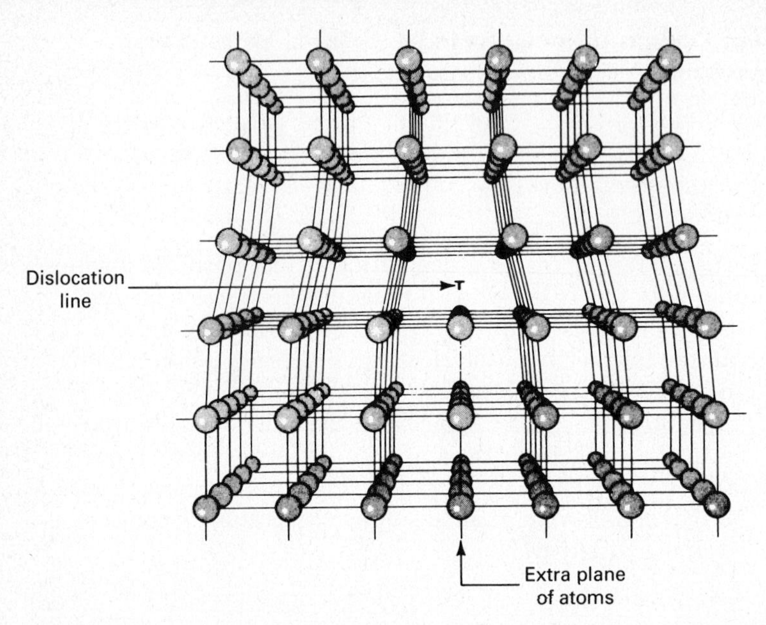

Figure 2.12 Schematic representation of an edge dislocation. An extra half-plane of atoms is wedged between two adjacent planes of atoms. The dislocation line is indicated and is normal to the plane of the page. [Reprinted from C. A. Wert and R. M. Thomson, *Physics of Solids,* 2nd ed. (McGraw-Hill, New York, 1970).]

placement is perpendicular to the edge dislocation line. Since the energy stored in a dislocation is generally *several* electron-volts per atom compared to about one electron-volt in the case of point defects, the dislocated solid is in a highly stressed state. Whereas point defects are in thermal equilibrium with the crystal lattice, dislocations are not. Hence dislocations cannot be removed from a solid by thermal **annealing,** although this possibility exists for removing point defects.

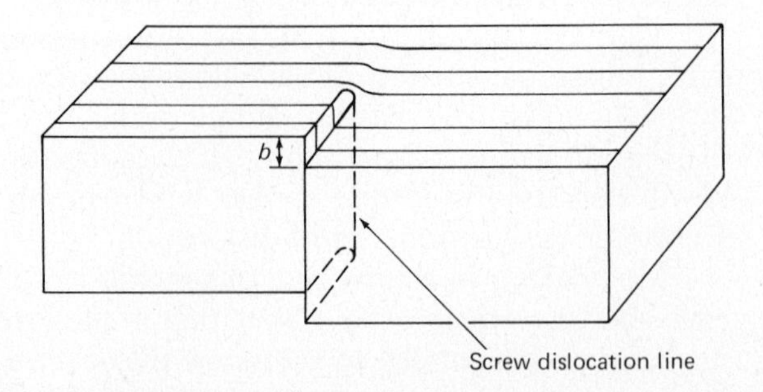

Figure 2.13 Structure of a crystal containing a screw dislocation. The height of the step on the top surface is usually one lattice spacing. The atom rows perpendicular to the dislocation are on a spiral ramp.

2.4.4 Effect of Crystal Defects on the Electrical Properties of Crystals with Diamond Structure

A vacancy in a diamond-type crystal yields four unsatisfied bonds (see Fig. 2.7b). Hence there is a tendency to the immobilization of electrons on these sites. These electrons would normally contribute to the electrical conductivity of the crystal. This is found to be the case since the resistivity of *n*-type silicon under atomic or nuclear radiation is observed to increase in value. Radiation of this type is known to produce point defects in crystals.

Line defects are observed to provide a focus for the precipitation of impurity atoms on the half-plane edge comprising the dislocation. Dislocation lines piercing *p-n* junctions tend to short-circuit these junctions since metallic impurity atoms often collect on such structural defects, providing a highly conducting path. These impurities can also scatter electrons in a uniform *n*-type crystal, impeding their motion and reducing their mobility.

Both point and line defects have been found to cause highly localized distortion of the crystal lattice leading to the formation of "trapping" sites where the recombination of positive (holes) and negative (electrons) carriers is enhanced. This may cause the electrons from the *n-p-n* transistor emitter to recombine with holes in the *p*-type base region before they can be collected at the *n*-type collector region, reducing the transistor current gain. This electron "lifetime" may be significantly reduced when as few as one out of 10^{11} silicon atoms are removed from their normal lattice sites by radiation.

□ *SUMMARY*

A definition was given of the single crystal, the form of starting material for the fabrication of most semiconductor devices; the methods of single-crystal growth were introduced. A periodic arrangement of points in space was defined as a space lattice. The primitive cell was defined as the smallest arrangements of atoms which when periodically stacked fills all of the crystal space; the cubic unit cell will also reproduce the whole crystal lattice but is not always the smallest volume that will do so. The length of an edge of the unit cell is called the lattice constant. Five simple basic crystal structures were defined, including the diamond lattice, which is the most common structure for popular semiconductor materials. The use of Miller indices in specifying crystal planes and directions was described. Crystal imperfections such as lattice vibrations, point defects, and dislocations were defined and their effect on the electrical properties of semiconductor crystals was introduced.

PROBLEMS

2.1 Distinguish between the terms
- **(a)** unit cell and primitive cell,
- **(b)** space lattice and crystal structure,
- **(c)** ionic bonding and covalent bonding.

2.2 Aluminum has a density of 2.7 g/cm^3 and an atomic weight 26.98.
(a) How many gram-atoms or moles are contained in 1.0 cm^3 of the solid? The dimensions of the unit cube (lattice constant) for this fcc metal as determined by X-ray techniques is 4.04 Å.
(b) Calculate the atomic radius of aluminum.
(c) How many atoms are contained in 1.0 cm^3 of an aluminum crystal?
(d) Determine Avogadro's number from these data.

2.3 Repeat Problem 2.2 for the metal sodium, whose crystal structure is bcc, unit-cell dimension 4.28 Å, density 0.97 g/cm^3, and atomic weight 22.99.

2.4 Assuming that the atoms are treated as rigid spheres, show that the ratio of the volume occupied by the atoms to the volume available in various crystal structures is
(a) $\pi/6$ (52%) for the simple cubic lattice,
(b) $\pi\sqrt{3}/8$ (68%) for the bcc lattice,
(c) $\pi\sqrt{3}/16$ (34%) for the diamond cubic structure.

2.5 Calculate the number of atoms/cm^2 for the following planes in a crystal of aluminum (see Problem 2.2):
(a) (1 0 0), **(d)** $(\bar{1}\,\bar{1}\,\bar{1})$,
(b) (1 1 0), **(e)** (1 2 0).
(c) (1 1 1),

2.6 Repeat Problem 2.5 for sodium (see Problem 2.3).

2.7 Determine the angle between the two crystallographic directions
(a) [1 0 0] and [1 1 0],
(b) [1 1 0] and [1 1 1],
(c) [1 1 1] and [1 2 0].

2.8 In connection with structure analysis using X-rays or elementary particles for radiation, determine
(a) the wavelength associated with an electron of kinetic energy 10 kilo-electron-volts (keV);
(b) the wavelength associated with a thermal neutron of 300°K temperature (kinetic energy = $\frac{1}{2}kT$);
(c) the minimum wavelength, in variable-wavelength (white) X-radiation, if the voltage applied to the X-ray tube anode is 30 kilovolts (kV).

***2.9** Distinguish between
(a) vacancies and interstitials,
(b) Schottky defect and Frenkel defect,
(c) point defect and dislocation.

***2.10** A vacancy of aluminum requires an expenditure of energy of about 0.75 eV for formation. Determine the ratio of the vacancies that exist in thermal equilibrium at 600°C relative to those present at 400°C.

CHAPTER 3

The Energy Band Structure of Crystals

3.1 THE NEW PHYSICS

To effectively explain the operation of solid-state devices it is necessary to develop the physics of the solid state, which, as currently understood, is based on quantum physics. First the historical manner in which a need for this new physics arose will be discussed.

By the end of the nineteenth century fabulous success had been achieved in explaining macroscopic phenomena in terms of the classical Newtonian laws of mechanics including problems ranging from the motion of the stars and planets to those involving interacting atoms in a gas. The elegance and simplicity of Newton's laws led some to believe that all physical phenomena could ultimately be explained within this framework. However, microscopic phenomena on a subatomic level were often incapable of being described by means of the classical laws of physics.

At the beginning of the twentieth century, Max Planck found these classical laws to be inadequate in accurately describing the dependence of blackbody radiation on wavelength. Also, soon after the turn of the century, Niels Bohr postulated a theory of structure of the hydrogen atom which abandoned some of the fundamental ideas of classical physics in order to explain available spectroscopic data on hydrogen gas. In addition, about that time it was realized that electric current was conducted in some metal and semiconductor solids by charge carriers similar to electrons but apparently positively charged. This surprising postulate of the existence of both positively and negatively charged current carriers in semiconducting materials is basic today in our understanding of modern bipolar junction transistor behavior.

It soon became clear that a general physical theory which could adequately account for the mounting number of phenomena unexplainable in classical terms was needed. A giant step in this direction was provided in 1926 by Erwin Schrödinger[1] when he published his mathematical theory for predicting the spectral lines of the hydrogen atom based on postulates which were previously stated by Planck and de Broglie. This new physics took the form of a second-order partial differential equation known today as Schrödinger's equation. This formulation still serves as the basis of our understanding of many microscopic phenomena including electrical conduction processes in metals and semiconductors. For example, the fact that the electrical conductivity of a class of materials known as semiconductors increases sharply as the temperature increases above a certain "intrinsic" temperature is not explained by the classical theory of electrical conduction in solids, but can be explained starting with Schrödinger's equation. **Wave mechanics** or **quantum physics** is the branch of science which is employed to describe electronic processes in solids by solving the Schrödinger (or matter-wave) equation.

This chapter begins with an explanation of the basic postulates of wave theory leading to a statement of the Schrödinger equation. This equation bears the same relationship to wave mechanics as Newton's laws do to classical physics. A result of the solution of the Schrödinger equation for the motion of an electron in a periodic potential field, characteristic of the array of ions in a crystal lattice, is then discussed. This leads to the **energy band theory of solids** which will be used to describe, in the remainder of this book, the electrical properties of semiconductors. This concept is essential for understanding the properties of most solid-state materials and devices such as transistors and integrated and optoelectronic circuits.

3.1.1 Energy Quanta

At the end of the nineteenth century, there were two well-known formulations of the problem of the spectral distribution of radiation from a blackbody[2]—one the Wien law and the other the Rayleigh–Jeans law. The former accurately

[1]E. Schrödinger, "Quantization as an Eigenvalue Problem," *Ann. Phys.*, **79**, 489 (1926).
[2]A blackbody may be defined as an ideally perfect emitter and absorber of radiation.

described the variation of radiated energy with wavelength of a blackbody for wavelengths shorter than the wavelength of maximum radiation, while the latter was good for longer wavelengths.[3] In 1901 Max Planck[4] found a formula which was correct at all wavelengths and reduced to the Wien and Rayleigh–Jeans expressions in the limits of short wavelength and long wavelength, respectively. Planck's empirical formulation fitted the precisely determined spectrographic data then available with amazing accuracy. The Wien and the Rayleigh–Jeans formulas were both based on a straightforward application of classical mechanics and thermodynamics. Planck was able to arrive at his formula only by postulating that the blackbody consisted of oscillators which radiated energies that were integral multiples of a certain amount of energy. This basic amount or **quantum** of energy was given by

$$E = h\nu, \tag{3.1}$$

where ν is the frequency of the radiated energy and h is a constant now associated with Planck's name. These discrete steps in energy values were in sharp contrast with classical radiation theory, which permitted all energy values.

The idea of discrete quanta of energy was successfully applied by Einstein in 1905[5] to explain the photoelectric effect. He related the energy threshold for electron emission from metals to the minimum frequency of the incident light according to the Planck formula. Light of frequency below this minimum would then be totally ineffective in producing photoelectrons, regardless of its intensity. The experimental evidence supporting this quantum theory of light radiation could not be explained by classical radiation theory derived from Maxwell.

About 1913 Niels Bohr[6] was able to accurately predict the major spectral lines emitted by excited hydrogen atoms by applying a pseudo-classical theory with the Planck quantum idea as a basic postulate. Although his physics was a strange mixture of the old and new physics, the prediction of the major sharp spectral lines was indisputably in excellent agreement with the accurate spectroscopic observations made at the time. Hence, in spite of some inherent contradictions in the theory, it appeared to have some elements of truth.

3.1.2 Wave–Particle Duality

Working in the totally unrelated area of relativity considerations, Louis de Broglie in 1924 proposed[7] that particles should be assigned a wavelength

$$\lambda = h/p, \tag{3.2}$$

[3]For a description of the dilemma presented by the purely classical explanation of blackbody radiation, see F. Richtmyer, E. Kennard, and T. Lauritsen, *Introduction to Modern Physics,* 5th ed. (McGraw-Hill, New York, 1955).

[4]M. Planck, "Distribution of Energy in the Spectrum," *Ann. Phys.,* **4**, 553 (1901).

[5]A. Einstein, "Generation and Transformation of Light," *Ann. Phys.,* **17**, 132 (1905).

[6]N. Bohr, "On the Constitution of Atoms and Molecules," *Philos. Mag.,* **26**, 1 (1913).

[7]L. de Broglie, "A Tentative Theory of Light Quanta," *Philos. Mag.,* **47**, 446 (1924).

where p is the particle momentum and h is again the Planck constant. De Broglie's reasoning was that in classical physics electromagnetic radiation was always considered a wave phenomenon, yet the work of Planck and Einstein dictated the consideration of light radiation in terms of discrete particles of energy or quanta. He suspected then that perhaps the inverse might also be true—that the classical particle might also be represented by a characteristic wavelength, according to his formula. The fact that particles such as electrons do, indeed, exhibit wave properties and undergo diffraction by a grating was demonstrated by Davisson and Germer in 1927.[8] The diffraction grating used was not the ruled grating commonly used for light, but a periodic arrangement of atoms in a nickel crystal provided by nature. This choice was necessitated by the small wavelength of the electron predicted by de Broglie and the necessity that the crystal periodicity be of the order of this wavelength to obtain a strong diffraction pattern. This observation finally resolved the nineteenth-century debate between Isaac Newton, who favored a corpuscular theory of light, and Christian Huygen, who emphasized its wave nature. There exists, in fact, a **wave–particle duality.** Light, or an electron, can be considered to have some of the properties of both particles and waves. Hence an electron ejected from the hot filament of an electron gun has a definite charge and mass, is deflected by the usual laws of electron ballistics in flight, and yet shows some diffraction properties when interacting with a periodic crystal structure. Matter and light can be considered as particles. However, the behavior of these corpuscles when interacting with atoms may not be described by Newton's mechanics but by a wave or quantum mechanics. Also, it is impossible to observe *simultaneously* the wave and particle properties of matter or light.

3.2 THE SCHRÖDINGER EQUATION

What has been discussed thus far is a series of ideas and concepts which are peculiar to the modern physics of quantum mechanics. What is needed now is a mathematical formulation that will permit us to predict and explain the many subatomic processes which govern the behavior of electronic materials of interest to us. As previously indicated, the Schrödinger wave equation expresses the essence of this modern physics. Obtaining solutions to this equation for a variety of physically interesting cases will permit us to analyze problems which may be subject to experimental verification. (In fact, the ability to stand up to experimental testing is, of course, the final test of a physical theory.)

The Schrödinger equation refers to the motion of an electron in various force fields. It replaces in microscopic phenomena the Newton expression $F = ma$ for the motion of particles on a larger scale and permits a description of effects not previously explainable in classical terms. Since we have already

[8] C. Davisson and L. Germer, "Diffraction of Electrons by a Crystal of Nickel," *Phys. Rev.*, **30**, 705 (1927).

indicated that the electron can be pictured as having a wavelength associated with it, it should not be surprising that the Schrödinger equation takes the form of a wave equation.

The **time-dependent** Schrödinger wave equation of an electron having a potential energy V, which in general can be expressed as a function of position and time, is[9]

$$\frac{h^2}{8\pi^2 \mathrm{m}_0}\frac{\partial^2\Psi}{\partial x^2} + \frac{\partial^2\Psi}{\partial y^2} + \frac{\partial^2\Psi}{\partial z^2} - V\Psi = \frac{h}{2\pi j}\frac{\partial\Psi}{\partial t} \qquad \text{(time-dependent Schrödinger equation)} \tag{3.3}$$

Here h is the Planck constant, m_0 the electron mass, $j = \sqrt{-1}$, and Ψ is the wave function, which depends in general on position and time; that is,

$$V = V(x, y, z, t)$$

and

$$\Psi = \Psi(x, y, z, t), \tag{3.4}$$

where Ψ is a complex quantity in general. This is a complicated second-order differential equation which can only be solved in closed form in a limited number of special cases. However, the equation may be greatly simplified in a large number of problems[10] by dealing *separately* with the position and time dependences. This is possible when *the potential energy depends only on position* and not on time. Let us introduce this condition into the one-dimensional Schrödinger equation, which we consider for additional simplicity, although the three-dimensional formulation may be handled similarly. The one-dimensional equation is

$$\frac{h^2}{8\pi^2 m_0}\frac{\partial^2\Psi(x,t)}{\partial x^2} - V(x)\Psi(x,t) = \frac{h}{2\pi j}\frac{\partial\Psi(x,t)}{\partial t}. \tag{3.5}$$

3.2.1 Separation of Variables

Let us postulate a solution of the form

$$\Psi(x,t) = \psi(x)\phi(t), \tag{3.6}$$

assuming that the space and time parts of the wave function Ψ can be separated. Substituting Eq. (3.6) into (3.5) gives

$$\frac{h^2}{8\pi^2 m_0}\frac{d^2\psi(x,t)}{dx^2}\phi(t) - V(x)\psi(x)\phi(t) = \frac{h}{2\pi j}\frac{d\phi(t)}{dt}. \tag{3.7}$$

[9] Note that V here refers to potential *energy* and in other chapters V is used for electric potential. The electric potential (in volts) is potential energy (in joules) per unit charge (in coulombs).

[10] We will deal only with such problems.

Dividing both sides of the equation by $\psi(x)\phi(t)$ we get

$$\frac{h^2}{8\pi^2 m_0}\frac{1}{\psi(x)}\frac{d^2\psi(x)}{dx^2} - V(x) = \frac{h}{2\pi j}\frac{1}{\phi(t)}\frac{d\phi(t)}{dt}. \tag{3.8}$$

Note that the left-hand side of the equation is a function of x only, while the right-hand side is a function only of t. A useful theorem of mathematical physics states that in such a case, where x and t are independent variables, both sides of this equation must be separately equal to a constant. In this case, when we arbitrarily choose a value of x, the left-hand side of Eq. (3.8) takes on a value which is not necessarily equal to the right-hand side, since t can be chosen arbitrarily as any value. Hence for the equality to hold, both sides must be equal to a constant, which we will arbitrarily take as $-E$.[11]

This then reduces Eq. (3.8) to two ordinary differential equations, one in position and one in time, whose solutions are more easily obtainable. That is,

$$\frac{h^2}{8\pi^2 m_0}\frac{d^2\psi}{dx^2} + (E - V)\psi = 0 \qquad \begin{array}{c}\textbf{(time-independent}\\ \text{Schrödinger equation)}\end{array} \tag{3.9a}$$

and

$$\frac{d\phi}{dt} = -\frac{2\pi j}{h}E\,\phi. \tag{3.9b}$$

After some manipulation the second equation integrates to

$$\ln \phi = -2\pi jEt/h + A. \tag{3.10}$$

We can take the integration constant $A = 0$ without loss of generality and introduce a constant later in reconstructing the product $\Psi = \psi\phi$. Then Eq. (3.10) becomes

$$\phi = e^{-(2\pi jE/h)t}, \tag{3.11}$$

which is oscillatory in time; $E\ (= h\nu)$ is recognized here from Planck's law as the total electron energy.

The problem of finding the solution for the motion of an electron having a potential energy V due to some force field now reduces to solving the second-order linear ordinary differential equation (3.9a), and the time-dependent solution can always be written by tacking on the results of Eq. (3.11).

3.2.2 Boundary Conditions on Ψ

We must now specify some boundary conditions on ψ so that we can determine the two arbitrary constants necessary for the solution of this second-order differential equation. Although we have not specified the physical significance of Ψ, ultimately some connection must be found between this quantity and the

[11]The reason for this peculiar choice of arbitrary constant will be seen later; E turns out to be the total electron energy.

real world, and hence Ψ must have intuitively acceptable behavior.[12] The usual conditions imposed are

$$\Psi(x,t) \quad \text{and} \quad \frac{\partial \Psi}{\partial x} \quad \text{must be continuous, finite, and single-valued for all } x \tag{3.12a}$$

and

$$\int_{-\infty}^{+\infty} |\Psi|^2 \, dx \quad \text{must be finite} \tag{3.12b}$$

Since these conditions are defined for $\Psi(x,t)$, they will apply to $\psi(x)$ as well. The physical meaning of these mathematical requirements will become clear in our subsequent discussion of the physical interpretation of Ψ.

3.2.3 Physical Interpretation of Ψ

Max Born[13] in 1926 proposed an important physical interpretation of the wave function. Since the Schrödinger equation refers to matter waves à la de Broglie, Ψ must represent the amplitude of these matter waves. Born reasoned that, by analogy with the theory of electromagnetic waves, the absolute value of Ψ squared gives the *probability* of finding the electron between x and $x + \Delta x$ at a given time.[14] This interpretation provides us with a means of comparing the predictions of the Schrödinger equation with experiment. A great deal of information about the motion of the electron is contained in the wave function Ψ. For example, the motion parameters such as momentum and energy are calculable once this function is known.

3.2.4 The Uncertainty and Correspondence Principles

This statistical interpretation of wave mechanics led to another important statement of the deviation of this new physics from Newtonian mechanics as applied to the subatomic realm. In classical mechanics the one-dimensional motion of a particle is described by a position function of time $x(t)$ and a momentum function of time $p(t)$. According to Newton, we can fully determine the motion of a particle such as an electron given its initial momentum and position and the mechanical laws that describe its motion. This is sometimes referred to as a deterministic or **causal** point of view. Wave mechanics, on the other hand, takes a probabilistic or **statistical** point of view. That is, it is *not* possible to predict a definite trajectory and motion of a particle but *only the probability* that it will behave in a specific way.

[12]Indeed, Schrödinger himself did not realize its full significance in his first paper.

[13]M. Born, "Quantum Mechanics of Collision," *Z. Phys.*, **37**, 863 (1926).

[14]Since mathematically Ψ in general is a complex quantity and must have physical significance, $|\Psi|^2$, which is always real and positive, is most useful. According to the definition, $\int |\Psi|^2 \, d\Omega = 1$, when taken over all space Ω.

This may be expressed in terms of the **uncertainty** or **indeterminacy** principle of Werner Heisenberg (1927), which states[15] that *it is not possible to measure simultaneously the position and momentum of a particle with arbitrary accuracy for the purpose of predicting future behavior.* If the position is precisely identified then the momentum will be somewhat uncertain, and vice versa. Mathematically, this can be expressed as follows: given that the position of a particle can be measured within Δx, its momentum uncertainty in the x-direction, Δp_x, cannot be ascertained to any greater accuracy than

$$\Delta p_x = h/\Delta x, \tag{3.13a}$$

where h is the ubiquitous Planck constant. This indeterminacy demonstrates the noncausal nature of quantum mechanics. It will also be useful in predicting the necessary separation of energy states in an energy band in a semiconductor (to be discussed in the next section), and hence constitutes another expression of energy quantization.

Another statement of this same principle can be shown to be

$$\Delta E \, \Delta t \geq h. \tag{3.13b}$$

Here ΔE can represent the uncertainty in the energy of a photon and Δt the accuracy limits of time estimation of its emission. Using the Planck formula this becomes

$$\Delta \nu \, \Delta t \geq 1. \tag{3.13c}$$

These relations are useful in estimating the sharpness of a semiconductor laser spectral line, where $\Delta \nu$ is the linewidth and Δt is the electron transition time. Another test of the validity of this quantum physics was proposed in 1923 by Bohr by pronouncing his **correspondence principle.** This requires that the laws of quantum mechanics, in the classical limit where many quanta are involved, lead to the classical physics equations, on the average. For example, in macroscopic problems the Schrödinger equation gives the same result as Newton's laws. This leads to the conclusion that the uncertainty in *macroscopic* particle motion is negligible.

EXAMPLE 3.1

The average time duration of an excited state of an atom in a gas laser is about 10^{-8} sec. This laser emits a spectral line whose central wavelength is 6328 Å.

(a) Compute the minimum frequency variation of the emitted red light, according to the uncertainty principle.

[15] W. Heisenberg, "The Actual Content of Quantum Theoretical Kinematics and Mechanics," *Z. Phys.*, **43**, 172 (1927).

(b) Express the possible sharpness of the spectral line by the ratio to the frequency variation to the fundamental radiated frequency in parts per million (ppm).

Solution

(a) From Eq. (3.13c),

$$\Delta\nu \geq \frac{1}{\Delta t} = \underline{\frac{1}{10^{-8}\ \text{sec}}}.$$

The longer the time that the atom spends in an excited state, the narrower the spectral line.

(b) The central frequency is given by

$$\nu_0 = \frac{c}{\lambda_0} = \frac{3.0 \times 10^{10}\ \text{cm/sec}}{6328 \times 10^{-8}\ \text{cm}}$$

$$= 4.7 \times 10^{14}\ \text{Hz}.$$

Hence the sharpness is expressed by

$$\frac{10^8}{4.7 \times 10^{14}} = 0.21 \times 10^{-6} \quad \text{or} \quad \underline{0.21\ \text{ppm}}.$$

3.3 MOTION OF AN ELECTRON IN A PERIODIC POTENTIAL

The problem of the transport of electrons in a crystal will now be considered from a quantum mechanical viewpoint. This is necessary since countless physical phenomena observed in solid materials have no classical explanation. The crystal considered consists of a periodic arrangement of atoms, some stripped of an electron (ions), with these electrons free to move through the crystal lattice. Later we will see that the properties of solid crystals such as metals, semiconductors, and insulators can be derived from this analysis.

The model that we will assume is drastically simplified. We will reduce this problem to that of a single electron traveling in one dimension, impinging on a series, infinite in extent, of periodically spaced potential wells. The potential energy function representing the force field created by the periodically spaced ions (nuclei plus immobile core electrons) in a sea of free electrons is shown in Fig. 3.1a. The simplified mathematical representation of this periodic potential function assumed by Krönig and Penney in 1930[16] is given in Fig. 3.1b.

[16]R. de L. Krönig and W. G. Penney, "Quantum Mechanics of Electrons in Crystal Lattices," *Proc. R. Soc. London,* **A130**, 499 (1930).

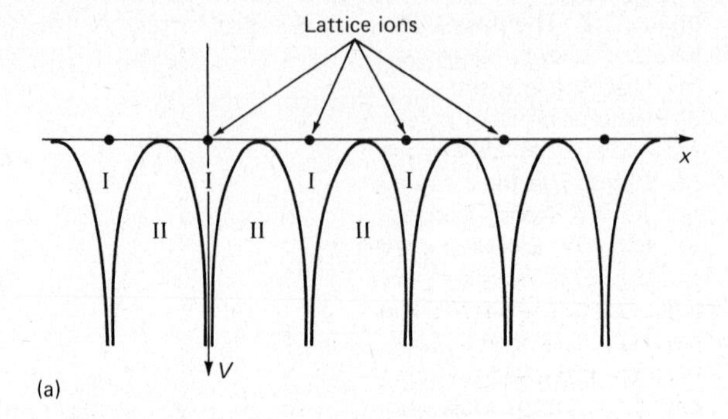

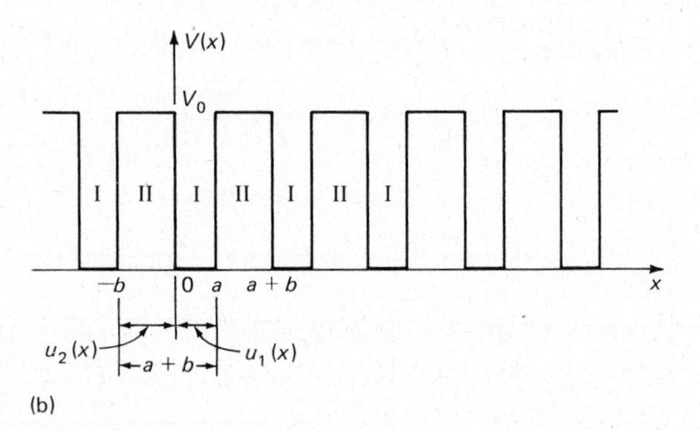

Figure 3.1 (a) A graph of the potential energy function representing the force field created by periodically spaced ions. (b) A simplified mathematical representation of the periodic function of (a) assumed by Krönig and Penney.

Note that the x-space is broken down into regions of width a (Region I) and b (Region II), with the distance between nuclei being $a + b$. In spite of the extreme simplicity of this formulation, the essential ideas of the quantum mechanical theory of the three-dimensional solid become clear when the one-dimensional Schrödinger equation is solved for this problem. Although the model is simple, it is nevertheless somewhat complicated mathematically to arrive at a solution in closed form.

3.3.1 The Schrödinger Equation for an Electron in a Periodic Potential

The time-independent Schrödinger equation for the motion of the electron in a periodic potential is, in one dimension,

$$\frac{d^2\psi}{dx^2} + \frac{8\pi^2 m_0}{h^2}[E - V(x)]\psi = 0, \tag{3.14}$$

where $V(x)$ is given in Fig. 3.1b as

$$\begin{aligned} V &= V_0 \quad \text{for} \quad -b < x < 0, \quad a < x < a + b, \quad \text{etc.;} \\ V &= 0 \quad \text{for} \quad 0 < x < a, \quad a + b < x < 2a + b, \quad \text{etc.} \end{aligned} \tag{3.15}$$

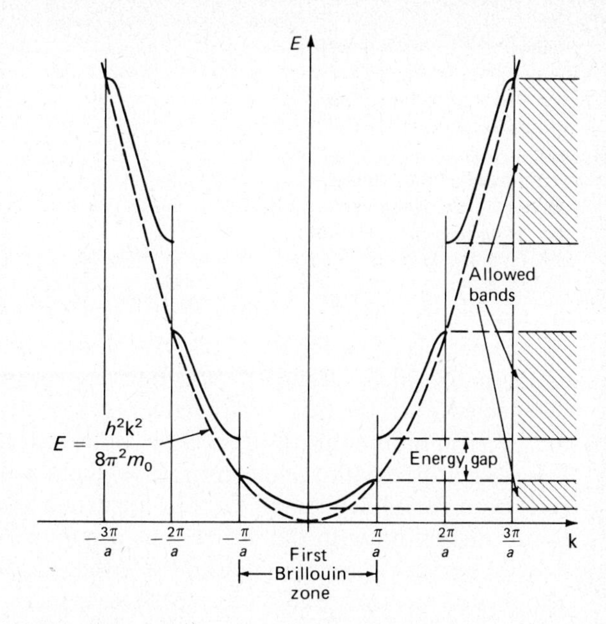

Figure 3.2 The permitted values of energy for an electron traveling in a one-dimensional periodic potential field are indicated by the solid lines. This indicates the results of a Krönig–Penney solution. The E versus k relationship for a free electron is indicated by the dashed-line parabola. The deviations from the free-electron motion caused by the periodic field give rise to the allowed energy bands shown crosshatched.

A theorem proved by Felix Bloch in 1928[17] states that all one-electron wave functions for problems involving periodically varying potential energy functions must be of the form

$$\psi(x) = u(x)e^{jkx}, \tag{3.16}$$

where k is a constant of motion[18] referred to as the electron wave vector. Here $u(x)$ is a **periodic function** (which repeats itself in the present case after every lattice distance $a + b$).[19] [For example, $u(x)$ may be a function of the form $\sin(2\pi x/d)$, d being the distance after which the function $u(x)$ repeats itself.]

Substitution of the Bloch function of Eq. (3.16) into the Schrödinger Eq. (3.14) and solving the resulting equation for $u(x)$, subject to the periodic boundary conditions provided by the periodic potential of Fig. 3.1b, is a straightforward but laborious task.[20] The resulting solution yields a relationship between the electron energy E and the electron wave vector k. This relationship is represented in the graph of Fig. 3.2, which shows a plot of E versus k.

[17]F. Bloch, "On the Quantum Mechanics of Electrons in Crystal Lattices," *Z. Phys.*, **52**, 555 (1928).

[18]In the Schrödinger solution discussed here $hk/2\pi$ can be shown to be identified with the electron momentum in the crystal.

[19]Note that when the time-dependent part of the wave function is tacked onto this function, it is recognized as that of a traveling wave, which is periodically modulated in space. Using Eqs. (3.11) and (3.16) with Eq. (3.6) yields $\Psi(x,t) = u(x)e^{j(kx-2\pi\nu t)}$, a traveling wave associated with the electron.

[20]For the details of this solution see Appendix C.

3.3.2 Brillouin Zones

A typical plot of E versus k for a periodic potential is given in Fig. 3.2. [The dashed curve is parabolic and corresponds to the E(k) relationship for a free electron, $E = p^2/2m_0 = h^2\mathrm{k}^2/8\pi^2 m_0$.] Indicated are energy discontinuities at values of

$$\mathrm{k} = n\pi/a, \tag{3.17}$$

where $n = \pm 1, \pm 2, \pm 3, \ldots$. These are referred to as the **Brillouin zone** boundaries and correspond to values of electron momenta at which a discontinuity in the energy versus wave vector relation occurs. Equation (3.17) describes the situation in which a standing wave due to an electron wave impinges upon and is reflected from the potential barriers created by the lattice ion charges. Since the electron momentum is $p = h\mathrm{k}/2\pi$ and from the de Broglie formula $p = h/\lambda$, k is identified with the electron wavelength as $\mathrm{k} = 2\pi/\lambda$. Then the Brillouin zone boundaries defined by Eq. (3.17) correspond to $\lambda = 2a/n$. That is, maximum interaction with the lattice ions occurs when the electron wavelength is twice the lattice spacing, equal to the lattice spacing, etc. This phenomenon is also related to the case of X-rays (electromagnetic waves) diffracted by a crystal lattice. The **Bragg condition** for reflection of X-rays is directly related to Eq. (3.17), which refers to electron wave diffraction in a crystal. This will now be shown. The one-dimensional Bragg condition ($\theta = 90°$) for reflection from crystal planes where the planes are spaced a distance a apart can be written as[21]

$$2a \sin\theta = 2a = n\lambda\,. \tag{3.18}$$

Dividing by 2π gives

$$\frac{a}{\pi} = \frac{n\lambda}{2\pi} = \frac{n}{\mathrm{k}} \quad \text{or} \quad \mathrm{k} = \frac{n\pi}{a}. \tag{3.19}$$

Hence, invoking wave–particle duality, we see that electrons in a periodic lattice and X-rays or photons incident on a periodic structure *cannot* be transmitted through this lattice when the electron or photon wave vector (or momentum) bears a specific relationship to the lattice spacing. This corresponds to standing electron waves and a maximum reflection (minimum transmission) of X-rays.

*3.4 THE CENTRAL FIELD PROBLEM

Another important solution of the Schrödinger equation which can be obtained in closed form corresponds to the problem of the simplest atom, that of hydrogen. This is the case of the motion of a single valence electron in the radial field created by one proton in the nucleus. The potential energy of the electron in the field of the proton is

$$V(r) = -q^2/4\pi\epsilon_0 r, \tag{3.20}$$

[21]See any first-year college physics text.

where q is the electronic charge, ϵ_0 the permittivity of free space, and r the distance from the central nucleus to the electron. This problem differs from the other examples discussed thus far in that the *three-dimensional* Schrödinger equation must be solved. The latter is normally written in spherical (r, θ, ϕ) coordinates owing to the spherical form of the potential energy function.

The solution is somewhat lengthy (but beautiful) mathematically, and will not be presented here.[22] However, the treatment of the one-electron atom forms the basis of the analysis of the many-electron atoms which comprise the electronic materials of interest to the engineer. The important results of this calculation will now be summarized in order to introduce some definitions which will be of use later in the discussion of electronic materials.

3.4.1 Results of the Solution of the One-Electron Atom Problem

Since the electron is confined to the vicinity of the nucleus (bound) by a Coulombic force, the uncertainty principle dictates that the possible electron energies must be quantized. The energy states of the electron in the hydrogen atom are given by

$$E_n = \frac{-m_0 q^4}{8h^2\epsilon_0^2}\frac{1}{n^2}, \tag{3.21}$$

where m_0 is the electron mass and h the Planck constant; the integer $n = 1, 2, 3, \ldots,$ is usually referred to as the **principal quantum number.** These quantized energy states are derived mathematically by requiring that the wave-function solutions of the Schrödinger equation are well behaved (see Section 3.2.2).

It can be shown for this case that the three-dimensional partial differential Schrödinger equation can be separated into three ordinary differential equations involving separately r, θ, and ϕ. One quantum number is derived from each of these equations, the quantum numbers n, l, and m_l corresponding to the spherical coordinates r, θ, and ϕ. As the quantum number n refers to energy quantization, the quantum number l refers to quantization of angular momentum of the electron and is called the **orbital momentum quantum number.** The quantum number m_l refers to the angular orientation of the orbital angular momentum vector and is called the **magnetic quantum number.** Here only certain discrete directions are permitted; this constitutes **spatial quantization.**

To explain certain features of atomic optical spectra, Uhlenbeck and Goudsmit (1925)[23] postulated in an ad hoc fashion that the electron, besides its known orbital angular momentum, has an additional **intrinsic** angular momentum. The fact that the electron acts like a spinning solid body leads to

[22] For a straightforward formulation and solution of this problem see R. M. Eisberg, *Fundamentals of Modern Physics* (Wiley, New York, 1961), Chap. 10.

[23] G. E. Uhlenbeck and S. Goudsmit, "Substitution of the Hypothesis of a Non-Mechanical Force Through a Consideration of the Intrinsic Behavior of Each Electron," *Naturwissenschaften,* **13**, 953 (1925).

the association of a quantum number m_s with this intrinsic angular momentum or **spin.** The Dirac formulation of the Schrödinger problem does not make an ad hoc introduction of electron spin necessary. Dirac (1928)[24] showed that the concept of spin quantization is derived directly from a relativistically corrected form of the Schrödinger equation. Four quantum numbers, n, l, m_l, and m_s, and their possible values are automatically shown to be necessary to fully define the state of an electron.[25] The following is a summary of the possible values of the quantum numbers which are used to define the possible states of an electron in any atom:

$$
\begin{aligned}
n &= 1, 2, 3, \ldots, \\
l &= 0, 1, 2, \ldots, n-1, \\
m_l &= 0, \pm 1, \pm 2, \ldots, \pm l, \\
m_s &= \pm \tfrac{1}{2}.
\end{aligned}
\qquad (3.22)
$$

The last number corresponds to the two spatial quantizations of the electron spin, **up** or **down.** A theorem due to Pauli (1927)[26] states that each electron state can be occupied by no more than one electron. This is the famous **Pauli exclusion principle.**

3.5 ELECTRICAL PROPERTIES OF CRYSTALS

Although the Krönig–Penney treatment discussed in Section 3.3 applies only to a hypothetical one-dimensional lattice, the basic results derived, which are indicated in Fig. 3.2, apply to the more general three-dimensional crystal model. This formulation was made in order to illustrate what the Schrödinger equation has to say about the electrons in crystals in order to better explain the vast difference, for example, between the electrical and thermal properties of metals compared to insulators. The electrical resistivity of solid materials varies from about 10^{-6} Ω-cm for metals to 10^{16} Ω-cm for some insulators. This fantastic range of more than 20 orders of magnitude is perhaps the largest for any known measurable physical parameter. Classical physics has no explanation for this phenomenon.

What will now be developed is the quantum mechanical explanation. These ideas will make it apparent that there is an intermediate class of electrical conductors called **semiconductors.** As it turns out, these materials are even more important for the construction of electronic devices than metals or insulators. This is the "stuff" from which transistors and microchips are made!

[24] P. A. M. Dirac, "The Quantum Theory of the Electron," *Proc. R. Soc. London,* **A117**, 610 (1928).

[25] Spectroscopists refer to the electron states defined by $l = 0, 1, 2, \ldots$, as s, p, d, . . . , states, respectively. Hence the 1s quantum state refers to an energy state corresponding to $n = 1$, $l = 0$; 2p refers to a state corresponding to $n = 2$, $l = 1$; etc.

[26] N. Pauli, "Quantum Mechanics of the Magnetic Electrons," *Z. Phys.*, **43**, 601 (1927).

A way of illustrating the results of the Krönig–Penney model for an electron wave traveling through a periodic array of ions is to plot the energy band diagram as shown in Fig. 3.2. Indicated here are the ranges of permitted and forbidden energy levels that an electron may occupy in a crystal.

A real crystal, of course, has many electrons. For example, a large number of metal crystals have approximately one free (valence) electron per atom. Since there are approximately 10^{22} atoms/cm^3 in a typical metal crystal, there are about the same number of electrons free to move and conduct current.[27] In addition, each atom in the crystal possesses many other electrons bound to the nucleus according to the Bohr–Rutherford picture of the atom. These electrons are securely attached to their nuclei and hence do not move through the crystal lattice and conduct electric current. Now the electrons in the crystal, whether bound or free to move, may possess energies only in the allowed bands shown in Fig. 3.2. The bound electrons have large *negative* potential energies relative to the energy of an electron far removed from the crystal because they are attached to their nuclei[28]; the electrons physically closer to the nucleus have larger negative energies. These latter electrons represent the *lower* energy electrons.

3.5.1 Energy Bands in Crystals from Atomic States

It turns out that within an energy band there are a *finite number of energy levels* which electrons can occupy. This can be seen by a qualitative argument making plausible the manner in which energy bands are formed for a crystal. First consider the electron energy levels characteristic of an isolated atom. These levels will be similar to those already discussed for the hydrogen atom, although, in general, the location and spacing of these levels will depend upon the particular atom chosen. A result of quantum mechanics is that the discrete energy levels of an atom in a perturbing electric or magnetic field will break up or split into two or more closely spaced levels as a result of this interaction. If a crystal is considered to be formed by bringing a number of atoms in close proximity to one another, the electrons of one atom will interact with the electric field due to another atom (and vice versa), causing a splitting of its energy states. If the interaction of the multitude of atoms in the crystal is taken into account, an energy band is formed consisting of many such closely spaced energy levels. This is shown in Fig. 3.3, which indicates the splitting and shift of atomic energy levels as the lattice spacing is reduced to that characteristic of equilibrium, r_0. Since the separation of these multiple levels is very small (of the order of 10^{-19} eV) and their number is large, this group of allowed levels is referred to as an energy band of the type derived in the Krönig–Penney problem. Bands derived from the atomic levels 1s and 2s are shown here; higher energy bands generate from 2p, 2d, etc., levels.

[27] This "free electron" theory of metals was already postulated by Drude in 1900: P. Drude, "Electron Theory of Metals," *Ann. Phys.*, **1–3**, 566 (1900).

[28] Energy must be *supplied* to remove them.

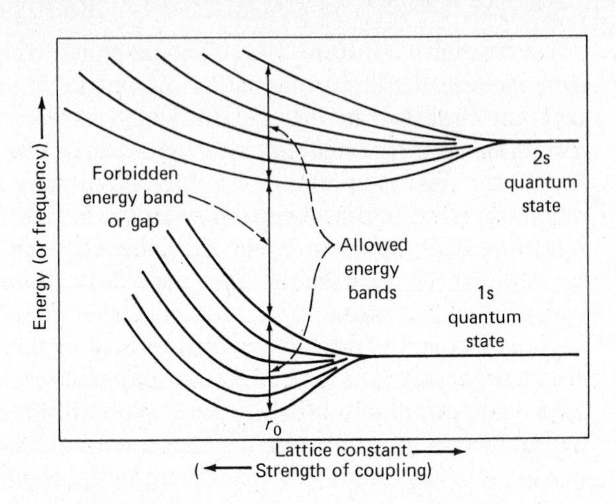

Figure 3.3 To the right are shown two of the energy levels characteristic of each isolated atom in a crystal. As the lattice constant or spacing between such atoms is reduced, a splitting of energy levels occurs due to their mutual interaction. At an equilibrium spacing r_0 two energy bands with a gap between them are indicated. Bands deriving from 2p and 2d levels will occur at higher energy values. Sometimes these bands can overlap to produce a single band containing a mixture of 2s and 2p states, for example. [Taken from W. Shockley, *Electrons and Holes in Semiconductors* (Van Nostrand Co., New York, 1950), p. 132.]

A theorem of quantum mechanics states that bringing atoms together leaves the total number of quantum states, totaling those corresponding to each of the atoms, unchanged. Hence in a crystal the band formed from the s atomic state can hold two electrons per atom, one for each direction of spin. The s-state has two spin states, $m_s = \pm 1/2$. This latter result derives from a postulate due to Wolfgang Pauli which states that *only one electron may occupy each quantum state available.*[29] This principle is essential to our understanding of the electrical properties of solids as well as the periodic table of the elements. Correspondingly, the band derived from the p atomic state can contain six electrons per atom and that from the d state can hold ten electrons per atom.

3.5.2 Electrical Conduction in Energy Bands

Now all of the electrons in the solid may be introduced into the available energy states, the lower energy levels being filled first according to the "least energy" principle of physics. The lower energy bands are usually filled with electrons and represent the innermost (closest to the nucleus) bound electrons. *No electric current can be carried by electrons in such a filled band.* The reason for this can be seen by referring to Fig. 3.2. For any value of electron energy, because of the symmetry of the E versus k curve, there exist two states having

[29]This postulate was proposed prior to the Schrödinger theory, just after the Bohr hydrogen atom model was developed. The quantum states here refer to the spatial quantum numbers as well as the spin quantum number (see Sec. 3.4).

wave vectors +k and −k. This corresponds physically to equal electron momenta in the $+x$ and $-x$ directions. Hence if all energy states are filled, in thermal equilibrium the electrons are paired with as many moving in one direction as in the other and the *net* momentum of all electrons is zero. When an electric field is applied, for a filled band there will *still* be just as many electrons traveling in one direction as in the opposite direction and hence the net current will be zero, for every available state in the band is filled and there is no way for the field to increase the total momentum of the electrons in the direction of the field.

Of course, the field can supply energy to these electrons, conceivably raising a few across a forbidden energy gap to the next higher band, which is perhaps not completely filled. Then the possibility does exist that these **excited** electrons may enter into electrical conduction. This, in fact, does happen in the presence of *very high* (10^6 V/cm) electrical fields under special conditions in semiconductor diodes and is known as the **Zener effect.**[30] Normally, however, electrical conduction will occur only if an energy band is *partially* filled. This is the case in a number of metals such as copper, silver, sodium, and aluminum, where the highest energy band containing s-state electrons is only half filled.

Figure 3.4 The band structure of a typical insulator (a), metal (b), and semiconductor (c) according to the energy band theory of solids. The crosshatched areas represent bands filled with electrons. The Fermi energy E_F is indicated for all three cases. Its significance will be discussed later.

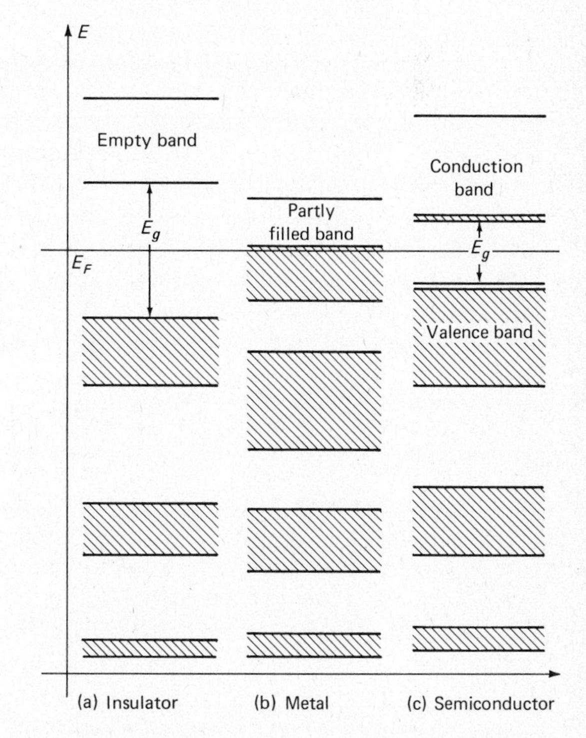

[30]C. Zener, "Non-Adiabatic Crossing of Energy Levels," *Proc. R. Soc. London,* **A137**, 696 (1932).

3.5.3 Electrical Conduction in Metals and Insulators

Figure 3.4 illustrates the distinction between a metal and an insulator in terms of the energy band theory of solids. In an insulator (Fig. 3.4a) the number of electrons present is *just sufficient* to fill some lowermost energy bands. There is an empty band just above the filled bands where electron conduction can take place. A forbidden energy gap of 3–7 eV exists separating the highest filled band from the next empty band. Perhaps at very high temperatures sufficient thermal energy may be present to cause a few electrons to be *raised* into conduction. However, the probability that an electron in the filled band will make a transition upward in energy to a **conduction band** is very small at room temperature. Hence at low temperatures insulators like mica, quartz, and diamond have electrical resistivities of greater than 10^{16} Ω-cm (see Table 3.1).

This is in contrast to resistivities of about 10^{-6} Ω-cm exhibited by some metals also shown in Table 3.1. Here the uppermost band is about half filled with electrons (Fig. 3.4b), accounting for the high electrical conductivity. (The lower filled bands correspond to tightly bound electrons near the atomic nucleus, whereas the conducting electrons correspond to the outer electrons normally responsible for chemical valence.)

TABLE 3.1 Electrical resistivity of metals, semiconductors, and insulators at room temperature.

Material	Resistivity (Ω-cm)
Metals	
Ag	1.59×10^{-6}
Al	2.66×10^{-6}
Au	2.35×10^{-6}
Cu	1.67×10^{-6}
Fe	9.70×10^{-6}
Semiconductors	
Ge	0.47
Si	2.3×10^{5}
GaAs	$>1 \times 10^{8}$
Insulators	
Mica	9×10^{16}
Quartz	3×10^{16}
Diamond	10^{16}

From R. E. Bolz and G. L. Tuve, *Handbook of Tables for Applied Engineering Science* (Chemical Rubber Company, Cleveland, 1970).

3.5.4 Electrical Conductivity of Semiconductor Materials

There is a class of materials called semiconductors which appear quite similar in band structure to insulators. In fact, the major difference is the width of the energy gap above the uppermost filled band. Figure 3.4c is a sketch of the band structure of an **intrinsic (pure) semiconductor.** The energy gap E_g is usually between 0.1 and 2.5 eV. It is sufficient for the description of a number of semiconductor phenomena to consider in detail the uppermost nearly filled energy band and the next highest nearly empty band separated from it by a **forbidden** energy gap.[31] At absolute zero temperature the lower band is completely filled, with all electron energy states occupied corresponding to complete electronic shells; the upper band is completely empty. This lower filled band is referred to as the **valence band** since it corresponds to the energy states of the valence electrons of the atoms. The empty band is called the **conduction band** representing the many energy states to which electrons may be excited by gaining energy from thermal excitation or a very high electric field. At absolute zero Si and GaAs are perfect insulators since the filled band can contribute nothing to electrical conduction because there are no energy states available for electrons to move into and the empty band contains no current carriers. When the electric field is moderate, there is not sufficient energy available to raise electrons across the forbidden energy gap into the conduction band and no electrical conduction takes place. As the temperature is raised, however, there is a small but definite probability that a given electron will be excited into the conduction band. Since there are many electrons in the valence band, a large number will go into conduction. This is shown in Fig. 3.4c. When this occurs an electric field can be applied to supply energy, and hence momentum, to these electrons in the conduction band, resulting in electrical conduction. Also, some states in the lower valence band which have been emptied by this excitation are available, so other electrons in the valence band may be raised in energy. Hence conduction can also take place in the valence band as a result of the availability of a few empty states, or **holes.** *For an intrinsic semiconductor it follows that the number of electrons in conduction equals the number of holes.* These materials have intrinsic resistivities intermediate between those of insulators and metals or about 10–10^7 Ω-cm. Some values for the electrical resistivity of typical metals, insulators, and pure semiconductors are given in Table 3.1.

The forbidden energy gap at room temperature is about 1.12 eV in the case of silicon and about 1.42 eV for gallium arsenide. The element carbon forms a crystalline substance known as diamond by tetrahedral covalent bonds; this material, however, is referred to as an insulator since the width of its energy gap is about 5.5 eV at room temperature. With this wide gap, thermal energy at normal temperatures is insufficient to excite any substantial number of electrons into conduction and the material has extremely low electrical conductivity. A result of the Krönig–Penney solution predicts that the energy gap width

[31]There are several other filled bands below these bands which correspond to complete inner electronic shells of the semiconductor atoms.

depends inversely on the interatomic spacing. It is interesting to note that the dimensions of the conventional unit cell[32] for diamond, silicon, and germanium are, respectively, 3.567, 5.430, and 5.646 Å and the gaps are 5.5, 1.12, and 0.66 eV. Of course, this problem is much more complex and other factors enter as well, such as the electron shell structure of the atoms.

The specification of the energy gap value at which a material passes from insulator to semiconductor is rather arbitrary. However, active electronic devices primarily utilize about 1-eV gap semiconductor materials. The reason for this is twofold: (i) the electrical conductivity is then sufficient so that the application of a few volts yields some milliamperes of current, and (ii) this conductivity can then be increased by the introduction of impurity atoms into the crystal. The latter modify the band structure somewhat by introducing energy levels into the forbidden energy gap, thereby enhancing electrical conduction. This will be discussed in some detail in the later sections on semiconductors. At that time it will be pointed out that there exist two types of **impurity** (as distinct from intrinsic) **semiconductor** materials, one which conducts electricity with negative charges, while the other conducts by positive charges. This fact will be of primary importance in describing the operation of bipolar (two-carrier) junction transistors of the type discussed briefly in Chapter 1.

□ *SUMMARY*

The prediction of the electronic properties of solids is based on solutions of the Schrödinger equation. A brief historical introduction was given of the developments that led to the formulation of the Schrödinger equation. For ease of solution this partial differential equation was separated into a time-independent ordinary differential equation and a time-dependent one. General boundary conditions were specified which can be applied for the solution of the Schrödinger equation. The solution of this equation for the motion of one electron in a periodic potential, as derived by Krönig and Penney, was presented to simulate the transport of an electron in a crystal material. This resulted in the derivation of the energy band theory of solids. Some results of the solution of the Schrödinger equation for the electron in a central electric field were also presented in order to give some definitions which will be of use later in the discussion of the electronic behavior of semiconductor materials. Next the differences between the electrical conduction properties of metals, insulators, and semiconductors were described by utilizing the energy band theory of crystals. The differences were vast and found to be due to the extent of the filling of the energy bands and the presence of energy gaps between bands. Metals were predicted to be excellent electrical conductors, insulators poor conductors, and semiconductors were shown to exhibit an intermediate conductivity.

[32]These are known as the lattice constants and are determined by X-ray diffraction studies.

PROBLEMS

3.1 Compare
(a) the wave nature of the electron and the particle nature of the electron,
(b) Planck's law and de Broglie's law,
(c) wave mechanics and Newtonian mechanics.

3.2 It is desired to produce X-ray radiation with a wavelength of 1 Å.
(a) Through what potential voltage difference must an electron be accelerated in vacuum so that it can, on colliding with a target, generate such a photon? (Assume that all the electron's energy is transferred to the photon.)
(b) What is the de Broglie wavelength of the electron of (a) just before it hits the target?
(c) What is the de Broglie wavelength of a 0.10-g mass moving at a velocity of 1 cm/sec?

3.3 An electron and a photon have the same energy. At what value of this energy (in eV) will their respective wavelengths be equal?

3.4 The work function of tungsten is 4.52 eV and that of cesium is 1.81 eV. The work function refers to the minimum energy necessary to remove an electron from the metal.
(a) Find the maximum wavelength of light for photoelectric emission of electrons from tungsten and cesium.
(b) Which, if any, of these materials can be used in designing a photocell for detecting the laser light of Example 3.1?

3.5 Assume the wave function $\Psi(x,t) = A[\sin(\pi x/L)]e^{-j\omega t}$; find $|\Psi|^2$. Choose A such that $\int |\Psi|^2 \, dx = 1$. Assume that the electron is confined in a region $-L < x < L$.

3.6 Due to the linearity of the Schrödinger equation, superposition is often employed to obtain its general solution. Assume $\Psi_1(x,t)$ and $\Psi_2(x,t)$ are each solutions of the one-dimensional time-dependent Schrödinger equation.
(a) Prove the superposition principle by showing that $\Psi_1 + \Psi_2$ is also a solution.
(b) Is $\Psi_1\Psi_2$ a solution of the Schrödinger equation in general?

3.7 Assume that the Bloch function of Eq. (3.16) is a solution of the one-dimensional Schrödinger equation for the motion of an electron in a periodic potential. By introducing the Bloch function into Eq. (3.14), derive the equation for $u(x)$ that must be solved to determine Ψ.

3.8 **(a)** How many possible 5d states exist for the electron in an atom of hydrogen?
(b) Which of the following are *not* electronic states of the hydrogen atom:

1s, 10s, 112p, 2d, 1d?

Only one electron can occupy each state, with the lower energy states being filled first.
(c) Designate by symbols the highest state occupied by an electron in an atom of the element silicon which contains 14 electrons.

CHAPTER 4

Introduction to the Quantum Theory of Semiconductors in Equilibrium

4.1 INTRODUCTION TO THE ELECTRON PHYSICS OF CRYSTALS

The description of the motion of an electron in a crystal according to the Krönig–Penney quantum mechanical model, which assumes an idealized potential function, was already seen to be rather complex, even in the one-dimensional case. Solutions of the three-dimensional problem—for example, those needed for such real crystals as sodium, silicon, and gallium arsenide—cannot be obtained in closed form. Approximation techniques are used and calculations need to be carried out with the aid of computers. However, as previously discussed, the general outline of the three-dimensional solution is already indicated by the one-dimensional analysis. In fact, if appropriate approximations are made, many of the electrical and thermal properties of crystal materials can be described simply from a quantum mechanical point of view. The trick is to select an approximation which is valid under the conditions of the problem under discussion. For example, the Krönig–Penney solution predicts a complicated relationship between the energy of the electron in a crystal lattice and its wave vector (wavelength or momentum), i.e., $E = E(\mathrm{k})$. However, when this function is plotted (as in Fig. 3.2), it is seen that for a small range of energies near $\mathrm{k} = 0$, at the bottom of a band, E is essentially proportional to k^2. This parabolic relationship suggests that the electron, with wavelength *other* than that corresponding to the Bragg condition [see Eq. (3.18)], is essentially free. This almost free electron near $\mathrm{k} = 0$ will have energy of motion given by

$$[E - E_0] = \frac{m_0 v^2}{2} = \frac{(p - p_0)^2}{2m_0} = \frac{h^2(\mathrm{k} - \mathrm{k}_0)^2}{8\pi^2 m_0}, \tag{4.1}$$

where E_0 and k_0 represent a reference electron energy and wave vector, respectively. Equation (4.1) indicates that this simple expression can be applied where strong interaction with (or scatter by) the lattice atoms is negligible. This results in the **nearly free electron** theory, which can be successfully applied to low energy electrons near the bottom of the conduction band of crystals of monovalent metals such as sodium, or semiconductors such as silicon. Physically these electrons correspond to outer-orbit, valence electrons somewhat loosely bound to the ions that occupy the crystal lattice. The monovalent nature of the atoms in a sodium crystal as well as the tetravalent crystal structure of silicon shown in Fig. 2.7 provide examples in which nearly free electrons are present.

For simplicity, a two-dimensional representation of tetravalent silicon is given in Fig. 4.1. Here the covalent bonds result in the sharing by a central atom of one electron from each of four surrounding atoms. Since the central silicon atom has four electrons in an outer shell, the sharing of these electrons tends to complement the original four, resulting in a completed shell of eight electrons. This is a very stable structure and, owing to the strength of the cova-

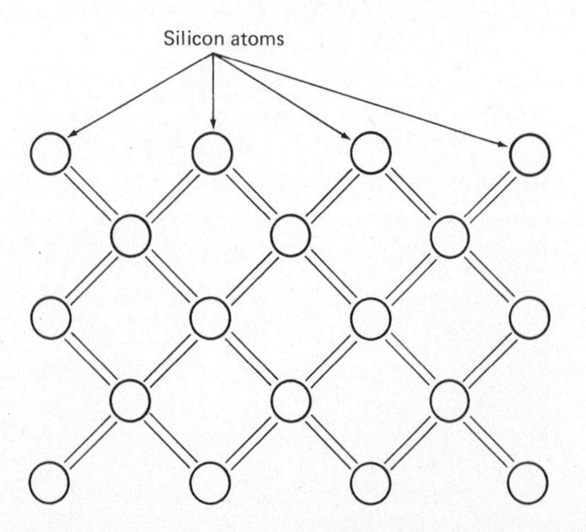

Figure 4.1 Two-dimensional schematic diagram of a silicon crystal. Each pair of lines represents a two-electron covalent bond.

lent bond, practically no electrons can move through the crystal lattice at low temperatures. The material is an electrical insulator at absolute zero temperature. However, as the temperature is raised, thermal energy tends to break a few electrons loose from their bonds and some electrical conduction can take place.

This is a qualitative description of the variation of the electrical behavior of an intrinsic semiconductor with temperature. For a quantitative discussion which will lead to a calculation of the electrical conductivity of semiconductors as a function of temperature, it is necessary to refer to the energy band description of these crystals.

4.2 NEARLY FREE ELECTRON THEORY OF SEMICONDUCTORS

Since the low energy electrons in the conduction band of a crystal can be treated as nearly free, a theory can be developed to predict the conduction properties of these crystals. This formulation must be consistent with the principles of quantum mechanics since subatomic physics is involved. Calculating the electrical conductivity of a material requires a knowledge of the number of electron carriers present. In quantum mechanics this means the calculation of the density of available electron energy states as well as the probability that these states are occupied by electrons. The calculation of the density of electron energy states will now be pursued using the concept of **momentum space,** coupled with the uncertainty principle. The probability of energy state occupancy will be determined by the use of **Fermi–Dirac statistics.**

4.2.1 Density of Electron Energy States

Let us extend Eq. (4.1) to three dimensions by writing the electron kinetic energy as

$$E = \frac{p_x^2 + p_y^2 + p_z^2}{2m_0}, \tag{4.2}$$

where p_x, p_y, and p_z are the x-, y-, and z-components of momentum; m_0 is the electron mass; and E_0 and p_0 are arbitrarily taken as zero. It is convenient now to define a momentum space in a very similar manner to the way in which configurational (x, y, z) space is normally defined. That is, consider three mutually perpendicular orthogonal axes labeled p_x, p_y, and p_z. Figure 4.2 shows an octant in this three-dimensional space. Also shown is a spherical surface in this space which is a surface of constant energy. This follows from Eq. (4.2) compared with the well-known analogous equation of a spherical surface in configurational space, $x^2 + y^2 + z^2 = r^2$, where r is a sphere radius. The radius of this sphere in momentum space is $\sqrt{2mE}$.

In classical theory a continuous set of energy values would be possible for these essentially free electrons. However, the uncertainty principle restricts our

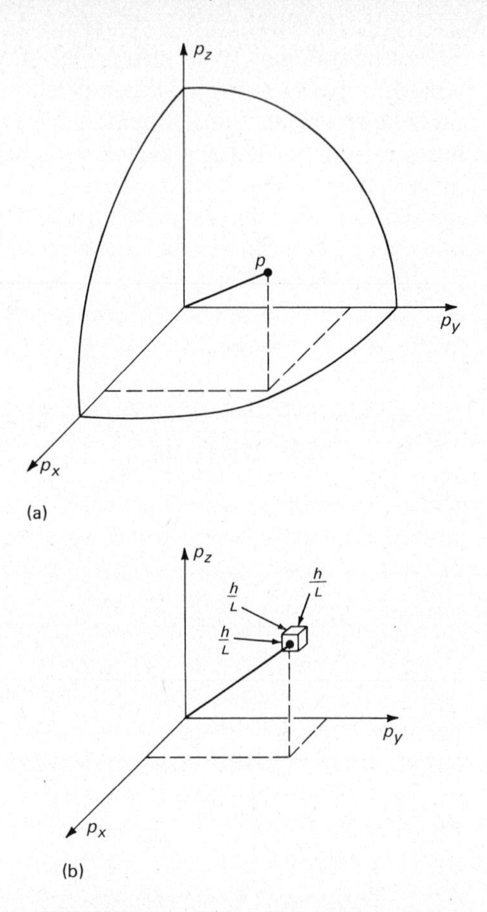

Figure 4.2 (a) An octant in three-dimensional momentum space. The spherical surface is one of constant energy for free electrons. (b) An elemental volume in momentum space of size h^3/L^3, as specified by the uncertainty principle.

ability to define both the electron momentum and position with arbitrary accuracy, which leads to a discrete number of energy values in a crystal of finite dimensions. Let us now determine the number of electron states possible in a crystal of a given size.[1]

If the volume of the crystal in which we are considering these electrons is assumed to be a cube whose sides are of length L, the location of the electron is given by

$$0 < x < L, \qquad 0 < y < L, \qquad 0 < z < L. \tag{4.3}$$

Hence $\Delta x = \Delta y = \Delta z = L$, since the electron can be found anywhere in this volume. The minimum uncertainty (fuzziness) of the electron momenta in each of the three directions is then

$$\Delta p_x = h/L, \qquad \Delta p_y = h/L, \qquad \Delta p_z = h/L. \tag{4.4}$$

[1]This problem may be treated by solving the Schrödinger equation for a free electron in a three-dimensional box; such a formulation *automatically* yields energy quantization corresponding to the uncertainty principle!

This follows from Eq. (3.13a). The product $\Delta p_x \, \Delta p_y \, \Delta p_z$ defines an elemental volume in momentum space (see Fig. 4.2) and represents a discrete electronic state. That is, it is not possible to define or resolve any electron state of momentum or energy any more precisely than given by this volume in momentum space. Hence, from Eq. (4.4), this elementary quantum state is defined by

$$\Delta p_x \, \Delta p_y \, \Delta p_z = h^3/L^3. \tag{4.5}$$

These cubes in p-space correspond to electronic states with energies below some value E. Hence all electron states are contained within a sphere of radius $\sqrt{2m_0E}$.

Consider the above description to refer to the electrons at the bottom of the conduction band in a semiconductor. The Pauli exclusion principle dictates that there are N possible electron energy states up to a value E. Since two electrons (with opposite spin) can be accommodated in each elemental momentum space volume, the number of these electron states can be expressed as twice the total volume under E divided by the elemental volume, or

$$N = 2\,\frac{(4\pi/3)(2m_0E)^{3/2}}{(h/L)^3}\,. \tag{4.6}$$

In this derivation of the number of electron states, the electron mass has been considered as that of an electron in free space, m_0. However, in our present problem the electron is actually moving through a crystal lattice. It is found convenient to take into account the effect of the lattice ions on the motion of the *nearly free* electrons by assigning them an **effective mass** m^*. The broader significance of this concept will be discussed later.

If this is done then the motion of the electrons, for example in an applied electric field, can be given a pseudo-classical description in terms of Newton's law $F = m^*a$. It can be seen that the degree of binding of the electron to the lattice ions will determine its effective mass. Strong binding would be indicated by a larger effective mass as the ionic electric field slows down the electron. However, the picture is not quite so simple and it is found from electron conductivity measurements that the ionic field in some cases is actually *aiding* the electron motion, or that the effective mass is lower than the free-space mass. A more rigorous and general definition of the effective mass will be given later in a more detailed discussion of electron transport. Table 4.1 gives values of the effective mass of electrons in some semiconductors and metals. Also given are the effective masses of holes, which are positively charged current carriers, to be discussed later.

Let us now write down the number of electronic states with energy between some value E and $E + dE$ (see Fig. 4.3). This number may be obtained by taking the elemental volume in momentum space between the spherical shells corresponding to energies E and $E + dE$, and dividing by the elemental volume of one electronic state, h^3/L^3. Letting $dE \to 0$ will yield the desired number. Mathematically this operation may be performed by taking the differential

TABLE 4.1 Energy gaps and effective masses[a] of electrons and holes in some crystal materials.

Crystal Material	Energy Gap (300°K, eV)	Electron Effective Mass m^*/m_0	Hole Effective Mass m^*/m_0
Sodium	—	1.2	—
Germanium	0.66	0.22	0.29
Silicon	1.12	0.33	0.55
GaAs	1.42	0.067	0.82
GaP	2.26	0.82	0.60

[a]The effective masses are the so-called density of states masses, which are defined in this chapter.

of the total number of states N with energy less than E, which from Eq. (4.6) is $N(E) = 8\pi(2m^*E)^{3/2}L^3/3h^3$.

This gives

$$dN = \frac{4\pi(2m^*)^{3/2}L^3}{h^3} E^{1/2}\, dE. \tag{4.7}$$

Hence the energy **density of states** function $\rho(E) = (dN/dE)/L^3$ is

$$\rho(E) = \frac{8\sqrt{2}\pi(m^*)^{3/2}}{h^3} E^{1/2}, \tag{4.8}$$

which represents the number of quantum states at any energy per *unit energy* per *unit volume;* it varies as the square root of energy. These energy states exist in the energy band and are available to electrons but are not necessarily occupied.

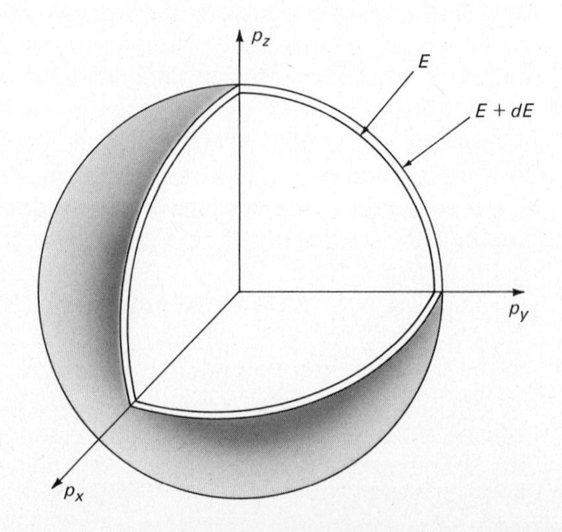

Figure 4.3 A figure useful for the determination of the density of electronic states versus the energy of nearly free electrons in a crystal. Two spherical surfaces of constant energy in momentum space are shown differing only infinitesimally in energy.

4.2.2 The Fermi–Dirac Distribution Function

The density of electronic energy states near the bottom of an energy band of a crystal has just been derived. In order to determine the number of electrons present in these states, an expression is needed which gives the probability that any energy state E is filled. A function for expressing the probability of an electron occupying any particular energy level E at an *absolute* temperature T is given by

$$f(E) = \frac{1}{e^{(E-E_F)/kT} + 1}. \tag{4.9}$$

This is the famous **Fermi–Dirac distribution function** which was derived by Enrico Fermi in 1926[2] for electrons obeying the Pauli exclusion principle. This derivation was accomplished by asking the question: Given a set of energy states, what is the statistically most probable distribution of electrons among these states assuming no *a priori* preferable state and no more than one electron per state (Pauli principle)? This result dictates that *the probability of occupancy of energy levels by electrons is one below E_F and zero above E_F*, at an

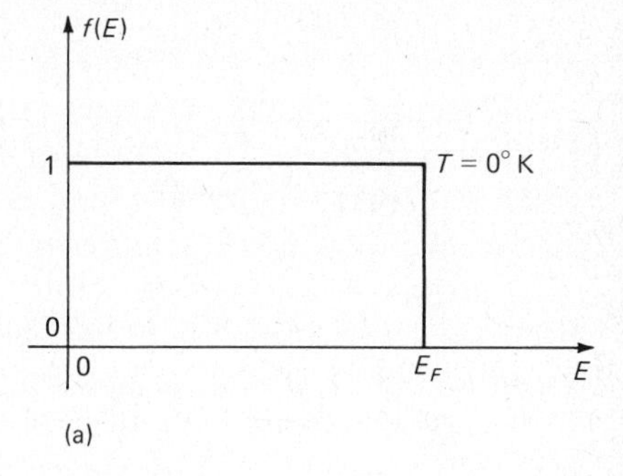

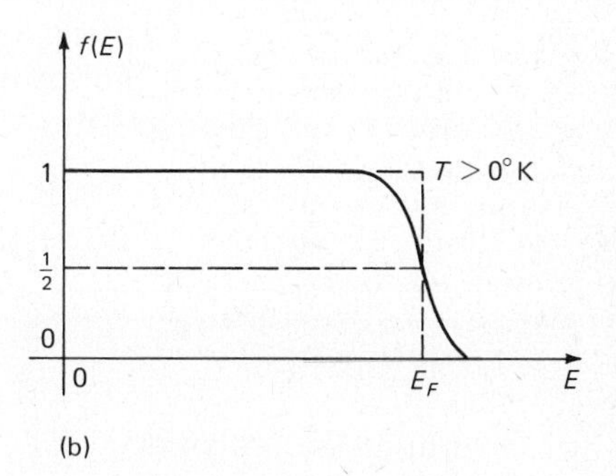

Figure 4.4 The Fermi function $f(E)$ for two temperatures, (a) $T = 0°K$, (b) $T > 0°K$. At $T = 0°K$ the probability of occupancy of energy levels by electrons is absolute (= 1) for energy values less than the Fermi energy E_F, and zero for $E > E_F$. At $T > 0°K$, electrons may occupy levels above E_F. At $E = E_F$ the occupation probability is always ½.

[2]This function was derived by Fermi and also independently by P. A. M. Dirac in the same year. See E. Fermi, "Quantization of the Ideal Monatomic Gas," *Z. Phys.*, **36**, 902 (1926). Here k represents the Boltzmann constant ($= 1.38 \times 10^{-23}$ J/°K). In this book the Boltzmann constant is normally followed by T, the absolute temperature.

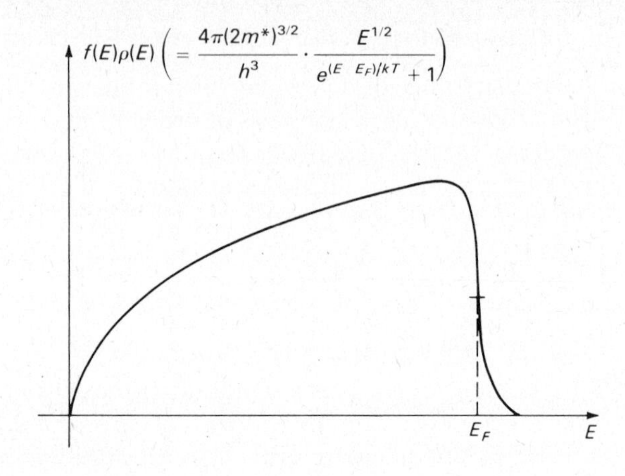

Figure 4.5 A graph of the number of electrons per unit volume in any energy state between E and $E + dE$ at a temperature $T > 0$°K. The Fermi energy where the occupation probability is ½ is indicated.

absolute temperature $T = 0$°K. Here E_F is the **Fermi energy level.** At any other temperature the thermal energy present will raise (or excite) electrons to higher energy levels, leaving some levels below empty. The probability of an energy level E being occupied at any temperature T is predicted by this Fermi function, plotted in Fig. 4.4. The effect of temperature is to make it possible for some electrons to have energies above E_F. In fact, there is a reasonable probability for some electrons to have kT units of energy in excess of E_F but not much in excess. Figure 4.4b and Eq. (4.9) indicate a very useful definition of the Fermi energy level: *The Fermi level is that energy for which the probability of occupancy by an electron is one-half.* This concept is universally true for all electrons which obey the Fermi statistics and will be especially helpful in the discussion of the electrical properties of semiconductors.

Expressions have now been given for the density of states and the probability of occupancy of these states by electrons. The product of these quantities given by Eqs. (4.8) and (4.9) yields an equation specifying the *number of electrons per unit volume* occupying an energy state between E and $E + dE$ at a particular temperature. A plot of this function is shown in Fig. 4.5; it will be used shortly to calculate the electrical conductivity of semiconductor materials.

EXAMPLE 4.1

Determine the probability that an energy level is occupied by an electron at 300°K if it is located above the Fermi level by

(a) 0.026 eV ($= kT$),
(b) 0.078 eV ($= 3kT$),
(c) repeat parts (a) and (b) at 600°K.

Solution

From Eq. (4.9)

(a)

$$f(E)|_{0.026\text{ eV}} = \frac{1}{e^{kT/kT} + 1} = \underline{0.27},$$

(b)

$$f(E)|_{0.078\text{ eV}} = \frac{1}{e^{3kT/kT} + 1} = \underline{0.05}.$$

The latter result indicates that at room temperature, any energy level only 0.078 eV above the Fermi energy level has a probability of 1 in 20 of being occupied. At the Fermi energy this probability of occupancy is 1 in 2.

(c) At 600°K, 0.026 eV = ½kT and 0.078 eV = $\frac{3}{2}kT$. Hence at 600°K

$$f(E)|_{0.026\text{ eV}} = \frac{1}{e^{(1/2)kT/kT} + 1} = \underline{0.38}.$$

$$f(E)|_{0.078\text{ eV}} = \frac{1}{e^{(3/2)kT/kT} + 1} = \underline{0.18}.$$

Hence the occupation probability increases substantially with temperature increase.

4.3 HOLES IN SEMICONDUCTORS

The previous discussion related to the nearly free low energy electrons at the bottom of the conduction band of a semiconductor which can conduct current. It was pointed out in Chapter 1 that electric current in semiconductors can be conducted by positively charged **holes.** These holes occur at the *top* of a nearly filled energy band, located below the conduction band, and result from the presence of some empty states near the top of this **valence** band. These electron energy states occur close to the value of energy (or electron wave vector) at which the Bragg condition applies, hence these electrons interact strongly with the crystal lattice ions and are certainly not "free." However, the energy variation with electron momentum or wave vector is nearly parabolic (see Fig. 3.2), which permits these carriers to be treated in a manner similar to free carriers. However, note that for such electrons the *rate* of energy increase *decreases* as its wave vector (momentum) *increases.* It is this peculiar property that indicates that these electrons exhibit a *negative effective mass.* In semiconductors it is important to account for the electrical conduction in a band completely filled with electrons except for a few empty states at the top of the band.

The conduction properties of this nearly filled band are most conveniently accounted for by treating the electrical conduction due to the *absence* of electrons (with negative charge and negative mass) in these few empty states. This may be accomplished even more conveniently by considering that these states are *filled* with particles of *positive* charge and *positive* mass called holes. Hence the electrical conduction in the nearly filled valence band of a semiconductor is described in terms of the flow of holes. The fact that semiconductor materials can conduct current both by negatively charged electrons and by positively charged holes makes bipolar, two-carrier transistor operation possible. This will be discussed in a later section on transistor devices.

4.4 THE DENSITY OF ELECTRONS AND HOLES IN PURE (INTRINSIC) SEMICONDUCTORS

The task is now to calculate the number of conducting electrons and holes in an **intrinsic** semiconductor (such as silicon) in order to be able to express the conductivity of the substance in terms of this and other physical parameters such as carrier collision relaxation time and effective mass. In this way the difference between the conducting properties of silicon, germanium, and gallium arsenide can be understood. It turns out to be convenient to give the number of current carriers in terms of a parameter which expresses the state of a material—the Fermi energy.

4.4.1 The Fermi Energy in an Intrinsic Semiconductor

The density of electrons and holes in an intrinsic semiconductor can be stated in terms of the density of states in the conduction and valence bands and the Fermi–Dirac distribution function for occupancy of these states. It has already been pointed out in Section 4.1 that the electrons at the very top and bottom of an energy band can be assumed (approximately) to follow the relationship $|E - E_0| \propto k^2$. Hence the density of states function per unit energy per unit volume at the *bottom* of a conduction band has the form [see Eq. (4.8)]

$$\rho_c(E)\,dE = \frac{8\sqrt{2}\pi}{h^3}\,m^{*3/2}_n(E - E_c)^{1/2}\,dE. \tag{4.10}$$

Similarly the density of states at the *top* of a valence band has the form

$$\rho_v(E)\,dE = \frac{8\sqrt{2}\pi}{h^3}\,m^{*3/2}_p(E_v - E)^{1/2}\,dE. \tag{4.11}$$

Hence $\rho_c(E)$ and $\rho_v(E)$ represent the density of states per unit energy per unit volume near the bottom of the conduction band and near the top of the valence band, respectively, which is of primary interest for an intrinsic semiconductor; m^*_n and m^*_p refer to the effective masses of electrons (negative charge) and holes (positive charge), respectively, and E_c and E_v correspond to

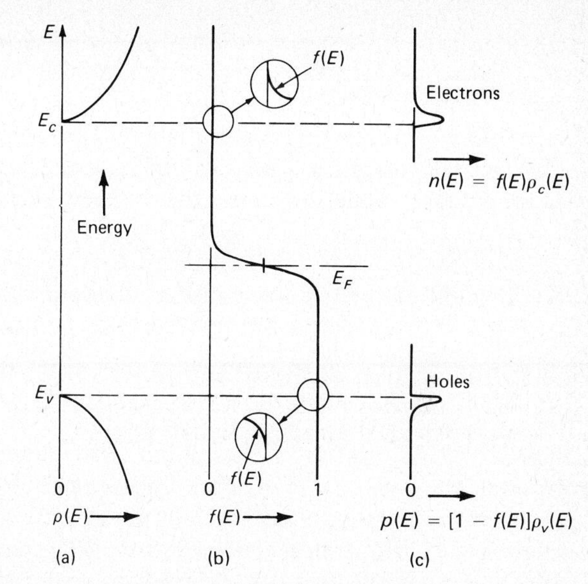

Figure 4.6 Calculation of the density of electrons and holes in an intrinsic semiconductor. (a) Density of states function at the bottom of the conduction band and top of the valence band.
(b) The Fermi function with $f(E)$ greatly magnified at the edges of the bands and shown in the circles.
(c) Density of electrons and holes.

the energies defining the bottom of the conduction band and top of the valence band. This is shown in Fig. 4.6.

Next we must consider the actual number of electrons *occupying* these states. This information is provided by the Fermi–Dirac distribution function of Eq. (4.9), which states that at a temperature of absolute zero any energy level below the Fermi energy is occupied whereas all levels above the Fermi energy are empty. Since it is known that at absolute zero temperature the valence band in a semiconductor is completely full and the conduction band completely empty, this locates the Fermi energy as somewhere between E_c and E_v. At any temperature above absolute zero some electrons fill energy levels in the conduction band corresponding to electrons which have been excited there from the valence band, leaving an equal number of empty states (holes) behind. The statement that the number of electrons within dE of E_c equals the number of holes within dE of E_v is expressed as

$$\rho_c(E)f(E)\,dE = \rho_v(E)[1 - f(E)]\,dE, \tag{4.12}$$

where $[1 - f(E)]$ represents the probability of an energy level in the valence band being empty. If the effective masses of electrons and holes are approximately equal, then $\rho_c(E) \simeq \rho_v(E)$, and using Eq. (4.9) for the Fermi–Dirac distribution function we get for electrons and holes just at the band edges, E_c and E_v,

$$\frac{1}{e^{(E_c - E_F)/kT} + 1} = 1 - \frac{1}{e^{(E_v - E_F)/kT} + 1}. \tag{4.13}$$

After some algebraic manipulations, solving for E_F we get

$$E_F = \tfrac{1}{2}(E_c + E_v), \tag{4.14}$$

and since $E_c - E_v = E_g$, the forbidden energy gap, then we have

$$E_F = E_v + \tfrac{1}{2}E_g. \tag{4.15}$$

Hence the Fermi energy level for an intrinsic semiconductor is at the middle of the energy gap if the electron and hole effective masses are equal. This is *essentially* true even if m_n^* doesn't equal m_p^*, for then

$$E_F = E_v + \tfrac{1}{2}E_g + kT \ln(m_p^*/m_n^*)^{3/4}, \tag{4.16}$$

and the additional term is usually small compared to $\tfrac{1}{2}E_g$.

4.4.2 Calculation of the Electron and Hole Densities

Now to calculate the density of conduction electrons in an intrinsic semiconductor, consider first the Fermi–Dirac distribution function under the condition that $E_c - E_F > kT$. Near room temperature kT is about 0.026 eV and $E_c - E_F > kT$ means that the Fermi energy is at least a multiple of kT removed from E_c. In practical cases involving pure or moderately doped silicon,[3] E_F is at least 0.1 eV below the conduction band edge E_c. Then the Fermi–Dirac function reduces to the Boltzmann function

$$f(E) \simeq e^{-(E-E_F)/kT} \tag{4.17}$$

Using this approximation we obtain the number of conduction electrons per unit volume at a temperature T by integrating the product of the occupation probability and the density of available states over the conduction band:

$$\begin{aligned} n(T) &= \int_{E_c}^{\infty} f(E)\rho(E)\, dE \\ &= \int_{E_c}^{\infty} \frac{8\sqrt{2}\pi}{h^3}\, m_n^{*3/2}(E - E_c)^{1/2} e^{(E_F-E)/kT}\, dE. \end{aligned} \tag{4.18}$$

Actually the top of the conduction band has a finite energy value, but since the Fermi–Dirac function falls off so rapidly as energy increases, extending the integral limit to infinity contributes little error but facilitates the integration. This means that electrons only occupy states near the bottom of the conduction band.

Since energy is the only variable in Eq. (4.18), it may be rewritten as

$$n(T) = \frac{8\sqrt{2}\pi}{h^3}\, m_n^{*3/2} e^{E_F/kT} \int_{E_c}^{\infty} (E - E_c)^{1/2} e^{-E/kT}\, dE. \tag{4.19}$$

Changing the variable of integration to $x = (E - E_c)/kT$ yields

$$n = \frac{8\sqrt{2}\pi}{h^3}\, (m_n^* kT)^{3/2} e^{(E_F-E_c)/kT} \int_0^{\infty} x^{1/2} e^{-x}\, dx. \tag{4.20}$$

[3]Moderately doped semiconductors are referred to as "nondegenerate."

Integrals in the form of Eq. (4.20) are tabulated and this one equals $\frac{1}{2}\sqrt{\pi}$. Hence we get

$$n = 2(2\pi m_n^* kT/h^2)^{3/2} e^{-(E_c - E_F)/kT}. \tag{4.21}$$

In certain semiconductor materials the conduction band has more than one parabolic-shaped minimum (see Fig. 5.8b). Then the right-hand side of Eq. (4.21) must be multiplied by the number of such minima. For silicon this number, M_c, is 6, for germanium the equivalent of 4, but for GaAs it is 1. In the case of the valence band for these semiconductors this number is 1. Hence in general Eq. (4.21) becomes

$$n = (M_c)2(2\pi m_n^* kT/h^2)^{3/2} e^{-(E_c - E_F)/kT} \equiv N_c e^{-(E_F - E_c)/kT}, \tag{4.21a}$$

where N_c is the total density of states near the edge of the conduction band, or

$$n = (M_c)4.82 \times 10^{15} \left(\frac{m_n^*}{m_0}\right)^{3/2} T^{3/2} e^{(E_F - E_c)/kT} \text{ electrons/cm}^3. \tag{4.21b}$$

The number of empty states near the top of the valence band can be calculated in a very similar manner. Here we must use the Fermi–Dirac function for *unoccupied* states, which is defined as

$$f'(E) = 1 - f(E) = 1 - \frac{1}{e^{(E - E_F)/kT} + 1} \tag{4.22}$$

or

$$f'(E) = \frac{1}{e^{(E_F - E)/kT} + 1}. \tag{4.23}$$

Repeating a similar argument as in the case of electrons in the conduction band, assume $E_F - E_v > kT$; that is, the Fermi energy is at least a few kT energy units above the top of the valence band. Then Eq. (4.23) reduces to

$$f'(E) \cong e^{-(E_F - E)/kT}. \tag{4.24}$$

The total number of empty energy states in the valence band then becomes

$$p(T) = \frac{8\sqrt{2}\pi}{h^3} m_p^{*3/2} e^{-E_F/kT} \int_{-\infty}^{E_v} (E_v - E)^{1/2} e^{E/kT}\, dE. \tag{4.25}$$

Hence again for mathematical convenience the lower integration limit is chosen as minus infinity although there exists a finite lowest energy in this valence band. The rationale again is the rapid falloff of the Fermi function for holes as the energy decreases. Introducing $y = (E_v - E)/kT$ as the integration variable in Eq. (4.25) yields

$$p = \frac{8\sqrt{2}\pi}{h^3} (m_p^* kT)^{3/2} e^{-(E_F - E)/kT} \int_0^\infty y^{1/2} e^{-y}\, dy. \tag{4.26}$$

Writing in the value of the integral as before gives

$$p = 2(2\pi m_p^* kT/h^2)^{3/2} e^{-(E_F - E_v)/kT} \equiv N_v e^{(E_v - E_F)/kT}, \tag{4.27a}$$

where N_v is the density of states near the edge of the valence band. This can be written as

$$p = 4.82 \times 10^{15} \left(\frac{m_p^*}{m_0}\right)^{3/2} T^{3/2} e^{(E_v - E_F)/kT} \text{ holes/cm}^3 \tag{4.27b}$$

for the density of holes in the valence band. As was already discussed, in interpreting the conduction properties of semiconductors these empty states or holes can be considered as positively charged current carriers with a positive effective mass m_p^*. It was pointed out previously that this is more convenient than treating the electrical conduction due to all of the occupied electron states in the valence band with the exception of a few empty states at the top of the band (see Section 4.3).

4.4.3 The Electron–Hole Product

Multiplying Eqs. (4.21a) and (4.27a) together yields an extremely useful expression, the electron–hole product, which is

$$np = 4M_c \left(\frac{2\pi(m_p^* m_n^*)^{1/2} kT}{h^2}\right)^3 e^{-E_g/kT}. \tag{4.28}$$

This is the so-called **law of mass action** of electrons and holes in a semiconductor. It states that the product of electrons and holes in a semiconductor in thermal equilibrium is a *function only of temperature* where the hole and electron effective masses and the energy gap are very mild functions of temperature.

The fact is that foreign atoms introduced substitutionally into a pure semiconductor crystal can also contribute current carriers. This will be referred to as an **impurity semiconductor** in contrast to an intrinsic semiconductor. In this case, the numbers of electrons and holes are no longer necessarily equal. Nevertheless, the assumptions made in the above derivation of the *np* product are also true for the impurity semiconductors. Hence Eq. (4.28) applies for *both* intrinsic and impurity semiconductors. Therefore, given the number of electrons in a semiconductor of the type being discussed, the number of holes can be determined if m_n^*, m_p^*, E_g, and kT are known. This fact is extremely useful in explaining the behavior of semiconductor devices and will be used later.

In the intrinsic semiconductor $n = p$ is usually referred to as the intrinsic number of carriers per unit volume, n_i, and Eq. (4.28) is given by

$$np = n_i^2 \tag{4.29}$$

where

$$n_i(T) = 2\sqrt{M_c} \left(\frac{2\pi(m_p^* m_n^*)^{1/2} kT}{h^2}\right)^{3/2} e^{-E_g/2kT}. \tag{4.30}$$

For calculation purposes Eq. (4.30) becomes

$$n_i = 4.82 \times 10^{15}\sqrt{M_c}(m_n^* m_p^*/m_0^2)^{3/4}T^{3/2}e^{-Eg/2kT}\ \text{cm}^{-3}. \tag{4.30a}$$

At 300°K, n_i is $1.5 \times 10^{10}/\text{cm}^3$ for Si and $1.8 \times 10^6/\text{cm}^3$ for GaAs. These values have been determined experimentally and should be used to work out the problems in this text.

EXAMPLE 4.2

Intrinsic semiconductor material A has an energy gap of 0.36 eV, while material B has an energy gap of 0.72 eV. Compare the intrinsic density of carriers in these two semiconductor materials at 300°K. Assume that the effective masses of all the electrons and holes are equal to the free electron mass.

Solution

Using Eq. (4.30) we have

$$\frac{n_{iA}}{n_{iB}} = \frac{e^{-Eg_A/2kT}}{e^{-Eg_B/2kT}} = e^{Eg_B - Eg_A/2kT}$$

$$= e^{(0.72-0.36)\ \text{eV}/0.052\ \text{eV}}$$

$$= \underline{1000}.$$

Hence, although the energy gap of these two intrinsic semiconductors differs only by a factor of 2, the intrinsic density of carriers in the narrower gap material is 1000 times greater than that in the wider gap semiconductor.

Noting that $n = n_i$ and $p = n_i$ for an intrinsic (pure) semiconductor in equilibrium, where E_F is defined as E_i, and using Eqs. (4.21) and (4.27), one obtains very useful, *generally applicable* expressions for the electron and hole densities as

$$n = n_i e^{(E_F - E_i)/kT} \tag{4.31a}$$

and

$$p = n_i e^{(E_i - E_F)/kT}. \tag{4.31b}$$

4.5 THE IMPURITY SEMICONDUCTOR

The introduction of foreign atoms into the lattice of a semiconductor crystal significantly affects the electrical conductivity of that material. Atoms of the elements antimony, arsenic, boron, or gallium can be introduced into a silicon or germanium crystal lattice by heating the crystal in contact with a vapor of these elements at temperatures near 1000°C. Because of the similarity of the atomic radius of these impurity atoms to that of the semiconductor, they are found to enter the lattice in place of silicon atoms, **substitutionally.** Similarly zinc, cadmium, selenium, or tellurium atoms can replace the gallium or arsenic atoms in GaAs in a substitutional fashion.

4.5.1 Donors

An arsenic atom can be made to occupy an empty lattice site normally occupied by a silicon atom (vacancy) in a pure silicon crystal. Now the unique semiconducting properties of silicon or germanium result from these elements having an electron valency of four. They appear in column IV of the periodic table of elements. When an arsenic atom substitutes for a silicon atom, it not only supplies the four electrons for bonding normally supplied by the silicon, but introduces an extra electron owing to its valency of five. The four arsenic electrons are strongly bound in the silicon lattice in the normal tetrahedral covalent bonds. However, the extra electron is very loosely bound to its nucleus, much like the valence electron in metals. At some moderate temperature this extra electron contributes to the electrical conductivity of the semiconductor, providing greater electrical conduction than pure silicon. This is shown schematically in Fig. 4.7.

For convenience in dealing quantitatively with these electron-contributing or **donor-type** impurities, some technique must be devised for fitting these impurity atoms into the energy band scheme. Let us assume (as is usually the case) that there are only relatively few of these impurities in a silicon lattice.[4]

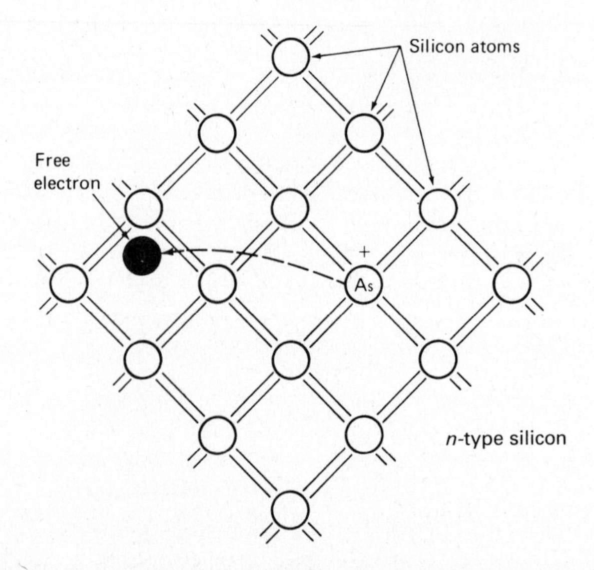

Figure 4.7 Free electron arising from ionization of a substitutional arsenic impurity atom which becomes positively ionized.

[4]Common impurity levels range from one to a million foreign atoms per billion atoms of silicon.

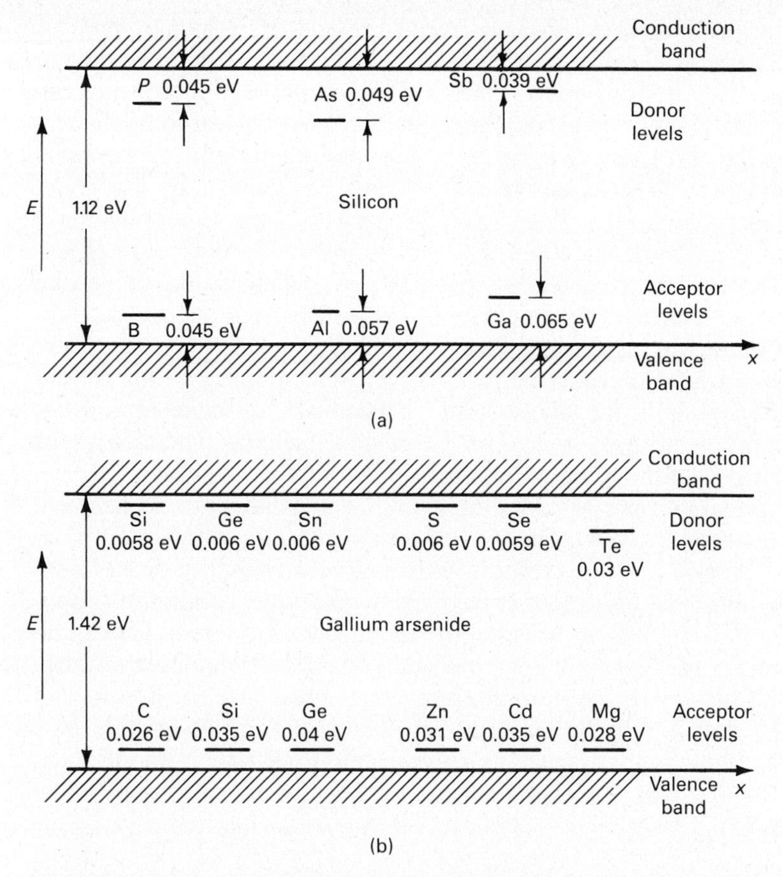

Figure 4.8 Energy levels of some impurity states in (a) silicon and (b) gallium arsenide. The levels are drawn as short dashes near the particular impurity to signify the localized nature of an electron in this energy level. The donor and acceptor energy values are measured *from* the conduction and valence band edges, respectively.

Then the extra donor electron of the arsenic atom is attracted by the unit net positive charge of its nucleus (the atom must be electrically neutral). This is taken as similar to the problem of the electron attached to the hydrogen atom nucleus with the exception that the arsenic atom is embedded in a matrix of silicon whereas the hydrogen atom is in vacuum. Now the binding energy of an electron to a hydrogen nucleus is 13.6 eV and can be obtained by the formula

$$E_B = -m_0 q^4/8\epsilon_0^2 h^2 = -13.6 \text{ eV}. \tag{4.32}$$

The binding energy of the electron to the arsenic nucleus in a silicon lattice can be roughly estimated by modifying this expression somewhat. The dielectric constant now should be 11.9 as for silicon compared to 1.0 for vacuum. Since the dielectric constant enters into the energy expression as an inverse square, the donor-electron binding energy becomes $-13.6/(11.9)^2 \simeq -0.1$ eV.[5] This indicates that a donor electron requires about 0.1 eV of energy to remove it from the influence of its nucleus so that it can participate in electronic conduction. Hence the donor energy state must be about 0.1 eV less than the conduction band edge of silicon. This can be indicated on the normal energy band diagram for intrinsic silicon as shown in Fig. 4.8a. The figure shows that about 0.049 eV of energy is required to raise an electron on an arsenic donor level just into the conduction band. When this occurs the donor atom becomes ionized and takes on a net positive charge. The 0.049 eV is a measured rather than a calculated value. Note that the impurity levels are localized in space as indicated by short lines, whereas the edges of the conduction and valence band are not. This corresponds to the fact that the relatively few impurity atoms present occur only occasionally in the silicon lattice. Silicon containing arsenic, phosphorus, or antimony atoms is said to be n-type **doped** silicon. These elements appear in column V of the periodic table of elements.

The situation for impurity-containing gallium arsenide crystals is somewhat similar to that for silicon. However, in this case either a gallium atom (column III) or an arsenic atom (column V) can be substituted for by an impurity atom. Following the argument given for silicon above, column VI impurities such as selenium, tellurium, or sulfur should act as donors in GaAs; column IV impurities such as silicon, germanium, or tin should supply an extra electron when replacing the gallium atom in GaAs and hence are donors and produce n-type conduction. Figure 4.8b shows the energy levels for these donors. Note that the donor levels for GaAs are very shallow (close to the conduction band edge) owing to the small effective mass of electrons in this material [see Eq. (4.32)].

4.5.2 Acceptors

A corresponding set of arguments can be given now for the case of an impurity atom like boron introduced substitutionally into a silicon crystal. The boron atom has a valence of only 3 and hence is one electron short of satisfying the covalent silicon atom bonding structure. It is sometimes convenient to view this as an acceptor nucleus with a positively charged hole orbiting around it. Should a free electron somehow be made available, it would tend to be captured on this boron **acceptor** site, completing the tetrahedral bond and *ionizing* the acceptor by taking on a negative charge. This is shown in Fig. 4.9. The electron may actually be one which comes from one of the valence electrons of some nearby atom. This provides an empty energy state or hole in the valence band structure of silicon.[6] Energy must be supplied to remove the elec-

[5]The actual calculated value should be even less since the "effective" electron mass in silicon is less than the free electron mass.

[6]An ionized acceptor contributes a positive hole to conduction in the valence band. This is identically equivalent to stating that the acceptor energy level has an electron excited into it from the valence band leaving a mobile hole behind.

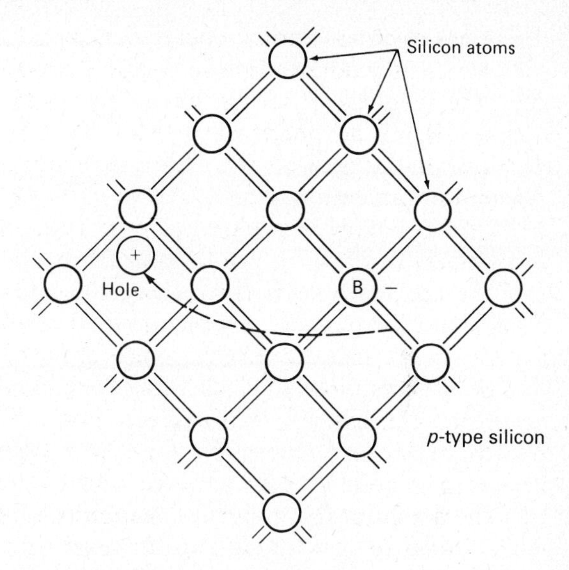

Figure 4.9 Free hole arising from ionization of a substitutional boron impurity atom which captures an electron, becoming a negative ion.

tron from the silicon covalent tetrahedral bond so that it can be captured by an acceptor atom. This energy is also of the order of 0.1 eV as for the donor electron ionization, and hence the acceptor level appears about 0.1 eV above the edge of the valence band. Some such acceptor levels are indicated in Fig. 4.8a.

The electrical conduction in the valence band of a semiconductor containing acceptor atoms can now be described. Under thermal equilibrium conditions in the crystal lattice, some valence electrons of silicon atoms come free from their normal positions and are immobilized or trapped on an acceptor impurity site. This action leaves behind an empty energy level in the semiconductor valence band. Now electrical conduction can take place in the valence band by another valence electron being raised in energy by an electric field so as to fill this empty hole. The electron so raised in energy itself creates a hole and another valence electron can be raised into this energy level by the field. This is pictured schematically in Fig. 4.10. If the excited valence electron moves to the left to fill an empty valence state there, leaving behind a hole, the empty state has *apparently moved to the right.* This leads to a simple but not strictly correct way[7] of visualizing the motion of these bound electrons by considering their corresponding hole moving in a direction opposite to that of the valence electrons. Thus in an electric field the holes in the valence band will move in a direction opposite to electrons in the conduction band and hence the concept of the hole as a particle with a positive charge. This is not strictly correct since it is the motion of the enormous number of electrons present in a nearly filled valence band that must be considered rather than the motion of

[7]See R. B. Adler et al., *Introduction to Semiconductor Physics,* S.E.E.C. (Wiley, New York, 1964), Vol. I, Sec. 1.3.2.

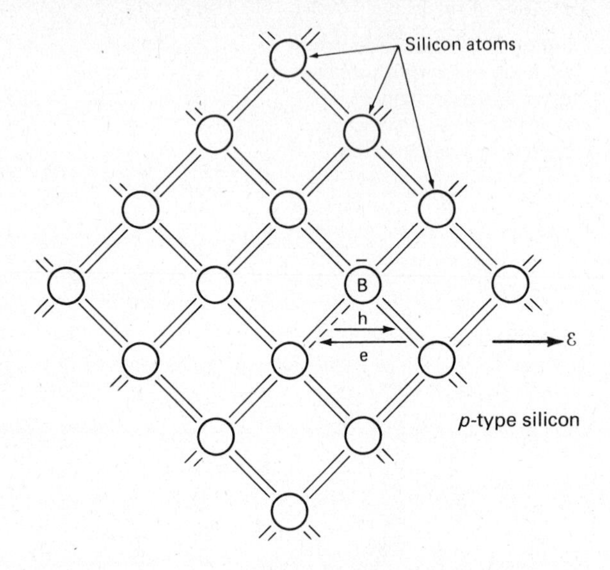

Figure 4.10 Free hole arising from ionization of a substitutional boron impurity atom is accelerated to the right by the electric field. Correspondingly, a valence electron moves to the left to fill the empty state, leaving an empty valence state behind.

the few holes. The concept of the mobile charge carrier called the hole is an expedient in discussing the electrical conductivity exhibited by all the electrons in the valence band of a semiconductor; this is similar to the way in which the effective mass concept permits the use of a nearly free electron analysis for the motion of an electron in the crystal field of a solid.

When a zinc atom (column II) replaces a gallium atom (column III) in a GaAs crystal, this represents a deficiency of one electron and hence gives rise to an acceptor state in this material. Hence column II atoms such as zinc, cadmium, or magnesium act as acceptor impurities in GaAs. Column IV impurities such as carbon, silicon, or germanium can replace the column V element arsenic, and hence they act as acceptors. Gallium arsenide crystals doped with these impurities are *p*-type and conduct by holes. Figure 4.8b shows the energy levels for these acceptors.

4.5.3 Impurity Semiconductor Doping

Note that the elements from group III in the periodic table (B, Al, Ga, and In) are acceptors in Ge and Si, and produce conduction by holes. With the addition of these materials Ge and Si conduct by positive charges and are termed *p*-type. The elements from group V in the periodic table (P, As, and Sb) are donors, produce conduction by electrons, and such semiconductors are referred to as *n*-type. The science of purposefully adding *p*- and *n*-type impurities to a semiconductor material is the essence of semiconductor device technology and is accomplished by subjecting the semiconductor to the impurity at high temperature, or implanting these atoms.

A semiconductor material can contain both *n*- and *p*-type impurities. If the impurities are primarily acceptors the material is *p*-type; if they are mainly donors it is *n*-type. Sometimes a semiconductor material can contain nearly

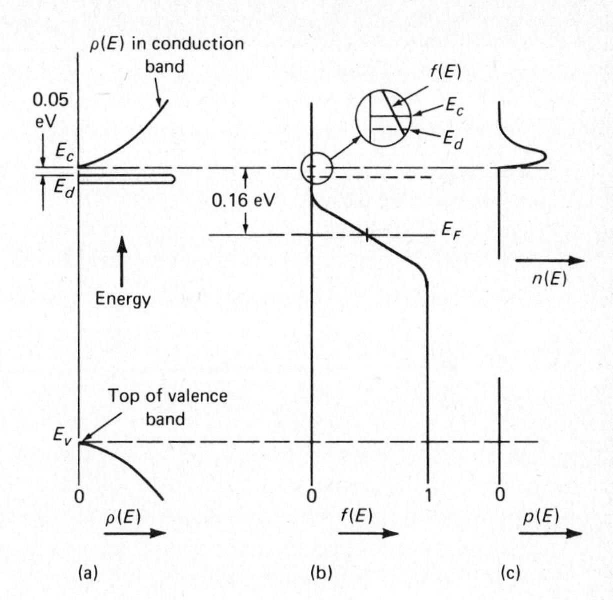

Figure 4.11 Calculation of the density of electrons in *n*-type silicon at room temperature. There are very few holes in the valence band and nearly all the donors are ionized. (a) Density of states function at the bottom of the conduction band and top of the valence band. (b) The Fermi function with *f*(*E*) greatly magnified at the edge of the conduction band shown in the circle. (c) Density of electrons and holes.

equal quantities of donors and acceptors; it is then said to be **compensated.** When this occurs the extra electrons from the donors become trapped on the acceptor sites and hence these two impurities compensate each other or cancel one another as far as electrical conductivity is concerned. Hence a material containing equal numbers of donors and acceptors "electrically speaking" is an intrinsic semiconductor, but hardly "pure."

In the above discussion all the donor atoms were considered to be "ionized"; that is, their extra electrons were considered to be contributing one electron per atom to conduction. At very low temperatures (below 200°K for Si) there is not sufficient thermal energy available to ionize these donors; some remain attached to their nuclei and hence, being immobile, do not contribute to the electrical conductivity. A very similar argument applies to acceptors. At very low temperatures some holes which contribute normally to conduction remain attached to their acceptor nuclei and hence do not contribute positive carriers to conduction. The acceptor atoms are no longer "ionized."

It is possible to calculate the number of ionized donors or acceptors in a semiconductor in thermal equilibrium if the Fermi energy level is known. The number of ionized donors is equivalent to the number of empty donor energy states. The probability of occupation of any energy state is defined in terms of the Fermi energy. If there are N_d donor atoms or states per cubic centimeter present in a crystal occupying an energy state E_d, the number of empty ionized states is given by

$$N_{\text{ionized donors}} = N_d[1 - f(E_d)] = N_d\left(1 - \frac{1}{\frac{1}{2}e^{(E_d - E_F)/kT} + 1}\right). \tag{4.33}$$

(The Fermi function differs by the factor ½ here owing to donor electron spin considerations.) It follows from this equation that if $(E_d - E_F) > kT$, then nearly *all* the donors are ionized. On the other hand, if $(E_F - E_d) > kT$, then practically all donor sites have electrons trapped or immobilized on them. This is illustrated in Fig. 4.11. Similar reasoning applies to acceptor or other energy levels as well. However, one must remember that an ionized acceptor is an acceptor level with an electron trapped on it, and that the Fermi function refers to the probability of occupation of the state by electrons.

4.6 CALCULATION OF THE FERMI ENERGY

The question often arises as to the position of the Fermi energy level in a given material. To calculate the Fermi level in a uniform semiconductor in thermal equilibrium, the number of donor and acceptor atoms, the energy gap, the temperature, and the density of state functions must be known. That is, the **state of the system** must be defined. Note that the Fermi level reflects *all* of these factors and any change in these quantities is reflected by a shift in the Fermi energy. Note too that the Fermi energy is defined *only* under thermal equilibrium conditions although sometimes a **quasi-Fermi level** is defined in non-equilibrium situations. This is done to enable one to express the number of electrons and holes present in a simple manner, and will be discussed later in connection with semiconductor devices.

For calculation of the Fermi energy the condition of **charge neutrality** can be applied to a homogeneous semiconductor. That is, it is nearly always assumed that no net charge can exist locally anywhere in a uniformly doped semiconductor containing the same number of impurities throughout its volume. This result is explained in terms of the Faraday "ice pail" experiment. That is, if any net charge occurs in the interior of a conductor, it will disperse itself to the surface in an extremely short time on the order of the dielectric relaxation time of the material.[8] Charge neutrality is expressed as

$$p + (N_d - n_d) = n + (N_a - p_a), \tag{4.34}$$

where p and n refer to hole density in the valence band and electron density in the conduction band, respectively; N_d and N_a are the density of impurity donor and acceptor atoms, respectively; and n_d and p_a refer to the density of electrons trapped on donor sites and holes trapped on acceptor sites, respectively. Note that the bracketed quantities on the left- and right-hand sides of Eq. (4.34) represent the concentration of ionized donors and acceptors, respectively, the ionized donors being positively charged and the ionized acceptors having a negative charge.

When expressions for all of the quantities in this equation are introduced from Eqs. (4.21), (4.27), and (4.33), the Fermi energy is determined.[9] Note that

[8] Of the order of 10^{-12} sec for semiconductors.

[9] An expression similar to Eq. (4.33) must be used for acceptors (see Problem 4.9).

the solution for the Fermi energy generally involves a complicated transcendental equation containing exponentials. Graphical techniques have been devised to solve for the Fermi energy.[10] However, it is often possible to obtain an algebraic solution by using very good approximations. For example, in a moderately doped donor impurity semiconductor at room temperature the number of conduction electrons may be assumed to be equal to the number of ionized donors, if $E_g \sim 1$ eV.

Equation (4.34) indicates that at 300°K, the Fermi level is somewhere above the middle of the energy gap for a donor-doped semiconductor since then many donor atoms are ionized; it is correspondingly below the middle of the gap for acceptor-doped semiconductors. The more donor impurities added, the closer the Fermi energy comes to the edge of the conduction band. The Fermi energy level approaches the edge of the valence band as more acceptor impurities are added.

EXAMPLE 4.3

Given the n-type semiconductor silicon at 300°K with an energy gap of 1.12 eV. The material contains only donor-type impurities, all of which are ionized. The donor density is $1.0 \times 10^{16}/\text{cm}^3$. Calculate the Fermi energy.

Solution

Beginning with the charge neutrality condition given in Eq. (4.34), since no acceptors are present and all donors are ionized, $n_d = 0$ and $N_a - p_a = 0$. This equation then reduces to

$$p + N_d = n.$$

Since the np product for silicon is $(1.5 \times 10^{10})^2/\text{cm}^6$ and there must be at least $10^{16}/\text{cm}^3$ electrons in conduction, the hole concentration must be less than $[(1.5 \times 10^{10})^2/10^{16}]/\text{cm}^3$. Hence $p \ll n, N_d$. So to a good approximation

$$N_d = n,$$

or essentially all the conduction electrons come from ionized donors. Using Eq. (4.21a) we have

$$N_d = M_c \times 2(2\pi m_n^* kT/h^2)^{3/2} e^{-(E_C - E_F)/kT}$$

$$= 6(4.82 \times 10^{15})(m_n^*/m_0)^{3/2} T^{3/2} e^{-(E_C - E_F)/kT}.$$

[10]See, for example, W. Shockley, *Electrons and Holes in Semiconductors* (Van Nostrand, New York, 1950), Sec. 16.3.

Hence, using Table 4.1, we have

$$1.0 \times 10^{16} = 6(4.82 \times 10^{15})(0.33)^{3/2}(300)^{3/2}e^{-(E_c-E_F)/0.026},$$

where $E_c - E_F$ is in electron-volts.

Taking the natural logarithm of this equation and using some algebra yield

$$E_c - E_F = \underline{0.21 \text{ eV}}.$$

Hence the Fermi energy is about 0.21 eV below the conduction band edge.

4.6.1 Temperature Dependence of the Fermi Level

At high temperature, when many electrons are thermally excited into the conduction band exceeding the electron concentration from donor atoms, the Fermi level tends toward the center of the gap. This is because thermally excited electrons leave behind an equal number of conducting holes and the material is sometimes said to become "intrinsic." Reducing the temperature causes the Fermi level to move toward the conduction band edge for n-type and toward the valence band edge for p-type semiconductors. The variation with temperature of the Fermi level of acceptor- and donor-doped crystals of Si is given in Fig. 4.12.

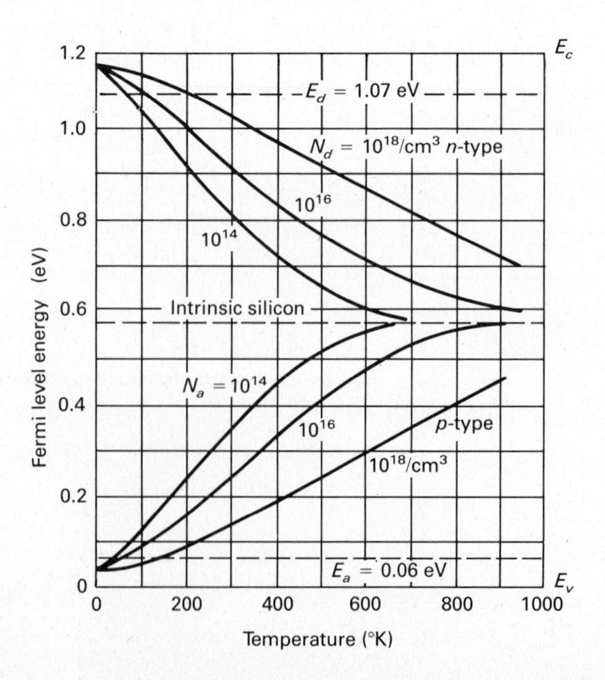

Figure 4.12 Calculated variation of the Fermi energy with temperature. This is for silicon, energy gap assumed constant at 1.12 eV, with impurity concentration as a parameter. A donor level ionization energy of 0.05 eV and an acceptor level of 0.06 eV is assumed. The energy at the top of the valence band, E_v, is here taken as zero.

☐ SUMMARY

The nearly free electron theory of the intrinsic semiconductor was developed and a quantum mechanical derivation of the density of electron energy states was given. This, coupled with the Fermi–Dirac function describing the occupancy probability of these electron states, was used to determine the electrical carrier densities in pure semiconductors. The concept of the "hole" representing any empty state near the top of the valence band was introduced in order to permit a representation of electron motion in the valence band. An expression was given for the important electron–hole density product, which was found to be a function of the energy gap of the semiconductor and the temperature. Semiconductors of *n*- and *p*-type were described which contain donor and acceptor impurities, respectively. Finally, a method for calculating the Fermi energy level of an impurity semiconductor was presented.

PROBLEMS

4.1 **(a)** Show that for an intrinsic semiconductor the Fermi energy

$$E_F = E_v + \tfrac{1}{2}E_g + \tfrac{3}{4}kT\ln(m_p^*/m_n^*).$$

(b) If $m_p^*/m_n^* = 9.5$, as in GaSb, which has an energy gap of 0.72 eV, find the deviation of the Fermi energy from the center of the energy gap at 300°K.

(c) Repeat (b) for a temperature of 600°K.

4.2 The effective mass for electrons and holes in silicon is given in Table 4.1.

(a) Determine the number of carriers in intrinsic Si at 300°K.

(b) Repeat (a) at 600°K.

4.3 Repeat Problems 4.2 for GaAs.

4.4 Using Eq. (4.8) calculate the energy level spacing in electron volts at an energy of 0.026 eV above the edge of the conduction band of a 1-cm^3 silicon crystal.

4.5 **(a)** Find the separation in electron volts of the Fermi energy from the intrinsic energy E_i at 300°K for *n*-type Si doped with 1.0×10^{14} donors/cm^3.

(b) Repeat (a) for *p*-type Si doped with 1.0×10^{14} acceptors/cm^3.

(c) Determine the density of holes in the *n*-type Si of (a).

4.6 Using Eq. (4.32), estimate the ionization energy of a typical donor atom in GaAs.

4.7 What is the average distance between dopant atoms (measured in parent-crystal interatomic spacings) for a donor density level of 10^{16}/cm^3 in silicon?

4.8 Calculate the position of the Fermi level at 300°K for

(a) silicon containing 10^{17} boron atoms/cm^3,

(b) silicon containing 10^{17} atoms of arsenic/cm^3 plus 5×10^{16} atoms of boron/cm^3,

(c) Repeat (a) at 600°K.

4.9 In analogy with Eq. (4.33), write an expression for the density of ionized acceptors, atoms which have captured an electron. Take the spin factor for holes as ¼.

4.10 Refer to Fig. 4.12 for the variation with temperature of the Fermi energy level in silicon doped with 10^{16} donors/cm^3.

(a) Determine the number of ionized donors at 300°K and at 30°K.

(b) Using the concept of charge neutrality, prove that the Fermi energy increases at 300°K as the number of donors is increased from 10^{16} to 10^{18}/cm^3.

(c) Prove that adding acceptors to this crystal will make the Fermi energy decrease.

CHAPTER 5

Transport of Carriers in Semiconductors

5.1 ELECTRON MOTION IN UNIFORM SEMICONDUCTORS

The electronic motion of current carriers in uniformly doped semiconductor crystals will now be considered from a quantum mechanical viewpoint, leading to a simple derivation of Ohm's law and an expression for the electron and hole effective mass. The previous chapter discussed the availability of carriers for conduction in a uniform semiconductor in equilibrium. This chapter will consider the speed with which they move. These two factors together determine the electrical conductivity of a crystal material. Considered here in addition will be the generation and recombination of *excess* carriers introduced into a semiconductor by optical excitation or by injection across a *p-n* junction. These carriers also participate in the electron conduction process. The movement of electrical carriers not only by an electric field as a driving force but also by a carrier gradient will be discussed. The concepts developed here will be useful later for the general description of current flow in semiconductor devices.

5.1.1 The Electron Velocity

In order to discuss the motion of an electron in a crystal, let us consider again the energy band structure for a simple crystal. The E versus k curve for the central Brillouin zone is redrawn in Fig. 5.1a from Fig. 3.2. In order to consider the velocity of an electron in such a periodic structure, we invoke the concept of the electron as a wave packet which travels with a group velocity given by $v = 2\pi\, d\nu/d\text{k}$.[1] Since quantum mechanics postulates that $E = h\nu$, the group velocity of the electron in the crystal is seen to be by differentiation

$$v = \frac{2\pi}{h}\frac{dE}{d\text{k}}. \tag{5.1}$$

Hence the electron velocity at a particular energy in the energy band depends on the slope of the E versus k curve[2] as indicated in Fig. 5.1b. Note that the electron velocity approaches zero at both the top and the bottom of the band; this must always be true regardless of the problem solved because of the periodic nature of the Schrödinger solution in a periodic crystal field. The zero velocity of electrons at the top and bottom of the band can be explained physically by using the wave concept of the electron. At the edges of the band the Bragg condition applies [see Eq. (3.18)]. Hence for electrons of this wave vector (wavelength), standing waves result, so no such electron waves can propagate through the lattice. Note, too, that the electron velocity reaches a maximum somewhere near the middle of the energy band and then decreases for increasing values of energy. This type of behavior is certainly drastically different from the motion of electrons in free space and is caused by the electrical interaction of the electrons with the ions of the crystal lattice.

5.1.2 The Effective Electron Mass

The motion of an electron in an accelerating electric field $\mathcal{E}$ applied to a crystal will next be discussed. Consider the energy gained by the electron when acted on for a short time dt by a small applied electric field. This is given classically by the product of the force F times the distance $dx = v\, dt$, or

$$dE = -q\mathcal{E}v\, dt; \tag{5.2}$$

$F\, dx$ represents the work done on the electron by the field, delivering energy to it. Now let us introduce the electron velocity in a crystal given by Eq. (5.1); then (5.2) becomes

$$dE = -2\pi \frac{q\mathcal{E}}{h}\frac{dE}{d\text{k}}\, dt. \tag{5.3}$$

[1] This concept of a combination of waves arises also in the treatment of other wave phenomena such as electromagnetic and light waves.

[2] The variation of E with k is always sought by means of solution of the Schrödinger equation for a particular crystal material, no matter how complex in structure. Equation (5.1) may also be derived by differentiating with respect to k, $E = p^2/2m_n^* = h^2\text{k}^2/8\pi^2 m_n^*$.

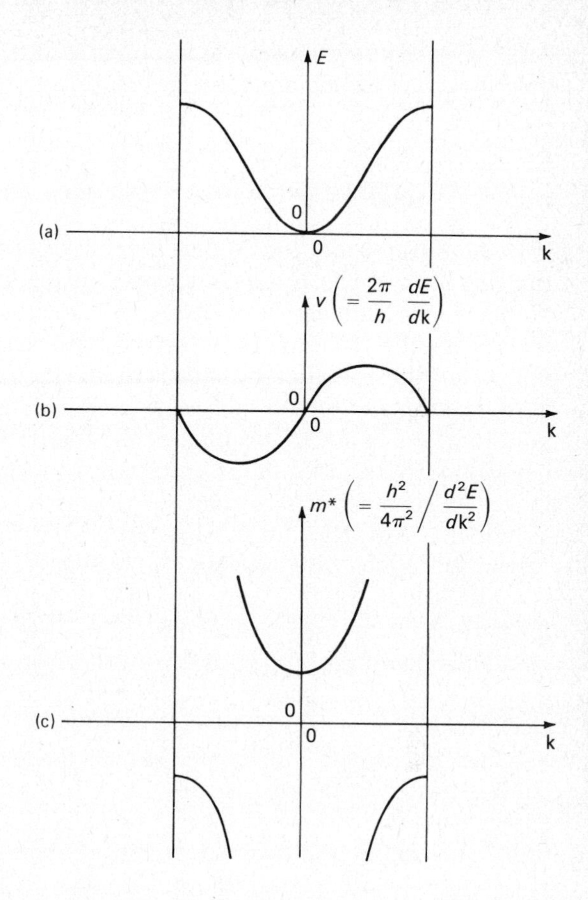

Figure 5.1 (a) Energy, (b) velocity, and (c) effective mass as a function of the electron wave vector k for an electron moving in a crystal.

By noting that $dE = (dE/d\text{k})\, d\text{k}$, this equation can be solved for $d\text{k}/dt$ as

$$\frac{d\text{k}}{dt} = -\frac{2\pi q\mathcal{E}}{h} = \frac{2\pi F}{h}, \tag{5.4}$$

where F is the external force exerted on the electron by the field. From Eq. (5.1) for velocity, we can obtain the electron acceleration by differentiation as

$$a = \frac{dv}{dt} = \frac{2\pi}{h}\frac{d^2E}{d\text{k}^2}\frac{d\text{k}}{dt}. \tag{5.5}$$

Introducing $d\text{k}/dt$ from Eq. (5.4) yields

$$a = -\frac{4\pi^2 q\mathcal{E}}{h^2}\frac{d^2E}{d\text{k}^2} = \frac{4\pi^2 F}{h^2}\frac{d^2E}{d\text{k}^2}. \tag{5.6}$$

The force F here is that due only to the electric field applied, excluding the forces of the crystal lattice ions. The interaction of the electrons with the lattice ions is taken into account by the introduction of the concept of an **effective mass.** Expression (5.6) for electron acceleration is strikingly different from the classical Newtonian law of motion $F = ma$. However, it is convenient to discuss the motion of an electron in a crystal in a similar manner to the way we describe a ball rolling down an inclined plane. This is possible if we compare Eq. (5.6) with Newton's law and define an effective electron mass as

$$m^* = \frac{h^2}{4\pi^2\, d^2E/d\mathrm{k}^2}\,. \tag{5.7}$$

Now mathematically the second derivative of E with respect to k represents the *curvature* of the E versus k curve of Fig. 5.1a. Zero curvature occurs at points of inflection and, as a result of the reciprocal relationship between effective mass and curvature, the effective mass becomes infinite. This is shown in Fig. 5.1c. Note, too, that the effective mass has a minimum absolute value both at the bottom and top of the energy band. However, this mass is *positive* at the bottom of the band but *negative* at the top. Hence for a quantum mechanical description of the motion of an electron in a crystal lattice the mass can no longer be considered a constant in the classical sense. In order to provide a pseudo-classical description, the particle is considered to move according to Newton's laws as a free particle, and the effect of the crystal lattice is taken into account by defining an essentially constant effective mass. This is only possible for particles with energy near the bottom or top of a band. In this way it is possible in many cases to consider the electron as nearly free and introduce the effect of the periodic potential by substituting a constant m^* for m, since the electrons are often near an extremum in $E(\mathrm{k})$.

The significance of the negative mass assigned to electrons at the top of the band is related to the discussion of electron states at the top of a nearly filled energy band in Section 4.3. There the concept of the "hole" was defined and it was indicated that the reduced rate of increase of energy with increasing momentum would lead to the concept of a negative electron mass. It was also pointed out that for convenience the motion of particles in a nearly filled band could be described by assigning to the empty states pseudoparticles (holes) with a positive mass and a positive charge. The concept of holes and electrons is particularly important in the description of semiconductor materials and devices.

5.2 OHM'S LAW

The task which still remains is to determine Ohm's law for electric field conduction in semiconductors. This derivation will predict the conditions under which a linear relationship exists between the current density in a bar of semiconductor and the electric field applied to it. A microscopic, elementary quantum mechanical treatment will be followed.

Under the influence of a small electric field $\mathcal{E}$, a randomly moving free electron would be accelerated to a value $a = -q\mathcal{E}/m$ in a direction opposite to the field and its velocity would continue to increase with time. The electron in a crystal lattice, however, will collide with some lattice ion a certain time after having been put in motion; the time between random collisions is taken as τ, which may be a function of the random electron velocity. The result of a collision is to randomize the motion or remove all of the electron velocity which is due to the accelerating field.[3] Hence after a collision the electron "relaxes" to its condition of random velocity as before acceleration. The time $\bar{\tau}$ is called the **relaxation time** or mean-free time between collisions. The average velocity increase of the electrons between collisions due to the field (drift velocity) is given by

$$\bar{v}_d = \tfrac{1}{2}at = -\tfrac{1}{2}q\mathcal{E}\bar{\tau}/m^*, \tag{5.8}$$

where the effective mass is introduced to take into account the fact that the electron is moving in a crystal lattice. Now the density of electrons n moving per second in a particular direction through a square centimeter of area perpendicular to that direction is $n\bar{v}_d$, the electron flux. The rate of charge flow is the particle charge times this flux, which constitutes the electric current density given by

$$J = -nq\bar{v}_d. \tag{5.9}$$

Introducing Eq. (5.8) into (5.9) gives

$$J = \frac{nq^2\bar{\tau}}{2m^*}\mathcal{E}. \tag{5.10}$$

This is the correct form of Ohm's law but for the factor of 2 in the denominator, which is eliminated if proper averaging of τ over the distribution of random electron velocities is performed.[4] Since the usual statement of Ohm's law is $J = \sigma\mathcal{E}$, where σ is the electrical conductivity of the crystal, this conductivity correctly expressed is

$$\sigma = \frac{nq^2\bar{\tau}}{m^*}. \tag{5.11}$$

The linear relationship between current density and electric field was derived assuming the linear variation of electron velocity with electric field in Eq. (5.8). Under conditions of high electric field this is no longer true (see Section 5.2.4).

5.2.1 The Electron Mobility

It is convenient, as a measure of the speed of an electron in a field, to specify a carrier **mobility** μ_n, which is defined numerically as the velocity per unit applied electric field; i.e.,

$$\mu_n = |\bar{v}_d/\mathcal{E}| = q\bar{\tau}/m_n^*, \tag{5.12}$$

[3]Note that the electron does not start from rest. In the absence of externally applied electric field, the electron will have a velocity, as previously described, according to its state in the energy band.

[4]See A. van der Ziel, *Solid State Physical Electronics,* 3rd ed. (Prentice-Hall, Englewood Cliffs, NJ, 1976), Chap. 6.

using Eqs. (5.9) and (5.11). Then the conductivity can be written as

$$\sigma = nq\mu_n. \tag{5.13}$$

EXAMPLE 5.1

Determine the collision relaxation time in lightly doped n-type silicon at 300°K.

Solution

From Eq. (5.12) and Tables 4.1 and 5.1

$$\bar{\tau} = \mu_n m_n^*/q$$

$$= \frac{(0.15 \text{ m}^2/\text{V-sec})(0.33 \times 9.11 \times 10^{-31} \text{ kg})}{1.6 \times 10^{-19} \text{ C}}.$$

But 1 kg-m²/sec² = 1 J. Hence $\bar{\tau} = \underline{2.8 \times 10^{-13} \text{ sec}}$.

In a semiconductor which conducts electrically by both electrons and holes, the conductivity becomes

$$\sigma = nq\mu_n + pq\mu_p, \tag{5.14}$$

where μ_n and μ_p correspond to the electron mobility of electrons near the edge of the conduction band and holes near the valence band edge, respectively. If the semiconductor is intrinsic (pure), we can then write

$$\sigma = \sqrt{M_c} \times 4.82 \times 10^{15} T^{3/2} q(\mu_p + \mu_n) e^{-E_g/2kT} (m_n^* m_p^*/m_0^2)^{3/4}, \tag{5.15}$$

taking into account that the number of electrons and holes is equal. The mobility of electrons and holes as measured in some typical pure semiconductor materials is given in Table 5.1. The high electron mobilities in GaAs, InAs, and InSb result from the small electron effective masses in these materials.

5.2.2 Determination of m^* and τ in Semiconductors

It is possible in principle to determine the effective mass of an electron or a hole in a semiconductor crystal by quantum mechanical calculations. If the crystal structure of the material is known, the ionic potential field can be estimated and introduced into Schrödinger's equation to yield an E versus k relation. Now the effective mass can be calculated from Eq. (5.7) as a property of the material.

At absolute zero temperature, where the ions in the crystal lattice have no thermal energy, in principle the electrons would move through the lattice without any net scattering. That is, the lattice ions would supply periodically an equal amount of acceleration and deceleration, there would be no effective

TABLE 5.1 The mobility of electrons and holes in some pure semiconductor materials at 300°K, in units of cm^2/V-sec.

Crystal Material	μ_n (Electrons)	μ_p (Holes)
Ge	3900	1900
Si	1500	480
InAs	33,000	460
InSb	80,000	1250
GaAs	8500	400
GaP	110	75
CdS	340	50

From S. M. Sze, *Physics of Semiconductor Devices,* 2nd ed. (Interscience, New York, 1981), Appendix G.

scattering, and the electrical conductivity would tend toward infinity. However, at any finite temperature the lattice vibrations would tend to scatter the electrons in motion since all ions at any particular time would not be in their precisely periodic locations. This has already been described in Chapter 2 as a type of crystal irregularity or crystal defect. At higher temperatures, the lattice vibration amplitude increases, shortening the time between collisions, or reducing the relaxation time.

The electrical conductivity of a semiconductor crystal material is determined by adding the effects of all crystal defects. For example, the presence of impurity doping atoms in the crystal, acting as scattering centers, also tends to reduce the relaxation time and hence the carrier mobility as well as the electrical conductivity. To obtain the carrier mobility in the presence of both lattice and impurity scattering, the number of scattering events must be summed. But the number of collisions is inversely proportional to the time between collisions, which is proportional to the mobility. Hence the net carrier mobility μ can be written as

$$\frac{1}{\mu} = \frac{1}{\mu_L} + \frac{1}{\mu_I}, \tag{5.16}$$

where μ_L is the lattice mobility if there is no scattering due to impurities and μ_I is the mobility corresponding strictly to scattering by impurities. Figure 5.2 gives the measured mobility values of electrons and holes in bulk silicon and gallium arsenide versus impurity concentration. As the temperature of a semiconductor crystal is increased, the lattice vibrations become more vigorous and μ_L decreases. The scattering by impurities on the other hand tends to decrease as the temperature increases, causing μ_I to increase. At temperatures above 300°K in silicon the lattice mobility factor dominates and the net mobility tends to fall off approximately as the inverse square of temperature.

The mobility of carriers moving near the surface of a semiconductor may

Figure 5.2 Carrier mobility versus impurity concentration in (a) silicon and (b) gallium arsenide at 300°K. Data for the resistivity of these materials versus impurity concentration are given in Appendix B.

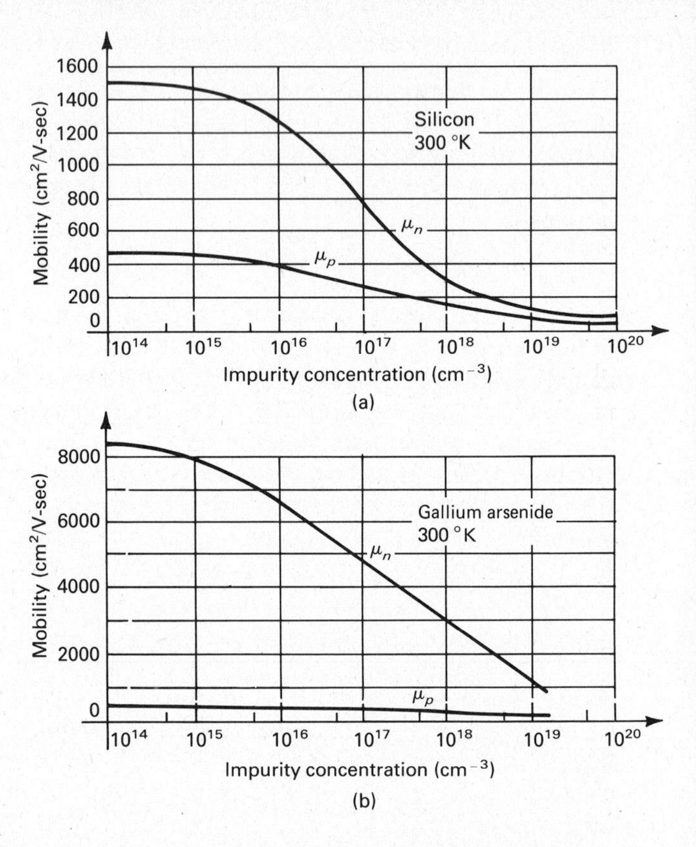

be significantly less than in the bulk owing to surface scattering. This is of importance in analyzing the performance of the surface field-effect transistor or MOSFET. The mobility of carriers can be enhanced in **superlattice** materials produced by growing thin (100 Å) alternating layers of two different semiconductor materials.[5] Very high mobilities have been measured in a direction parallel to these layers.

The quantum mechanical calculation of the lattice relaxation time by considering these different scattering mechanisms is quite difficult. It is often much easier to measure the electron mobility[6] in an electric field and the effective mass by a method such as cyclotron resonance,[7] and then calculate the relaxation time from Eq. (5.12).

[5]R. Dingle et al., "Electron Mobilities in Modulation-Doped Semiconductor Heterojunction Superlattices," *Appl. Phys. Lett.*, **33**, 665 (1978).

[6]The mobility in silicon can be measured directly by a time-of-flight measurement, or else it can be determined by measuring the Hall coefficient for the carrier density plus the electrical conductivity which gives the mobility–carrier density product.

[7]The effective masses of electrons and holes in a few semiconductors are listed in Table 4.1. Here the "density of states" effective masses are given as determined by electrical conductivity measurements. These values do not agree precisely with those found by cyclotron resonance measurements carried out in a magnetic field. In fact, the latter values depend on the direction of the magnetic field. For electrical calculations values in Table 4.1 should be used.

5.2.3 The Hall Effect for Measuring Carrier Density

The carrier density can be determined directly by measuring the Hall effect (E. H. Hall, 1890). Here a voltage is measured in the y-direction in a sample while an electric field is applied in the x-direction and a magnetic field in the z-direction, as shown in Fig. 5.3. The Hall voltage measured in the y-direction is given by

$$V_H = IB/Wnq, \tag{5.17}$$

where I is the electric current in the x-direction, B is the magnetic induction in the z-direction, W is the sample thickness in the z-direction, n is the carrier density, and q is the carrier charge. If the Hall voltage, the current, and the magnetic field are measured, then the carrier density can be calculated directly. Taking into account the polarity of the voltage, the sign or polarity of the charged carriers in the sample can be determined, for the Hall voltage derives from the magnetic field deflection of charge carriers in the y-direction. They are deflected to one edge of the sample; this charge pileup creates an electric field. The direction of the field depends on the polarity of the carrier charge.

5.2.4 The Electron Velocity at High Fields

As was previously discussed, the simple form of Ohm's law, derived as Eq. (5.10), depends on the presence in the semiconductor crystal of a small electric field such that the electron velocity varies linearly with field. However, at high electric field values (in the kV/cm range) this linear dependence is no longer observed. In fact, at very high fields in semiconductors the electron velocity in silicon and germanium tends toward a constant velocity, independent of field, known as the **saturation velocity.** This is illustrated in Fig. 5.4, which shows the velocity–field dependence for electrons and holes in the semiconductor silicon and indicates that both saturate at about 10^7 cm/sec. Similar behavior is observed for germanium with a saturation velocity of 6×10^6 cm/sec and gallium arsenide with a peak velocity of 2×10^7 cm/sec.

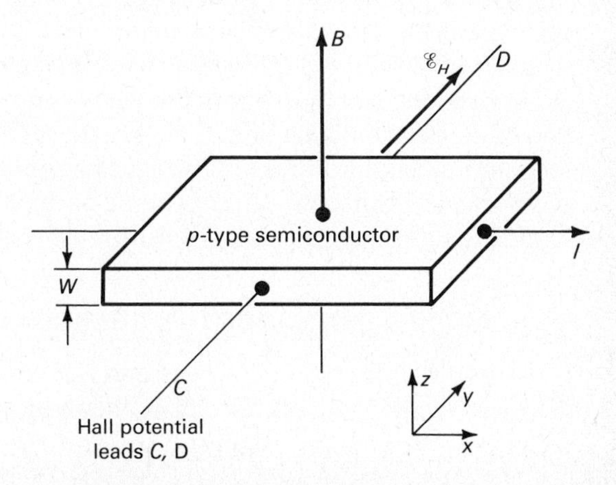

Figure 5.3 The directions of current, magnetic field, and Hall field for determining the carrier density in a rectangular sample.

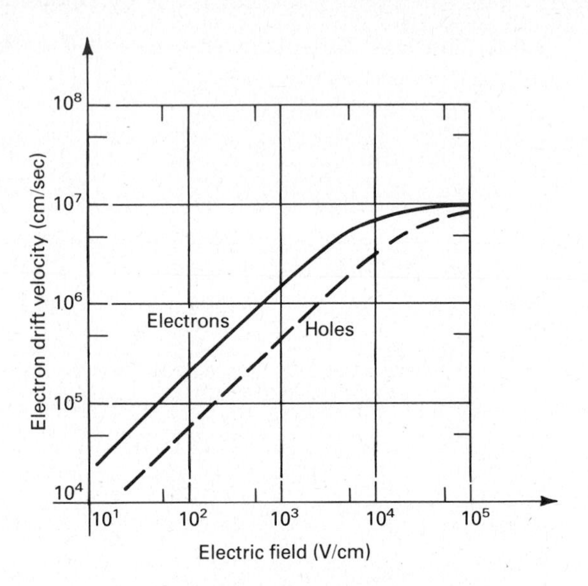

Figure 5.4 The velocity–field dependence for electrons and holes in silicon. The velocity achieves a constant value, independent of field, for electric field values above about 50 kV/cm.

At values of electric field in excess of 2 kV/cm, some of the energy delivered by the field excites higher energy lattice vibrations, and hence increases the scattering rate of the electrons by the crystal lattice atoms. Now not all of the increasing energy delivered by the raised electric field results in increased electron velocity; some of this energy is delivered to the crystal atoms. These high energy carriers are termed **hot electrons.** Hot electrons generally tend to reach a *scattering-limited* saturation velocity at approximately the average thermal velocity of electrons in their random motion in the semiconductor crystal in thermal equilibrium. At 300°K this thermal velocity is 10^7 cm/sec, which is approximately the measured saturation velocity of electrons in silicon (see Fig. 5.4). In gallium arsenide the electron velocity at high electric fields has an even more complex behavior. This will be discussed in connection with GaAs Gunn-effect devices in a later chapter.

At electric fields in excess of 200 kV/cm in silicon the hot electrons have sufficient energy to knock out valence electrons from their bound states in the crystal lattice. This is equivalent to exciting an electron from the valence band to the conduction band across the energy gap, and corresponds to the production of an **electron–hole pair.** These carriers can in turn become *hot* and **ionize** other atoms and produce additional electron–hole pairs. This process is called **avalanche multiplication** and provides a high field (or voltage) limitation of semiconductor devices and will be discussed in this connection later.

5.3 CARRIER TRANSPORT IN SEMICONDUCTORS

In Section 5.2 an expression was derived connecting the current density and the electric field in a uniform semiconductor crystal. This relationship is known as Ohm's law and is related to the **drift** of electrical carriers under the influence of an electric field. The picture here is that there is a net movement of electrons in the direction opposite to the electric field superimposed on the equilibrium random motion of the carriers which they possess by virtue of their position in the energy band and the temperature.

In a semiconductor such as silicon at room temperature, there are electrons available for movement at the bottom of the conduction band plus holes which are available for motion at the top of the valence band. These latter carriers behave as positive charges and the manner in which they participate in electrical conduction has already been described. Hence in general the net current in a semiconductor is given by the sum of the electron and hole currents $J_n + J_p$, so in one dimension Ohm's law becomes

$$J_x = J_{n_x} + J_{p_x} = \sigma_n \mathcal{E}_x + \sigma_p \mathcal{E}_x = nq\mu_n \mathcal{E}_x + pq\mu_p \mathcal{E}_x. \tag{5.18}$$

Here n and p refer to the density of electrons and holes, respectively; μ_n and μ_p are the electron and hole mobilities, respectively. In addition to this conduction current, the transport of carriers in a semiconductor can also occur by a process known as *diffusion.*

5.3.1 The Diffusion Current Density

The motion of carriers by diffusion occurs when there is a nonuniform distribution of these particles. This does not occur to any appreciable extent in metals since for the most part the conduction in these materials is due to one type of carrier and any nonuniformity would be quickly dispersed. In a uniform semiconductor there can be considerable differences spatially in the distribution of electrons; of course space-charge neutrality requires that these charges must be balanced locally by comparable numbers of holes or ionized donors.

Consider a distribution of electrons in distance as shown in Fig. 5.5. If we take an incremental region Δx about x_1, there are more carriers to the left of Δx than to the right. Hence there exists a tendency, *not* due to Coulombic

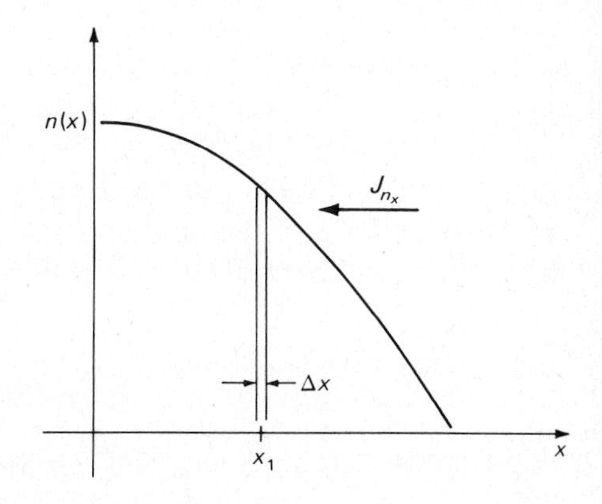

Figure 5.5 Nonuniform distribution of electron carrier density $n(x)$ resulting in an electron diffusion current density J_{nx}.

repulsion of these charges, for a net flux of carriers to move to the right to minimize this nonuniformity. This results from the random motion of the electrons, some moving to the right and others to the left. Statistically it is evident that more electrons in any given time interval will cross to the right of Δx than to the left of this region. The effect is to even out the concentration nonuniformity, similar to the dispersion of an ink drop in a tank of water—and is termed **diffusion.** No electric field is required for this motion, which results from the net to and fro movement of the carriers.

In a semiconductor when the concentration of carriers varies with distance, this constitutes a concentration gradient. This concentration gradient in one dimension is measured by the change in carrier concentration Δn in a distance Δx in the limit as $\Delta x \to 0$. This concentration gradient is denoted by dn/dx in the case of electrons and by dp/dx in the case of holes. A reasonable assumption is that the magnitude of this diffusive flux of electrons at x_1 is proportional to the extent of this nonuniformity, which is expressed by the electron gradient there. That is,

$$\text{diffusion electron flux} \propto \left.\frac{dn}{dx}\right|_{x=x_1}, \tag{5.19}$$

and is expressed in electrons per cm[4]. The diffusion electric current density for electrons may be expressed by multiplying the flux by the electron charge. This can then be written in the form of an equation as

$$J_n = qD_n \frac{dn}{dx}, \tag{5.20}$$

where D_n is a proportionality constant called the **diffusion constant for electrons** and has dimensions of square centimeters per second. A corresponding expression for holes in a semiconductor is

$$J_p = -qD_p \frac{dp}{dx}, \tag{5.21}$$

where D_p is the diffusion constant for holes. Note that the negative sign is necessary if the electron charge q is considered to have just a numerical value. For if the concentration of holes decreases in the direction of the $+x$-axis, dp/dx is negative and J_p represents a positive current density which travels in the $+x$-direction. For electrons, a decreasing carrier concentration in the $+x$-direction means that dn/dx is negative and hence Eq. (5.20) predicts a current density in the $-x$-direction. For the same type of gradient the hole and electron currents are in opposite directions owing to the opposite sign of their respective electrical charges.

5.3.2 Current Flow with Drift and Diffusion

In the general case of current flow in a semiconductor, movement of charge occurs as a result both of the presence of an electric field (drift) and of concentration gradients (diffusion). The mathematical treatment of problems involving such a case is normally difficult. However, fortunately a good number of practical problems exist in which one or the other transport process predominates; we shall only concern ourselves with these cases. This reduces the complexity enormously. However, for completeness the total current flow in a semiconductor in one dimension is expressed as

$$J_x = J_n + J_p = q\left(n\mu_n\mathscr{E}_x + D_n\frac{dn}{dx}\right) + q\left(p\mu_p\mathscr{E}_x - D_p\frac{dp}{dx}\right). \tag{5.22}$$

5.3.3 The Einstein Relation

The two material parameters which express the ability of carriers to drift or diffuse in a semiconductor are the mobility μ and the diffusion constant D. The concept of mobility, previously developed, derives from the net carrier motion resulting from random collisions of these charges with the lattice atoms under the action of an applied **potential gradient** or electric field. Similarly, the diffusion process can be described in terms of a net motion of carriers superimposed on their random thermal motion, under the effect of a **concentration gradient** and involving collisions with the lattice. (This is analogous to the manner in which heat flows in a material under the influence of a thermal gradient.)

Since drift and diffusion both result from similar statistical mechanisms it can be shown that the parameters μ and D for a moderately doped semiconductor material are not independent. They are related by an equation known as the **Einstein relation.** This relation for electrons and holes, respectively, is given by

$$D_n = (kT/q)\mu_n \tag{5.23a}$$

and

$$D_p = (kT/q)\mu_p. \tag{5.23b}$$

Here k is the Boltzmann constant, q the electronic charge, and T the absolute temperature. Since the units of the diffusion constant are square centimeters per second and the mobility is square centimeters per volt-second, kT/q must be expressed in volts and equals 0.026 V at room temperature.

EXAMPLE 5.2

The p-type base region of the n-p-n bipolar silicon transistor has a width of 2.0×10^{-4} cm and is doped with 1.0×10^{15} acceptors/cm^3. Electrons are injected into this region from the emitter at x_E, producing a uniform gradient of electrons there with the electron concentration dropping to zero at the collector at x_C. If 2.0×10^{14} electrons/cm^3 are present at the emitter edge of the base region (x_E), (a) calculate the diffusion current density of electrons through this base region under steady-state conditions. (b) What electric field must be present in this base region to yield an electron drift current density exactly equal to the diffusion current density just calculated? (c) Determine the voltage drop across this base width corresponding to this field.

Solution

(a) From Table 5.1, $\mu_n = 1500\ \text{cm}^2/\text{V-sec}$, and from the Einstein relation (5.23a), $D_n = (0.026\ \text{V}^{-1})(1500\ \text{cm}^2/\text{V-sec})$ or $D_n = 39\ \text{cm}^2/\text{sec}$. From Eq. (5.20)

$$(J_n)_{\text{diffusion}} = qD_n \frac{dn}{dx}$$

$$= (1.6 \times 10^{-19}\ \text{C})\left(39 \frac{\text{cm}^2}{\text{sec}}\right)\left(\frac{2.0 \times 10^{14}}{2.0 \times 10^{-4}}\ \text{cm}^{-4}\right)$$

$$= \underline{6.2 \frac{\text{A}}{\text{cm}^2}}.$$

(b) From Eq. (5.18)

$$(J_n)_{\text{drift}} = qn\mu_n\mathscr{E} \qquad \text{(use average electron density);}$$

$$6.2 \frac{\text{A}}{\text{cm}^2} = (1.6 \times 10^{-19}\ \text{C})(1.0 \times 10^{14}\ \text{cm}^{-3})\left(1500 \frac{\text{cm}^2}{\text{V-sec}}\right)\left(\mathscr{E} \frac{\text{V}}{\text{cm}}\right)$$

$$\underline{\mathscr{E} = 2.6 \times 10^2 \frac{\text{V}}{\text{cm}}}.$$

(c) Voltage drop:

$$V = \mathscr{E}W = (2.6 \times 10^2)(2.0 \times 10^{-4}) = \underline{5.2 \times 10^{-2}\ \text{V}}.$$

5.4 GENERATION AND RECOMBINATION OF MINORITY CARRIERS IN SEMICONDUCTORS

The discussion of homogeneous semiconductors thus far has been confined to thermal equilibrium conditions. However, semiconductor devices in general operate under nonequilibrium conditions. For example, a bar of n-type silicon acting as a photodetecting device is not in equilibrium in the presence of light. When the bar is illuminated **excess charge carriers** (above the equilibrium number) are produced in the material and the electrical conductivity of the bar increases. This can be demonstrated by an experiment illustrated in Fig. 5.6. Energy has been absorbed from the light by the silicon in producing these excess carriers. Hence the semiconductor is no longer in thermal equilibrium.

The extra carriers are produced by the light energy $h\nu$ breaking electrons free from their covalent tetrahedral bonds in the valence band and raising

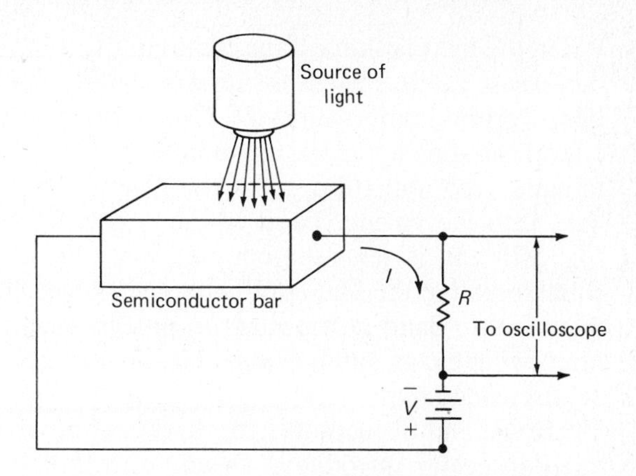

Figure 5.6 Apparatus for observation of the photoconductivity of a semiconductor and excess carrier recombination. The effect of the light is to increase the conductivity of the semiconductor bar, increasing the circuit current and hence causing an increased voltage output as viewed across the resistor *R*, on the oscilloscope.

them into the conduction band. A conducting hole is left behind for every excess electron carrier so produced. This is referred to as electron–hole **pair production** or **generation** and is illustrated in Fig. 5.7a. The electrical neutrality of the bar, of course, is maintained. Generally, a donor impurity semiconductor has many orders of magnitude more electrons than holes. When weak light introduces electron–hole pairs, the number of electrons is hardly increased above the equilibrium number already present, but the number of holes in this respect normally increases significantly. Hence this process is usually referred to as **minority carrier hole injection,** since this positively charged carrier is greatly outnumbered by the electrons or **majority carriers** present in thermal equilibrium. In transistors minority carrier introduction or injection takes place by the supply of electrical energy instead of by light energy, via a *p-n* junction. Nevertheless, this type of carrier injection is basically the same and charge neutrality is normally maintained (i.e., for every hole injected an electron must be introduced). The mobility of minority carriers is generally taken as approximately the same as that of majority carriers in a particular material.

5.4.1 Minority Carrier Recombination

Electron–hole pairs are only produced by light of sufficiently high frequency ν, such that $h\nu > E_g$, where E_g is the semiconductor forbidden energy gap.

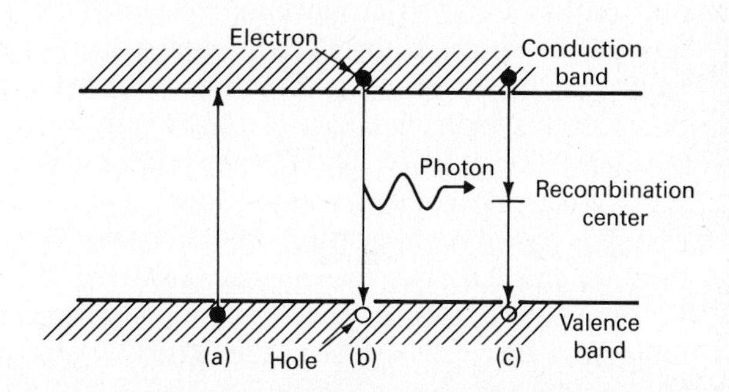

Figure 5.7 (a) Generation of an electron–hole pair. (b) Direct recombination of electron and hole with emission of photon. (c) Electron recombining with a hole via a recombination center.

When the light is shut off the material must return to thermal equilibrium and the excess carriers must disappear. This occurs by electron–hole **recombination.** Recombination can take place by one of the conducting electrons being bound again on a normal covalent tetrahedral site. This is referred to as band-to-band recombination since the electron drops from the conduction band back into the valence band in the energy band picture. This is illustrated in Fig. 5.7b and is generally coupled with the emission of photons with energy equal to that of the semiconductor energy gap. It has been found, however, that this band-to-band recombination process is not favored in silicon and germanium, whereas in gallium arsenide, for example, it is. The reason for this will be discussed later.

In Ge and Si recombination is generally found to take place via a **recombination center** introduced by some impurity atom such as copper or gold. Physically, these impurity atoms tend to immobilize a hole which immediately recombines with a conduction electron. This speeds up the process, and the recombination center thus "catalyzes" the electron–hole "reaction." This is illustrated by the energy band diagram in Fig. 5.7c. Atoms such as copper or gold are impurities which tend to introduce energy levels at about the center of the energy gap. The recombination process may be described as follows: When a hole jumps up to the recombination energy level, an electron promptly drops into this level and annihilation takes place. A completely analogous way of describing the same process is: An electron from the conduction band drops back into a hole in the valence band, using a recombination level as a "stepping stone" (see Fig. 5.7c).

The main point is that the one hole and one electron disappear; the time taken for this recombination process is called the **minority carrier lifetime.** In silicon and germanium for carriers recombining through an impurity recombination center, a lifetime can range from less than 1 ns to 1 ms. The value of the recombination time is shorter in samples with more impurity centers and also depends on the facility with which a center catalyzes the reaction, which is referred to as the **capture cross section.** The band-to-band recombination time in GaAs is much shorter, less than a nanosecond.

5.4.2 Light Emission from Semiconductors

Since energy is required to raise an electron from the valence band to the conduction band and so produce an electron–hole pair, energy must be released on recombination. If the electron–hole pair is produced by energy supplied by an electric field and most of the recombination energy is released via light or photons, this forms the basis of a **light-emitting diode** (LED) as well as an **injection *p-n* junction laser.** The former produces incoherent light, while the latter yields coherent light as described later. In Si and Ge most of the recombination energy is absorbed in heating up the crystal, whereas in gallium arsenide photons are emitted. The reason that GaAs emits photons and hence is intrinsically a better material for fabricating semiconductor lasers will be discussed soon. However, let us first mathematically formulate the recombination process, which will be of importance later in the discussion of transistors and integrated-circuit microchips as well as light-emitting diodes and lasers.

5.4.3 Minority Carrier Lifetime

Under thermal equilibrium conditions it has been shown that the product of electrons and holes in any semiconductor must be constant at any temperature. This constant will depend on such semiconductor parameters as the energy gap [see Eq. (4.29)]. When carriers are injected via the energy of an electric field or light, equilibrium is disturbed and carriers in excess of the equilibrium np product are introduced. For reasons of electrical neutrality, the number of electrons and holes so introduced must be equal; we shall also assume that their number is small compared to the equilibrium number so that **quasithermal equilibrium** occurs. This possibility of introducing excess carriers, yet maintaining electrical neutrality, distinguishes semiconductors from metals and results from the two-carrier nature of the semiconductor. This accounts for the fact that transistors utilizing two electrical carriers—**bipolar transistors**—are fabricated from semiconductor materials and not metals.

It should be understood that in thermal equilibrium generation and recombination of carriers are constantly taking place, but on the average for every electron excited into conduction one recombines or drops back into the valence band. If the equilibrium generation rate is G carriers per cubic meter per second and the recombination rate is R, then this fact may be expressed as $R - G = 0$. In the case of band-to-band recombination, a reasonable assumption is that the recombination rate is linearly proportional to the number of both electrons and holes present. Hence at equilibrium we can write

$$rn_0p_0 - G = 0, \tag{5.24}$$

where r is a constant of proportionality and n_0 and p_0 refer to the equilibrium density of electrons and holes, respectively. If now the crystal is slightly disturbed from equilibrium by a small excess hole density Δp and an equal electron density Δn, the net rate of excess hole carrier recombination (or loss) at any time will be given by

$$-\frac{d\,\Delta p}{dt} = r[(n_0 + \Delta n)(p_0 + \Delta p)] - rn_0p_0. \tag{5.25}$$

This assumes that r is unchanged since the density of excess carriers introduced is small. Then Eq. (5.25) reduces to

$$\frac{d\,\Delta p}{dt} = -r(n_0 + p_0)\,\Delta p, \tag{5.26}$$

if we neglect second-order terms. The quantity $1/r(p_0 + n_0)$ is usually defined as the **minority carrier lifetime** τ, and Eq. (5.26) is then written as[8]

[8]Here τ is not to be confused with the relaxation time of Eq. (5.8).

$$\frac{d\,\Delta p}{dt} = -\frac{\Delta p}{\tau}. \tag{5.27}$$

Integrating this equation yields

$$\Delta p = Ae^{-t/\tau}. \tag{5.28}$$

To evaluate the integration constant A, let the initial injected density of holes be $(\Delta p)_0$ at time $t = 0$ and the source of injection be removed at that time. Then

$$\Delta p = (\Delta p)_0 e^{-t/\tau}, \tag{5.29}$$

which indicates that the excess holes (and electrons) disappear exponentially with time, so after a time τ only a fraction $1/e$ are left. Note that the speed of annihilation of excess holes varies inversely as τ, which in Si and Ge depends on the density of recombination centers and the capture cross section of these centers. In the fabrication of transistors and integrated circuits impurities such as Cu and Au are selectively introduced into the semiconductor material by high temperature processing to establish the lifetime at some value predetermined by design.

EXAMPLE 5.3

A bar of n-type silicon at 300°K contains 5×10^{15} donor impurity atoms/cm^3 and exhibits a minority hole lifetime of 1.0 μsec. The sample is illuminated with light of wavelength $\lambda = 8000$ Å which introduces 1.0×10^{14} excess electron–hole pairs/cm^3.

(a) Prove that this illumination will excite electron–hole pairs in the sample.
(b) How long after the light is shut off will it take for the excess hole density to fall to 10% of its initial value?

Solution

(a) The photon energy for $\lambda = 8000 \times 10^{-8}$ cm is

$$E = \frac{hc}{\lambda} = \frac{(6.63 \times 10^{-34}\ \text{J-sec})(3 \times 10^{10}\ \text{cm/sec})}{8000 \times 10^{-8}\ \text{cm}}$$

$$= 2.5 \times 10^{-19}\ \text{J} \quad \text{or} \quad \frac{2.5 \times 10^{-19}\ \text{J}}{1.6 \times 10^{-19}\ \text{J/eV}} = \underline{1.55\ \text{eV}}.$$

Since this photon energy exceeds the silicon gap width of 1.12 eV, electron–hole pairs will be excited across the gap.

(b) $\Delta p = \Delta p_0 e^{-t/\tau}$ or $t = -\tau \ln(\Delta p/\Delta p_0) = -(1.0 \times 10^{-6}\ \text{sec}) \ln(0.10)$ and $t = \underline{2.3\ \mu\text{sec}}$ for the excess hole density to reduce to 10% of its original value.

5.4.4 The Direct Transition

Let us return to the question of the important difference between recombination processes in Si and Ge compared to GaAs. The answer must be sought in terms of the band theory of solids. The solution of the Schrödinger equation for a crystal has already been shown to yield a relationship between the energy E of an electron in the solid and its wave vector k. In the simple case discussed, this relationship was indicated by the graph of Fig. 3.2, which indicates that there exist certain forbidden bands of energy.

A theorem of solid-state physics states that each segment of the E versus k curve may be translated in the k-direction by $\pm n\pi/a$ without loss of generality (n being an integer); this follows from the periodicity of the Schrödinger solution, which results from the periodicity of the potential due to the lattice ions in the crystal. In this way all segments of the E versus k curve of Fig. 3.2 may be included between k $= +\pi/a$ and $-\pi/a$, in a **reduced-zone** representation. The shape of the E versus k curves now becomes symmetric about k = 0; portions of these curves representing two adjoining energy bands are shown in Fig. 5.8a. In this way the E versus k structure near k = 0 for adjoining bands may be conveniently studied. For example, for a semiconductor material, the lower band may be the valence band and the upper band the conduction band (see Fig. 5.8a). The uppermost parabola represents the energy states of electrons in the conduction band. The inverted parabola corresponds to electrons in the valence band; the negative curvature near the top of the band is indicative of a negative effective mass of these electrons (see Section 5.1.2), leading to the concept of holes.[9] The range of energies between the top of the valence band at k = 0 and the bottom of the conduction band at k = 0 constitutes the forbidden energy gap.

As an example of the usefulness of this diagram, consider a photon of light striking the semiconductor represented by the energy band diagram of Fig. 5.8a.[10] The excitation of an electron into the conduction band occurs most probably for electrons with zero momentum (i.e., k = 0). Note that the transition takes place directly upward, since this requires minimum energy expenditure, as shown in Fig. 5.8a, and is referred to as a **direct transition.** That is, the photon hardly alters the momentum of the electron on the scale of k shown. This can be proved by a simple calculation which is the subject of one of the exercises at the end of this chapter.

5.4.5 The Indirect Transition

The E versus k solutions obtained from the Schrödinger equation do not always result in curves as shown in Fig. 5.8a but may instead yield behavior as indicated in Fig. 5.8b. Although the maximum energy in the valence band

[9]Note that the magnitude of the curvature is different for electrons and holes since in general their effective masses are different.

[10]This diagram is valid only for electron motion along one crystalline direction.

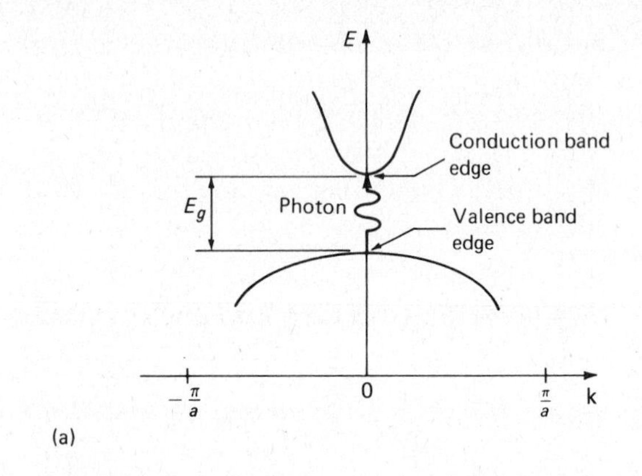

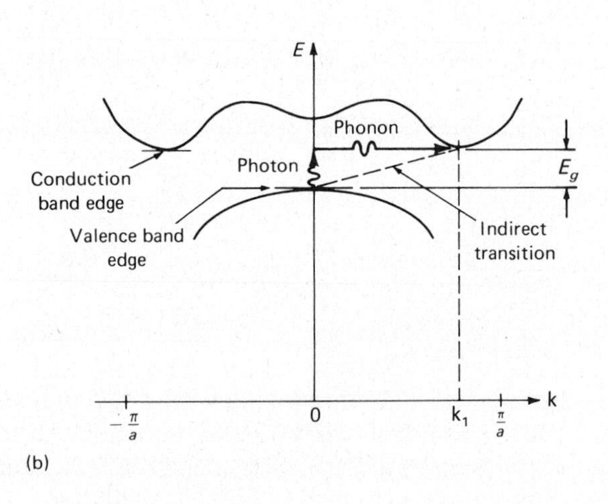

Figure 5.8 (a) Photon-induced direct transition of an electron from the valence band to the conduction band. Here the lowest energy of the conduction band occurs at the same value of wave vector k as the highest energy of the valence band. The energy gap E_g corresponds to the separation of this minimum and maximum. Note that in this illustration the hole effective mass at the top of the valence band is much larger than the electron effective mass at the bottom of the conduction band since the E versus k curvature is smaller in the former case. (b) Indirect transition of an electron from the valence band to the conduction band. Here the lowest energy of the conduction band occurs at a larger value of k, k_1, than the highest energy of the valence band. This transition requires the participation of both a photon (vertical) and a phonon (horizontal) to provide respectively the energy and momentum for the transition. The energy gap still corresponds to the separation of the top of the valence band from the bottom of the conduction band as shown.

here corresponds to k = 0, the minimum electron energy in the conduction band occurs at a value of k greater than zero. Now a direct upward transition is not favored since the minimum energy for excitation of an electron into the conduction band does not occur at k = 0. The energetically favored transition, in fact, is from k = 0 in the valence band to k = k_1 in the conduction band. Hence this process does not take place at constant momentum. As already stated, the additional momentum cannot come from the photon; hence it must be supplied otherwise. It is contributed by collisions with the atoms on the crystal vibrating in their lattice sites. The quantization of the lattice vibrational energy gives rise to the concept of the **phonon** particle, which is useful in describing electron–lattice atom collisions.

This type of **indirect transition** is indicated in Fig. 5.8b. Here again the minimum energy principle is illustrated as well as the law of conservation of momentum. The participation of a phonon as well as a photon is necessary for this transition. Note that the energy gap for the semiconductor represented by this energy band diagram corresponds as always to the **minimum** separation of the valence and conduction bands, which in this case does not occur at k = 0.

5.4.6 Band Structure and Semiconductor Devices

The two different types of semiconductor band structures, sometimes referred to as "direct" and "indirect," as illustrated in Figs. 5.8a and 5.8b, correspond to gallium arsenide and silicon, respectively. Although the crystal structure of both of these semiconductors is basically diamond cubic, subtle differences in the bonding forces separate these materials into dramatically different electrical categories. For example, excellent semiconductor light-emitting diodes can be constructed from GaAs but not from Si. Because the direct excitation is favored energetically, the inverse reaction, direct recombination, is also highly probable. This causes reduced current gain in GaAs transistors since the injected carriers recombine (are lost) in transit between emitter and collector. However, the directness of this recombination results in efficient energy emission in terms of photons and hence light emission, or laser action.

The indirect transition typical of silicon is inefficient since momentum must be exchanged with the crystal lattice atoms. Hence most of the recombination energy goes into the vibrational energy of the crystal atoms, which "heats up" the crystal solid. Little energy is left for the production of photons and hence Si is a poor light emitter. However, because of the inefficient recombination, this process is slow and the lifetime correspondingly long, accounting for the general use of Si for transistors and integrated circuits.

These concepts will be useful later in the detailed description of semiconductor devices.

*5.5 SEMICONDUCTOR SURFACES AND SEMICONDUCTOR DEVICE STABILITY

All of the discussion of the physics of semiconductor materials thus far has been confined to a treatment of the properties of the crystal bulk or volume. The electrical characteristics of *all* the semiconductor devices to be discussed in succeeding chapters, however, are affected to a greater or lesser extent by the properties of the *surface* of the semiconductor crystal. For example, success or failure in the quality fabrication of insulated-gate field-effect transistors (Chapter 10) and the charge-coupled devices (Chapter 12) is almost totally determined by the control of the surface characteristics of the semiconductor material, since all the electronic action in these devices takes place within a few micrometers of the surface. Current conducted across the *p-n* junctions in the diode and bipolar transistor may take place at the surface of the semiconductor crystal, where the *p-n* junction intersects the surface, instead of through the semiconductor bulk. The control of surface properties requires an understanding of the semiconductor crystal's surface and its interaction with the surrounding ambient. The intent of this section is to gain insight into the electrical behavior of semiconductor surfaces.

The outstanding stability and reliability of solid-state devices results from the fact that the structure of the crystal, as well as metal contacts to it, is *time invariant* at moderate temperatures and in somewhat hostile environments. In normal use there are few short-term burnout mechanisms for semiconductor devices and any aging problems are usually very long term. Radiation, of an atomic, nuclear, optical, or thermal variety, can penetrate the bulk of the solid-state material and cause permanent damage therein. But this usually requires comparatively high energy radiation. However, the outer surface of the solid is particularly vulnerable to low energy interaction with its environment. Since the bulk electrical properties of semiconductor materials are sensitive to minute quantities of impurities (donors and acceptors), it would be expected that a small amount of contaminant on the semiconductor surface will significantly affect the electrical properties of devices fabricated from these materials. This is the case and accounts for the fact that all semiconductor devices are encapsulated in some way to protect them from the surrounding ambient. The primary task is to stabilize the semiconductor surface so that the inherent reliability of solid-state semiconductor devices can be maintained. For this purpose these devices are sometimes sealed in vacuum or more often coated with various electrically insulating materials such as pure glasses, oxides, nitrides, and silicones or epoxies. Silicon devices specifically are most often protected by the material's **natural oxide,** which forms on the silicon crystal surface on exposure to pure oxygen at high temperatures, or which may be deposited onto its surface. It is the excellent electrical insulating properties of the uncontaminated form of this material and its ease of application that make it particularly desirable to protect silicon transistors and integrated circuits using the planar technology illustrated in Fig. 8.2a; here all the *p-n* junctions intersecting the surface are covered with silicon dioxide, greatly enhancing the electrical stability of these devices. Gallium arsenide does not have a stable oxide; silicon dioxide is often deposited to protect *p-n* junction devices in this material.

Some *p-n* junctions which terminate at the crystal surface exhibit spurious electrical conduction due to current caused by the mobility of ion contaminants on the surface, across the junction. Even **fixed** charges on the surface can attract mobile holes and electrons from the semiconductor bulk to the surface, where they may undergo electrical conduction due to a voltage applied across the *p-n* junction. Without some form of protection these effects could be cumulative, increasing with time or varying in time, causing unpredictable changes in device electrical behavior. Time-invariant electrical characteristics are a requisite for the modern semiconductor devices used in vast numbers in complex electronic systems. Let us now turn our attention to the physics of semiconductor surfaces in order to better understand the influence of surface phenomena on the devices to be discussed in subsequent chapters.

5.5.1 Semiconductor Surface Physics

Consider a semiconductor material like silicon freshly cleaved or cut in high vacuum to expose an ultraclean surface. Although the crystal structure described in Chapter 2 was generally considered to be a perfectly regular arrangement of atoms, this can no longer be the case at the cleaved surface of the crystal. From a symmetry viewpoint one would expect that two of the four covalent bonds which hold the silicon diamond-type lattice together would be broken, and hence unsaturated, at the crystal surface in contact with the vacuum (see Fig. 2.7). These broken covalent bonds are lacking two electrons per atom and hence about 10^{15} atoms/cm^2 are in a position to attract negative charges which might alight on the crystal surface and give rise to possible energy states in the forbidden energy gap. This represents an enormous quantity of acceptor electronic states which, if they existed in practice, would dominate the behavior of most semiconductor devices.

Fortunately any silicon surface when exposed to air will quickly form an oxide layer about 40 Å thick. Apparently the silicon surface atoms become bound to the oxygen atoms in the air, forming a silica (SiO_2) polyhedron structure. This satisfies the silicon "dangling" bonds at the surface and reduces these surface states under clean conditions to 5×10^{10}/cm^2 or less. These are the so-called **fast surface** or **interface states** which are responsible for electron and hole recombination and generation effects at the silicon surface similar to those already described in Section 5.4 for the semiconductor bulk. The time constants are correspondingly within the nanosecond to microsecond range. This is in contrast to the much slower states in the range of seconds to days or even months observed in some heavily oxidized silicon surfaces. The surface densities of these slow states are comparable to those of the fast states and seem to be located in the oxide, within 20 Å of the silicon–oxide interface. Both of these apparently act like donor states, are positively charged when ionized, and hence induce a tendency toward an n-type skin just below the silicon surface due to the electrons in the semiconductor bulk being attracted to the silicon surface. The slow surface states have been attributed to excess silicon ions in the oxide near the silicon interface which have not yet reacted with the negative oxygen ions which diffuse through the silicon dioxide at high temperature to form a stable SiO_2 tetrahedral structure.[11]

Figure 5.9a shows an equilibrium energy band diagram for the interface between a p-type silicon crystal and a vacuum, assuming a perfect silicon crystal structure out to the surface, ignoring the effects of broken surface bonds; this may be referred to as the **flat-band** surface situation and is taken as a reference, but obviously doesn't occur normally. Figure 5.9b shows the equilibrium energy band structure for the same p-type silicon crystal coated with a 1000-Å-thick oxide layer, including the effect of fast and slow interface state charges. The electric field lines from these positive charges[12] terminate on some

[11]A. S. Grove, *Physics and Technology of Semiconductor Devices* (Wiley, New York, 1967), Chap. 12.

[12]These positive charges have been attributed partly to unoxidized silicon ions.

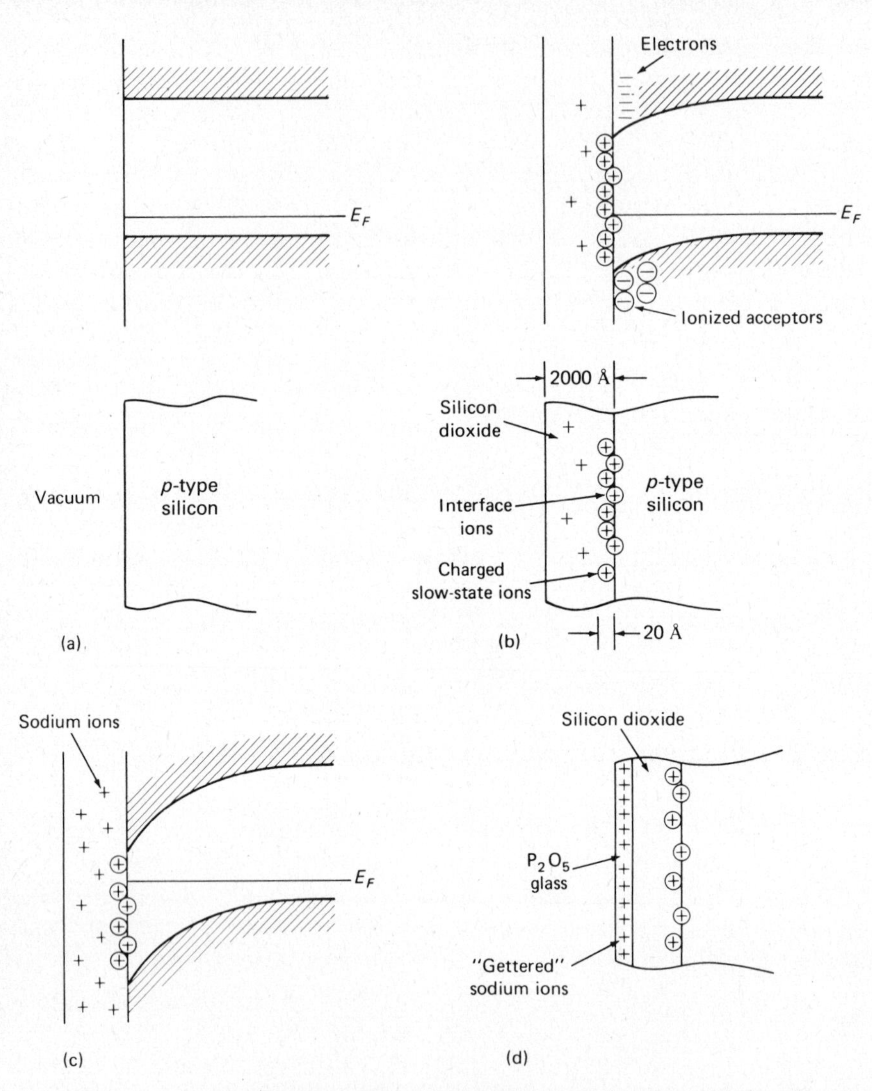

Figure 5.9 (a) Interface between *p*-type silicon and vacuum, ignoring broken surface bonds. (b) Equilibrium energy band diagram for *p*-type silicon showing fast and slow states at the interface and in the oxide. Also shown are electrons collected in the silicon at the interface as well as ionized acceptors.
(c) Equilibrium energy band diagram for *p*-type silicon with its surface inverted to *n*-type by positive ions in the silicon dioxide as well as at the oxide–silicon interface.
(d) Sodium ions gettered in the phosphosilicate glass formed on the silicon dioxide surface.

electrons collected at the silicon surface as well as on some negatively charged ionized acceptors fixed in the silicon crystal close to the surface. Hence the silicon surface just under the oxide tends toward *n*-type as indicated by the closer proximity of edge of the conduction band to the Fermi level. This **band bending** near the silicon surface represents a potential gradient and hence an electric field in the silicon due to the electric dipole at the interface. The extent of this band bending (and the tendency of the original *p*-type surface to become less *p*-type) depends on the chemical treatment of the silicon surface prior to oxidation as well as the care taken during the oxidation process.[13] The crystal plane orientation of the silicon surface also affects the density of surface states. Cleaving the crystal along (111) planes gives rise to the rupturing of the greatest number of bonds per square centimeter, which accounts for the observation of the largest density of surface states in this situation. The lowest density of surface states is observed when the oxide is grown onto the (100) surface planes which include a minimum number of broken bonds per square centimeter.

5.5.2 Charges in Silicon Dioxide

Surface effects in semiconductors may also be due to ionic charges in the bulk of the silicon dioxide layer which in turn induce charges in the silicon, near the silicon–oxide interface. For example, positively charged sodium and other alkali metal ions are known to have a high solubility as well as high mobility in SiO_2 or quartz. These positive charges imbedded in the oxide will tend to drive the silicon near the interface even more toward *n*-type, causing additional band bending. This is shown in Fig. 5.9c. Note that the *p*-type silicon crystal has now a surface which has been **inverted** to *n*-type; this is indicated by the fact that the Fermi level is closer to the conduction band edge than the valence band edge near the surface. (Positively charged oxygen vacancies have been reported to be another source of mobile charge in the oxide.) The high mobility of these ions in silicon dioxide accounts for their motion even at temperatures of 150°C and lower, particularly when an external voltage is applied across the oxide, giving rise to an electric field therein. The motion of these charges in the silicon dioxide causes corresponding changes in the electrical nature of the silicon surface just under the oxide. How the surface layer on a *p*-type silicon crystal may be inverted to *n*-type by charges in the coating oxide has already been indicated. The extent of this conversion depends not only on the number of such contaminant charges but also on their proximity to the silicon–oxide interface (the nearer the interface, the larger the effect). The normal fields present in device operation can cause the motion of these charges and hence time-variable changes in device characteristics that are surface sensitive. Methods are needed to reduce this contamination to significantly less than one part per million.[14]

Techniques have been developed to minimize the effect of the sodium ion, which is a primary contaminant and difficult, if not impossible, to eliminate during semiconductor device processing. The final processing of an *n-p-n*

[13]Indeed, this treatment is a carefully guarded secret of individual semiconductor device manufacturers.

[14]E. H. Nicollian, "Surface Passivation of Semiconductors," *J. Vac. Sci. Technol.*, **8**, 539 (1971). This paper provides a good brief summary of surface effects in silicon.

planar transistor often involves exposure of the oxidized silicon to phosphorous pentoxide (P_2O_5) to form the n^+-type emitter region (see Fig. 8.2). This reacts with upper surface of the SiO_2 layer already on the silicon to form a phosphosilicate "glass." Sodium is much more soluble in this phosphorus glass than the pure silicon dioxide and hence tends to collect or be "gettered" by this layer, far from the silicon–oxide interface, minimizing the electrical effect in the silicon under the oxide. This is shown in Fig. 5.9d. Another effect of this P_2O_5 treatment is for this superoxidant to supply excess oxygen to fill the oxygen vacancies in the SiO_2 structure which can also act as mobile positive charges or else enhance the rate of sodium drift. The introduction of the chlorine ion in the oxidation process is also generally used to minimize the oxide and interface charge states.

Another way to inhibit the drift of sodium ions toward the silicon–oxide interface is to provide an intervening layer of silicon nitride, Si_3N_4. This material may be deposited onto an oxidized silicon wafer by reacting silicon tetrachloride, $SiCl_4$, and ammonia, NH_3, at a temperature of about 800°C to form this compound. The mobility of sodium or alkali metal ions in silicon nitride is known to be significantly slower than in silicon dioxide owing to the denser nature of the nitride and hence can help "seal off" the device so coated from sodium contamination from the environment.

Finally, it has been observed that hole traps may be introduced into a silicon dioxide layer by high-energy electron or ultraviolet radiation. Again these immobilized positive charges can induce an n-type layer at the silicon surface under the oxide. In fact the effect of all these oxide–silicon surface phenomena is invariably to drive the silicon toward n-type. The limitation of semiconductor device behavior due to surface effects will be mentioned in ensuing chapters.

□ *SUMMARY*

In order to describe the electrical carrier motion in a semiconductor, expressions were obtained for calculating the carrier velocity and effective mass from the band structure. Ohm's law describing the relation between the electric current density and the electric field, at low fields, was derived in terms of the carrier intercollision (or relaxation) time τ and the carrier effective mass m^*. The useful concept of the carrier mobility, the velocity per unit electric field, was introduced and expressed in terms of τ and m^*; experimental methods were described for determining these latter parameters. Measurements of the Hall coefficient and the electrical conductivity permit determination of the majority carrier mobility and the carrier concentration in a semiconductor. High electric field effects were described. A component of electric current due to a carrier concentration gradient in a semiconductor, termed "diffusion" current was defined; this is in addition to the electric-field-dependent component called "drift." The Einstein relation, which expresses the relation between the diffusion coefficient and the carrier mobility, was introduced. The generation and recombination processes of electrical carriers in semiconductors were described and the concept of "minority carrier lifetime" was introduced. The differences between these processes in semiconductors which exhibit "direct" or "indirect" carrier transitions from the valence to the conduction band were explained. Finally, the relation between semiconductor surface physics and semiconductor device stability was discussed.

PROBLEMS

5.1 Derive Eq. (5.1) from the definition of group velocity and Planck's law.

5.2 The energy of an electron at the bottom of the conduction band of a semiconductor is given by $E = Ak^2$, where k is the electron wave vector.
- **(a)** Plot E versus k if A is a positive constant independent of k. On the same graph, plot E versus k for a value of A three times greater.
- **(b)** What is the effective mass for electrons for A and $3A$?
- **(c)** Repeat (a) for electrons near the top of a valence band where $E = E_{max} - Bk^2$, and where E_{max} is the energy at the top of the valence band and B is a positive constant.
- **(d)** What is the effective mass for electrons near the top of the valence band?

5.3 The Hall voltage for N-type silicon is 110 mV measured at $I = 10.0$ mA, $B = 2.0 \times 10^{-4}$ webers/cm^2, and $W = 0.025$ cm. (*Note:* A **weber** dimensionally is a volt-second.)
- **(a)** Calculate the number of carriers/cm^3 in the silicon.
- **(b)** Calculate the mobility of the electrons in silicon, if the electrical conductivity is 1.0 $(\Omega\text{-cm})^{-1}$.

5.4 **(a)** Using Eqs. (5.9) and (5.11), derive an expression relating the relaxation time $\bar{\tau}$ and the mobility and effective mass of a carrier in a semiconductor.
- **(b)** Find $\bar{\tau}$ for nearly pure n-type gallium arsenide.

5.5 Compute the average drift velocity of electrons in a bar of silicon 1.0 mm^2 in cross section, containing an electron density of 4.5×10^{15}/cm^3 and carrying a current of 100 mA.

5.6 Using the Einstein relation, determine the carrier diffusion constant for electrons and holes at 300°K in pure
- **(a)** germanium,
- **(b)** gallium arsenide.

5.7 **(a)** Determine the maximum value of the energy gap which a semiconductor, used as a photoconductor, can have if it is to be sensitive to yellow light $(6000 \times 10^{-8}$ cm).
- **(b)** A photodetector whose area is 5.0×10^{-2} cm^2 is irradiated with yellow light whose intensity is 2.0×10^{-3} W/cm^2. Assuming that each photon generates one electron–hole pair, calculate the number of pairs generated per second.

5.8 **(a)** From the known energy gap of the semiconductor gallium arsenide, calculate the primary wavelength of photons emitted from this crystal as a result of electron–hole recombination.
- **(b)** Is this light visible?
- **(c)** Will a silicon photodetector be sensitive to the radiation from a GaAs laser? Why?

5.9 **(a)** Determine the magnitude of the wave vector k for yellow light ($\lambda = 6 \times 10^{-5}$ cm).

(b) Calculate the width of the first Brillouin zone as indicated in Fig. 3.2 and partly duplicated in Fig. 5.8a, for a material with a lattice constant $a = 5.0 \times 10^{-8}$ cm.

(c) Using the results of (a) and (b), show that the momentum (or wave vector) of a yellow photon is negligibly small compared to the momentum of nearly all of the electrons in the first Brillouin zone.

5.10 Explain with the aid of a simple calculation why the crystal gallium phosphide ($E_g = 2.26$ eV) is transparent to red light ($\lambda = 7 \times 10^{-5}$ cm), while the crystal silicon is opaque to this light. (*Hint:* When a photon incident on a crystal has sufficient energy to raise an electron from the valence band to the conduction band, its energy will be absorbed by the crystal.)

CHAPTER

6

Introduction to Semiconductor Junctions

6.1 THE *p-n* JUNCTION DIODE IN EQUILIBRIUM

In Chapter 5 the properties of semiconductor materials containing a uniform distribution of impurity atoms have been considered in some detail. There are a number of electronic devices which employ uniformly doped semiconductors such as germanium or silicon. Some of these devices are thermistors, photoconductive cells, and strain gauges; these are all linear devices.[1] However, the devices most extensively in use today contain at least one *p-n* junction and exhibit a *nonlinear* variation of current with voltage.

A typical current–voltage (*I-V*) characteristic for such devices is given in Fig. 6.1. There is extensive general use of this type of device in electronic circuitry. For example, semiconductor diodes are employed as voltage limiters and references, current rectifiers, signal demodulators, etc. Perhaps the most

[1]In linear devices the electric current through the device is directly proportional to the applied voltage; i.e., Ohm's law applies.

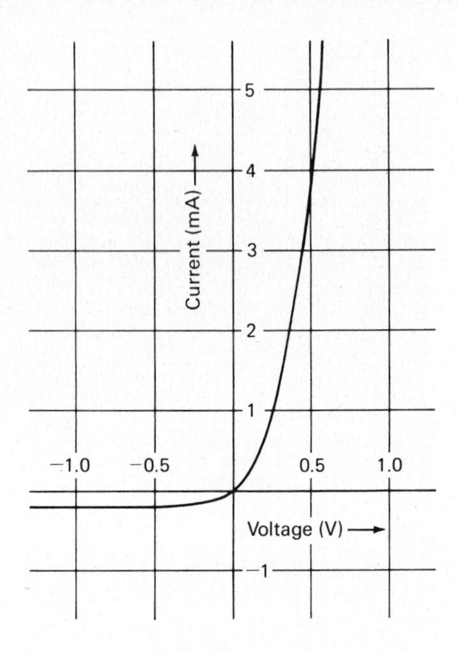

Figure 6.1 Current–voltage characteristic of a semiconductor *p-n* junction diode. The forward bias direction is specified when the *p*-region is positive relative to the *n*-region and the voltage and current are positive. Reverse bias is indicated where the voltage and current are negative, indicating that the *p*-region is negative relative to the *n*-region. The forward bias condition gives low impedance and the reverse bias yields high impedance.

important application of the *p-n* junction is as the emitter of the bipolar transistor. Hence the following diode considerations will have application in the description of transistors.

This chapter will present a derivation of the DC and low frequency current–voltage characteristic of the *p-n* junction diode by considering charge carrier flow through the junction. The temperature behavior and the voltage limitation of the *p-n* junction will also be described. In the next chapter, the higher frequency properties of the junction diode will be treated as well as the transient switching behavior.

6.1.1 A Physical Description of the *p-n* Junction

A schematic diagram for a *p-n* junction diode grown in the form of a semiconductor single-crystal bar is shown in Fig. 6.2a along with the circuit symbol representing this device. The more common diode geometry is similar to that shown in Fig. 6.2b. This **planar** configuration can be formed by selectively diffusing a *p*-type impurity (such as boron) into the *n*-type semiconductor crystal by exposure to the impurity at a very high temperature, or by ion implantation. Similarly, a junction can be formed by diffusing an *n*-type impurity (such as arsenic) into a *p*-type semiconductor.[2] In either case, the structure consists of a single crystal of semiconductor material with a *p-n* junction, shown schematically in Fig. 6.3a. On one side of the junction are found predominantly *p*-

[2]The impurity diffusion technology will be described in Chapter 11 in connection with the fabrication of integrated circuits.

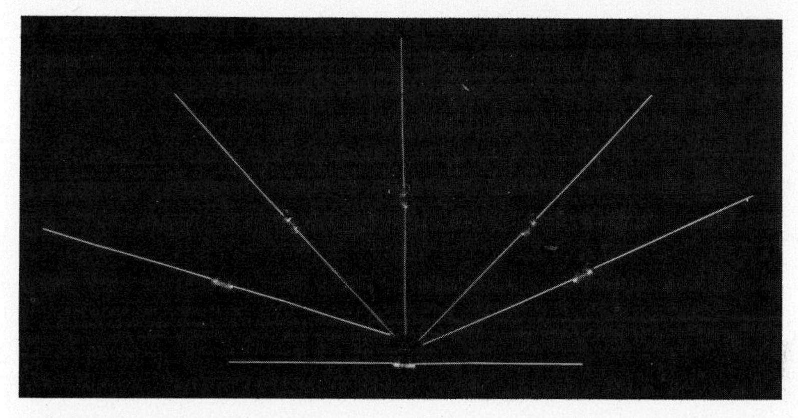

Small-signal, glass-encapsulated p-n junction diodes for commercial applications. (Courtesy of the Unitrode Corporation, Lexington, MA.)

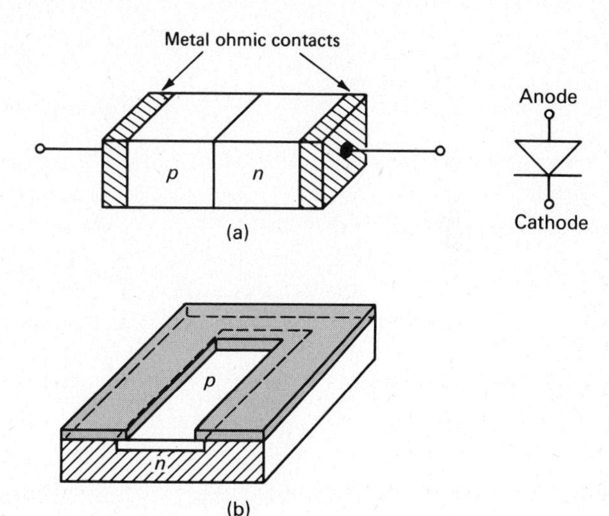

Figure 6.2 (a) A p-n junction diode grown in the form of a single-crystal bar and circuit symbol for this diode. (b) Planar p-n junction formed by selectively diffusing a p-type impurity into an n-type semiconductor crystal. The darkened region represents the insulating layer protecting the junction from the environment.

type impurities, whereas the other side has mainly *n*-type doping atoms. In this analysis it will be assumed that the doping is uniform on each side of the *p*-*n* junction as indicated in Fig. 6.3b.

The electrical characteristics of this diode depend mainly on the properties of the junction.[3] Figuratively, one can consider the formation of the *p*-*n* junction by bringing together two uniformly doped bars of semiconductor material,

[3]It will be shown that the voltage drops in the uniform semiconductor material away from the junction and across the metal contacts are normally small at moderate current levels. Also, surface effects can influence somewhat the diode characteristics, as discussed in Section 5.5 and later in this chapter.

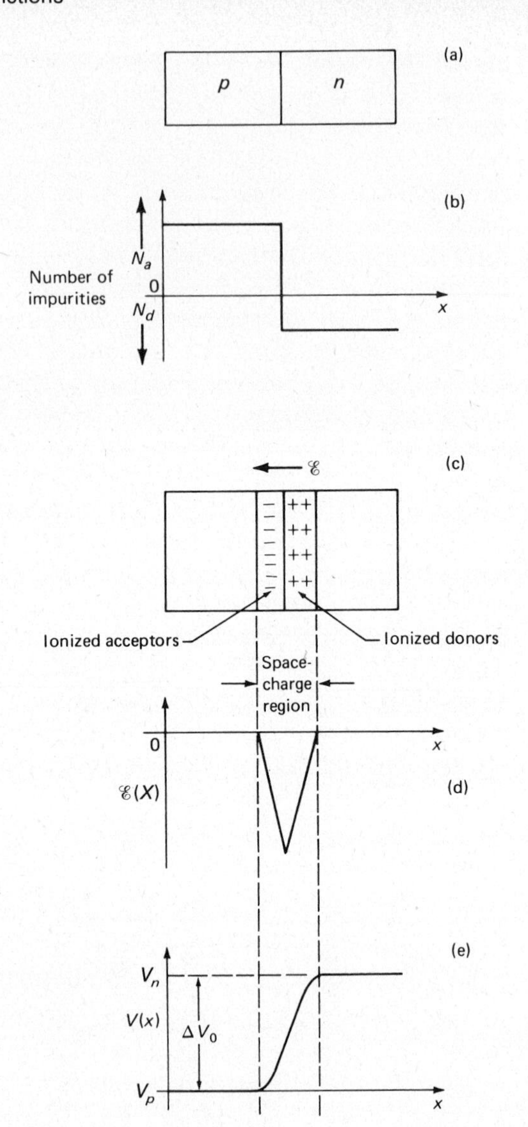

Figure 6.3 The *p-n* junction in thermal equilibrium.
(a) Schematic of *p-n* junction diode.
(b) Uniform distribution of acceptor and donor impurities on each side of an abrupt *p-n* junction.
(c) Space-charge region under thermal equilibrium conditions showing ionized acceptor and donor charges in a region otherwise depleted of mobile carriers.
(d) The variation of electric field with distance in the depletion layer, reaching a maximum negative value at the junction.
(e) The variation of electric potential with distance in the diode, showing that all the potential drop is across the space-charge region; ΔV_0 is the barrier height or contact potential.

one *p*-type and the other *n*-type. The *p*-side contains holes as the majority carriers coming mainly from acceptors and very few minority carrier electrons.[4] The *n*-side contains electrons derived mainly from the donors as the majority carriers and very few minority holes. Before contact each side is electrically neutral because of the balance of the mobile carriers and fixed ion charges. However, as these two sections are brought together, one can imagine some

[4]The greater the number of acceptors and hence holes, the fewer the number of electrons since the product is constant at any particular temperature [see Eq. (4.29)].

charge transfer taking place across the junction, for the sharp carrier concentration gradient cannot be sustained in the same sense as was discussed in Section 5.3.1 and the mobile charges begin to diffuse, causing a tendency to smooth out this drastically nonuniform distribution of mobile carriers in the neighborhood of the junction. That is, the holes tend to move toward the *n*-region, where there are few holes, leaving behind negatively charged acceptor ions; the electrons tend to flow toward the *p*-side, where there are few electrons, leaving behind positively charged donor ions.[5] This tendency to redistribution is opposed by an electric field which is set up in the neighborhood of the metallurgical *p-n* junction. This retarding field results from the fixed electric charges due to the positively charged donor ions and the negatively charged acceptor ions in the proximity of and on opposite sides of the junction, left behind by the mobile carriers.[6] This very field, caused by the migration of the mobile carriers across the junction, tends to limit this migration and establish an equilibrium situation. This is shown schematically in Fig. 6.3c.

One can visualize the creation of an electric field at the *p-n* junction in a descriptive manner as resulting from the transfer of electrons from the electrically neutral donor atoms in the *n*-region to the electrically neutral acceptor atoms in the *p*-region which takes place in close proximity to the *p-n* junction. This redistribution of charge occurs only in a **space-charge** region within a few micrometers of the metallurgical junction, which is **depleted** of mobile carriers and contains only fixed charges. The exact distribution of electric field near the *p-n* junction can be obtained by the solution of the Poisson equation applied to this region, and is graphed in Fig. 6.3d. This formulation will be presented later. Figure 6.3e gives the potential variation through the diode obtained by integrating the electric field over distance.

6.1.2 The *p-n* Junction Contact Potential

Another way of expressing the effect of the electric field which is "built-in" by ionized donors and acceptors at the junction is in terms of a **potential barrier** there. The height of the barrier in volts is termed the **contact potential** and is of the order of a volt for silicon and germanium *p-n* junctions. The maximum value of the built-in potential is determined by the energy band gap of the material; it can be as large as 1.12 eV for Si and only 0.66 eV for Ge. This built-in potential is somewhat reminiscent of the electrochemical potential that arises at the junction between an electrode and electrolyte in a battery. However, no current can be drawn from the *p-n* junction under thermal equilibrium conditions; there is no chemical reaction and so power derived from such a

[5]If concentration gradients occurred in a uniformly doped bar of semiconductor crystal, the carrier gradients would be reduced quickly to zero in a period equal to the dielectric relaxation time of the semiconductor material, this being of the order of 10^{-12} sec.

[6]It can be shown that a very abrupt transition from *p*-type to *n*-type (in less than a few micrometers) must occur to support such a field.

device would be contrary to the second law of thermodynamics since the device would then comprise a perpetual motion machine. Hence, if a metal wire is connected between the two ends of the p-n junction, contact potentials must be set up at the metal–semiconductor end connections to just balance the contact potential appearing at the p-n junction. Now the net voltage in the loop is zero and so no current can flow, as required on theoretical grounds.

The junction contact potential can be estimated indirectly by extrapolation of junction capacitance measurements as will be described later. This is possible because the capacitance effectively measures the charge distribution near the junction which in turn determines the barrier potential. The contact potential will now be analytically calculated for a semiconductor p-n junction in thermal equilibrium at a temperature T in terms of the number of acceptors and donors on each side of the p-n junction, respectively .

The analysis is based on the concept that in equilibrium not only is the total current zero, but the electron and hole currents must *separately* be zero. (This derives from a very fundamental and useful physical concept which is called the principle of **detailed balance.**) The hole current in one dimension generally can be written [see Eq. (5.22)] as

$$J_p = q\left(p\mu_p\mathscr{E}_x - D_p\frac{dp}{dx}\right), \tag{6.1a}$$

and the electron current as

$$J_n = q\left(n\mu_n\mathscr{E}_x + D_n\frac{dn}{dx}\right). \tag{6.1b}$$

Here $\mathscr{E}_x$ represents the field at any place x and dn/dx and dp/dx are the carrier gradients there which cause diffusive current flow. Setting each of these equations separately to zero yields

$$-p\mu_p\frac{dV}{dx} = D_p\frac{dp}{dx} \tag{6.2a}$$

and

$$n\mu_n\frac{dV}{dx} = D_n\frac{dn}{dx}, \tag{6.2b}$$

where V is the electric potential, which is related to the electric field by definition as

$$\mathscr{E} \equiv -\frac{dV}{dx}. \tag{6.3}$$

Our attention is directed mainly at this point to the space-charge region, where the electric field is not zero. Now integrating Eq. (6.2a) yields

$$\ln\left(\frac{p}{p_c}\right) = -\frac{\mu_p}{D_p}(V - V_c) = -\frac{q}{kT}(V - V_c), \tag{6.4}$$

where p_c and V_c are arbitrary hole concentration and voltage reference values and the Einstein relationship [Eq. (5.23)] between μ_p and D_p is used. Let us call V_p the potential on the *p*-side of the junction, far to the left and remote from the junction, as shown in Fig. 6.3e. The potential on the *n*-side, far to the right and remote from the junction, is taken as V_n. Using Eq. (6.4) we then have

$$\ln\left(\frac{p_{p_0}}{p_c}\right) = -\frac{q}{kT}(V_p - V_c), \tag{6.5a}$$

where p_{p_0} represents the *equilibrium* density of majority holes in the *p*-region remote from the junction. Similarly for the p_{n_0} minority holes on the *n*-side, remote from the junction,

$$\ln\left(\frac{p_{n_0}}{p_c}\right) = -\frac{q}{kT}(V_n - V_c). \tag{6.5b}$$

Subtracting Eq. (6.5b) from (6.5a) yields

$$\ln\left(\frac{p_{p_0}}{p_{n_0}}\right) = -\frac{q}{kT}(V_p - V_n). \tag{6.6}$$

Then

$$\Delta V_0 = \frac{kT}{q}\ln\left(\frac{p_{p_0}}{p_{n_0}}\right) \quad \text{or} \quad \frac{p_{p_0}}{p_{n_0}} = e^{q\,\Delta V_0/kT}, \tag{6.7}$$

where ΔV_0 is the potential difference (or contact potential) between the two sides of the *p-n* junction, remote from the junction, under thermal equilibrium conditions.

In Section 4.5.3 it was pointed out that at room temperature in germanium and silicon all donors and acceptors are ionized, so essentially $p_{p_0} = N_a$ and $n_{n_0} = N_d$. Also, Eq. (4.29) yields $n_{n_0}p_{n_0} = n_i^2$. Hence Eq. (6.7) can be written as

$$\Delta V_0 = \frac{kT}{q}\ln\left(\frac{N_aN_d}{n_i^2}\right). \tag{6.8}$$

This expression relates the semiconductor *p-n* junction built-in voltage to the impurity levels on each side of the junction. An analogous treatment could be done beginning with Eq. (6.2b) for electrons. The identically same result would be obtained, indicating the strong interdependence between electron and hole flow in a semiconductor.

The same value of the contact potential for a *p-n* junction may be obtained in quite another manner. This is accomplished with the aid of the energy band diagram for electrons and holes in semiconductors along with the concept of the Fermi energy. In Fig. 6.4a the energy band diagrams for separate *p*-type and *n*-type semiconductor crystal segments are shown. Note the position of the Fermi level E_{F_p} near the valence band edge for the *p*-type semiconductor and E_{F_n} near the conduction band edge for the *n*-type semiconductor. Now consider the two segments to be brought toward each other and finally brought into

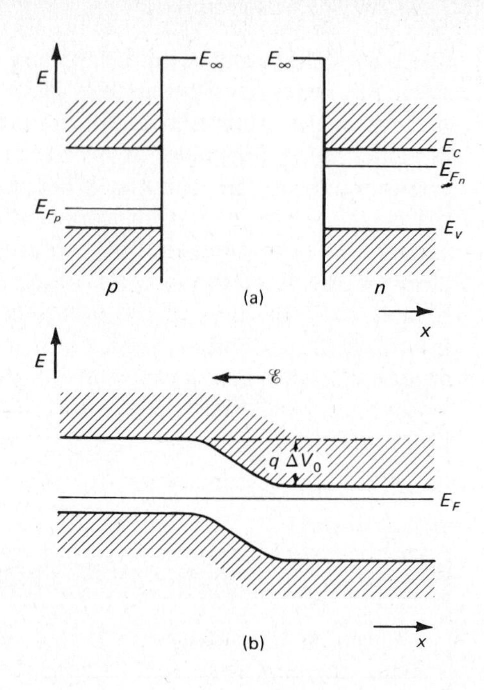

Figure 6.4 (a) Energy band diagrams for separate *p*- and *n*-type crystal segments. (b) Energy band diagrams for a *p-n* junction diode under thermal equilibrium conditions where the Fermi energy is constant throughout the device.

contact in thermal equilibrium. Once this has occurred the Fermi energies on the *p*- and *n*-sides must agree. This, in fact, defines **thermal equilibrium** for a crystal, namely that the Fermi energy is the same throughout the sample.[7] This requirement causes a potential barrier to be formed between the *p*- and *n*-sides of this junction as shown in Fig. 6.4b. That is, the *n*-side must be depressed in energy relative to the *p*-side to permit the respective Fermi levels to line up. (Or vice versa the *p*-side is raised relative to the *n*-side.) Since this energy diagram is for electrons and the electron energy E is given by $E = -qV$, decreasing energy means a numerically greater potential V. Hence the electrical potential on the *n*-side of the *p-n* junction is higher than on the *p*-side. This, of course, is in agreement with the picture obtained by consideration of carrier flow as shown in Fig. 6.3e. The potential barrier indicates the existence of an electric field in the depletion region directed from the *n*-to-*p* side of the junction which prevents the large number of electrons in the *n*-material from diffusing to the *p*-region. It also keeps the large quantity of holes on the *p*-side from entering the *n*-region of the crystal.

It should be pointed out that although the *p*-region contains predominantly holes, it also includes some minority electrons. The electric field of the *p-n* junction tends to transport any such electrons that diffuse into the space-charge region across this region and over to the *n*-side. Also, any electrons

[7]This is analogous to two objects, each at a different temperature, being brought into contact and coming to a uniform temperature after some time has elapsed as thermal equilibrium is established.

residing in this space-charge region will be correspondingly transported. Although the number of such electrons is small relative to the other charges present in materials like extrinsic silicon at room temperature, this number becomes quite substantial as the temperature is raised. (A corresponding argument applies for the minority holes on the n-side of the junction being transported to the p-side by the field present at the junction.) Of course, the electron carrier flow in thermal equilibrium must balance to zero. That is, any minority electron flow from the p-region to the n-side of the junction must be balanced by a comparable flow of the more energetic majority electrons on the n-side mounting the potential barrier and traveling to the p-side. A corresponding argument applies also to hole flow. However, when an external reverse voltage is applied (as will become clear later) this minority carrier flow through the junction is then the source of current flow through a reverse-biased silicon p-n diode.

EXAMPLE 6.1

Determine the change in barrier height of a p^+-n junction diode at 300°K when the doping on the n-side is changed by a factor of 1000, if the doping on the p-side remains unchanged.

Solution

From Eq. (6.8)

$$(\Delta V_0)_1 = (kT/q)\ \ln[(N_a)_1(N_d)_1/n_i^2]$$

and

$$(\Delta V_0)_2 = (kT/q)\ \ln[(N_a)_2(N_d)_2/n_i^2],$$

where the subscript 1 refers to the lightly doped case and 2 to the heavily doped case. Subtracting the first equation from the second gives

$$\begin{aligned}(\Delta V_0)_2 - (\Delta V_0)_1 &= (kT/q)\ \ln[(N_a)_2(N_d)_2/(N_a)_1(N_d)_1] \\ &= (0.026\ \text{V})\ \ln(1000) = \underline{0.18\ \text{V}}.\end{aligned}$$

Hence a 1000-fold change in doping alters the barrier height by only 180 mV.

6.1.3 The Space-Charge Region in Equilibrium

A quantitative calculation of the electric field distribution in the space-charge region will now be given. In a practical p-n junction diode one side of the junction is usually much more heavily doped with impurities than the other side.

The reason for this is mainly explained by certain practical considerations in the fabrication technology for the semiconductor *p-n* junction diode. If this junction is the emitter junction of a bipolar transistor, there are also important electrical reasons for requiring an unsymmetrically doped *p-n* junction.

For our calculation here assume that the *p*-region of the junction contains many more impurity atoms than the *n*-side. This junction will now be referred to as a p^+-n junction. Again we will assume that there is a uniform distribution of impurities on each side of the junction. The total charge density on each side is shown schematically in Fig. 6.5a. Note that the negative charges on the *p*-side and the positive charges on the *n*-side originate from ionized acceptors and donors, respectively. Note too that the region containing the positive charges is considerably wider than the negatively charged region. This results

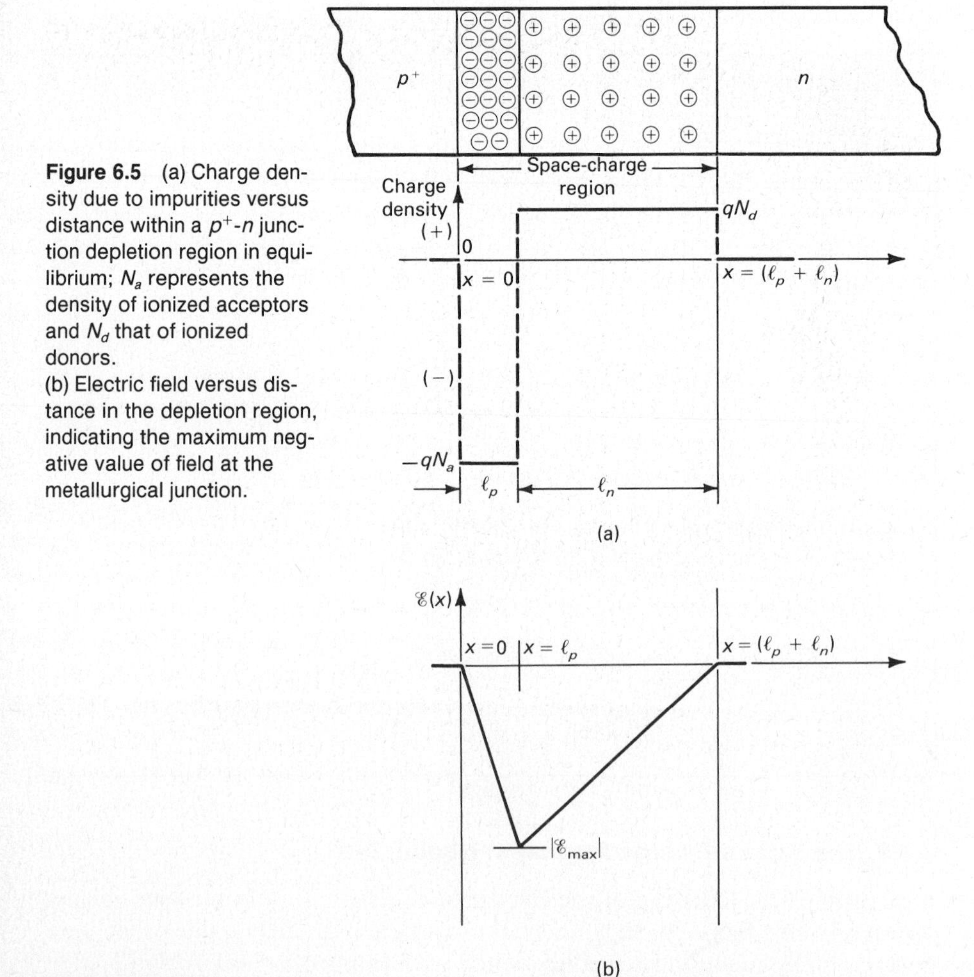

Figure 6.5 (a) Charge density due to impurities versus distance within a p^+-n junction depletion region in equilibrium; N_a represents the density of ionized acceptors and N_d that of ionized donors.
(b) Electric field versus distance in the depletion region, indicating the maximum negative value of field at the metallurgical junction.

from the assumption made that the *density* of impurity atoms is much greater on the *p*-side than on the *n*-side and that the total number of positive ion charges must equal the number of negative ion charges. The latter results from the fact that the field flux lines which originate on the positive charges in the *n*-region must all terminate on the negative charges of the *p*-region, since the electric field is everywhere zero in the uniformly doped material outside this **space-charge** region. (Charge neutrality in semiconductors dictates that there can be no space-charge region in a crystal unless a relatively sharp impurity gradient occurs in a region of the order of a few micrometers as at a *p-n* junction.[6]) Equating the number of charges on each side of the junction of Fig. 6.5a yields

$$N_a l_p A = N_d l_n A, \tag{6.9}$$

so the ratio of the space-charge width on the *n*-side to that on the *p*-side is

$$l_n / l_p = N_a / N_d. \tag{6.10}$$

In this discussion it has been tacitly assumed that there are no electrons or holes in the space-charge region, that the only charges there are ionized donors or acceptors. This is a good approximation since the high electric field present in the space-charge region tends to sweep out mobile electrons and holes, and this volume is sometimes referred to as the **depletion region.** At the outer edges of the space-charge layer the field is small and some mobile carriers are present. However, this causes only a small correction to this space-charge calculation and so the effect of mobile charges will be neglected here and only the fixed charges considered, in a so-called **depletion-layer** approximation.[8]

Gauss's law applied to the *p*-side of the *p-n* junction depletion region yields

$$\epsilon_0 \epsilon_s \frac{d\mathscr{E}}{dx} = -qN_a. \tag{6.11}$$

Here $\mathscr{E}$ is the electric field intensity, N_a the acceptor concentration, ϵ_0 the permittivity of free space, and ϵ_s the semiconductor dielectric constant. Integrating this equation gives

$$\mathscr{E} = -\left(\frac{qN_a}{\epsilon_0 \epsilon_s}\right) x + B. \tag{6.12}$$

The electric field vanishes at the edge of the space-charge layer; i.e., $\mathscr{E} = 0$ at $x = 0$. Using this boundary condition requires that $B = 0$. So the electric field decreases linearly with x and reaches a maximum negative value at $x = l_p$. This is illustrated in Fig. 6.5b. Hence

$$|\mathscr{E}_{\max}| = qN_a l_p / \epsilon_0 \epsilon_s. \tag{6.13a}$$

[8]See P. E. Gray et al., *Physical Electronics and Circuit Models of Transistors* (Wiley, New York, 1969), p. 20, for a discussion of the accuracy of this approximation.

The edge of the space-charge region of the n-side of the junction is at $x = l_p + l_n$. Integrating Gauss's equation for the n-side gives

$$|\mathcal{E}_{\max}| = qN_d l_n/\epsilon_0\epsilon_s\,. \tag{6.13b}$$

The total potential drop across the space-charge layer is the sum of the drops across the p- and n-sides and is obtained by integrating the electric field across the depletion region. This gives the equilibrium barrier height

$$\Delta V_0 = \left|\int_0^{l_p+l_n} \mathcal{E}(x)\,dx\right|, \tag{6.14}$$

which represents the area under the field curve of Fig. 6.5b. Integrating gives the barrier height

$$\Delta V_0 = \left(\frac{q}{2\epsilon_0\epsilon_s}\right)(N_a l_p^2 + N_d l_n^2). \tag{6.15}$$

With Eqs. (6.13a) and (6.13b) this becomes

$$\Delta V_0 = |\mathcal{E}_{\max}|\left(\frac{l_p + l_n}{2}\right). \tag{6.16}$$

Note that this equation may be deduced from inspection of Fig. 6.5b, by determining the area under the curve geometrically. This can be rewritten using Eq. (6.10) as

$$\Delta V_0 = \frac{|\mathcal{E}_{\max}|\,l_n}{2}\left(1 + \frac{N_d}{N_a}\right). \tag{6.17}$$

If we use the previously derived expression for the equilibrium barrier height given in Eq. (6.8), Eqs. (6.13) and (6.17) now permit calculation of the space-charge widths in terms of device design parameters as

$$l_p = \left(\Delta V_0 \frac{2\epsilon_0\epsilon_s}{qN_a}\frac{N_d}{N_d + N_a}\right)^{1/2}, \tag{6.18a}$$

$$l_n = \left(\Delta V_0 \frac{2\epsilon_0\epsilon_s}{qN_d}\frac{N_a}{N_d + N_a}\right)^{1/2}. \tag{6.18b}$$

For a p^+-n junction where $N_a \gg N_d$, then

$$l_n \simeq \left(\Delta V_0 \frac{2\epsilon_0\epsilon_s}{qN_d}\right)^{1/2} = \left(\frac{2\epsilon_0\epsilon_s kT}{q^2 N_d}\ln\frac{N_a N_d}{n_i^2}\right)^{1/2} \tag{6.18c}$$

and $l_p \ll l_n$. Hence the depletion-layer width is approximately inversely proportional to the square root of the impurity level in the most *lightly* doped side of the p-n junction and penetrates mainly into this latter region.

Precisely the same procedure may be followed in cases in which the transition from p- to n-type is not abrupt, as assumed here. Gauss's law can be integrated easily if, for example, a linear variation in impurity concentration is assumed as one proceeds from the *p*- to the *n*-region. Then the space-charge width under equilibrium conditions varies inversely as the *cube root* of the impurity gradient. In the case of this linear impurity gradient a, in atoms/cm^4, the space-charge width l is given by[9]

$$l = \left(\frac{12\epsilon_0\epsilon_s\,\Delta V_0}{qa}\right)^{1/3}. \tag{6.19}$$

EXAMPLE 6.2

Determine the space-charge width of the *n*-region of an abrupt silicon p^+-n junction in thermal equilibrium at 300°K if the doping on the *n*-side is 1.0×10^{14} donors/cm^3 and that on the *p*-side is $5.0 \times 10^{19}/cm^3$. The relative dielectric constant of silicon is 12.

Solution

From Eq. (6.10)

$$l_n/l_p = N_a/N_d = (5.0 \times 10^{19})/10^{14} = 5.0 \times 10^5.$$

Therefore the space-charge width in the *p*-region is negligibly narrow compared to that in the *n*-region. Hence using Eq. (6.18c) we have

$$l_n = \left[\frac{2\epsilon_0\epsilon_s kT}{q^2 N_d}\ln\frac{N_a N_d}{n_i^2}\right]^{1/2}$$

$$= \left[\underbrace{\frac{kT}{q}\ln\frac{N_a N_d}{n_i^2}}_{\Delta V_0}\frac{2\epsilon_0\epsilon_s}{qN_d}\right]^{1/2}$$

$$= \left[\left((0.026\text{ V})\ln\frac{(5.0 \times 10^{19})(1.0 \times 10^{14})}{(1.5 \times 10^{10})^2}\right) \times \frac{2(8.85 \times 10^{-14}\text{ F/cm})12}{(1.6 \times 10^{-19}\text{ C})(10^{14}\text{ cm}^{-3})}\right]^{1/2}$$

$$= [0.80(1.33 \times 10^{-7})]^{1/2} = \underline{3.3 \times 10^{-4}\text{ cm}}.$$

This derivation has assumed no externally applied voltage. If an inverse voltage bias is applied to the junction (*p*-side negative relative to *n*-side), this

[9] S. M. Sze, *Physics of Semiconductor Devices,* 2nd ed. (Wiley, New York, 1981), p. 81.

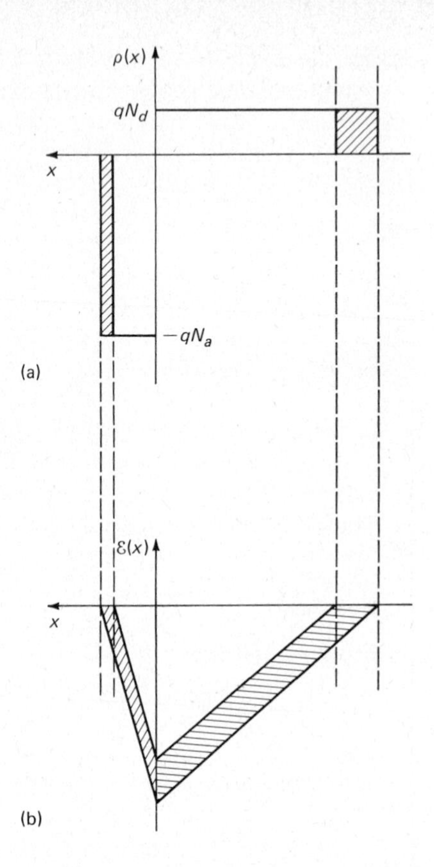

Figure 6.6 (a) Schematic diagram showing the additional fixed charge (above equilibrium) of ionized donors and acceptors (crosshatched areas) introduced into the p^+-n junction space-charge region under reverse bias conditions. Space-charge widening results. (b) Graph showing the electric field increase in the space-charge region under reverse bias conditions. The crosshatched area represents the applied reverse voltage.

creates an additional electric field in the space-charge region in the same direction as the built-in field. Hence, according to Eq. (6.18), the space-charge width will tend to increase until $\int \mathcal{E}\, dx$ over this space-charge width is equal to the sum of the built-in voltage plus the absolute value of the applied voltage. In order for this to occur, additional ionized acceptors must be added to the p-side of the depletion layer and an equal number of ionized donors to the n-side. A schematic representation of this is shown in Fig. 6.6.

This method of calculating the space-charge width is of interest in the design of a **charge-coupled device**[10] as well as the **varactor diode,** which is a variable-capacitor-type semiconductor device. The capacitive character of the junction diode is indicated by the fact that an incremental applied voltage results in an incremental change in charge on each side of the depletion layer, in analogy with the charging of a parallel-plate capacitor. It will be shown quantitatively later how the capacitance of the junction can be adjusted by applying an external voltage, thereby altering the space-charge width.

[10]This device produces a simple type of shift register and will be described in Chapter 12.

6.2 STEADY-STATE CURRENT FLOW IN A *p-n* JUNCTION DIODE AT LOW FREQUENCIES

The previous discussion was a description of a semiconductor *p-n* junction under equilibrium conditions with no external voltage applied. Consider now the case of a *p-n* junction under applied voltage conditions and hence current flow. If the *p*-side of the junction is brought to a higher or more positive potential, V_{pF}, than the *n*-side, V_{nF}, this is referred to as **forward** bias. **Reverse** or **inverse** bias refers to the *n*-side being raised to a higher potential than the *p*-side. It will next be shown analytically that the current I through the *p-n* junction and the voltage drop $V(= V_{pF} - V_{nF})$ across the junction are related generally by a diode equation of the form

$$I = I_0(e^{qV/kT} - 1). \tag{6.20}$$

That is, when the junction is forward biased (V is made positive), the current increases rapidly for small changes in voltage. In reverse bias the current is much smaller and becomes essentially equal to I_0 as the voltage is increased. Physically this is explained in terms of reduction of barrier height due to forward bias, causing appreciable current to flow; in the reverse direction, the increased barrier height limits carrier flow across the junction. This is illustrated in Fig. 6.7. Note that increasing the potential in a positive sense

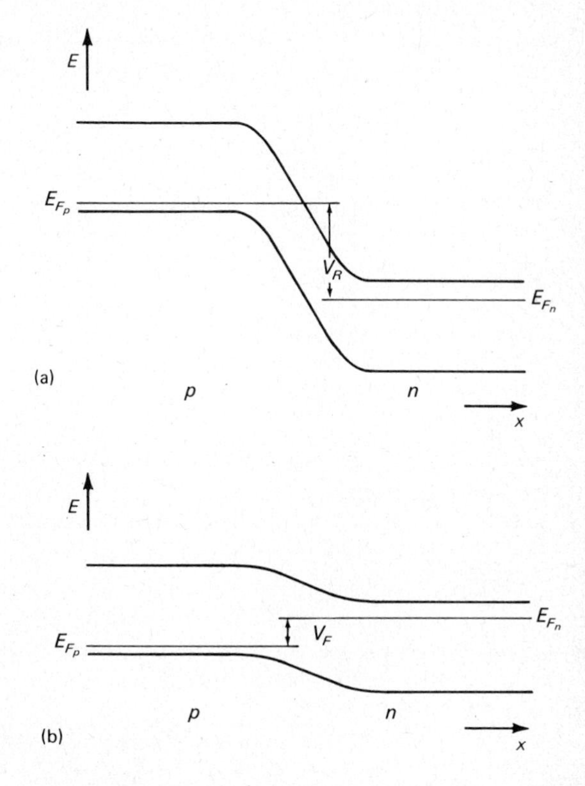

Figure 6.7 (a) Energy diagram for a p^+-n junction under reverse bias. The barrier height is increased over equilibrium conditions. (b) Energy diagram for a p^+-n junction under forward bias. The barrier height is decreased over equilibrium conditions.

depresses the energy levels in the electron energy band diagram. Again this results from the negative charge on the electron. The Fermi energy is not defined under nonequilibrium conditions, but levels of E_{F_p} and E_{F_n} are called the **quasi-Fermi levels.**

6.2.1 Steady-State Minority Carrier Flow

Consider a simple one-dimensional model of the p^+-n junction as shown in Fig. 6.8a. A forward voltage bias causes the reduction of barrier height and permits many holes from the p-region to enter the n-region as minority carriers. This process is known as **minority carrier injection.** Correspondingly, electrons from the n-side cross the lowered barrier and enter the p-side as minority carriers. This is shown in Fig. 6.8b. Although the current through the junction consists of the sum of both the electron and hole components, the hole flow dominates because of the preponderance of holes on the heavily doped p-side compared to electrons on the n-side of the p^+-n junction considered. In addition, although the total hole current in the n-region consists of both a drift component due to electric field and a diffusion component [see Eq. (5.22)], the latter current dominates completely in the n-region *near* the junction under low level injection conditions. (Low level injection refers to the case in which the injected minority holes are smaller in density by at least an order of magnitude compared with the density of majority electrons in the n-region at equilibrium.) Dominance of diffusion current occurs in this case since the applied voltage appears across the space-charge region primarily and creates

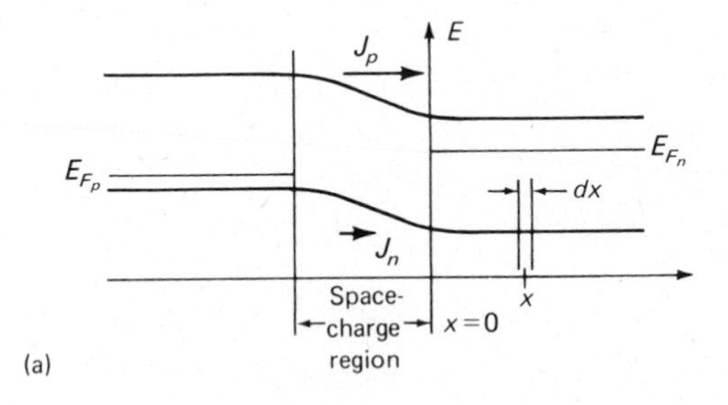

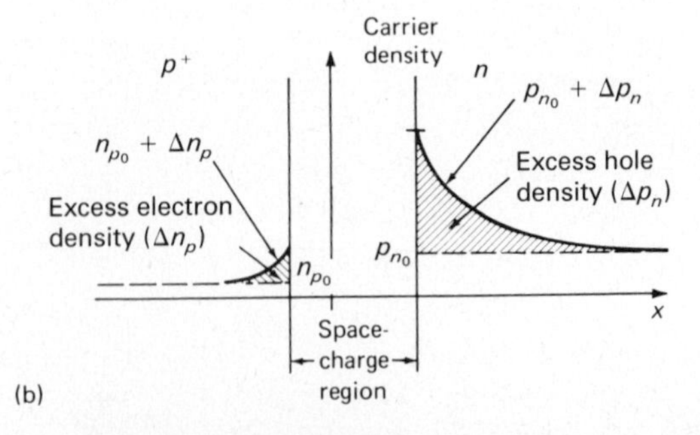

Figure 6.8 (a) Energy diagram for a p^+-n junction diode under forward bias causing injection of electrons and holes; x is taken as zero at the n-side edge of the space-charge region. The n-region is assumed to extend to infinity in the x-direction (long diode approximation). (b) Injection of excess hole density into the n-region (Δp_n) and electrons (Δn_p) into the p-region of the p^+-n junction diode; p_{n0} and n_{p0} correspond respectively to the equilibrium concentrations of holes in the n-type and electrons in the p-type regions. The exponential fall-off of excess holes and electrons with distance is due to minority carrier recombination.

an electric field there; very little of the field penetrates the n-region to produce a drift component of the current there.[11] This is illustrated by Fig. 6.8a since the slope of energy band edge is nearly zero everywhere except in the space-charge region. Since $E = -qV$, $dE/dx = -q\,dV/dx \simeq 0$ and the electric field is nearly zero outside the space-charge region.

Now consider a very small volume of width dx and unit area, in the n-region located a distance x from the edge of the space-charge region as shown in Fig. 6.8a. Under steady-state conditions the current of carriers entering the region dx differs from that leaving the region by the amount of minority hole loss by recombination in that volume. This statement is an expression of the **continuity** of carrier flow and can be expressed mathematically as

$$-\frac{1}{q}\frac{dJ_p}{dx} = \frac{\Delta p_n}{\tau_p}, \tag{6.21}$$

where $(1/q)(dJ_p/dx)$ is the change of hole flux with respect to x in dx and $\Delta p_n/\tau_p$ represents the rate of recombination of *excess* injected hole density Δp_n in the n-region, within dx, where τ_p is the hole lifetime [see Eq. (5.27)]. Since only hole diffusion current is considered, this is given by

$$J_p = -qD_p\frac{dp_n}{dx}. \tag{6.22}$$

Combining Eqs. (6.21) and (6.22) gives the steady-state **continuity equation** for hole density in the n-region:

$$D_p\frac{d^2p_n}{dx^2} = \frac{\Delta p_n}{\tau_p}. \tag{6.23a}$$

In terms of injected carriers above the equilibrium number, the excess hole density continuity equation becomes, since $p_n = p_{n_0} + \Delta p_n$,

$$D_p\frac{d^2\,\Delta p_n}{dx^2} = \frac{\Delta p_n}{\tau_p}. \tag{6.23b}$$

The solution of this equation may be written as

$$\Delta p_n(x) = Be^{-x/L_p} + Ce^{x/L_p}, \tag{6.24}$$

where $\sqrt{D_p\tau_p}$ is defined as the **minority carrier diffusion length** L_p. Just as τ_p represents the *time* taken for minority holes to decrease in number by a factor of $1/e$ by recombination, L_p represents the *distance* that holes can diffuse on the average before their number decreases by a factor of $1/e$ as a result of recombination. Since recombination occurs all along the path of motion of the hole carriers in the n-region, then Δp_n must approach zero for large positive values of x. (We assume in this derivation that the diode thickness is much greater than L_p; this is the **long** diode approximation.) This boundary condition

[11] The demonstration that at low injection levels the current in the n-region near the junction is almost exclusively diffusion current is given in Appendix D.

requires that C in Eq. (6.24) be identically zero, for otherwise Δp_n would grow with x.

6.2.2 The *p-n* Junction Current–Voltage Relation

Making use of Eq. (6.22) for the hole diffusion current density and introducing Eq. (6.24) gives for this current density at x

$$J_p(x) = -qD_p \frac{d\,\Delta p_n(x)}{dx} = \frac{qD_p B e^{-x/L_p}}{L_p}. \tag{6.25}$$

This expression is true for any distance x from the edge of the space-charge region. Hence it must also be true at the n-side edge of this region, where $x = 0$. There Eq. (6.25) may be written as

$$J_p(0) = qD_p\,\Delta p_n(0)/L_p\,, \tag{6.26a}$$

where $\Delta p_n(0)$ represents the excess minority hole density present at the n-side edge of the space-charge region under forward voltage bias conditions. Combining Eqs. (6.25) and (6.26a), we can write the hole current density at any value of x in the n-region as

$$J_p(x) = [qD_p\,\Delta p_n(0)/L_p]\, e^{-x/L_p}. \tag{6.26b}$$

This implies that the hole current falls off with x, and so reaches $1/e$ of its value at $x = 0$ in a diffusion length L_p (see Fig. 6.9a). However, under the steady-state conditions assumed, the current density must be the same at all values of distance x and time t. Hence the loss of hole current with distance due to hole recombination must be made up by the generation of electron current just to compensate this effect. This is shown diagrammatically in Fig. 6.9b. Note that at a distance x, far (several L_p) from the junction, the current is nearly entirely electron drift current, as one would expect in an n-type semiconductor lacking injected holes; but the electric field is quite small there.

It is now necessary to determine a relationship between $\Delta p_n(0)$ and the voltage V applied across the space-charge layer in order to derive the p-n junction current–voltage characteristic sought. It is a reasonable assumption that, because of the large hole density in the p-region under normal forward voltage bias conditions, the junction is intrinsically capable of supplying very much more current than it actually supplies. This means that the balance of drift and diffusion currents in the space-charge region, which is absolute under equilibrium conditions, is not much disturbed by the applied voltage. In fact, the applied bias raises the diffusion component of current only slightly above the drift component. This is demonstrated in Example 6.3. Then, in analogy with Eq. (6.7), since the equilibrium is hardly disturbed, the density of holes on the n-side of the depletion region and the holes on the p-side are related by

$$\frac{(p_p)_{p\text{-side edge}}}{(p_n)_{n\text{-side edge}}} = e^{q(\Delta V_0 - V)/kT}, \tag{6.27}$$

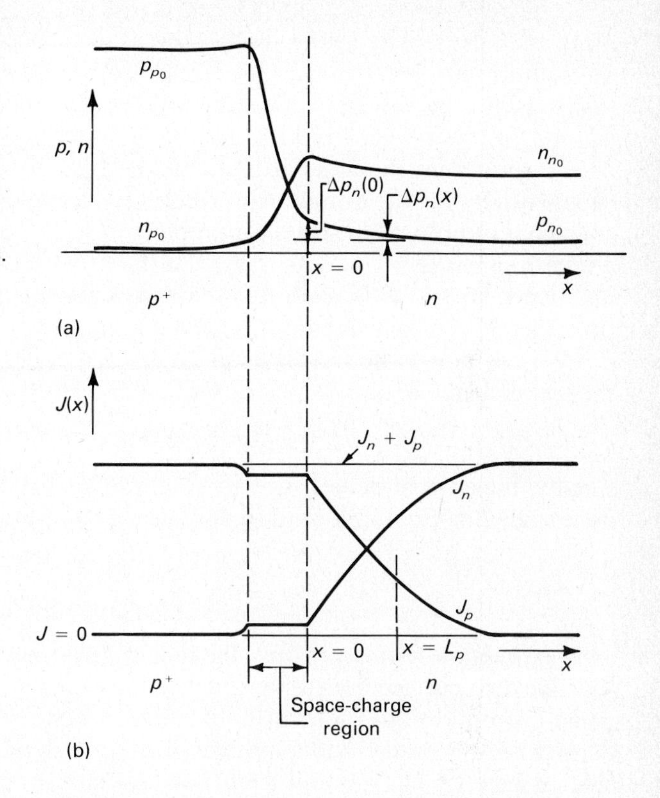

Figure 6.9 (a) Plot of electron density n and hole density p in a long p^+-n junction diode in forward bias. (b) Plot of the electron current density J_n and the hole current density J_p in a long p^+-n junction diode versus distance, under forward bias.

where ΔV_0 is the equilibrium built-in barrier potential and V represents the forward voltage (p-side positive with respect to n-side) applied across the space-charge layer. Using Eq. (6.7) to eliminate ΔV_0, we rewrite Eq. (6.27) as

$$\frac{(p_n)_{n\text{-side edge}}}{(p_p)_{p\text{-side edge}}} = \frac{p_{n0}}{p_{p0}}\, e^{qV/kT}. \tag{6.28}$$

For low level injection conditions the injected minority hole density is assumed to be much smaller than the density of the equilibrium number of majority electrons present in the n-region; similarly, the injected minority electron density is considerably less than the hole density present in the heavily doped p-region. Since the condition of charge neutrality requires that electrons and holes be injected into the p-region in equal numbers, this latter statement requires that $(p_p)_{p\text{-side edge}} \simeq p_{p0}$. Then Eq. (6.28) reduces to

$$p_n(0) = p_{n0}e^{qV/kT}. \tag{6.29}$$

This is the extremely useful **junction law.** Multiplying both sides by $n_n(0)$ and noting that $n_n(0) \simeq n_{n_0}$ (the equilibrium majority carrier concentration), we obtain $p_n(0)n_n(0) = p_{n_0}n_{n_0}e^{qV/kT} = n_i^2 e^{qV/kT}$, which is the nonequilibrium version of $p_n n_n = n_i^2$ [see Eq. (4.29)].

When the voltage applied to the junction is zero, Eq. (6.29) shows that the minority hole density at the n-side edge of the space-charge region reduces to the equilibrium density, as expected. It also predicts that the density of holes injected into the n-region increases exponentially with applied voltage. That is, the density of holes from the heavily hole-populated p-side of the junction, which can mount the junction potential barrier and enter the n-side, depends exponentially on the amount by which the barrier is reduced. This is analogous, for example, to the classical problem of gas atoms in free space mounting a retarding barrier. This was treated in the nineteenth century by Boltzmann, and hence Eq. (6.29) is often referred to as the Boltzmann relation; it is sometimes taken as a basic assumption, even at moderate injection levels.

We can now write for the excess holes just appearing at the n-side edge of the space-charge region

$$\Delta p_n(0) = p_n(0) - p_{no} = p_{no}(e^{qV/kT} - 1). \tag{6.30}$$

Finally, the current density due to holes crossing the p-n junction is obtained by introducing Eq. (6.30) into (6.26b) as

$$J_p(x) = \left(\frac{qD_p p_{no}}{L_p}\right)(e^{qV/kT} - 1)e^{-x/L}. \tag{6.31}$$

This expression represents only the hole diffusion current density. In general, the hole drift current density must be added to this to get the total hole current density. In addition, the drift and diffusion components of electron current density must be added too. However, for the analysis of the p^+-n junction considered here, it has already been argued that the hole diffusion component dominates near the junction at $x = 0$. Hence in this case Eq. (6.31) at $x = 0$ approximately represents the relation between the *total* junction current density and applied voltage

$$I = \frac{qAD_p p_{no}}{L_p}(e^{qV/kT} - 1) = \frac{qp_{no}(L_p A)}{\tau_p}(e^{qV/kT} - 1), \tag{6.32a}$$

where A is the p-n junction cross sectional area; this is the form predicted in Eq. (6.20).

The coefficient $qp_{n0}(L_p A)/\tau_p$ in Eq. (6.32a) represents the diffusion component of the **saturation** or **leakage** current through the junction when a few volts of negative or reverse potential is applied to the device. For then $e^{qV/kT} \rightarrow 0$ and $I \rightarrow qp_{n0}(L_p A)/\tau_p$. Physically this means that the diffusion component of the inverse leakage current in such a p^+-n junction is made up of the flow across the junction of the equilibrium number of minority holes on the n-side within a diffusion length of the junction, $p_{n0}(L_p A)$, which are hence "collected" during their lifetime τ_p. These carriers flow "downhill"[12] from the n-side to the p-side of the potential barrier (see Fig. 6.7a). Minority electrons from the p-side also can flow downhill through the barrier to the n-side. However, because of the

[12]Note that "downhill" for holes is "uphill" for electrons as pictured in Fig. 6.7a.

high acceptor concentration on the *p*-side, the number of majority holes is very large and so the number of minority electrons is small there, owing to the constancy of the *np* product. Therefore in a p^+-n junction, although the diffusion component of the leakage current consists of both diffusing holes and electrons, the hole component dominates. In the more general case of a *p-n* junction which has comparable doping on both the *p*- and *n*-sides, the diode current–voltage equation becomes

$$I = qA\left(\frac{D_p p_{n0}}{L_p} + \frac{D_n n_{p0}}{L_n}\right)(e^{qV/kT} - 1)$$

$$\simeq qA\left(\frac{D_p}{L_p N_d} + \frac{D_n}{L_p N_a}\right) n_i^2 (e^{qV/kT} - 1), \tag{6.32b}$$

where A is the area of the *p-n* junction. Note that p_{n0}, n_{p0}, and n_i are strongly temperature dependent as given by Eqs. (4.21), (4.27), and (4.29); D_p, D_n, L_p, and L_n are normally only mildly temperature dependent. Hence the temperature variation of the diffusion component of the junction saturation current is mainly dependent on the change of the semiconductor minority carrier densities and n_i with temperature. This leakage current is contributed mainly by the diffusion of minority carriers from the least-doped side of the *p-n* junction.

EXAMPLE 6.3

Assuming the silicon p^+-n junction diode design as in Example 6.2, show that the balance of drift and diffusion currents in the space-charge region, which is absolute under equilibrium conditions, is not much disturbed by the application of a normal forward bias voltage, say 0.60 V, at 300°K.

Solution

First let us roughly approximate the hole diffusion current in the space-charge region (which is equal and opposite to the hole drift current) under equilibrium conditions.

$$(J_{\text{diff}})_p = -qD_p \frac{dp}{dx}$$

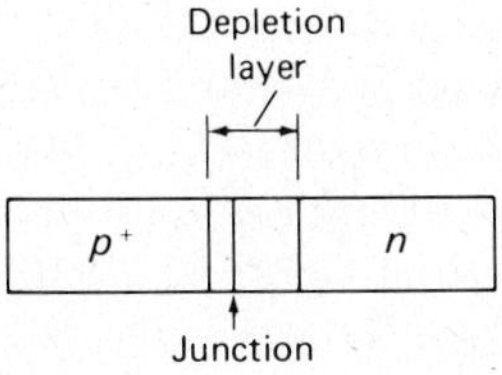

Depletion layer

p^+ n

Junction

From Example 6.2 the density of holes at the left-hand edge of the space-charge region is 5.0×10^{19} cm^{-3}. Since the *pn* product in equilibrium at 300°K is $(1.5 \times 10^{10})^2$ cm^{-6}, the hole density at the right-hand edge of the space-charge region is $(1.5 \times 10^{10})^2/10^{14}$ cm^{-3} or 2.25×10^6 holes/cm^3. Since the depletion-layer width from Example 6.2 is 3.3×10^{-4} cm,

$$-\left(\frac{dp}{dx}\right) \simeq \frac{5.0 \times 10^{19} - 2.25 \times 10^6}{3.3 \times 10^{-4}} \text{ cm}^{-4}$$

$$= 1.5 \times 10^{23} \text{ cm}^{-4}$$

and

$$(J_{\text{diff}})_p \simeq (1.6 \times 10^{-19} \text{ C})(1.25 \times 10^1 \text{ cm}^2/\text{sec})(1.5 \times 10^{23} \text{ cm}^{-4})$$

$$= \underline{3.0 \times 10^5 \text{ A/cm}^2}.$$

So $(J_{\text{drift}})_p \simeq \underline{3.0 \times 10^5 \text{ A/cm}^2}$.

Let us now calculate the hole current at the right-hand edge of the depletion layer, assuming in advance that the equilibrium drift and diffusion current balance is little upset by the application of a forward bias of 0.60 V. Using Eq. (6.31) and assuming that $L_p = 1.0 \times 10^{-2}$ cm, we have

$$J_p = \frac{qD_p p_{n0}}{L_p}(e^{qV/kT} - 1)$$

$$= \frac{(1.6 \times 10^{-19} \text{ C})(1.25 \times 10^1 \text{ cm}^2/\text{sec})(2.25 \times 10^6 \text{ cm}^{-3})}{1.0 \times 10^{-2} \text{ cm}}$$

$$\times (e^{0.60/0.026} - 1)$$

$$= \underline{4.7 \text{ A/cm}^2}.$$

This represents only $4.7/(3.0 \times 10^5) \times 100$ or $\underline{1.6 \times 10^{-3}\%}$, or a minute upset of balance of equilibrium drift and diffusion currents, and thus the proposition is proved.

6.2.3 The Short Diode Approximation

In practical p^+-n junction diode rectifiers the diode thickness is normally much less than L_p. That is, the minority carrier diffusion length greatly exceeds the thickness of the lightly doped n-region. This tends to reduce the series resistance of the rectifier in the forward direction and greatly diminishes the power loss at high forward currents due to this parasitic element. This **short diode** device will now be modeled by slightly altering the analytic treatment which resulted in Eq. (6.32a).

Assume that the n-region of the diode pictured in Fig. 6.8 is short; that is, its thickness W is much less than the minority hole diffusion length L_p. A metal ohmic contact will be taken to terminate the right extent of the n-region, at $x = W$. Because of the high density of electrons in a metal, the metal is a region of high minority hole recombination and hence the boundary condition to be assumed at $x = W$ is $\Delta p_n(W) = 0$. Under this condition Eq. (6.24) becomes

$$C = -Be^{-2W/L_p}, \tag{6.33a}$$

and

$$\Delta p_n(x) = B(e^{-x/L_p} - e^{(x-2W)/L_p}). \tag{6.33b}$$

In the short diode approximation $W \ll L_p$ and hence also $x \ll L_p$. Using a series expansion for the exponential functions in Eq. (6.33b) in the case of small values of the exponents, one can reduce that equation to

$$\Delta p_n(x) = 2B(W - x)/L_p, \tag{6.34}$$

which at $x = 0$ gives

$$B = (L_p/2W)\,\Delta p_n(0). \tag{6.35}$$

So

$$\Delta p_n(x) = (1 - x/W)\,\Delta p_n(0). \tag{6.36}$$

Hence $\Delta p_n(x)$ decreases linearly with x, equals $\Delta p_n(0)$ at $x = 0$, and equals zero at $x = W$. Introducing Eq. (6.36) into Eq. (6.22) yields

$$J_p(x) = qD_p\,\Delta p_n(0)/W. \tag{6.37}$$

In this approximation the hole current density is constant, independent of x, since hole recombination is neglected in the bulk of the device. Now using Eqs. (6.30) and (6.37) gives the diode current–voltage law

$$I = (qAD_pp_{n0}/W)(e^{qV/kT} - 1). \tag{6.38}$$

Note that for short diodes ($W \ll L_p$) the current increases as the diode width decreases.

6.2.4 Space-Charge-Generated Current

In the above calculation it was assumed that essentially all the current was derived from the diffusion of minority carriers through the junction. This is a valid assumption for germanium and semiconductor materials with smaller energy gaps. In the case of wider energy gap semiconductors such as silicon and gallium arsenide there exists still another important component of the current. This component derives from carriers that are **generated and recombine in the space-charge region,** and was neglected in the diffusion current treatment. In the space-charge region in equilibrium, hole–electron pairs are continuously being created thermally and recombining in a time called the lifetime. By definition of equilibrium, for every carrier generated one must recombine. Under forward bias, minority current carriers from the p-side of the junction can recombine in the space-charge region before reaching the n-side and this recombination current must be supplied. Also, a large electric field is present in the reverse-biased junction space-charge region, which causes mobile electrons and holes generated within it to be swept out before recombining, constituting a current. A formulation of this problem of space-charge-generated junction current has been carried out,[13] yielding approximately the following current–voltage characteristic

$$I = (qAW_{SC}n_i/\tau)(e^{qV/2kT} - 1), \tag{6.39}$$

where τ is the effective carrier lifetime in the space-charge width W_{SC}, A the junction area, and n_i the density of intrinsic carriers in the semiconductor. Note the factor of 2 difference in the exponential term in this formulation compared to the expression that considers only diffusion current [see Eq. (6.32)]. Note too that the coefficient of the exponential term represents the space-charge-generated junction leakage current under reverse bias. The form of this coefficient can be made plausible by identifying qn_i/τ as the charge density generated per unit time in the space-charge volume AW_{SC}. The charge density n_i is involved since the space-charge region is depleted of mobile carriers and hence effectively is "intrinsic." Since the space-charge width increases with increased reverse potential bias, the space-charge-generated leakage current rises with increased inverse voltage. This is in contrast to the diffusion component of leakage current which is voltage independent above a small voltage and hence is termed "saturation" current.

In general, both the diffusion and space-charge-generated components of the junction leakage current should be considered. However, by comparing the magnitude of the currents given by Eqs. (6.39) and (6.32b) it can be ascertained which is dominant in any particular case. Note that the diffusion current dominates over space-charge-generated current for narrow energy gap semiconductors since the former varies as n_i^2 whereas the latter follows n_i. Hence in germanium only diffusion current normally needs to be considered. However, in silicon (and wider energy gap materials) space-charge-generated current dominates at low forward currents and for reverse voltage bias. At relatively large forward currents in silicon junctions, diffusion controls again since the diffusion component increases rapidly with voltage whereas the space-charge-generated component is mainly unchanged.

*6.2.5 Surface Leakage Current

The reverse current of a real p^+-n junction diode includes a surface leakage component in addition to the bulk leakage terms due to diffusion and space-charge generation derived above. This may be of the ohmic type simply caused by the electrical conduction properties of spurious impurity ions which appear

[13] C. T. Sah, R. N. Noyce, and W. Shockley, "Carrier Generation and Recombination in p-n Junctions and p-n Junction Characteristics," *Proc. IRE,* **45,** 1228 (1957).

on the diode surface to a greater or lesser extent depending on manufacturer cleaning techniques. Surface ionic conduction is particularly enhanced in the presence of moisture or under high humidity conditions. This type of leakage current can be reduced many orders of magnitude by depositing a protective insulating layer on the p^+-n junction surface. Silicon planar junctions are commonly protected by depositing a thin, **passivating** silicon dioxide layer on the junction surface as shown in Fig. 6.2b. This is standard practice in the fabrication of integrated circuits and will be discussed in Chapter 11. (Some limitations connected with oxide passivation have been explained in Section 5.5.) The surfaces of *p-n* junctions are alternatively cleaned by chemically etching away contaminated and disturbed surface layers of silicon with a combination of hydrofluoric and nitric acids; a thorough washing in ultrapure water and drying then follows. Junctions so prepared are uniformly flat in contrast to the curvature at the edges of the planar junctions of Fig. 6.2b and are referred to as **mesa-type** junctions. When *p-n* junctions with blocking voltage capability of a thousand volts or more are required, mesa junctions are often fabricated; for a given applied voltage a portion of the space-charge region of a planar junction reaches a higher electric field than a comparably doped mesa junction since the highest field is developed at the point of greatest junction curvature.

Still another type of junction leakage is referred to as junction **channeling.** This refers to the conduction of current in the surface of the semiconductor crystal adjoining the junction due to mobile charges induced in the semiconductor by fixed charges on the crystal surface, or even under the oxide "protecting" the surface.[14] This produces a leakage current essentially independent of voltage in contrast to ohmic ionic leakage, in which the current and voltage are linearly related. An n^+-p planar junction consists of a small n^+-region imbedded in a *p*-type semiconductor substrate. As has already been discussed in Section 5.5, the oxide and oxide–surface interface charges are always positive, so an *n*-type surface skin is induced which can invert the *p*-type surface covered by the oxide. This effectively increases the *n-p* junction area, thereby increasing the leakage current. Since this induced *n*-layer is thin, a current-limiting action similar to that observed in a field-effect transistor (see Chapter 10) occurs, so the leakage current is constant, although higher, above a certain voltage and remains fixed as the voltage is raised. This will obviously not occur in the case of a p^+-n junction; instead the positive surface charges will tend to reduce the blocking voltage of this type of junction (see Section 6.3.3C).

6.3 OTHER ELECTRICAL PROPERTIES OF *p-n* JUNCTION DIODES

The successful circuit use of the *p-n* junction diode requires that some additional properties of the junction and electrical contacts to the device be under-

[14]For a full discussion of this phenomenon, see A. Grove, *The Physics and Technology of Semiconductor Devices* (Wiley, New York, 1965), pp. 298ff. See also Section 5.5 of this text.

stood. For example, successful circuit applications of the diode require a knowledge of the temperature behavior of the device. The electrical characteristics of semiconductor devices in general are quite temperature sensitive [see Eq. (6.32)]. Self-heating caused by the current flowing through the junction diode provides a high current limitation for the device. A high voltage limitation results from excessive junction leakage current due to current multiplication in the reverse-biased space-charge region. High frequency or high speed performance is limited by the time necessary to charge the space-charge capacitance of the junction through the bulk series resistance of the semiconductor body and ohmic contacts. High speed switching performance can also be limited by injected-minority-carrier transit time as well as storage effects in the semiconductor bulk. These various types of diode limitations will be discussed in the remainder of this chapter and in the following chapter.

6.3.1 Temperature Behavior of the *p-n* Junction

The electrical characteristics of the semiconductor *p-n* junction are quite temperature sensitive. This becomes apparent when the current–voltage relationship for the junction is examined. In addition to the explicit exponential variation with temperature of the junction current as indicated in Eq. (6.20), the leakage or saturation current is also temperature sensitive. In order to illustrate this we can write the saturation current for a germanium *p-n* junction from Eq. (6.32b) as

$$I_0 = qA\left(\frac{D_p}{L_pN_d} + \frac{D_n}{L_nN_a}\right)n_i^2. \tag{6.40}$$

Here n_i^2 represents the square of the density of intrinsic carriers and can be written as [see Eq. (4.29)]

$$n_i^2 = CT^3e^{-E_g/kT}, \tag{6.41}$$

where C is a constant independent of temperature. This drastic temperature dependence of n_i^2 completely dominates the leakage current indicated in Eq. (6.40) over a wide temperature range since the minority carrier diffusion constant and lifetime vary rather weakly with temperature in comparison. The approximate fractional change in the saturation current in germanium diodes with temperature is then

$$\frac{1}{I_0}\frac{\partial I_0}{\partial T} = \frac{3}{T} + \frac{E_g}{kT^2} \tag{6.42}$$

or about 0.1/°K at room temperature.

For moderate forward bias voltages such that $e^{qV/kT} \gg 1$, the fractional change in junction current for a fixed applied voltage is obtained by differentiating Eq. (6.20) as

$$\begin{aligned}\frac{1}{I}\left(\frac{\partial I}{\partial T}\right)_V &= \frac{1}{I_0}\frac{\partial I_0}{\partial T} - \frac{qV}{kT^2} \\ &= \frac{3}{T} + \frac{E_g - qV}{kT^2}\,.\end{aligned} \tag{6.43}$$

This temperature variation is somewhat less than that of the saturation current, as given by Eq. (6.42).

Consideration of temperature variation of the diode voltage drop for a fixed bias current yields, for current values $I \gg I_0$,

$$\left(\frac{\partial V}{\partial T}\right)_I = \frac{V}{T} - \frac{kT}{q}\frac{1}{I_0}\frac{\partial I_0}{\partial T} = \frac{V - E_g/q}{T} - \frac{3k}{q}\,. \tag{6.44}$$

This coefficient averages about -1.5 mV/°K at room temperature for germanium junctions.

In the case of silicon the temperature dependence of the junction leakage current as given by Eq. (6.39) is dominated by the temperature variation of n_i. Hence the fractional change in leakage current of a reverse-biased silicon junction is given by

$$\frac{1}{I_0}\left(\frac{\partial I_0}{\partial T}\right)_V = \frac{1}{2}\left(\frac{3}{T} + \frac{E_g}{kT^2}\right), \tag{6.45}$$

or about 0.08/°K at room temperature. For moderate forward bias voltages such that $e^{qV/kT} \gg 1$, the fractional change in silicon diode current for a given applied voltage is

$$\begin{aligned}\frac{1}{I}\left(\frac{\partial I}{\partial T}\right)_V &= \left(\frac{1}{I_0}\frac{\partial I_0}{\partial T} - \frac{qV}{2kT^2}\right) \\ &= \frac{1}{2}\left(\frac{3}{T} + \frac{E_g - qV}{kT^2}\right).\end{aligned} \tag{6.46}$$

This temperature variation is somewhat less than that of the leakage current given by Eq. (6.45). The temperature variation of the diode forward voltage drop at a fixed bias current for current values $I \gg I_0$ is given by

$$\left(\frac{\partial V}{\partial T}\right)_I = \frac{V - E_g/q}{T} - \frac{3k}{q}\,. \tag{6.47}$$

This coefficient averages about -2 mV/°K at room temperature for silicon diodes.

These junction temperature coefficients are of particular importance in the case in which the junction is the emitter of a transistor. It is generally desired that transistor circuits be unaffected by temperature fluctuations. Temperature variations are often compensated for in temperature-sensitive circuit applications by balancing the emitter voltage against the voltage drop across a similar diode in thermal proximity to the emitter junction. This is referred to as a **differential** arrangement. It is in general use in microchip integrated-circuit amplifiers and will be discussed in Chapter 11.

6.3.2 Bulk and Contact Resistance in Diodes

Thus far the discussion has been restricted to the electrical properties of the *p-n* junction of the semiconductor diode. The applied voltage is assumed to appear completely across the junction space-charge layer. However, in a practical diode rectifier structure there is bulk semiconductor material on both the *p*- and *n*-sides of the junction. This provides resistance in series with the junction and in general there is an ohmic voltage drop in these regions. In addition, there may be additional potential drops at the ohmic metal–semiconductor contacts at the ends of the diode which provide electrical connections to an external circuit. In a practical silicon or germanium diode, these potential drops are negligible except at high current densities. A demonstration of this is provided in Fig. 6.10, which shows a semilog plot of the current in a typical silicon diode versus applied voltage. From Eqs. (6.32) and (6.38) it may be seen that when a diode is biased with only a few kT/q units of voltage in the forward direction, the diode equation becomes strictly exponential, i.e., $e^{qV/kT} \gg 1$. Hence a semilog plot of I versus V should yield a straight line. This is often found to be true over a range of eight or more orders of magnitude in current, which verifies the diode equation and confirms that nearly all of the applied

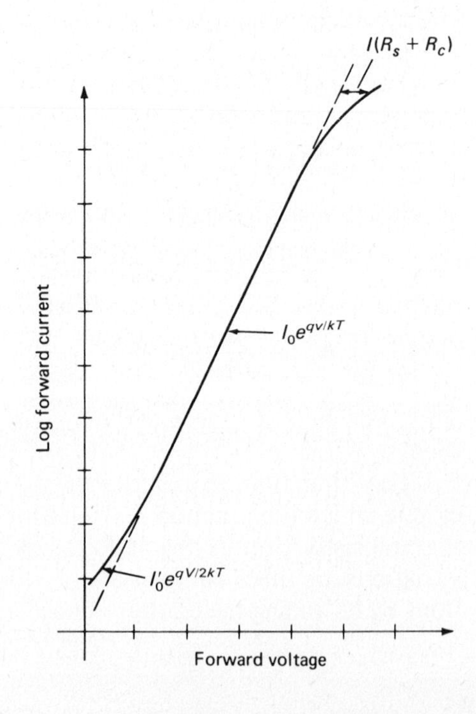

Figure 6.10 Graph of the logarithm of forward current versus forward voltage for a typical silicon *p-n* junction diode rectifier. Deviation from the simple exponential law occurs at high current where a contribution from the potential drop across the *n*-type semiconductor bulk and the metal–semiconductor end contacts, $I(R_s + R_c)$, is observed. Deviation at low voltage occurs as a result of charge generation in the space-charge region which supplies an additional component to the diffusion current.

voltage appears across the *p-n* junction. At some high current value, though, the voltage drop across the diode series resistance is no longer negligible and deviation from strictly logarithmic behavior is observed. This is seen in Fig. 6.10, and the extent of the deviation from exponential behavior is a measure of the diode series resistance as shown in the figure.

The value of diode current at which some of the applied voltage begins to appear across the diode series resistance can be estimated by comparing the effective junction resistance with the diode series resistance, at increasing values of current. The effective junction resistance is current dependent since the diode is basically a nonlinear device. The junction resistance at any current, usually termed the **dynamic** or **incremental** resistance, is defined as the derivative dV/dI and is obtained from Eq. (6.20), neglecting I_0 compared to the current I, as

$$\frac{dV}{dI} = \frac{kT}{qI}. \tag{6.48}$$

As an example, the dynamic resistance of a semiconductor *p-n* junction at 1 mA is 26 Ω.[15] At 26 mA this resistance reduces to 1 Ω. Now in a p^+-n junction the bulk series resistance on the *p*-side of the junction is usually negligible owing to the high acceptor content and high electrical conductivity. The *n*-side resistance may not be negligible. Consider a silicon junction of the form illustrated in Fig. 6.2a. Assume that the diode cross sectional area is 10^{-2} cm^2 and the 1-Ω cm-resistivity *n*-region is 2×10^{-2} cm thick. Hence the bulk series resistance of the diode can be calculated by Ohm's law to be 2 Ω.[16] This is nearly negligible compared to the 26-Ω junction resistance at 1 mA current, but certainly not at the 26-mA level of diode current. It should be pointed out, however, that the bulk series resistance of the diode at 26 mA of current would typically be much less than 2 Ω owing to the increased number of electrical carriers on the *n*-side due to minority hole injection. This increases the conductivity of that region and is termed **conductivity modulation.**

Account must also be taken of any potential drops at the metal–semiconductor contacts at the ends of the diode. Significant contact potentials may be obtained at metal–semiconductor junctions; this will be discussed in the next chapter under the subject of **Schottky barriers.** In practical devices these end contact drops are made negligible by heavily doping the ends of the semiconductor junction crystal with acceptor impurities on the *p*-side and donors on the *n*-side to the extent that they are nearly metallic. Then the metal–semiconductor contacts become nearly metal–metal contacts and hence yield very small contact potential drops. Since the current in a *p*-type semiconductor which is carried almost entirely by holes must convert completely to the electron current characteristic of most metals at the *p*-semiconductor–metal contact, this must be a region of high hole carrier recombination. There are plenty

[15]Note that the dynamic resistance is independent of junction area.

[16]$R = \rho l/A$, where ρ is the resistivity, l the *n*-region thickness, and A the cross sectional area.

of electrons in the metal available for recombination with these holes. In fact, an ohmic contact is sometimes defined as a region of infinitely high recombination.

Now the total voltage drop across a silicon p^+-n junction diode rectifier can be written as

$$V = \frac{kT}{q} \ln\left(\frac{I}{I_0} + 1\right) + I(R_s + R_c), \tag{6.49}$$

making use of Eq. (6.20). Here R_s represents the modulated bulk series resistance of the diode structure and R_c the metal–semiconductor contact resistance. The latter term may dominate the forward voltage drop in the case of diode rectifiers which operate at high current levels.

6.3.3 Reverse Voltage Limitation of the *p-n* Junction

Thus far, the components of leakage current in a reverse-biased junction diode due to minority carriers diffusing through, and generated within, the space-charge layer have been described. An additional source of inverse leakage current in a *p-n* diode is attributed to **impact** ionization of the semiconductor atoms in the space-charge layer by high energy carriers moving through this region under the effects of the high electric field there. This is termed current multiplication or **avalanching.** Another source of leakage current is due to carriers penetrating the junction potential barrier by quantum mechanical tunneling.[17] This is known as **Zener** tunneling after the physicist who predicted the effect. Both of these sources of inverse leakage current are extremely voltage sensitive and hence occur suddenly when a sufficiently high electric field is reached in the depletion layer. Above the value of applied voltage corresponding to this critical field the current increases drastically for a small increase in voltage. This rather well-defined voltage is termed the junction **breakdown voltage.** The complete current–voltage characteristic of the junction diode including avalanching is shown in Fig. 6.11. This breakdown sometimes limits the voltage blocking ability of the *p-n* junction but also often has important circuit uses in voltage limitation or regulation. The excessive leakage current in the diode in the avalanche region must be limited or else thermal runaway can occur as a result of the high voltage *and* high current and can destroy the device.

A. AVALANCHE BREAKDOWN

The phenomenon of avalanching results when electrons or holes, on diffusing into the junction space-charge region, acquire sufficient energy from the elec-

[17]The solution of Schrödinger's equation for a potential barrier of a given height predicts a finite probability for an electron with energy less than the barrier energy to penetrate this barrier. The probability can become substantial when the barrier width is less than 100 Å. This effect is known as quantum mechanical tunneling.

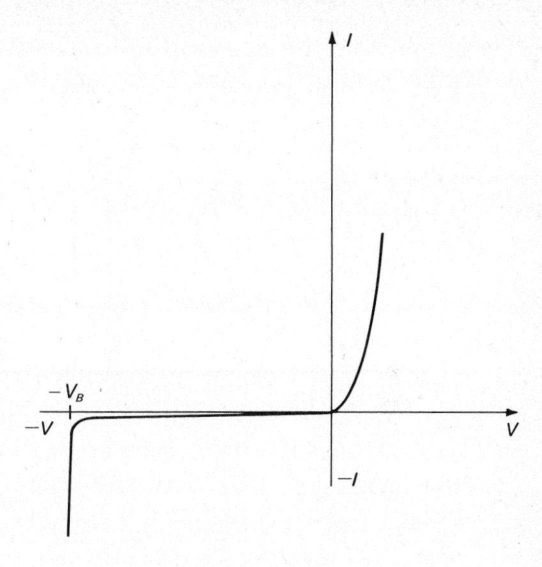

Figure 6.11 Complete current–voltage characteristic of a typical junction *p-n* diode showing avalanching at high reverse negative voltage. A complete set of specification sheets for a typical commercial silicon *p-n* junction diode is included in Appendix E.

tric field there to knock out bound valence electrons from the lattice atoms contained in that region. This high electric field usually results from a large inverse voltage applied to the diode. The junction depletion layer must be wide enough that the mobile carriers can gain sufficient energy from the field to cause ionization. If one electron or hole produces on the average less than one additional carrier, then the junction leakage current is hardly increased. If, however, on the average one additional carrier is produced and these extra carriers each produce one additional carrier, etc., this impact ionization provides a voltage limitation for diodes. However, this phenomenon can also be used to generate microwave power, as in the IMPATT diode described in Chapter 12. The avalanche process is, in many ways, identical to the one that causes inert gases to ionize in a fluorescent light tube. The mathematics for describing this effect is already known from the treatment of gaseous breakdown phenomena. When this formulation is applied to the semiconductor breakdown case, the current multiplication factor M can be explicitly derived. If the ionization rate of electrons and holes is known, the integration of this energy-dependent factor over the maximum-field path in the space-charge region yields the breakdown voltage.[18] A simple empirical expression for M is

$$M = \frac{1}{1 - (V/V_B)^n}, \tag{6.50}$$

where V_B is the junction breakdown voltage, V is the applied potential, and n is a numerical factor which depends on whether the diode is p^+-n or n^+-p and the doping of the lightly doped region (n = 3–6 for silicon). This factor mul-

[18]For a discussion of this ionization integral see S. M. Sze, *Physics of Semiconductor Devices*, 2nd ed. (Wiley, New York, 1981), p. 99.

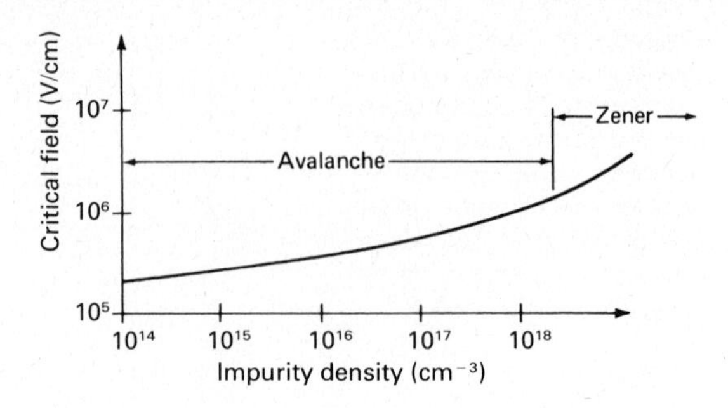

Figure 6.12 Graph of critical field for junction breakdown in *n*-type silicon versus impurity density indicating both the avalanche and Zener regions.

tiplies the normal junction leakage current denoted by I_0 in Eq. (6.20). Note that as $V \rightarrow V_B$, $M \rightarrow \infty$ and avalanche breakdown occurs. In silicon p^+-n junctions the breakdown voltage is approximately proportional to the resistivity of the lightly doped n-region. This result may be derived by first combining Eqs. (6.13b) and (6.17) to give

$$|\mathscr{E}_{\max}| = \left[\frac{2qN_d(\Delta V_0 + |V|)}{\epsilon_0\epsilon_s}\right]^{1/2}. \tag{6.51}$$

Note that the fields produced by the barrier voltage and the applied voltage are in the same direction and their effects are additive. Now for an extrinsic n-region, the resistivity $\rho_n = 1/N_d q\mu_n$, so Eq. (6.51) becomes

$$|\mathscr{E}_{\max}| = \left[\frac{2(\Delta V_0 + |V|)}{\rho_n\mu_n\epsilon_0\epsilon_s}\right]^{1/2}. \tag{6.52}$$

Since normally at breakdown $|V|_B \gg \Delta V_0$,

$$|V|_B \approx \frac{\rho_n\mu_n\epsilon_0\epsilon_s|\mathscr{E}_{\max}|_B^2}{2}. \tag{6.53}$$

In high resistivity (greater than 0.1 Ω-cm) silicon, the breakdown field is appreciably constant at about 3×10^5 V/cm.[19] Hence the breakdown voltage increases in proportion to resistivity. For example, a 100-V-breakdown silicon diode requires a resistivity of a little over 1 Ω-cm. For lower resistivity, the breakdown field increases somewhat. The breakdown field versus impurity density for n-type Si is given in Fig. 6.12, which shows this field decreasing with increasing resistivity. The avalanche breakdown voltage versus impurity concentration for one-sided, abrupt p^+-n semiconductor junctions in Ge, Si, GaAs, and GaP is given in Fig. 6.13.

B. ZENER BREAKDOWN

Excessive leakage current may also result when the electric field in the semiconductor becomes so high that there is a finite probability that electrons in their covalent bonds can be directly excited into conduction. The quantum

[19] S. L. Miller, "Ionization Rates for Holes and Electrons in Silicon," *Phys. Rev.*, **105**, 1246 (1957).

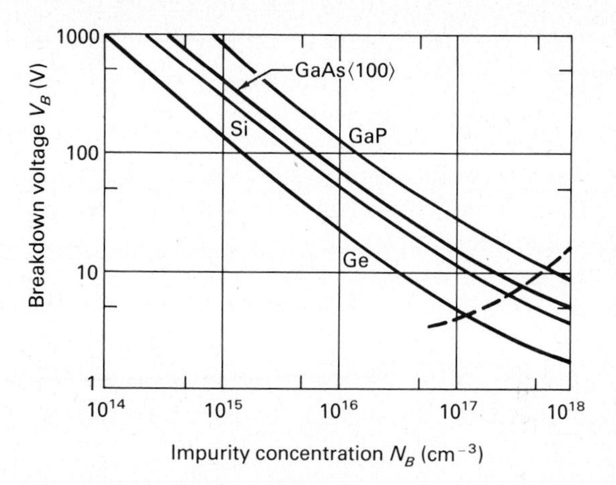

Figure 6.13 Avalanche breakdown voltage versus donor impurity concentration for one-sided, abrupt p^+-n junctions in Ge, Si, GaAs, and GaP. [After S. M. Sze and G. Gibbons, *Appl. Phys. Lett.*, **8,** 111 (1966).]

mechanical calculation of this probability was performed by Clarence Zener. No carrier acceleration or collision is required for this breakdown mechanism. Hence it is only observed in p^+-n^+ junctions where the space-charge region is so narrow owing to heavy doping that the carriers do not have a sufficiently long path to gain enough energy for impact ionization. Under these conditions the depletion region is so thin that tunneling of the type discussed in footnote 17 takes place. In fact even heavier doping results in the **tunnel diode** device.

A Zener breakdown field of about 10^6 V/cm is observed in silicon junctions when the doping level is about 10^{18} donors/cm^3. This is illustrated in Fig. 6.12 and corresponds to a breakdown voltage of about 6 V. Hence silicon diodes break down by the avalanche mechanism above approximately 6 V and by the Zener mechanism below 6 V. Because of the different mechanisms involved, it is interesting to note that the breakdown voltage of diodes which avalanche at greater than 6 V increases somewhat with temperature whereas those which undergo Zener breakdown below 6 V have breakdown voltages which decrease with increasing temperature. Hence the breakdown voltage has essentially a zero temperature coefficient of variation for approximately 6-V breakdown diodes. Diodes of this type are often used as temperature-stable voltage references.

When both sides of a p-n junction are doped so heavily that the Fermi level enters the conduction and valence bands, Esaki tunneling[20] results and a tunnel diode is produced. This device may be used as a high speed switch or to generate high frequency oscillations.

*C. VOLTAGE BREAKDOWN DUE TO SURFACE EFFECTS

In addition to the bulk mechanisms for p-n junction breakdown just described, there are surface effects which often limit the reverse blocking voltage of a junction diode. Contamination and imperfections on the semiconductor surface

[20] L. Esaki, "New Phenomenon in Narrow Ge p-n Junctions," *Phys. Rev.*, **109,** 603 (1958). See also S. M. Sze, *Physics of Semiconductor Devices,* 2nd ed. (Wiley, New York, 1981), Chap. 9.

near where the *p-n* junction intersects the surface can result in premature voltage breakdown, i.e., breakdown at a voltage lower than that predicted by the bulk considerations discussed in the previous sections. Hence mesa-type junctions must be heavily etched to remove the damaged semiconductor surface layer and then thoroughly washed and dried to maintain surface cleanliness. Surfaces so prepared are then usually coated with a pure, highly insulating silicone coating to stabilize the surface over a long period of time. Breakdown voltages in excess of 1000 V are achieved in mesa *p-n* junctions fabricated from silicon crystals.

For an oxide-protected planar p^+-n junction the positive charge normally present in the oxide and at the silicon–oxide interface tends to induce additional electron charges at the surface of the lightly doped *n*-substrate. This effectively causes a p^+-n^+ junction to occur at the silicon surface. Previous arguments have indicated that the breakdown voltage of a junction with heavy doping on both sides is lower than the case of low doping on at least one side of the junction. Hence reduced breakdown voltage can occur at the semiconductor surface of such a device, to the extent of the density of charges in the oxide.

For an oxide-protected planar n^+-p junction the positive oxide charge will tend to induce an *n*-layer on the lightly doped *p*-substrate. As discussed before, this will cause the *n-p* junction to extend over the entire silicon upper surface, causing higher leakage current and even reduced breakdown voltage due to a surface defect somewhere on the upper surface. This tendency can be reduced by introducing a heavily p^+-doped annular "guard ring" around the n^+-p junction to limit its extension. Sometimes this induced *n*-layer can tend to reduce the sharp curvature at the edge of a planar junction, which will in fact raise the junction breakdown voltage. A field-limiting ring, consisting of a doped region of the same type as the highly doped region of a junction but electrically floating, is used to raise the breakdown voltage of a sharp curvature planar diode.[21]

EXAMPLE 6.4

Determine the thickness in micrometers of the space-charge region of a p^+-n^+ junction, reverse biased by 6.0 V, near the onset of Zener breakdown. Also find the peak field in the space-charge region under this condition. The p^+-region is doped with 5.0×10^{19} acceptor/cm^3 and the n^+-region contains 2.0×10^{18} donor impurities/cm^3.

Solution

The calculation is identical to that carried out in Example 6.2 with the exception that an external voltage is applied in addition to the built-in

[21] For an excellent summary of methods used to raise the breakdown voltage of planar diodes see A. Blicher, *FET and Bipolar Power Transistor Physics* (Academic Press, New York, 1981), Chap. 4.

voltage. Calculation of the built-in voltage by Eq. (6.8) yields $\Delta V_0 = 1.06$ V. The space-charge width is then given by Eq. (6.18c) as

$$l_n = \left((\Delta V_0 + |V|)\frac{2\epsilon_0\epsilon_s}{qN_d}\right)^{1/2}$$

$$= \left((1.06 + 6.0)\frac{2(8.85 \times 10^{-14})12}{(1.6 \times 10^{-19})(2.0 \times 10^{18})}\right)^{1/2} \text{cm}$$

$$= 6.9 \times 10^{-6} \text{ cm} \quad \text{or} \quad \underline{0.069\ \mu\text{m}}.$$

Equation (6.16) gives the peak electric field as

$$|\mathscr{E}_{max}| = \frac{2(\Delta V_0 + |V|)}{l_n} = \frac{2(1.06 + 6.0 \text{ V})}{6.9 \times 10^{-6} \text{ cm}}$$

$$= \underline{2.1 \times 10^6 \text{ V/cm}}.$$

☐ SUMMARY

The *p-n* junction in equilibrium was analyzed, yielding the built-in field distribution in the space-charge region and an expression for the junction barrier height. This was repeated for a reverse voltage-biased junction. Next the steady-state current–voltage characteristic for a *p-n* junction diode operating at a low frequency was derived by using carrier diffusion theory. Here the junction law was determined and the concept of minority carrier injection through the *p-n* junction was introduced. Space-charge-generated and surface leakage currents for a reverse-biased junction were described. The temperature variations of the junction current and voltage were derived. The resistances in series with the junction such as the bulk semiconductor resistance and the electrode contact resistance were discussed. The voltage blocking limitations of the reverse-biased *p-n* junction provided by avalanche and Zener breakdown and surface breakdown were described.

PROBLEMS

6.1 A silicon *p-n* junction has 1.0×10^{18} *p*-type impurities/cm^3 uniformly distributed on the *p*-side and 9.4×10^{13} *n*-type impurities/cm^3 on the *n*-side.

(a) Calculate the junction contact potential under thermal equilibrium conditions at 300°K.

(b) If the impurity concentration on the *n*-side is increased to 1.7×10^{15}/cm^3, what is the new contact potential?

(c) Determine the width of the depletion layer on the *n*-side for (a). Repeat for the *p*-side. The relative dielectric constant for silicon is 11.9.

(d) Calculate the maximum electric field in the depletion region for (a).

(e) Repeat (a) at 450°K.
(f) Repeat (a) for a similarly doped GaAs p^+-n junction.

6.2 Beginning with Eq. (6.2b), derive Eq. (6.8).

6.3 Show that Eq. (6.15) follows from Eq. (6.14).

6.4 The cross sectional area of a long silicon p^+-n junction diode is 1.0×10^{-2} cm^2. The n-region is 200 μm wide and is doped with 5.0×10^{14} donor atoms/cm^3. The p^+-region is 100 μm wide and is doped with 5.0×10^{19} acceptors/cm^3. The minority hole lifetime in the n-region is 0.10 μsec. The minority electron lifetime in the p-region is 0.005 μsec. If the diode current is 1.0 mA,
(a) Determine the injected hole density at the n-side edge of the space-charge region at 300°K.
(b) Roughly estimate how far these holes penetrate the n-region.
(c) Determine the ratio of the density of minority holes injected to the density of majority electrons which are present in the n-region.
(d) Calculate the total amount of excess charge in the n-region due to injected holes.

6.5 Show that the expression of Eq. (6.24) is a solution of Eq. (6.23b).

6.6 For a p-n junction with very wide p- and n-regions, starting with expressions of the form of Eq. (6.26a), show that the ratio of hole to electron minority carrier current injected through the junction is given by $\sigma_p L_n/\sigma_n L_p$, where σ_p and σ_n are the electrical conductivities on the p-side and n-side, respectively, and L_p and L_n are the minority carrier diffusion lengths on the n- and p-sides, respectively.

6.7 Derive Eq. (6.37) beginning with Eqs. (6.22) and (6.36).

6.8 Assume that the p^+-n diode described in Problem 6.4 has its n-region reduced in length to 5 μm, but otherwise is unchanged in design. If a forward voltage of 0.60 V is applied to this redesigned diode, calculate the current in the device at 300°K.

6.9 Calculate the junction leakage current for the silicon p^+-n junction of Problem 6.4 due to
(a) Minority carriers in the p- and n-regions diffusing through the junction.
(b) Current generated in the space-charge region when the junction is reverse biased with 10 V. (Take the average lifetime in the depletion region to be 0.10 μsec and relatively temperature insensitive.)
(c) Derive an expression for the ratio of the diffusion component of the saturation current to the space-charge-limited component. Is this ratio larger for germanium or silicon?
(d) Determine the ratio of the diffusion component of the saturation current at 300°K to that at 450°K for a germanium p^+-n junction of the same design as the silicon p^+-n junction of Problem 6.4. Neglect the variation of D_p and L_p with temperature.
(e) By what factor does the space-charge-generated leakage current of the silicon p^+-n junction of (b) change as the reverse voltage is increased from 1.0 to 100 V?

6.10 Two germanium p-n junction diodes are connected in series with a reverse voltage of 100 V applied to the combination through a resistance of 1000 Ω. Diode A has a reverse saturation current of 0.10 μA, while diode B has a saturation current of

0.01 μA at 300°K. Assume that the diodes have negligible series and contact resistances.

(a) Estimate the voltage supported by diode A and diode B. What is the voltage drop across the 1000-Ω resistor?

(b) If the applied potential of 100 V is now reversed, determine the current through the 1000-Ω resistance and the voltage drop across this resistor.

(c) Calculate the ratio of the power dissipated in diode B in reverse bias to that in the forward-biased state.

6.11 Consider two diodes of the type described in Problem 6.10 in parallel. If 50 V is applied to this parallel combination, in the forward direction through a 1000-Ω resistor, calculate the current in each diode.

6.12 Find the temperature variation of the forward drop of a silicon *p-n* junction at 0.5 V and a temperature of 450°K. Assume that the energy gap width varies only slightly with temperature.

6.13 A silicon *p-n* diode rectifier has a leakage current at 300°K of 1.0×10^{-12} A. The forward drop across this rectifier when carrying 100 mA is 0.80 V. Compute the total bulk series and contact resistance of this device.

6.14 **(a)** Calculate the approximate breakdown voltage for the *p-n* junction diode of Problem 6.4.

(b) What is the width of the space-charge region at the voltage at which the junction avalanches?

(c) Repeat (a) if the doping on the *n*-side of the junction is increased to 1.0×10^{16} donors/cm^3.

6.15 **(a)** Plot the current multiplication factor M versus reverse voltage for the *p-n* junction diode described in Problem 6.4. Assume that the factor n in Eq. (6.50) is equal to 3.

(b) Repeat (a) for $n = 6$.

CHAPTER 7

Semiconductor Diode Devices and Frequency/Speed Behavior

7.1 INTRODUCTION TO SEMICONDUCTOR DIODE DEVICES

In Chapter 6, the theory of the semiconductor *p-n* junction was developed, leading to a derivation of the device's DC current–voltage characteristic. For this discussion the junction was assumed to lie entirely within a single crystal of a semiconductor material, albeit nonuniformly doped. In this chapter a few important examples of device applications of this *p-n* junction diode will be described. In particular, diode photodetectors and photoemitters employed in optical communication will be analyzed. The operation of *p-n* junction solar energy converters will also be discussed. Next an analysis will be presented of the factors which limit the switching speed and high frequency operation of semiconductor *p-n* diodes. In this connection the variable-capacitance **varactor** diode used for electronic frequency tuning applications will be discussed. A description of two different types of semiconductor diodes will follow next. First the theory of the metal–semiconductor junction or **Schottky** barrier will be presented. A diode device utilizing this type of junction is extensively employed to increase the switching speed of microchip digital integrated circuits. Then the theory of the **heterojunction** will be presented. This junction forms the boundary between two different energy gap semiconductor materials. Heterojunction diode devices offer the promise of overcoming some of the limitations of more conventional *p-n* junction diodes.

7.2 PHOTODETECTORS AND PHOTOEMITTERS

Great strides recently have been made in the use of **optoelectronics** for communication purposes.[1] Here an audio- or video-produced electrical signal applied to a *p-n* junction **light-emitting diode** (LED) produces a light signal which is transmitted along a hairlike glass optical fiber to a receiver in the form of a *p-n* junction diode detector which converts the received light signal back into an electrical signal. After a process of demodulation this electrical signal is changed back into speech or a TV picture. The semiconductor LED is an extremely small device which for its size also produces large amounts of light energy efficiently. The light is emitted from a semiconductor *p-n* junction (often made in GaAs or another compound semiconductor) by the application of an applied voltage. The receiver is in the form of a *p-n* junction photodiode (often fabricated of silicon) which converts the light signal into an electrical signal. Figure 7.1 shows a schematic diagram of such a communication system.

The tremendous advantage of optical communication in contrast to normal radio or television transmission is that light is an electromagnetic radiation much higher in frequency than radio or television waves. A theorem of communication theory states that the *higher the frequency of transmission, the greater the bits of information that can be transmitted.* For example, the wavelength λ of yellow light is about 6000 Å (6×10^{-5} cm). Since the velocity c of light is 3×10^{10} cm/sec, the frequency of light ν can be calculated from

$$\nu = c/\lambda. \tag{7.1}$$

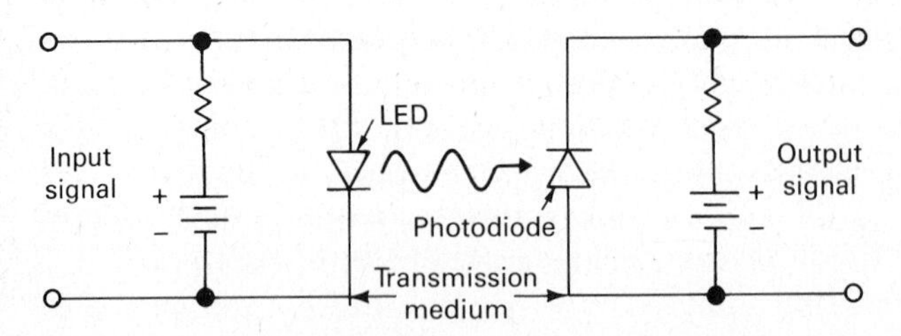

Figure 7.1 Light-wave communication system using an LED photoemitter as a transmitter and a photodiode detector as a receiver. The transmission medium is normally a glass optical fiber.

[1]See R. H. Saul, "Recent Advances in the Performance and Reliability of InGaAsP LED's for Lightwave Communication Systems," *IEEE Trans. Electron Devices,* **ED-30,** 285–295 (1983).

This comes out to be 5×10^{14} Hz, which is about a million times higher than the frequency of the television carrier waves, and hence accounts for the enormous excitement about light-wave communication systems.

In recent years, a perhaps less interesting but commercially exciting light-emitting solid-state device has come into general use. This is the gallium arsenide phosphide LED. When about a volt and 20 mA is applied to a *p-n* junction of this material, it emits visible red light and hence constitutes a miniature solid-state lamp which has no fundamental burnout mechanism. Owing to the LED's very high electric power-to-light conversion efficiency, only milliwatts of power are required to produce a significant amount of light. A 5×7 matrix of these devices is used to provide a small numerical readout, visible in daylight, of the type used in the digital clock and other numerical displays.[2]

In addition to the use of the photodiode as a receiver for light signals in communication systems, there has been a significant increase in the employment of a form of this device, known as the **solar cell,** in the conversion of solar energy to electrical power. Silicon solar cells with efficiencies of up to 14% are available commercially; GaAs cells have demonstrated conversion efficiencies of up to 25% in the laboratory.

7.2.1 The *p-n* Junction Photodiode

The *p-n* junction as a light **detector** or photodiode will now be discussed. If a reverse-biased semiconductor *p-n* junction is illuminated with photons having energy in excess of the forbidden energy gap, hole–electron pairs may be generated in the vicinity of the junction.[3] Those generated in the junction space-charge region will be swept through the junction by the electric field there, constituting an additional source of reverse current. In addition, hole–electron pairs generated in the *p*- and *n*-regions on either side of the junction region can diffuse to the space-charge region and then be collected there by the electric field. Because of the direction of the electric field only minority holes generated on the *n*-side and minority electrons generated on the *p*-side will be swept through the junction, constituting a current which can be detected at the terminals of the diode. The magnitude of this current will be shown to be proportional to the number of incident photons per unit time. Consider light impinging on a simple *p-n* junction bar as shown in Fig. 7.2. Since only the holes generated on the *n*-side within a hole diffusion length L_p of the space-charge region can be collected by the junction, giving rise to a photocurrent, this hole current can be estimated as

$$I_p = qAL_pG. \tag{7.2}$$

[2]For a survey of numeric readout devices, including the GaAsP LED, see "Special Report: Numeric Readout Displays," *Electronics,* **44,** 65 (1971).

[3]See Section 5.4 for a discussion of carrier generation by light.

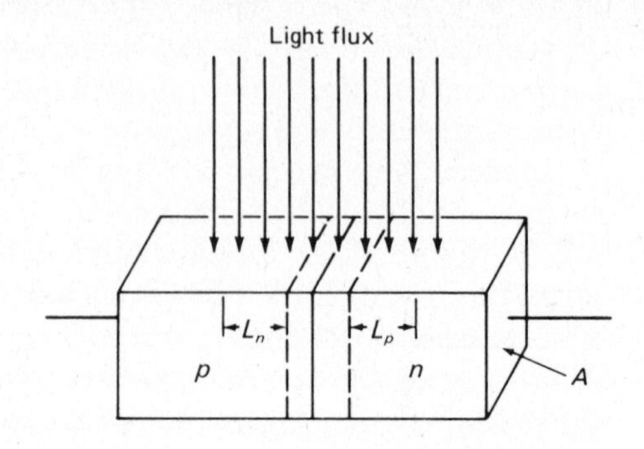

Figure 7.2 Schematic diagram of an illuminated *p-n* junction photodiode. Only photons with energy greater than the semiconductor energy gap produce electron–hole pairs. Only those produced within a diffusion length of the junction are collected.

Here A is the junction cross sectional area and G is the hole-electron generation rate, which is proportional to the incident light flux. A corresponding expression can be written for the electrons generated on the p-side. Physically this expression can be interpreted as follows: G is the rate of production of holes per cubic centimeter and the product AL_p represents the volume in which generated holes are produced from which these carriers can diffuse to the space-charge region before recombination. Hence, when the magnitude of the hole charge q is taken into account Eq. (7.2) can be interpreted as the total hole charge produced by the light that is collected in the space-charge region per unit time, which when swept through the junction constitutes the photocurrent due to the illumination.

Since each photon with energy in excess of the semiconductor energy gap can produce one electron–hole pair, the photocurrent is proportional to the number of these incident photons per unit time. The total photodiode current I can then be written as

$$I = I_0(e^{qV/kT} - 1) - (qAL_pG + qAL_nG), \tag{7.3}$$

where the first term on the right refers to the ordinary p-n junction dark current when a voltage V appears across the device and the second term on the right corresponds to the photocurrent due to hole generation on the n-side and electron generation on the p-side of the p-n junction.

Using Eq. (7.3) we can derive the equivalent circuit of this diode; it is shown in Fig. 7.3a. The effect of illumination on the diode is indicated by the current source I_L, in parallel with the ordinary dark or light-free p-n junction diode characteristic. The I-V characteristic of the photodiode in the dark and in the presence of light is shown in Fig. 7.3b. The photodetector aspects of the device are indicated by noting that the change in reverse bias current of the device is proportional to the incident light flux. For when the diode is short circuited ($V = 0$) the first term in Eq. (7.3) vanishes, and the current is negative and just given by the optically generated current I_L. If the diode is open circuited ($I = 0$) in the presence of illumination, a photovoltage develops across its terminals marked V_{OC} in Fig. 7.3b and with polarity as shown. Operated in this mode the device is known as a **photovoltaic** cell. The magnitude of this

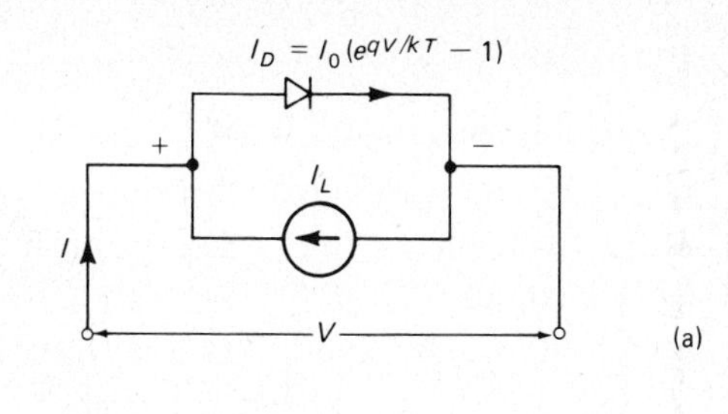

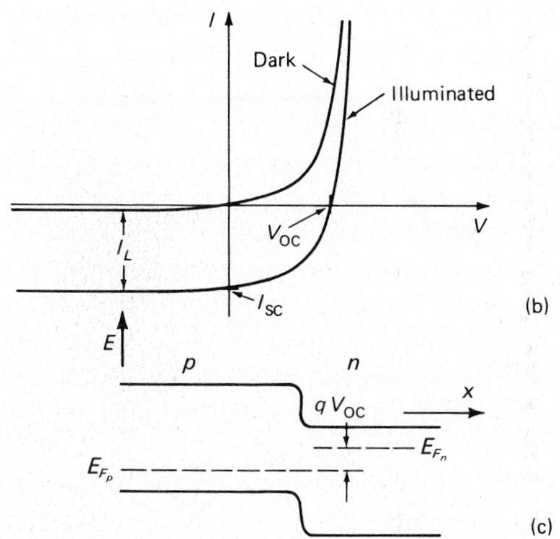

Figure 7.3 (a) Equivalent circuit of a *p-n* junction photodiode showing the polarity of photovoltage produced by illumination.
(b) *I-V* characteristic of a *p-n* junction photocell both in the dark and under illumination. The open-circuit voltage V_{OC} and the short-circuit current I_{SC} under illumination are indicated.
(c) Energy band diagram for a photodiode open-circuited and illuminated. The shift of the Fermi level on the *n*-side relative to the *p*-side, far from the junction, represents the open-circuit voltage.

photovoltage may be calculated by setting $I = 0$ in Eq. (7.3). This gives for the open-circuit voltage

$$V_{OC} = \frac{kT}{q} \ln\left(\frac{qAG(L_p + L_n)}{I_0} + 1\right). \tag{7.4}$$

A practical value of this quantity is about one-half the energy gap or about 0.5 V for silicon photodiodes. The manner in which this voltage is developed by using the energy band concept is diagrammed in Fig. 7.3c. The separation of the photogenerated holes and electrons by the junction built-in field tends to collect holes on the *p*-side and electrons on the *n*-side since no net current can flow. These charges produce a field which opposes the built-in field, reduce the barrier height, and give rise to a photovoltage which may be determined from Eq. (7.4).

A. The Photodiode Detector

The figure of merit of a photodetector in general is measured in terms of (i) the quantum efficiency or gain, (ii) the response time or speed, and (iii) the sensitivity. The quantum efficiency η is defined as the number of electron–hole pairs generated per incident optical photon. The primary light current I_{ph} can then be expressed as

$$I_{\mathrm{ph}} = q\eta P_{\mathrm{opt}}/h\nu, \tag{7.5}$$

where P_{opt} is the incident light power and $h\nu$ represents the photon energy. In order to collect a maximum number of these light-generated carriers it is advantageous to have a thick region in which an electric field exists to separate the electron–hole pairs. A device to satisfy this need is the ***p-i-n*** diode, which consists of a very lightly doped (near intrinsic) thick central region surrounded by thin, heavily doped p^+- and n^+-regions. This device is sketched in Fig. 7.4a. In this structure the separating electric field occupies a large fraction of the cell. However, the carrier transit time through this region will be proportional to its thickness, which will increase the response time. Hence there is a design compromise between speed of response and output current. Of course the speed of the photodetector also depends on the junction capacitance since this must be charged to provide output. This dictates the use of small-area photocells for high speed operation. An antireflecting coating on the cell surface will also improve the quantum efficiency of the photodiode. Use can also be made of the avalanche process to obtain improved diode photosensitivity. If the electric field in the collecting space-charge region is high enough, carrier multiplication will occur for those electrons and holes generated optically there, resulting in increased output current. This can be accomplished by supplying sufficient reverse bias to the photodiode. Photodiodes can also be constructed from metal–semiconductor junctions (Schottky barriers) and heterojunctions, both of which will be described later in this chapter. Table 7.1 gives some typical operating characteristics for different types of photodiodes.

The spectral response of the photodiode is important both for communication system detectors and for photocells for solar energy conversion. In the case of light-wave communication systems using wavelengths in the 0.8–0.9 μm range, silicon photodiodes are commonly used as detectors, since photons of this wavelength have energy in excess of the energy gap width of silicon. However, in systems utilizing a wavelength of 1.3 μm, sufficient photon energy is not available to excite electron–hole pairs. This suggests then the use of ger-

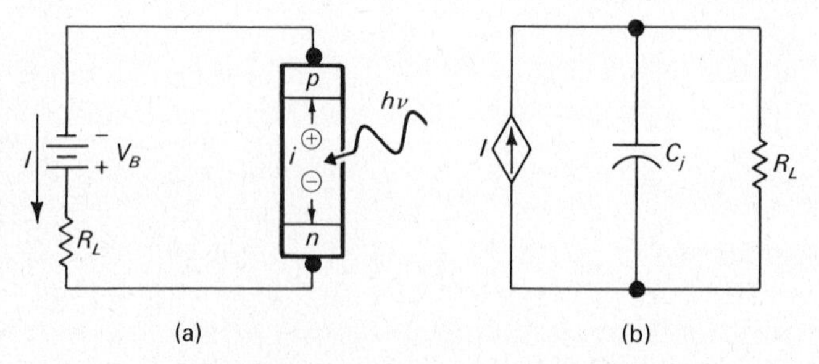

Figure 7.4 (a) Sketch of a *p-i-n* photodiode in an operating circuit, delivering an electrical signal to a load resistor R_L.
(b) Equivalent circuit useful in determining the speed of this photodetection system.

TABLE 7.1 **Typical values of gain and response time for different types of photodiodes.**

Photodiode Type	Gain	Response Time (sec)
p-n junction	1	10^{-11}
p-i-n junction	1	10^{-8}–10^{-11}
Avalanche photodiode	10^2–10^4	10^{-10}
Schottky photodiode	1	10^{-11}

manium photodetectors, since good *p-n* junctions can be fabricated of this material and it has a sufficiently small energy gap (0.66 eV). However, the high leakage currents and temperature sensitivity of narrow gap semiconductor diodes dictate the use of a wider gap material. InGaAs and InGaAsP are other materials currently used as photodetectors at 1.3 μm wavelength. Naturally the technology for fabricating cells of these compound semiconductors is much more complex than that for producing germanium cells.

B. The Solar Cell

The semiconductor solar light energy converter in general has more rigorous spectral requirements than a simple photodiode. Highest efficiency conversion is obtained when there is an optimum match between the photocell spectral sensitivity and the solar energy emission spectrum. This requirement is similar to matching of the output impedance of an amplifier with the load that it is driving to obtain optimum power transmission. The solar energy spectrum extends in wavelength from about 0.3 μm in the ultraviolet to more than 5 μm in the infrared, peaking at about 1.8 μm.[4] The large amount of infrared energy would seem to dictate the use of *small* energy gap materials so that even these long wavelength (low energy) photons will have sufficient energy to excite electrons across the semiconductor gap E_g and create electron–hole pairs. From Eq. (7.1) and Planck's law the carrier-producing photons must have a wavelength

$$\lambda < \frac{hc}{E_g} \quad \left(= \frac{1.24}{E_g\,\text{eV}}\,\mu\text{m} \right). \tag{7.6}$$

However, the voltage generated by a solar cell in general will increase as the energy gap increases. This voltage is related to E_g as indicated by Eq. (7.4) and its maximum is the junction barrier height. In this equation it is seen that the open-circuit voltage is a strong inverse function of I_0, which in turn depends reciprocally on the energy gap. Hence the most efficient material for use in converting solar energy has an optimum, medium value gap width. This turns

[4]This corresponds approximately to the radiation expected from a blackbody heated to 5800°K.

TIROS I (Television Infrared Observation Satellite) showing its two hemicylindrical solar arrays used for converting solar energy into electrical energy. (Photo courtesy of National Aeronautics and Space Administration.) The drawing is a closeup of an individual solar cell.

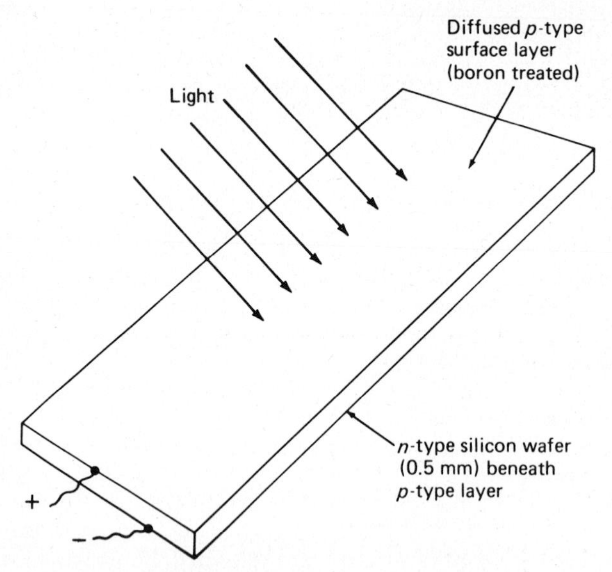

out to be between 1.1 and 1.4 eV, the energy gap widths of silicon and GaAs, respectively. This is shown in Fig. 7.5.

The current–voltage (I-V) characteristic of a solar cell can be represented as shown in Fig. 7.6. The maximum power delivered to an electrical load by this cell occurs when the product IV is maximum. This product, $I_m V_m$, is proportional to the crosshatched rectangle shown in Fig. 7.6. A method for cal-

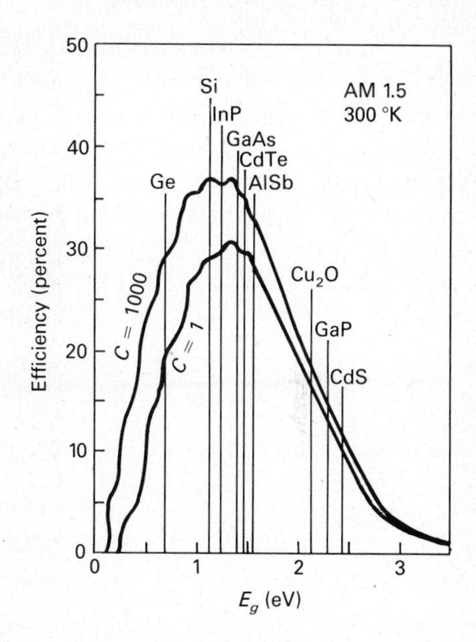

Figure 7.5 Ideal solar cell efficiency at 300°K for irradiation from 1 sun and concentrated illumination corresponding to 1000 suns versus solar cell material energy gap. AM 1.5 refers to air mass 1.5, indicating some radiation absorption by the earth's atmosphere; AM 0 refers to zero atmospheric absorption. [From S. M. Sze, *Physics of Semiconductor Devices,* 2nd ed. (Wiley, New York, 1981), p. 798.]

culating I_m and V_m is illustrated in Problem 7.3 at the end of this chapter. The ratio $I_m V_m / I_{SC} V_{OC}$ is always less than one and is referred to as the **fill factor.** A typical value for this fill factor (FF) for a practical solar cell is 0.8. Hence the maximum power output of a solar cell may be written as

$$P_{max} = FF\ I_{SC} V_{OC}. \tag{7.7}$$

A typical 2-cm^2 silicon solar cell at 300°K can deliver about 10 mW in noonday sun and converts solar to electrical energy at about 12% efficiency. Series and parallel connection of thousands of these devices yields a power supply of the type used to energize space vehicles, typically capable of 28 V and a few kilowatts. Solar cells are extremely lightweight direct energy converters and hence are used extensively in space projects.

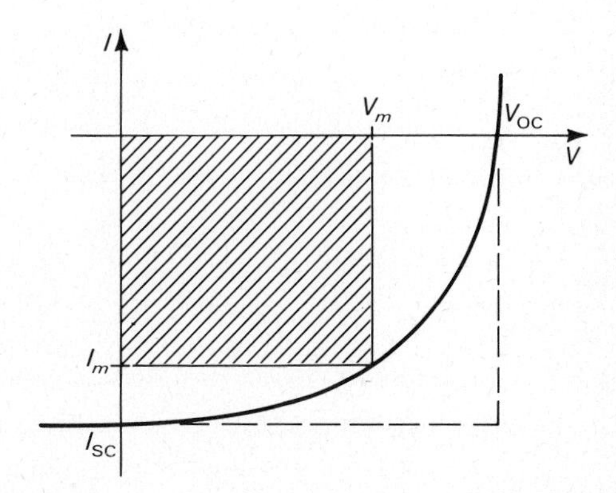

Figure 7.6 Current–voltage characteristic of an illuminated solar cell. I_{SC} and V_{OC} represent the cell short-circuit current and open-circuit voltage, respectively. The crosshatched area $I_m V_m$ represents the maximum operating power output when the cell is matched by an optimum load resistor.

The large cell area required to collect a large amount of solar energy and generate a significant amount of electrical power makes the solar cell industry material intensive. This is in contrast to the small amount of material needed for each microchip in the microelectronics industry. Owing to the abundance of silicon in the earth's crust, this material is relatively inexpensive and hence is in general use in manufacturing solar cells. Although GaAs has resulted in more efficient cells which can operate at higher temperatures than Si, the high cost of GaAs restricts it to use where solar concentrators are employed.

EXAMPLE 7.1

The leakage current I_0 of a 2.0-cm^2 silicon p^+-n junction solar cell at 300°K is 0.05 nA. The short-circuit current of this device exposed to noonday sun is 20 mA and the electron–hole pair generation rate in the silicon is then 3.0×10^{18}/cm^3-sec.

(a) What is the lifetime of minority holes in the n-region of this device? (*Hint:* Assume that the electron lifetime in the p-region is very small because of the high impurity level there.)
(b) What resistance value must be connected across the cell in order to ensure that 10 mA of load current is delivered to this load?
(c) Calculate the power delivered to the load of (b).

Solution

(a) From Eq. (7.3)

$$I_{SC} = I_L = qAG(L_p + L_n),$$

so

$$L_p = I_{SC}/qAG,$$

where it is assumed that $L_n \ll L_p$.
Hence

$$L_p = \frac{20 \times 10^{-3}\ \text{A}}{(1.6 \times 10^{-19}\ \text{C})(2.0\ \text{cm}^2)(3.0 \times 10^{18}\ \text{cm}^{-3})}$$

$$= 2.1 \times 10^{-2}\ \text{cm},$$

and

$$\tau_p = \frac{L_p^2}{D_p} = \frac{(2.1 \times 10^{-2})^2\ \text{cm}^2}{12.5\ \text{cm}^2/\text{sec}}$$

$$= \underline{35 \times 10^{-6}\ \text{sec} \quad \text{or} \quad 35\ \mu\text{sec}}.$$

(b) Equation (7.3) is

$$I = I_0(e^{qV/kT} - 1) - I_L.$$

Solving for V gives

$$V = \left(\frac{kT}{q}\right) \ln \left(\frac{I_L + I}{I_0} + 1\right)$$

$$= (0.026 \text{ V}) \ln \left[\frac{(20 \times 10^{-3}) + (-10 \times 10^{-3} \text{ A})}{0.05 \times 10^{-9} \text{ A}} + 1\right]$$

$$= 0.50 \text{ V};$$

$$R = V/I = 0.50/10 \times 10^{-3} = \underline{50 \ \Omega}.$$

(c) Power $= IV$

$$= (10 \times 10^{-3} \text{ A})(0.50 \text{ V}) = \underline{5.0 \times 10^{-3} \text{ W}}.$$

7.2.2 The *p-n* Junction LED

As was pointed out in Section 5.4, direct-gap semiconductor materials make better light emitters than indirect-gap materials. In the former materials the transition of an electron from the conduction band to an empty state (hole) in the valence band (recombination) at a constant value of k (momentum) results in an efficient release of light energy or photons; the frequency of these photons as given by the Planck law is centered about $\nu = E_g/h$. The semiconductor device for light emission usually takes the form of a *p-n* junction diode (LED). When a forward voltage is applied to this device, carriers are injected into the space-charge region and the neutral regions near the junction. Their subsequent recombination results in light-energy emission corresponding to the energy loss of recombination. If the shape of this device takes the form of a block of semiconductor crystal material with parallel, carefully polished mirrorlike faces, coherent light may be emitted by **laser**[5] action. Laser light emission is highly directional, within an angle of about 10°, and hence can provide for more efficient coupling to an optical fiber used for transmission of this light signal. In addition these injection lasers provide increased light output and a significantly narrower emission frequency band. However, fabrication, reliability, and operational difficulties have restricted the general use of injection lasers as yet for emitters in optical communication systems.

The choice of semiconductor material to be utilized in fabricating LEDs for light-wave communication systems depends on the transmission properties

[5]Laser is an acronym for "light amplification by stimulated emission of radiation," which will be further discussed in Chapter 12.

of the optical fiber materials available to carry the light signal. Since glass is the most appropriate material in use for such an application transmission is restricted to the 0.8–0.9 μm range or else a band centered at 1.3 μm. These frequency "windows" correspond to regions in which light energy absorption is at a minimum. For example the loss in the 0.8–0.9 μm range is presently about 2–3 dB/km. This wavelength corresponds to the light emission spectrum of AlGaAs LEDs; AlGaAs is a mixture of the binary compound semiconductors AlAs and GaAs. There exists a minimum in the absorption spectrum of glass fibers at 1.3 μm. This wavelength corresponds to the emission obtained from InGaAsP LEDs and the parent material is a four-component compound including the elements In, Ga, As, and P. The substantial improvement (0.6 dB/km) in fiber loss that can be obtained in shifting to the absorption minimum at 1.3 μm is presently being delayed until the reliability problem connected with InGaAsP LEDs is solved.

The modulation frequency for transmission of optical signals by LEDs is limited mainly by the turnon time needed to charge the junction capacitance of the diode and to establish the injected carrier distribution, plus the fall time corresponding to the decay of injected carriers. The short carrier lifetime of the direct-gap semiconductors used to produce efficient light emitters improves their switching speed. Nanosecond switching time is possible. Experiments have demonstrated data transmission rates of 274 Mbit/sec for a glass fiber transmission line 23 km long.

The most common material used to fabricate light-emitting diodes which luminesce in the visible range is GaAsP; red-light-emitting LEDs used in digital displays are fabricated from this mixture of GaAs and GaP. For alloys containing 0–45% mole fraction of GaP the energy gap is **direct.** For 0% GaP (all GaAs), the energy gap is 1.42 eV; for 45% mole fraction of GaP in GaAsP, the energy gap is 1.98 eV. In between these compositions the energy gap is a nearly linear function of composition. For GaP percentages in excess of 45%, the energy gap is **indirect.** Hence materials containing less than 45% GaP make naturally more efficient photoemitters.[6]

EXAMPLE 7.2

(a) Find the energy gap of a mixture of a 30% mole fraction of GaP in GaAsP.
(b) Determine the central wavelength emitted by an LED fabricated from the material in (a).
(c) Will the light emitted in (b) be visible? What color?
(d) Comment on the emission efficiency of the LED of (b).

[6]It has been found that the photoemission efficiency of "indirect" (GaAs)P can be significantly enhanced by incorporating nitrogen in this crystal.

Solution

(a) $(E_g)_{45\%\ \text{GaP}} = 1.98$ eV, $(E_g)_{0\%\ \text{GaP}} = 1.42$ eV. Approximate linearity gives $(E_g)_{30\%\ \text{GaP}} = (0.30/0.45)(1.98 - 1.42) + 1.42$ eV $=$ 1.80 eV.

(b)
$$\lambda = \frac{hc}{E_g} = \frac{(6.63 \times 10^{-34}\ \text{J-sec})(3 \times 10^{10}\ \text{cm/sec})}{(1.80\ \text{eV})(1.6 \times 10^{-19}\ \text{J/eV})}$$
$$= 6.9 \times 10^{-5}\ \text{cm} \quad \text{or} \quad 0.69\ \mu\text{m}.$$

(c) The visible light spectrum is 0.4–0.7 μm. The emitted light will be deep red.

(d) Since less than 45% mole fraction GaP in GaAsP yields a direct-gap semiconductor, 30% GaP should give *good emission efficiency*.

7.3 SMALL-SIGNAL, HIGH SPEED/FREQUENCY DIODE PERFORMANCE

Chapter 6 was confined to the discussion of the *p-n* junction electrical characteristic under DC or low frequency, **quasistatic** conditions.[7] One of the important applications of the semiconductor diode is its use as a high speed switch. Hence it is of interest to investigate those physical device characteristics which limit high speed performance so that very fast diodes can be designed. The small-signal, high frequency limitation of the diode used as a varactor diode or RF detector is also of interest. Hence the *p-n* junction diode frequency cutoff will also be discussed.

The time-dependent response of a long p^+-n junction diode operating in the forward direction, of the type illustrated in Fig. 6.8, is governed partly by the flow of minority carriers in the lightly doped *n*-region of the diode. This charge flow may be investigated by solving the time-dependent minority carrier flow equation in the *n*-region,

$$D_p \frac{\partial^2 \Delta p_n}{\partial x^2} = \frac{\Delta p_n}{\tau_p} + \frac{\partial \Delta p_n}{\partial t}. \tag{7.8}$$

The above equation is derived in the same manner as Eq. (6.23b) except that in the latter, steady-state case $\partial \Delta p_n/\partial t = 0$. The low level hole current density is then obtained from $J_p = -qD_p[\partial \Delta p_n(0)/\partial x]$, where $\Delta p_n(0) = p_{n0}(e^{qv/kT} - 1)$ from Eqs. (6.26a) and (6.30). Note that the time-dependent voltage is indi-

[7]**Quasistatic** means that a time-dependent problem is treated as a series of DC solutions at incremental steps in time.

cated here by the lower-case v, and that quasiequilibrium conditions are assumed.

When a time-varying voltage is applied to the junction, the space-charge width and hence fixed charge content of that region also varies. This variation of charge with time constitutes a displacement current through the junction, indicating the capacitorlike behavior of the junction space-charge region. The speed of response of the p^+-n diode to a time-varying voltage depends both on the time necessary for charge redistribution in the space-charge region and on the time for charge redistribution of injected carriers in the n-region in series with the junction. The p-n junction space-charge capacitance will first be considered.

7.3.1 Junction Space-Charge Capacitance and the Varactor

In Section 6.1.3 the p-n junction space-charge region in equilibrium was considered for an abrupt junction. The increase in the space-charge width with *reverse* applied voltage was illustrated in Fig. 6.6. This causes more fixed charge in the form of ionized donors and acceptors not neutralized by electrons and holes to appear within the depletion region. Forward bias reduces the number of un-neutralized ions in this region. The incremental charge increase dq_{SC} of ionized donors (or acceptors) taken per incremental voltage change dv is termed the **depletion-layer capacitance** and is given by

$$C_j = -\frac{dq_{SC}}{dv} . \tag{7.9a}$$

Here the negative sign results from the fact that the reverse voltage must increase in a *negative* sense for a charge increase. The additional current passed by this junction when the inverse voltage applied to it changes by dv/dt is given by the displacement current $i = C_j \, dv/dt$, which is in excess of the I_0 of Eq. (6.20).

In order to calculate this depletion-layer capacitance in terms of device design parameters for an abrupt p-n junction refer back to Eqs. (6.9) and (6.18). The total charge on either the p- or n-side of the space-charge region in the depletion-layer approximation is given by

$$q_{SC} = qAN_d l_n = qAN_a l_p = A\,[2\epsilon_0\epsilon_s qN_dN_a(\Delta V_0 - v)/(N_a + N_d)]^{1/2}. \tag{7.10}$$

Here v represents the applied voltage, negative for reverse bias, which may be time dependent, and A is the junction cross sectional area. Performing the differentiation as indicated in Eq. (7.9a) we express the junction space-charge capacitance as

$$C_j = A\left(\frac{\epsilon_0\epsilon_s q}{2(1/N_a + 1/N_d)(\Delta V_0 - v)}\right)^{1/2} . \tag{7.9b}$$

Hence the reciprocal of the capacitance squared for an abrupt junction is seen to vary linearly as the applied voltage and the resistivity of the most lightly doped side of the diode. A graph of the measurement of $1/C_j^2$ versus v should yield a straight line whose intercept at $1/C_j^2 = 0$ gives ΔV_0. In the case of a *p-n* junction which is not abrupt but has a linear grading of impurities, the exponent in Eq. (7.9b) has the value ⅓ [see Eq. (6.19)]. This represents a less rapid change of capacitance with voltage. A very sensitive variation of capacitance with voltage is obtained with a diode containing a **hyperabrupt** junction. This is designed with a very steep donor impurity gradient so that the exponent in Eq. (7.9b) is 2.[8]

Having derived an expression for the space-charge capacitance of a *p-n* junction, we are now in a position to point out an important application of this device. A reverse-biased *p-n* junction diode is a voltage-variable capacitor (varactor) which may be used for frequency modulation or as a tuning capacitor whose capacitance can be varied electronically rather than mechanically. The device also has applications in harmonic generator and parametric amplifier circuits because of its nonlinear variation of capacitance with voltage. The high frequency limitation f_{max} of the varactor diode can be written as

$$f_{max} = 1/2\pi r_s C_j, \tag{7.11}$$

where r_s is the resistance in series with the reverse-biased junction capacitance, which in a p^+-n diode is mainly the bulk resistance of the *n*-region. This cutoff frequency is defined as the frequency at which the capacitive reactance of the junction just equals the diode resistance. The impedance of this device is mainly capacitive below this frequency and the electrical energy is then stored in the capacitor and not dissipated in the series resistance. An important factor in the design of a varactor diode is the doping of the *n*-region. Light doping yields lower capacitance but also results in higher series resistance. Heavy doping gives higher capacitance but much lower series resistance. The design of a varactor device is the subject of a problem at the end of this chapter.

7.3.2 Diode Diffusion Capacitance

The p^+-n junction space-charge capacitance under forward bias can be obtained from Eq. (7.9b). However, this forward bias also introduces an additional injected charge distribution in the *n*-region which can be described by a capacitance known as the **diffusion capacitance.** This capacitance is of special interest when the junction is the emitter of a transistor. In fact, the diffusion capacitance in forward bias may be numerically greater than the depletion-layer capacitance and hence in this case can limit the device high frequency

[8]This requires the impurity density to vary as the −⅔ power of distance x (see footnote 9 to Chapter 6).

performance. This capacitance will now be calculated in terms of device parameters.

Under steady-state forward bias conditions in a long p^+-n junction diode, there exists an excess charge in the form of injected holes in the bulk n-region (see Fig. 6.7b). The total of these excess minority carrier charges may be obtained by integrating the expression for the excess hole distribution under forward bias, over this whole region. Using Eq. (6.24), we thus obtain for the total excess minority hole charge injected into the n-region

$$q_{\text{diff}} = \int_0^\infty qA\,\Delta p_n(0)e^{-x/L_p}\,dx = qA\,\Delta p_n(0)Lp. \tag{7.12}$$

If now the forward bias is changed incrementally, this charge stored in the n-region will change by

$$dq_{\text{diff}} = \frac{dq_{\text{diff}}}{dv}\,dv = C_D\,dv, \tag{7.13}$$

where C_D represents the diffusion capacitance. The charging current required to accomplish this charge change is given by

$$i_{\text{diff}} = \frac{dq}{dv}\frac{dv}{dt} = C_D\frac{dv}{dt}. \tag{7.14}$$

The minority charge distribution in the n-region before and after the application of an incremental voltage dv, in a quasistatic manner, is indicated in Fig. 7.7. From Eqs. (7.12) and (7.13) the diffusion capacitance is given by

$$C_D = \frac{dq_{\text{diff}}}{dv} = qAL_p\frac{d\,\Delta p_n(0)}{dv}. \tag{7.15}$$

Now $\Delta p_n(0)$ can be obtained from Eq. (6.26a) as

$$\Delta p_n(0) = i_pL_p/qD_pA. \tag{7.16}$$

Then, on differentiating with respect to voltage, one obtains

$$\frac{d\,\Delta p_n(0)}{dv} = \frac{L_p}{qD_pA}\frac{di_p}{dv}. \tag{7.17}$$

Introducing the diode law [Eq. (6.20)] into Eq. (7.17) and substituting into Eq. (7.15) gives for the diffusion capacitance of a long diode

$$C_D = qI\tau_p/kT, \tag{7.18a}$$

which is seen to be directly proportional to the DC bias current. A similar result can be obtained in the case of the short diode approximation, $2W \ll L_p$, where W is the width of the lightly doped n-region. Then the diffusion capacitance is given by

$$C_D = W^2qI/2D_pkT. \tag{7.18b}$$

This is the subject of a problem at the end of this chapter.

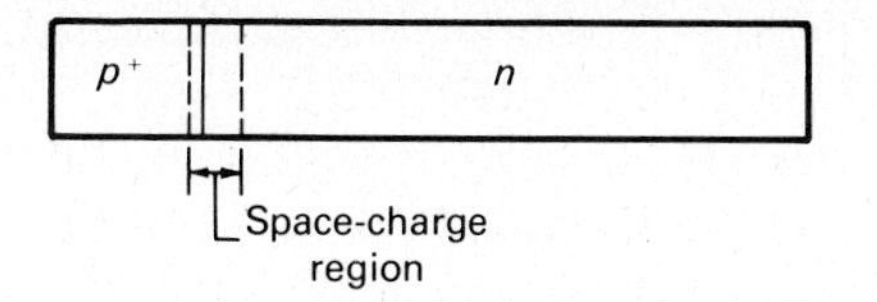

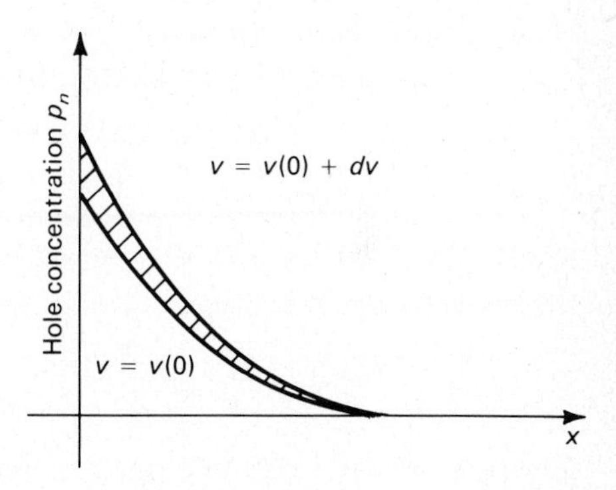

Figure 7.7 Minority carrier charge distribution in the n-region of a long p^+-n junction diode before and after the application of an incremental voltage dv. The crosshatched area represents the additional hole charge stored in the n-region as a result of the additional applied voltage dv.

EXAMPLE 7.3

The capacitance of an abrupt, long p^+-n junction diode 1.0×10^{-4} cm^2 in area measured at -1.0 V reverse bias is 5.0 pF. The built-in voltage ΔV_0 of this device is 0.90 V. When the diode is forward biased with 0.50 V, a current of 10 mA flows. The n-region minority hole lifetime is known to be 1.0 μsec at 300°K.

(a) Calculate the depletion-layer capacitance of this junction at 0.50 V forward bias.
(b) Calculate the diffusion capacitance of the diode operating as in (a) at 300°K.

Solution

(a) The depletion-layer capacitance of a p-n junction is given by Eq. (7.9b). Hence in the forward direction at 0.50 V, the depletion-layer capacitance is

$$C_j = (5.0 \text{ pF}) [(-1.0 - 0.90)/(0.50 - 0.90)]^{1/2}$$

$$= \underline{11 \text{ pF}}.$$

(b) From Eq. (7.18a)

$$C_D = \frac{I\tau_p}{kT/q}$$

$$= \frac{(10 \times 10^{-3}\ \text{A})\,(1.0 \times 10^{-6}\ \text{sec})}{0.026\ \text{V}}$$

$$= 38 \times 10^{-8}\ \text{F}$$

$$= \underline{380{,}000\ \text{pF}}.$$

This example illustrates that the diffusion capacitance of a forward-biased junction diode can be four or more orders of magnitude greater than the depletion-layer capacitance of a *p-n* junction. Then the *p-n* junction minority carrier charging time depends on the product of the junction dynamic resistance and the capacitance. This product for a long diode is essentially given by

$$r_{\text{jct}} C_D = \left(\frac{dv}{di}\right)_{\text{jct}} (C_D) = \frac{kT}{qI} \frac{qI\tau_p}{kT} = \tau_p \tag{7.19}$$

from Eqs. (6.20) and (7.18). Hence the time necessary to incrementally change the injected charge distribution in a forward-biased *p-n* junction is about equal to the minority carrier lifetime in the *n*-region, when the width of this region is much greater than L_p. For a short diode this charging time is $W^2/2D_p$.

When both the time for charge redistribution in the junction depletion layer and the distribution of charge in the semiconductor bulk are considered, a small-signal equivalent circuit can be constructed to approximate the high frequency diode performance. This is shown in Fig. 7.8. Note that the diffusion capacitance is negligible for a reverse-biased diode and usually dominates in forward bias. The total current i through a *p-n* junction with a DC bias voltage V and a small impressed AC voltage v is given by

$$i = I_0(e^{qV/kT} - 1) + (C_j + C_D)\frac{dv}{dt}. \tag{7.20}$$

7.4 LARGE-SIGNAL SWITCHING OF A *p-n* JUNCTION DIODE

In circuit applications the junction diode is often used as a switch which is "ON" when the diode is conducting in the forward direction and "OFF" or

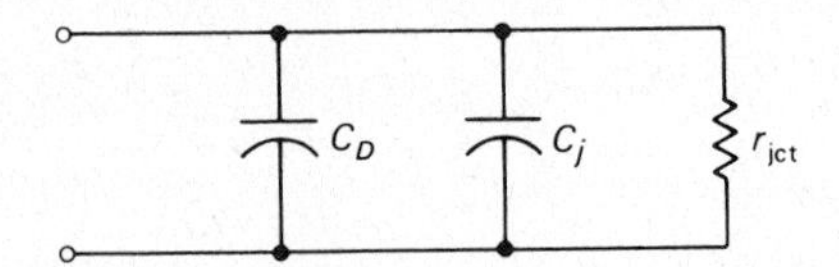

Figure 7.8 Approximate high frequency equivalent circuit of a *p-n* junction diode including the diffusion capacitance C_D, the depletion-layer capacitance C_j, and the dynamic junction resistance r_{jct}.

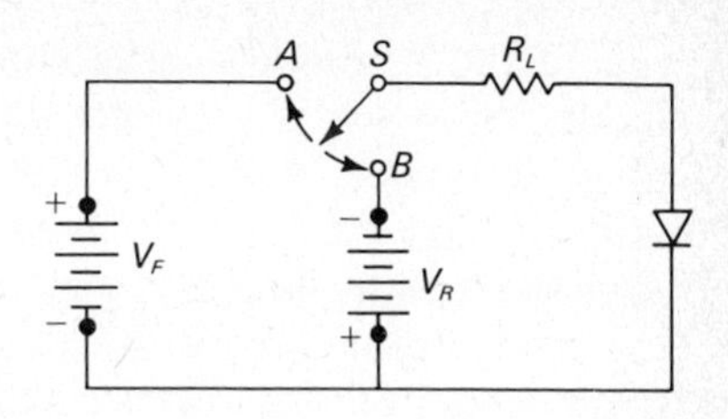

Figure 7.9 Circuit for demonstrating large-signal switching of a *p-n* junction diode. The diode is initially forward biased by having the switch S at terminal A. The diode is biased in the reverse direction by suddenly switching S to terminal B.

nonconducting when a reverse potential is applied to the device. Consider that a voltage V_F is applied to the circuit so that the diode is in its normal conducting ON state carrying a current $I_F \simeq V_F/R_L$, where the current is limited by a circuit resistance R_L much greater than the junction diode forward resistance.[9] Now the potential is suddenly reversed in an attempt to place the diode into the nonconducting OFF state. A circuit for demonstrating this type of large-signal switching is shown in Fig. 7.9.

Figure 7.10 gives a plot of the current through the load resistor R_L and the voltage across the diode as a function of time. Note that the current persists for a time longer than t_s after turnoff is attempted at $t = 0$. This effect is

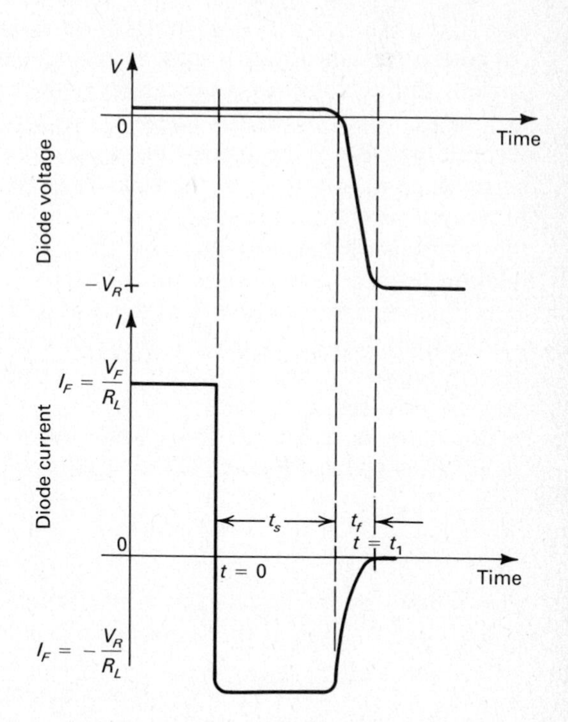

Figure 7.10 Graph of the voltage across and the current through the diode in Fig. 7.9 versus time. The moment of switching from the forward to the reverse direction is at $t = 0$. A time greater than t_s is required for the diode to recover to its normal, very low reverse current.

[9]Here we assume that the junction contact potential is small compared with V_F.

referred to as diode **charge storage** and is characteristic of all *p-n* junction diodes; however, it normally does not occur in Schottky barrier diodes, soon to be discussed, which turn off more rapidly. This fact demonstrates that minority carriers, injected while the device is forward biased, persist for a time after the forward bias is removed because of their finite lifetime. Reference to the previous section indicates that a finite time is also required to establish the steady-state minority carrier distribution characteristic of the forward current I_F. Schottky devices do not exhibit these phenomena normally since the current carriers in such diodes are majority carriers. A formulation will now be presented for the large-signal switching problem. This will point up the importance and the simplicity of the **charge-control** description of diode switching.

7.4.1 Charge Control and *p-n* Junction Diode Switching

In the steady-state situation, the distribution of holes in the *n*-region of a forward-biased long p^+-*n* diode [see Fig. (6.7b)] is given by Eq. (6.24). The integral of these charges over the whole *n*-region yields a value for the number of excess injected charge carriers stored in this region.[10] If we take the ratio of this charge to the current flowing through the junction [Eqs. (7.12) and (6.26a)], we obtain

$$\frac{Q_p}{I_p} = \frac{qAL_p\,\Delta p_n(0)}{qAD_p\,\Delta p_n(0)/L_p} = \frac{L_p^2}{D_p} = \tau_p. \tag{7.21a}$$

An interpretation of this simple relation is obtained by noting that the hole current through the junction simply serves the purpose of maintaining the exponential distribution of holes in the *n*-region, where they are constantly recombining. Since the average hole recombination time is τ_p, a charge Q_p must be supplied in a time τ_p by the current I_p. According to the normal definition of current as charge transmitted per unit time, Eq. (7.21a) results. Note that the relationship between junction current and charge is linear, whereas the relation between current and junction voltage is exponential [see Eq. (6.20)]. This illustrates the simplicity of the so-called charge-control model in dealing with current flow in the *p-n* junction diode. It is also of considerable use in the description of transistor switching when applied to the emitter *p-n* junction; this will be discussed later.

In the case of a short p^+-*n* diode, where the width of the *n*-region is W ($\ll \frac{1}{2}L_p$), the ratio of Eq. (7.21a) becomes

$$\frac{Q_p}{I_p} = \frac{W^2}{2D_p}. \tag{7.21b}$$

This follows by integrating the minority hole charge stored in this diode and using Eq. (6.37); this is the subject of a problem at the end of the chapter.

[10]See Eq. (7.12). Note that only the holes stored in the *n*-region are considered since very few electrons are injected by this p^+-*n* junction.

7.4.2 Diode Turnoff Time Calculation

The p^+-n diode, used as an electronic switch, has a current I_F flowing through it at time $t < 0$ under steady-state ON conditions, as indicated in Fig. 7.9 with the switch moved to terminal A. Now the diode is to be turned OFF by reversing the potential on the device by moving the switch to terminal B at $t = 0$. After some transient recovery time during which appreciable reverse current flows the diode finally attains the high impedance state with only a very small leakage current flowing through it, typical of a reverse-biased junction diode. The finite time required for the diode to recover to its high impedance state is the period necessary for the excess minority holes introduced into the n-region of the diode under forward bias conditions to disappear by recombination with majority electrons or by diffusing out of this region. The minority carrier distribution in the n-region of the diode just before and during switching is shown in Fig. 7.11.

Under transient conditions the minority hole current in the diode n-region

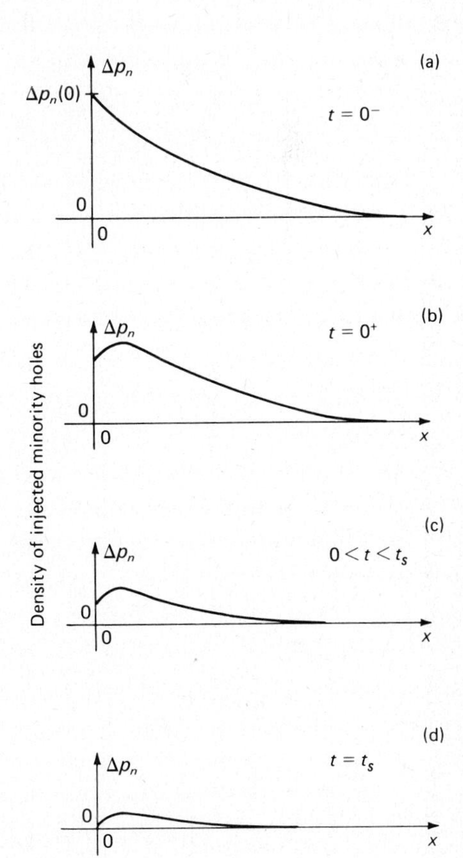

Figure 7.11 (a) The minority injected carrier distribution in the n-region of the p-n diode biased in the forward direction just prior to switching. The excess hole density at the edge of the space-charge region at $x = 0$ is $\Delta p_n(0)$.
(b) The minority carrier distribution just after switching at $t = 0$. Note that the slope of the carrier density at the edge of the space-charge region is proportional to the reverse current $-I_R$ since it is a diffusion current.
(c) Carrier distribution at a later time, less than t_s. Note that the carrier slope at $x = 0$ is the same as in (b), indicating that the reverse current is still $-I_R$.
(d) Carrier distribution at t_s, where the excess hole density at the edge of the space-charge region becomes zero but there are still some residual holes remaining in the n-region.

is given by[11]

$$i_p(t) = \frac{Q_p(t)}{\tau_p} + \frac{dQ_p(t)}{dt}, \tag{7.22}$$

where $Q_p(t)$ represents the total hole charge integrated over x, at a time t. The first term on the right-hand side of the equation refers to the supply of current necessary to compensate for the recombination of holes in the n-region under steady-state conditions. However, under transient conditions the net number of charges in the n-region must change with time; this is represented by the second term in Eq. (7.22). If the diode is being switched OFF, the net number of charges there must decrease with time and hence dQ_p/dt must be negative. For the circuit shown in Fig. 7.9, before turnoff (switch in position A), a steady-state current $I_F \simeq V_F/R_L$ flows in the diode when V_F is much greater than the forward diode drop. At $t = 0$ the switch is suddenly moved to B. Now the diode is reverse biased with a potential $-V_R$ and a current $I_R \simeq -V_R/R_L$ flows from the time $t = 0$ to $t = t_s$. During this period the junction is so swamped with minority injected carriers that its impedance is negligibly small compared to the circuit resistance R_L, which then limits the current. After a time t_s a sufficient number of excess holes have recombined in or flowed out of the n-region, and so the junction begins to return to its high impedance state. The recovery is essentially exponential in character and is effectively completed at time $t = t_1$ (see Fig. 7.10), when approximately the DC reverse steady-state leakage current I_0 is all that flows.

Equation (7.22) can now be written for the period $0 < t < t_s$ as

$$-I_R = -\frac{V_R}{R_L} = \frac{Q_p}{\tau_p} + \frac{dQ_p}{dt}. \tag{7.23}$$

The general solution of this equation is

$$Q_p(t) = -I_R\tau_p + Be^{-t/\tau_p}, \tag{7.24}$$

where B is an integration constant which will now be evaluated. Just prior to $t = 0$, the steady-state hole current is given by $I_F = Q_p/\tau_p$, where Q_p is the total charge stored in the n-region which cannot change instantaneously at $t = 0$. Hence introducing this charge into Eq. (7.24), taken at $t = 0^+$, yields $B = \tau_p(I_F + I_R)$. Now Eq. (7.24) becomes

$$Q_p(t) = \tau_p[-I_R + (I_F + I_R)e^{-t/\tau_p}]. \tag{7.25}$$

At $t = t_s$ the excess hole density at the junction becomes zero (see Fig. 7.11d), the diode voltage starts to reverse, and the junction impedance begins to regain control of the current. Depending on the details of the diode doping design, the excess junction current will then decay more or less rapidly and become

[11]It is assumed here that the junction capacitance charging current $i_c = C_j\, dv/dt$ is negligible compared to bulk charge storage terms since the voltage change across the diode during switching is usually small during most of the turnoff time.

small compared to I_R at a time $t = t_1$ when the stored charge $Q_p(t_1)$ becomes negligibly small. From Eq. (7.25), setting $Q_p(t) = 0$, we get as an approximation, assuming that $t_1 \frown t_s$ and hence falls within the interval in which I_R is essentially constant,

$$t_1 \simeq \tau_p[\ln(1 + I_F/I_R)]. \tag{7.26}$$

In some diode designs $t_1 \frown t_s$ ($t_f \approx 0$),[12] and so this latter equation represents approximately the charge storage time t_s, which is proportional to the lifetime τ_p. Hence the storage time can be reduced by decreasing the lifetime or by increasing I_R at the expense of additional energy expenditure. Note that the approximation of Eq. (7.25) is not useful for $t_1 \gg t_s$. The exact equation for obtaining t_s for a uniformly doped long *p-n* junction is[13]

$$\text{erf}(\sqrt{t_s/\tau_p}) = I_F/(I_F + I_R), \tag{7.27}$$

where erf is a tabulated function called the **error function.** Since erf(0) = 0, $t_s \rightarrow 0$ as $I_R \rightarrow \infty$, and since erf(0.45) = 0.5, $t_s = 0.2\tau_p$ when $I_R = I_F$. See Fig. 11.4 for other values of the error function. This relationship between t_s and τ_p can be used to determine the hole lifetime in the *n*-material of a long p^+-*n* diode.

For the case of a short diode, $W \ll L_p$, we can again solve Eq. (7.23), noting from Eq. (7.21b) that $I_F = Q_p/(W^2/2D_p)$. Proceeding as before we can show that

$$t_1 \approx \tau_p \ln\left[1 + \frac{1}{2}\left(\frac{W}{L_p}\right)^2 \frac{I_F}{I_R}\right]. \tag{7.28}$$

If I_F is of the same order as I_R and since $W \ll L_p$, Eq. (7.28) can be approximated by

$$t_1 \approx \frac{\tau_p}{2}\left(\frac{W}{L_p}\right)^2 \frac{I_F}{I_R}. \tag{7.29}$$

Hence, although the turnoff time is still proportional to the minority carrier lifetime, this time decreases as the *square* of the diode thickness W. This is because the total stored charge that must be removed and the charge gradient that produces a diffusion current to remove these charges are both proportional to W.

To speed turnoff the lifetime of the semiconductor material used for diode fabrication may be reduced, but this in general will reduce the diode forward ON current for a given diode voltage drop. Alternatively, the reverse current I_R can be increased to extract the stored charges. For large ratios of I_R/I_F, the

[12]The "snapoff" diode is designed with an impurity gradient which creates a built-in electric field that tends to rapidly sweep out minority carriers still remaining after time t_s, making $t_1 \approx t_s$. This very rapid turnoff is used for harmonic generation.

[13]R. H. Kingston, "Switching Time in Junction Diodes and Junction Transistors," *Proc. IRE*, **42,** 829 (1954).

turnoff time is given by the more exact solution[13] for the total turnoff time t_1 as

$$t_1 = \frac{\tau_p}{2} \frac{I_F}{I_R} \tag{7.30a}$$

for $W \gg L_p$, or

$$t_1 = \frac{W^2}{2D_p} \frac{I_F}{I_R} \tag{7.30b}$$

for $W \ll L_p$.

This time delay for the diode to return to its blocking state plus the time to turn ON the diode of course lengthens the total diode switching time and hence reduces the maximum switching rate of the device in switching applications. In silicon junction diodes designed for rapid switching, gold impurity atoms are introduced into semiconductor bulk in order to reduce the lifetime there and hence improve the switching speed. Gold atoms introduce energy levels near the middle of the energy gap which serve as recombination centers. However, there is a limit to the density of gold atoms which can be introduced metallurgically into the silicon and hence the recovery time cannot be reduced much below a nanosecond. (Also, gold levels trap out or immobilize electrons and can cause the silicon resistivity to increase.) The variation of hole lifetime with the gold impurity density in a silicon crystal is plotted in Fig. 7.12.

For subnanosecond switching the Schottky barrier diode is employed since the forward current in this device is not carried by injected minority carriers

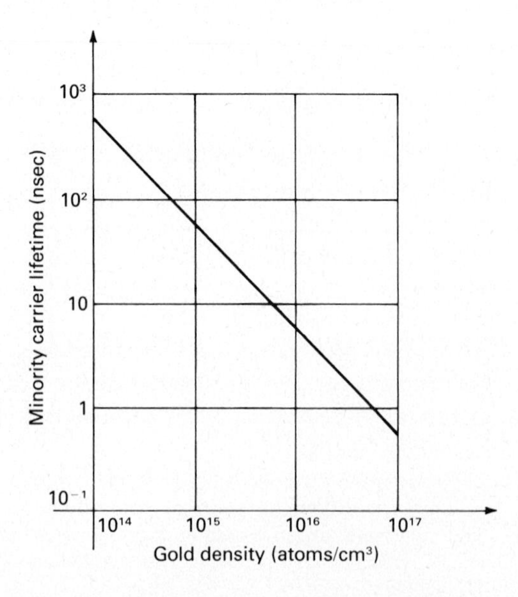

Figure 7.12 Logarithmic plot of the variation of minority hole lifetime with gold impurity density in a 0.1 Ω-cm *n*-type silicon crystal doped with gold atoms, at 300°K.

and the problem of minority carrier storage does not arise. It will be seen in the next chapter that the minority carrier storage effect also limits the switching speed of bipolar junction transistors. In microchip integrated circuits, Schottky barrier devices are sometimes used to speed up the switching of bipolar transistors. (See Section 11.5.4.)

EXAMPLE 7.4

The steady-state current I_F carried by a long silicon p^+-n junction diode biased in the forward direction is 10 mA. The hole lifetime in the 0.1 Ω-cm n-region is adjusted by introducing 10^{16} atoms of gold per cubic centimeter of silicon. The diode is switched ON and OFF in the circuit of Fig. 7.9.

(a) Determine the integrated excess hole charge in the n-region that must be eliminated to turn this diode off.

(b) How long will it take to essentially eliminate the charge calculated in (a) if the reverse turnoff current I_R is 10 mA?

(c) Calculate the storage time under the conditions of (b).

Solution

(a) From Fig. 7.12, 10^{16} gold atoms/cm^3 corresponds to a lifetime of 6×10^{-9} sec. Hence from Eq. (7.21a) the total excess hole charge in the n-region is

$$Q_p = I_p\tau_p = (10 \times 10^{-3}\text{ A})(6 \times 10^{-9}\text{ sec})$$

$$= \underline{6 \times 10^{-11}\text{ C} \quad \text{or} \quad 60\text{ pC}}.$$

(b) Using Eq. (7.26) we obtain the time to remove this charge:

$$t_1 \simeq \tau_p \ln(1 + I_F/I_R)$$

$$= (6 \times 10^{-9}\text{ sec}) \ln[1 + (10 \times 10^{-3}\text{ A})/(10 \times 10^{-3}\text{ A})]$$

$$= \underline{4 \times 10^{-9}\text{ sec}}.$$

(c) The storage time from Eq. (7.27) is given by

$$\text{erf}(\sqrt{t_s/\tau_p}) = I_F/(I_F + I_R),$$

$$\text{erf}(\sqrt{t_s/\tau_p}) = (10 \times 10^{-3}\text{ A})/(10 \times 10^{-3} + 10 \times 10^{-3}\text{ A}),$$

$$\text{erf}(\sqrt{t_s/\tau_p}) = \tfrac{1}{2}.$$

Hence

$$\sqrt{t_s/\tau_p} = 0.45,$$

and

$$t_s = \tau_p(0.45)^2 = (6 \times 10^{-9}\text{ sec})(0.45)^2$$

$$= \underline{1.2 \times 10^{-9}\text{ sec.}}$$

Note that this is less than the minority carrier recombination time (lifetime) since the reverse current extracts holes from the *n*-region.

7.5 THE METAL–SEMICONDUCTOR CONTACTS

The metal–semiconductor contacts to the *p-n* junction diode are normally made as low resistance and ohmic[14] as possible. This is by way of ensuring that the electrical characteristics of the device are solely those of the carefully designed junction. In general, though, the metal–semiconductor contact is a rectifying junction with a nonlinear *I-V* characteristic. This is due to the presence of a potential barrier which is present at a contact between a metal and the carefully prepared surface of a semiconductor[15] which has a substantially different **work function** than the metal. Here the more general definition of the work function is implied; that is, the energy necessary to remove an electron from a level of energy equal to the Fermi energy, to a distance where the material no longer exerts any force on the electron. In the following sections the theory of both ohmic and rectifying metal–semiconductor contacts will be discussed.

7.5.1 The Metal–Semiconductor Barrier (Schottky) Diode

The energy band diagram at thermal equilibrium for the contact between a particular metal and an *n*-type semiconductor crystal is shown in Fig. 7.13b. In Fig. 7.13a, the conduction band of the metal and the semiconductor conduction and valance bands are pictured for the case of materials completely separate from each other. For the situation illustrated, bringing the metal and semiconductor crystal into intimate contact in thermal equilibrium means that the energy levels of the semiconductor must be depressed relative to those of

[14]Here "ohmic" refers to the electrical symmetry and linearity of the contact; the resistance is the same, independent of the direction of current flow (see Section 6.3.2).

[15]Since the electrical characteristics of this diode depend very sensitively on the metal–semiconductor interface, the surface states discussed in Section 5.5 must be reduced to a very low density to avoid their significant influence on the properties of this device. Otherwise these states will trap or immobilize charge carriers attempting to transverse this junction. The presence of an interface layer (such as SiO_2 on Si) can also inhibit current flow. Charge carriers can tunnel through this layer efficiently only if it is a few angstroms thick.

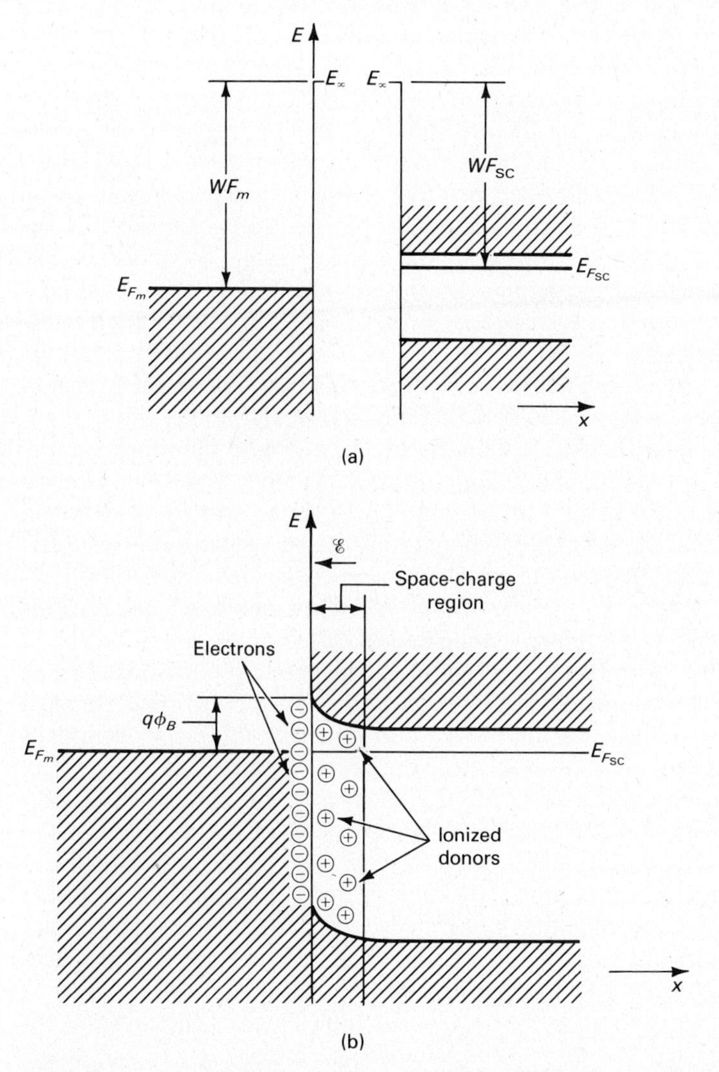

Figure 7.13 (a) Energy band diagram of a metal (left) and an *n*-type semiconductor crystal (right) when separate from each other. The metal work function is indicated as the energy necessary to carry an electron in the metal to a location far from the influence of the metal, to an energy E_∞.
(b) Energy diagram for a metal in intimate contact with an *n*-type semiconductor under thermal equilibrium conditions. Note that "band bending" in the semiconductor provides a field which keeps additional electrons in the conduction band of the semiconductor from spilling over into the metal. The barrier height for electron injection from the metal into the semiconductor is $q\phi_B$.

the metal so that their respective Fermi levels can line up. This means that the semiconductor is now at a higher potential energy than the metal in this electron energy diagram. This can be considered to occur as the result of a transfer of negative charge from the semiconductor to the metal on contact. Before contact there are electrons in the conduction band of the semiconductor which are in energy states above those of the electrons in the conduction band of the metal. Hence, on contact, the electrons from the semiconductor "spill over" into the metal, leaving behind fixed positive donor charges. This sets up an electric field which tends to discourage additional electrons from leaving the semiconductor and an equilibrium potential barrier results as shown in Fig. 7.13b. The electric field lines originate on the ionized donors in the semiconductor bulk and terminate on electrons in the metal attracted to the metal–semiconductor interface.

Note that the space-charge or depletion region occurs only in the semiconductor here, and not in the metal. This is due to the greater ability of the semiconductor to support space-charge than the metal, owing to the much smaller density of charges in the semiconductor compared to the metal. This creates a so-called "bending of the bands" in the semiconductor near the metal contact. It should be clear that the height of the potential barrier depends on the difference in work function between the metal and semiconductor.[16] A similar consideration may be applied to a metal and *p*-type semiconductor contact. This analysis is the subject of a problem at the end of this chapter.

The asymmetry of current flow through this metal–semiconductor or **Schottky** junction can be understood by considering the application of forward and reverse potential biases to the diode. Consider that the *n*-type semiconductor is biased positively with respect to the metal (reverse bias). The electrons in the metal are urged toward the barrier by the field but few have sufficient energy to overcome the barrier potential. Similarly, the holes in the *n*-type semiconductor are moved toward the barrier by the applied field but there are few of these minority holes available. Hence these carriers constitute a small leakage current through the barrier. Also, electrons in the semiconductor are repelled from the junction region and this depletion causes the space-charge region of ionized donors to widen. This is illustrated in Fig. 7.14a and may be compared to reverse bias in a *p-n* junction which is the low current direction and which also produces space-charge widening.

When the metal is now biased positively with respect to the semiconductor (forward bias), the potential barrier is reduced and electrons in the conduction band of the semiconductor can now flow more easily into the metal. This is shown in Fig. 7.14b. The electrons crossing the potential barrier to enter the metal have energies much greater than the Fermi energy in the metal. These are quite energetic and hence are termed **hot electrons.** Note that the majority

[16]An "image force" barrier lowering should strictly also be considered. This occurs when for example an electron leaves the metal, leaving a positively charged metal behind whose field in turn affects the electron transport and hence the current.

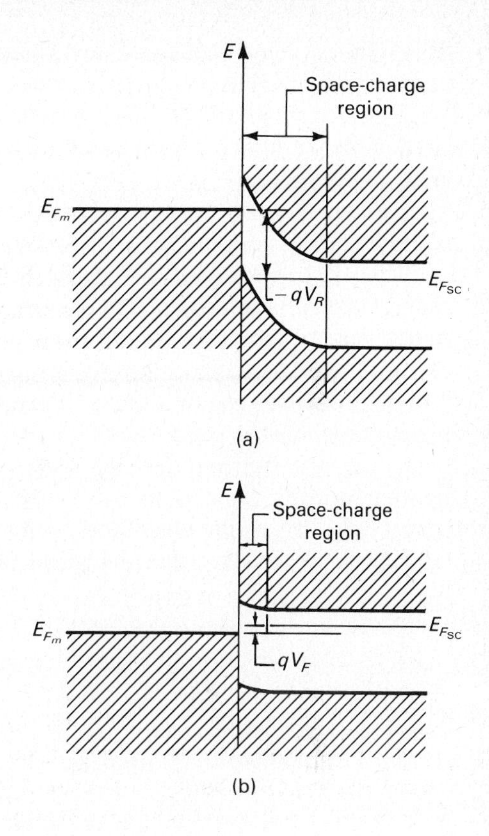

Figure 7.14 (a) The metal–semiconductor junction energy diagram under reverse bias conditions; V_R is the reverse voltage applied. (b) The metal-semiconductor junction energy diagram under forward bias conditions; V_F is the forward voltage applied.

carrier electrons carry the current in this case compared to the minority injected carriers that carry the main part of the forward current in the *p-n* junction.[17] This fact expresses perhaps the most important practical difference between a Schottky barrier diode and a *p-n* junction diode. As was shown in Section 7.4.2, the speed with which a *p-n* diode can be switched from the forward (conducting) mode to the reverse (blocking) mode may be limited by the minority carrier lifetime, which can be of the order of 10^{-6} sec. Since the Schottky diode conducts by majority carriers, the charge redistribution or dielectric relaxation time is less than 10^{-12} sec and so only the charging time of the space-charge capacitance normally limits switching speed. An expression such as that given in Eq. (7.11) may be used to calculate the maximum operating frequency of this device. The depletion-layer capacitance may be calculated by using an expression identical to that used in connection with a p^+-n junction assuming the space-charge region penetrates only into the *n*-type semiconductor [Eq. (7.9b) with $N_a \gg N_d$]. A maximum operating frequency of

[17]Under high voltage bias conditions there can be a small component of minority carrier current. For a gold/*n*-silicon diode ($N_d = 10^{15}/\text{cm}^3$) only 5% minority carrier current is predicted at a current density of 350 A/cm^2.

350 GHz has been measured for 10-μm-diameter gallium arsenide Schottky diodes and 120 GHz for silicon devices. The higher frequency capability of the GaAs device results from the higher electron mobility in this material compared with Si, hence lower diode series resistance, and consequently higher operating frequency.

The current–voltage characteristic of the Schottky diode can be approximated by an expression similar to Eq. (6.20) for the *p-n* junction, viz., $I = I_0(e^{qV/kT} - 1)$. However, I_0 for the Schottky barrier is usually substantially greater than the leakage current for Ge and Si *p-n* junctions. Hence substantial current can be obtained at moderately low voltages compared with the *p-n* junction.[18] This aspect of the device coupled with its high speed or high frequency properties make it particularly suitable for a high frequency, low level detector such as the RF stage of a radio or TV receiver. It is also used to great advantage in increasing the switching speed of transistor transistor logic (TTL) digital computer circuits, as will be described in Chapter 11. Sketches of some typical Schottky diode structures are shown in Fig. 7.15.

To compute I_0 for the ideal Schottky contact illustrated in Fig. 7.13b an

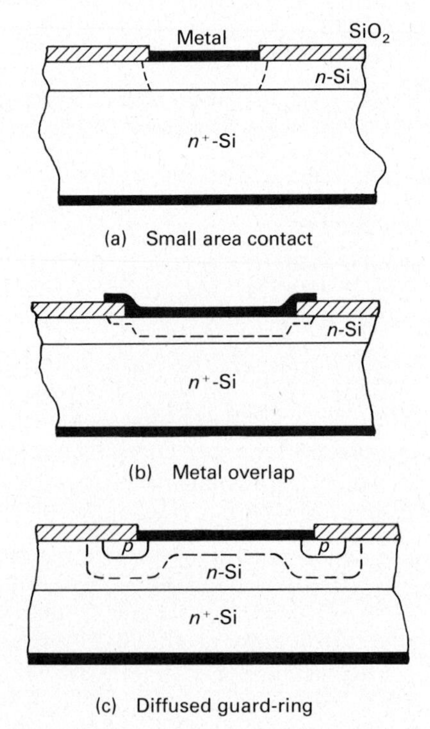

Figure 7.15 Typical metal–semiconductor (Schottky) device structures. [After Sze, *Physics of Semiconductor Devices*, 2nd ed. (Wiley, New York, 1981), p. 299.]

[18]About 0.3 V for a silicon Schottky diode in contrast to about 0.6 V for a silicon *p-n* junction device.

assumption must be made concerning the mechanism for electron transfer through this barrier. Thermionic electron emission theory assumes that the probability of an electron in the metal surmounting this potential barrier $q\phi_B$ is given by a Boltzmann factor similar to that assumed in the derivation of the *p-n* junction *I-V* characteristic [see Eq. (6.29)], and hence the reverse bias current is

$$I_0 \propto e^{-q\phi_B/kT}. \tag{7.31}$$

This theory yields as a final expression for the Schottky junction current density

$$J = A^*T^2 e^{-q\phi_B/kT}(e^{qV/kT} - 1), \tag{7.32}$$

where A^* is the effective Richardson constant for thermionic emission, which is relatively temperature insensitive. A carrier diffusion theory can be added to supplement the thermionic emission theory. This corrected theory only results in a revised expression for the Richardson constant. Some barrier height values for typical metal Schottky contacts to silicon and gallium arsenide are given in Table 7.2.

TABLE 7.2 **Schottky barrier heights at 300°K for various metals deposited onto Si and GaAs.**

Semiconductor	Metal	Schottky Barrier Height ϕ_B (V)
n-Si	Al	0.72
	Au	0.80
	Ti	0.50
	Pt-Si	0.84
p-Si	Al	0.58
	Au	0.34
	Ti	0.61
n-GaAs	Al	0.80
	Au	0.90
	Ag	0.88
	$PtAs_2$	0.88
p-GaAs	Au	0.42
	Ag	0.63

From S. M. Sze, *Physics of Semiconductor Devices,* 2nd ed. (Wiley, New York, 1981), p. 291.

7.5.2 The Ohmic Contact

In the previous section the properties of a rectifying metal–n-semiconductor contact were discussed. Because of the potential barrier that was produced at the junction of the metal and semiconductor owing to the work function difference, a rectifying, nonlinear I-V characteristic resulted. The width of this potential barrier can be determined by employing an expression identical to the one derived for a p^+-n junction [see Eq. (6.18c)]. For a metal contact made to heavily doped material ($>10^{19}/\text{cm}^3$ impurity atoms) the barrier is of the order of tens of angstroms. Solution of the Schrödinger equation (quantum mechanics) predicts a high probability of tunneling through such narrow barriers so that they can pass through these barriers without having sufficient energy to surmount them. Hence the Boltzmann expression given in Eqs. (6.32) and (7.31) no longer applies and the junction current will increase substantially with small applied-voltage increments. The resulting I-V characteristic will then tend to be ohmic and low resistance. A common technique used to produce low resistance contacts to the ends of a p-n junction or the emitter, base, and collector regions of a transmitter is to dope the areas to be contacted with a large concentration of impurities of the same semiconductor type as that to be contacted. Hence metal/n^+-n and metal/p^+-p contacts make low resistance, ohmic connections. For example, alloying aluminum (a p-type dopant) to p-silicon provides a good ohmic contact. Another way to enhance this carrier tunneling is to disrupt the crystal structure of the surface to be contacted.

In principle it is also possible to produce an ohmic contact to a semiconductor by choosing a metal with a particular work function relative to the semiconductor. Such a contact to an n-type semiconductor is sketched in Fig. 7.16b. If the metal is positively biased relative to the semiconductor, there is no barrier to flow from metal to semiconductor. If the metal is negatively biased, the large accumulation of electrons at the interface in the semiconductor, caused by the Fermi level entering the conduction band there, will be easily transported into the metal over the small barrier at the interface.

*7.6 THE HETEROJUNCTION

A **heterojunction** is a junction formed between two semiconductors with different energy gaps.[19] If the semiconductors each contain the same species of carrier type the junction is called an **isotype** heterojunction; otherwise it is an **anisotope.** Important applications of this type of junction are the room temperature operating semiconductor laser, improved LEDs, photodetectors and solar cells (see Section 7.2), and the wide gap emitter transistor (see Section 8.1.2). Also, the forming of thin (<100 Å), periodically alternating layers of two different semiconductor materials can result in a superlattice structure which has very high carrier mobility properties parallel to the layer interfaces. The equilibrium band diagram for a p-n anisotope heterojunction is shown in

[19]This is in contrast to the **homojunction,** which is formed in a single semiconductor material.

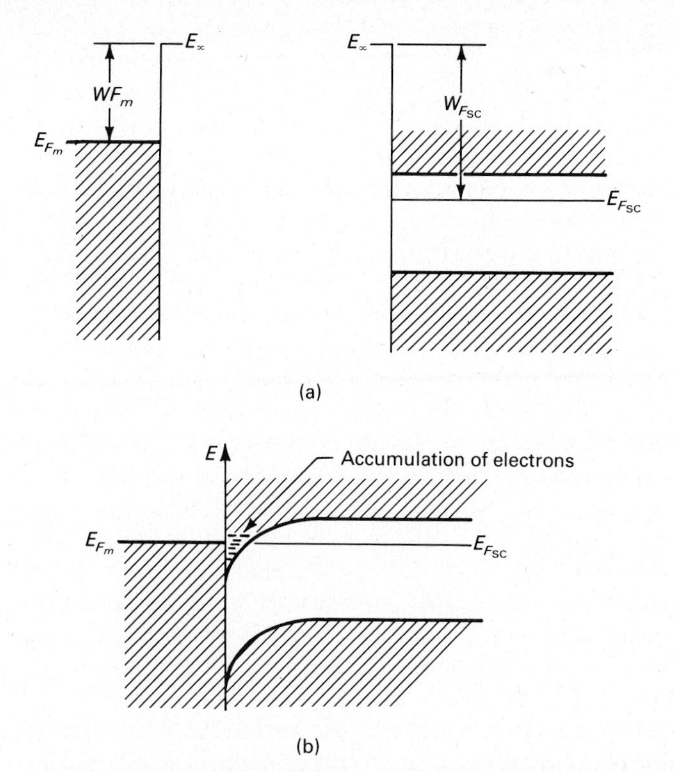

Figure 7.16 (a) Energy band diagram of a metal (left) and an *n*-type semiconductor when separate from each other. Note that in this case $E_{Fm} > E_{Fsc}$.
(b) Equilibrium energy diagram for a metal–semiconductor ohmic contact made by joining together the materials illustrated above. Note that the lining up of the Fermi levels causes band bending, providing an easy flow of electrons in both directions.

Fig. 7.17b. Here the minority hole diffusion from the *n*-type side to the *p*-type side under reverse voltage bias is suppressed by the barrier in the valence band, reducing its contribution to the leakage current I_0 even if the *n*-region is lightly doped. In the forward direction electron back-injection from the *n*-side to the *p*-side is suppressed to a greater extent than the injection of holes from the *p*-side to the *n*-side, resulting in almost exclusive hole current. Since this is the case almost independent of doping on either side, a highly efficient hole emitter for a high gain *p-n-p* bipolar transistor may be fabricated without requiring a heavily doped emitter p^+-n junction. Also, the *n*-base region of a *p-n-p* transistor can be heavily doped to provide low base resistance without sacrificing injection efficiency.

When a wide energy gap semiconductor such as AlGaAs is deposited onto GaAs[20] to form a heterojunction photodiode or solar cell, improved device

[20]These materials are metallurgically compatible in the single-crystal form since their lattice constants are nearly perfectly matched.

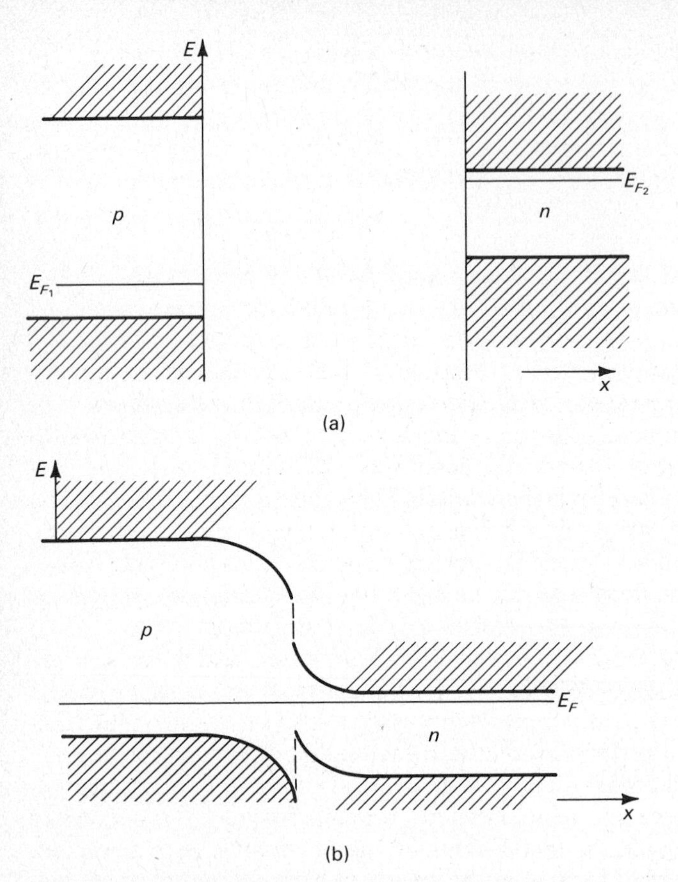

Figure 7.17 (a) Energy band diagrams for wide and narrow band gap semiconductors when separated from each other.
(b) Equilibrium energy band diagram for the two semiconductor materials sketched in (a) when placed in contact with each other.

properties result. This combination of materials can give rise to photocells which have improved quantum efficiency as well as good response speed for a given incident optical input. The wide gap semiconductor material acts as a "window" for transmission of incident light. The large energy gap prevents lower energy photons from generating electron–hole pairs in this material. Such photons transmitted to the substrate can have sufficient energy, however, to create electron–hole pairs there for collection by the *n-p* heterojunction. Some advantages of the heterojunction solar cell over the conventional *p-n* junction device are (i) enhanced ultraviolet response, (ii) lower cell series resistance to reduce power loss, and (iii) high tolerance to radiation in outer space. Because of the high absorption coefficient in semiconductors of ultraviolet light, UV-created electron–hole pairs are generated mainly near the semiconductor surface, where significant minority carrier recombination can take place, reducing the cell efficiency. The window in wide gap heterojunctions will cause these carriers to be generated in the material bulk, resulting in higher collection efficiency. The window material can be thick enough to provide low series resistance to reduce power loss. The thick window material can also absorb radiation and prevent lifetime reduction in the substrate material.

☐ *SUMMARY*

In this chapter several important device applications of *p-n* junction diodes were described. First the use of this device as a photodetector (receiver) in an optical communication system to convert light signals into electrical impulses was discussed; in the same sense the operation of large-area diodes for the conversion of solar energy into electrical power was analyzed. The application of the LED and semiconductor laser as an efficient light emitter (transmitter) in an optical communication system was described; the unidirectional nature of the light output of the laser was explained. The diode as a voltage-variable capacitor or "varactor" was analyzed. The concept of the *p-n* junction diffusion capacitance due to minority carrier injection was introduced and will be used later in determining the frequency limitation of the bipolar transistor provided by its emitter diode. The concept of charge control was employed to analyze the switching speed of the *p-n* junction diode; minority carrier storage was found to play a major role in determining the device's recovery time. Gold and other impurities can be used to improve the switching speed of silicon junction diodes. The physics of operation of the metal–semiconductor ("Schottky") diode was described, leading to the conclusion that minority carrier injection is minimal in this device and hence does not limit its frequency of operation. Methods for producing ohmic, metal–semiconductor contacts were discussed. Finally, the heterojunction formed at the interface between two semiconductors with different energy gaps was described; the intrinsic property of this junction in suppressing either hole or electron injection was stressed.

PROBLEMS

7.1 The *p-n* junction described in Problem 6.4 is employed as a photodetector. Illumination causes a hole–electron generation rate of 3.0×10^{21} pairs/cm^3-sec. Determine the photocurrent due to the incident light.

7.2 Two milliwatts of light power of wavelength 0.85 μm is incident on a photodiode and induces a light current of 1.0 mA. Determine the quantum efficiency of this photodetector.

7.3 **(a)** Derive an expression which will give the voltage delivered by the solar cell of Example 7.1 when it is delivering maximum power to a matched load. (*Hint:* Set $d(IV)/dV = 0$.)

(b) If the fill factor for this cell is 0.8, determine the maximum power delivered to a matched load at 300°K.

7.4 Determine the energy gap in eV for an LED which emits 1.3-μm radiation.

7.5 Starting with Eqs. (7.9a) and (7.10), derive Eq. (7.9b).

7.6 **(a)** Calculate the junction depletion-layer capacitance at a reverse bias of -1.0 V of an abrupt p^+-n silicon junction diode 1.0×10^{-4} cm^2 in area, whose n-region resistivity is 0.10 Ω-cm. The p^+-region resistivity is 0.01 Ω-cm.
(b) How much charge is stored on the n-side of the space-charge region in (a) and what is the nature of this charge?
(c) Repeat (a) and (b) at a reverse applied voltage of -10 V.
(d) If the n-region is only 5.0×10^{-4} cm thick, calculate the maximum operating frequency of the diode of (a) used in a varactor diode application.
(e) Determine the effect on the frequency cutoff of the diode in (d) if the doping in the n-region is reduced so that the resistivity there is 1.0 Ω-cm.

7.7 For a hyperabrupt p-n junction with a barrier height of 0.8 V the capacitance variation with voltage is given by Eq. (7.9b) except that the exponent is 2 instead of ½. It is desired to use this diode in a varactor circuit to tune the entire AM radio band from 550 to 1600 kHz. The diode is reverse biased with -1.0 V at 550 kHz. What is the voltage required to tune the circuit for 1600 kHz?

7.8 **(a)** Starting with Eq. (6.36) for the minority hole distribution in the n-region of a short p^+-n diode, derive Eq. (7.18b).
(b) Calculate the diffusion capacitance at 300°K for the diode in part (a) of Example 7.3 with a 50-μm-wide n-region. Compare this result with that of Example 7.3.

7.9 Explain the approximation made to obtain Eq. (7.26).

7.10 Show that the expression of Eq. (7.24) is a solution of Eq. (7.23).

7.11 Derive Eq. (7.21b) by calculating the stored charge for the short diode and using Eq. (6.37).

7.12 A p^+-n diode like that shown in Fig. 7.7 is used in a switching application in a circuit similar to that shown in Fig. 7.9. The minority hole lifetime in the n-region of this diode is determined by the gold doping there of 10^{15} gold atoms/cm^3. The minority electron lifetime in the p^+-region is very low compared to that in the n-region owing to the high impurity content there.
(a) Why is the switching speed of this diode limited by the minority hole lifetime in the n-region?
(b) If a forward current of 10 mA is initially flowing through the diode, determine the excess hole charges stored in the n-region of the diode.
(c) If the diode is suddenly reverse biased so that a reverse current of 10 mA is drawn out of the diode, compute the time necessary to reduce the excess stored hole charge to a negligible amount.
(d) Repeat (c) for 100 mA of reverse current extracted.
(e) Compute approximately the maximum switching rate possible with the diode operated as in (c).

7.13 Using Eq. (7.21b) for the short diode, derive an expression for the turnoff time given by Eq. (7.28).

7.14 Repeat Example 7.4, parts (a) and (b), for a short diode with n-region thickness 10 μm and hole lifetime 0.6 μsec.

7.15 Calculate the maximum operating frequency for a Schottky diode whose capacitance is 2.0 pF and series resistance 1.0 Ω.

7.16 A long p^+-n junction diode biased in the forward direction is conducting a steady-state current of 100 mA. A reverse bias of 50 V is suddenly applied to the diode through a 5.0-kΩ resistor. The avalanche breakdown voltage of the diode is 100 V, it has a saturation current of 0.10 nA, and the minority carrier lifetime in the n-region of the diode is 1.0 μsec.

(a) What is the current through the diode just after switching?
(b) What is the current in the diode 1 msec after switching?
(c) How long does it take for the diode current to reach an essentially steady-state value in this reverse direction? (Assume that the fall time is small compared to the storage time.)

7.17 Determine the barrier width near a metal–semiconductor contact in a semiconductor doped with 10^{20} impurities/cm^3. The barrier height from semiconductor to metal is 0.8 V.

7.18 Determine the current density in an Al/n-Si Schottky diode at 0.4 V forward bias at 300°K. Assume a Richardson constant of 264 A/cm^2-°K^2.

7.19 A Schottky barrier is formed by depositing a thin film of gold onto a surface of clean n-type 1 Ω-cm silicon. If the work function of gold is 4.8 eV and that of silicon is 4.2 eV, calculate the height of the potential barrier preventing the electrons in the conduction band of the semiconductor from entering the metal under equilibrium conditions.

7.20 Draw the energy band diagram for a metal–p-type semiconductor contact in thermal equilibrium, starting with Fig. 7.13a.

7.21 Find the minimum energy gap width for a heterojunction "window" in a solar cell so that ultraviolet photons of wavelength 0.4 μm are transmitted to the substrate to produce electron–hole pairs there.

CHAPTER

8

The Bipolar Junction Transistor

8.1 INTRODUCTION TO THE BIPOLAR JUNCTION TRANSISTOR (BJT)

The analysis and modeling of the *p-n* junction diode provide a background for a discussion of the physics and modeling of the **bipolar semiconductor junction transistor.** The transistor device consists of two back-to-back junctions in series, connected by a thin region of semiconductor material. Although the individual diodes are passive devices, when they are coupled by a pure, single-crystal, and nearly structurally perfect thin semiconductor region this device becomes active and exhibits good power gain. The amplification results from the nearly total transfer of minority current carriers from the emitter junction to the collector junction with the former forward biased (low impedance) and the latter reverse biased (high impedance). This is the source of power gain when the device is operating in the active region.

In the following sections the physics of operation of the bipolar junction transistor as an electronic device will be developed; first DC or low frequency operation will be described and later some high frequency limitations will be considered. The device will be modeled to permit the analysis of electronic circuits utilizing bipolar transistors. **Computer-aided** device design and circuit analysis will be discussed. The application of the BJT as an amplifying element and as an electronic switch will also be presented. A description of field-effect transistors will be reserved for the next chapter.

8.1.1 The Bipolar Junction Transistor Structure

The *n-p-n* **bipolar junction** transistor contains two *n*-type regions of semiconductor crystal coupled to a central *p*-region; a *p-n-p* junction transistor contains two *p*-type regions coupled by an *n*-region. In fact, all three regions are part of a single crystal of semiconductor material. Since the description of current flow in such a device involves the motion of both holes and electrons, it is referred to as a **bipolar** transistor. If all three regions are included in a single semiconductor material it is known as a **homojunction** transistor.

Figure 8.1 shows schematically a model of an *n-p-n* bipolar junction transistor as invented by Shockley. The bar of silicon pictured has two *p-n* junctions in a single crystal of silicon, one on the left- and one on the right-hand side of the bar. The left-hand junction will be called the **emitter** junction, while the one to the right will be called the **collector** junction. Since the device picture as shown is symmetric, either side can be called the emitter or collector. This, in fact, indicates the physical construction of the junction transistor, which was fabricated from the semiconductor element silicon in the early years of transistor development. However, more recently the **planar** transistor has been adopted as the industry standard for individual or discrete devices and is used as well in all bipolar microchip integrated circuits. A sketch of a discrete bipolar *n-p-n* planar junction transistor is shown in Fig. 8.2a. In this structure the collector junction is larger in area than the emitter junction, primarily because of the method used to fabricate this device. Also, in general, the impurity con-

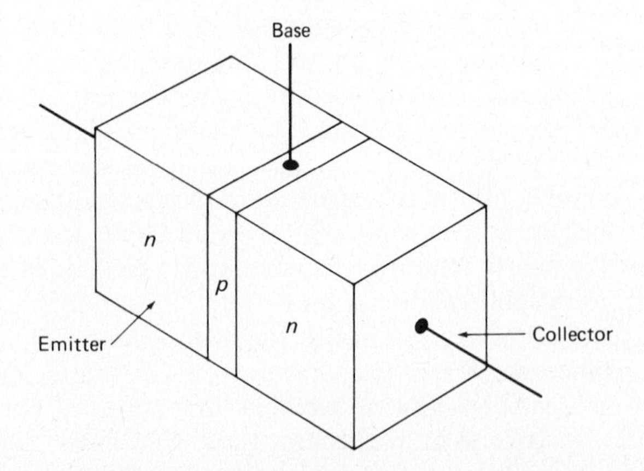

Figure 8.1 Schematic diagram of an *n-p-n* bipolar junction transistor. The actual device is constructed of a bar of single-crystal semiconductor containing the emitter, base, and collector regions.

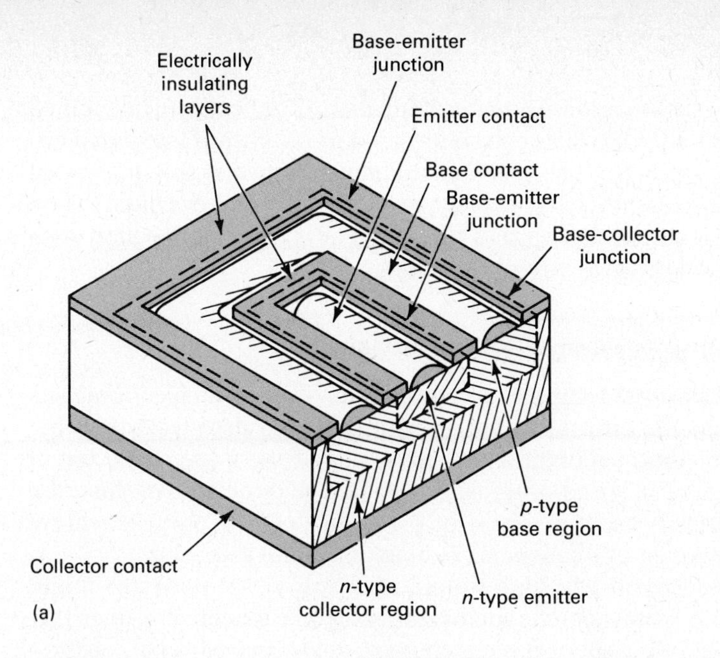

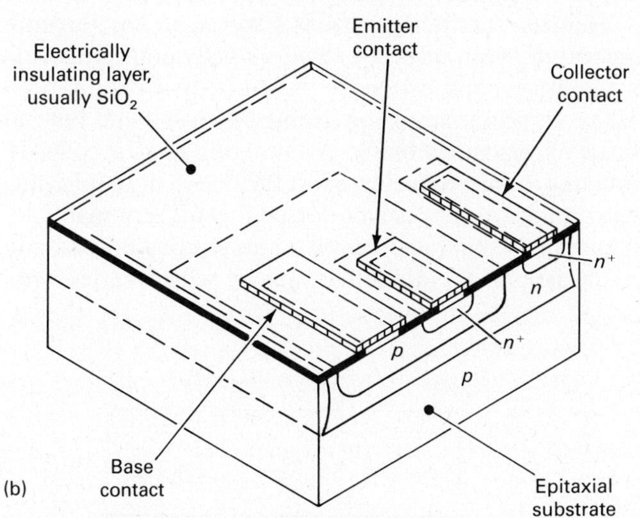

Figure 8.2 (a) Schematic drawing and cross section of a discrete bipolar *n-p-n* planar junction transistor. The emitter, base, and collector contacts are normally metallic (aluminum or gold) and are alloyed to the semiconductor body. The emitter, base, and collector regions are normally silicon, of a *p*- or *n*-variety. This structure is the simplest type used for discrete or individual transistors, such as those used in power circuits. Such devices are protected from contamination from the surrounding environment by a coating of electrical insulator such as silicon dioxide or glass.
(b) Schematic drawing and cross section (not to scale) of a typical bipolar *n-p-n* microchip integrated circuit transistor. The emitter, base, and collector contacts are aluminum or a refractory metal and are sintered to the silicon crystal. The transistor device is isolated from the *p*-type epitaxial substrate and other devices on the substrate chip by a high impedance *p-n* junction. The thick epitaxial *p*-substrate acts as a foundation material onto which the device is fabricated.

centration in the emitter region is greater than that in the collector region. A sketch of a bipolar planar transistor structure used in microchip integrated circuits is shown in Fig. 8.2b. The primary difference between this integrated structure and the discrete device is that the collector terminal or contact is on the upper surface of the silicon chip. This permits easy interconnection with other devices in the integrated circuit (see Chapter 11).

8.1.2 Bipolar Transistor Amplifier Power Gain

The transistor is intrinsically capable of power gain. That is, an electronic signal when applied to the emitter (input) junction can be detected somewhat later at the collector (output) junction, as an analog of the input signal but at a higher power level. The speed of transmission can be quite fast, of the order of 10^{-9} sec or less. This rapid response of the device can be used to achieve signal amplification even at frequencies as high as several gigahertz.

In order to understand why this simple structure yields power gain and why it is constructed from semiconductor materials like silicon, consider that the transistor device is **biased,** or has electric potentials applied to it by batteries as shown in Fig. 8.3. Here the emitter junction is biased in the forward direction (*n*-type region, negative, with respect to the *p*-type region, positive); the collector junction on the other hand is reverse biased (*n*-type region, positive, with respect to the *p*-type region, negative). Hence the input and output sections of the device consist of oppositely biased *p-n* junction diodes, coupled by a thin *p*-type base region and is said to be biased in the active or amplifying region. These diodes exhibit nonlinear *I-V* behavior (see Fig. 6.1) compared to a linear resistor since the latter's *I-V* characteristic is a straight line of constant slope, while the *I-V* characteristic of the diode has a slope which varies with

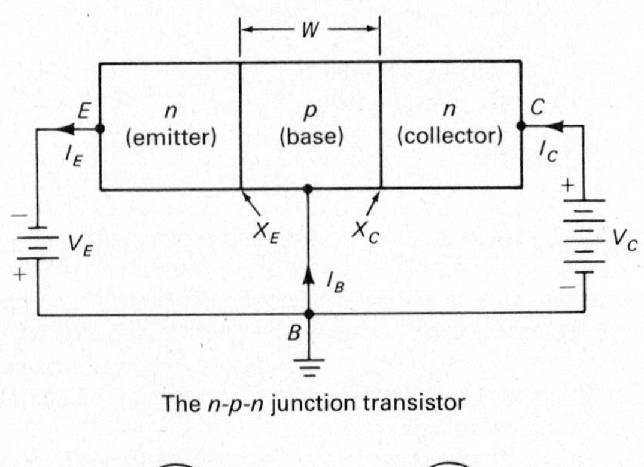

Figure 8.3 Electrical biasing of an *n-p-n* transistor. The emitter–base junction X_E is forward biased since the *p*-type base is positive relative to the negative *n*-type base. The collector–base junction X_C is reverse biased since the *n*-type collector is positive relative to the negative *p*-type base. The direction of current flow in the emitter, base, and collector leads is specified. The base-width thickness *W* must be narrow for good power amplification. Also shown are circuit symbols for an *n-p-n* transistor and a *p-n-p* transistor.

current as well as voltage polarity. The value of the slope (**dynamic conductance** dI/dV) of the diode characteristic is generally much less in the reverse ($-I$, $-V$) direction than in the forward direction ($+I, +V$). (In fact, the slopes in these two regions can differ by as many as 12 orders of magnitude.) It is this difference in input and output diode conductance that provides a possibility for power gain. This nonlinear type of diode I-V behavior is typical of semiconductor p-n junctions.

Figure 8.4 shows a simple transistor amplifying circuit with a small AC input voltage signal superimposed on the DC voltage bias V_B which is applied in the emitter circuit by battery; the output voltage is detected by the potential difference change, induced by the input signal, across the output resistor R_{out}. Consider an incremental increase ΔV_{in} in the voltage supplied to the transistor emitter by the input signal. Because of the steep slope of the emitter diode I-V curve, this small change in applied voltage gives rise to a large change in emitter electron current, ΔI_E. Since the input and output circuits are coupled by **transistor action** (to be explained later) this emitter current of electrons will be transmitted nearly undiminished through the p-type base region to the collector region, which is positively charged to attract electrons. Hence a ΔI_E increase in emitter current will result in a ΔI_C rise in collector electron current.

The collector junction acts like a current source because of the high resistance nature of the reverse-biased collector p-n junction. The output resistance R_{out} can be within an order of magnitude of this high collector resistance without limiting the collector current rise ΔI_C. The voltage gain of this device is then

$$V_{gain} = \frac{\Delta V_{out}}{\Delta V_{in}} = \frac{\Delta I_C R_{out}}{\Delta I_E r_e}, \tag{8.1a}$$

where r_e is the incremental or dynamic resistance (reciprocal of the dynamic conductance) of the emitter input junction.

In a well-designed transistor the coupling between the emitter and collector circuits is improved by making the p-type base region (W in Fig. 8.3) very narrow, of the order of a fraction of a micrometer. Under this condition ΔI_C is

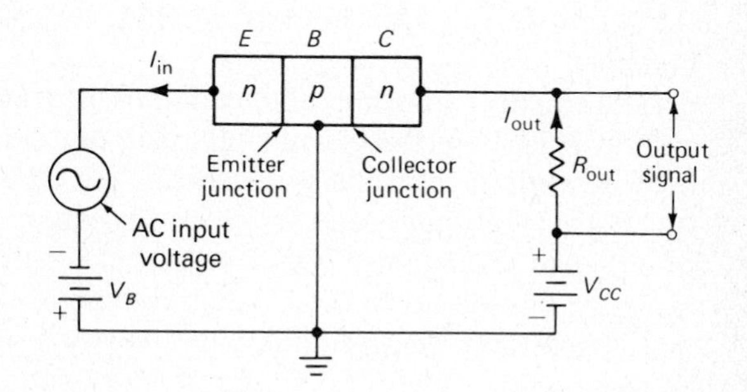

Figure 8.4 Simple transistor circuit for demonstrating power gain. V_B is a battery used to establish proper emitter DC bias. The collector bias V_{CC} is such that the collector resistance is much greater than R_{out}. In this **grounded-base connection** the power gain is $(\Delta I_C/\Delta I_E)^2(R_{out}/r_{in})$, where normally $\Delta I_C \approx \Delta I_E$ and $R_{out} \gg r_{in}$.

A variety of commercial bipolar transistors ranging from small-signal devices in plastic packages (lower section) to power devices in metal packages (upper section). (Photo supplied by the Unitrode Corporation, Lexington, MA.)

only a few percent less than ΔI_E. Hence the above equation indicates substantial voltage gain since $R_{out}/r_e \gg 1$. In a similar manner the power gain of this device can be written as

$$P_{gain} = \frac{\Delta P_{out}}{\Delta P_{in}} = \frac{(\Delta I_C)^2 R_{out}}{(\Delta I_E)^2 r_e}\,. \tag{8.1b}$$

Again the power gain can be substantial because $R_{out} \gg r_e$. Hence the gain of a transistor results from the facts that

1. the dynamic resistance of the reverse-biased collector junction is much greater than that of the forward-biased emitter junction; and
2. electron current can be transmitted from the emitter to the collector through the base region practically undiminished in magnitude owing to the high value of the minority carrier lifetime in the base region.

Hence the transistor acts as an amplifier, in that the output signal power exceeds that of the input signal power. Conservation of energy requires that the source of this gain be derived from the chemical energy supplied by the batteries.

An equally important application of the transistor is its use as an electronic switch. This is especially true in the case of modern digital computers, in which literally millions of transistor switches are used. The computational processes in a digital computer are carried out by circuit elements which can be placed in either of two states, "ON" or "OFF." The ON state, in analogy with an ordinary household light switch, corresponds to the joining together of two points in an electronic circuit with a very low resistance, causing a substantial current to flow. The OFF state refers to opening the circuit or placing a very high resistance between two points in an electronic circuit, practically cutting off current flow. In the transistor switch, these two terminals are normally the emitter and collector terminals. A small voltage signal applied to the third terminal or base lead can cause the normally nonconducting transistor to convert to the conducting state.

The fact that the transistor can easily be put in either of two states, ON or OFF, quite separate from each other, accounts for the device's significant use in digital computers. Digital computations are carried out in the binary system with only two numbers, 0 or 1. The transistor in the ON or "high" state can represent the number 1, while the OFF or "low" state corresponds to 0. Since the transistor can be switched from one state to the other in less than 10^{-9} sec, more than one billion computations can be carried out in a second. This accounts for the enormous computational speed of the modern digital calculator.

8.1.3 Current Flow in an *n-p-n* Transistor

The *n-p-n* transistor is the most common type of bipolar transistor used in microchip integrated circuits today. An analysis of the operation of this device involves following the electron charges injected by the emitter in their transport through the base region to the collector. The *n-p-n* transistor description is identical to that of the *p-n-p* if the roles played by holes and electrons are reversed. A schematic diagram for this device is given in Fig. 8.5 with currents and voltages specified in a conventional manner. Figure 8.6 shows the circuit symbol and typical bias arrangements for the transistor as an amplifier. Note that the directions of current and junction potential biases normally assumed are independent of whether the device is *n-p-n* or *p-n-p*. The transistor has

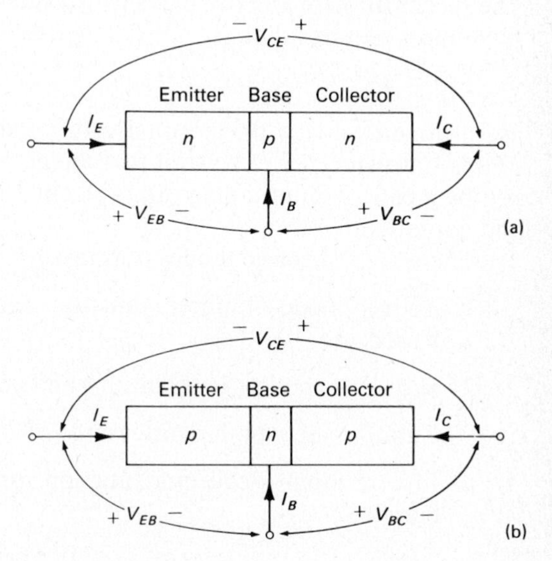

Figure 8.5 (a) Schematic diagram of an *n-p-n* transistor with externally applied voltages and currents defined. Note the order of the double-letter subscripts. The first represents the electrode which is more positive (higher) in potential relative to the second. For example, V_{CE} is the collector-to-emitter potential and is positive if the collector is positive (higher) in potential relative to the emitter. If the emitter electrode is more positive than the collector, V_{CE} is then a negative quantity ($= -V_{EC}$). (b) As (a), except the transistor is *p-n-p*.

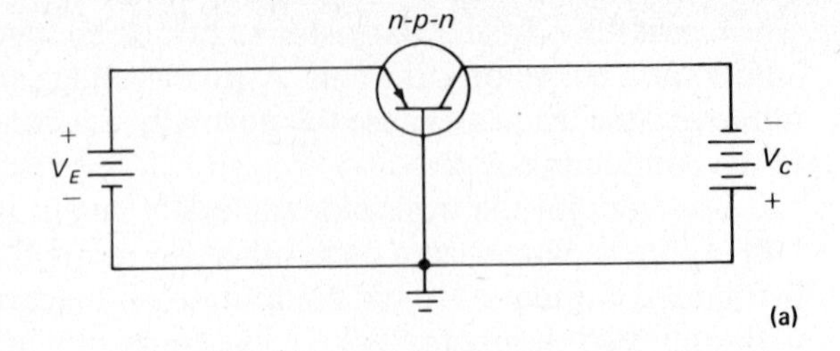

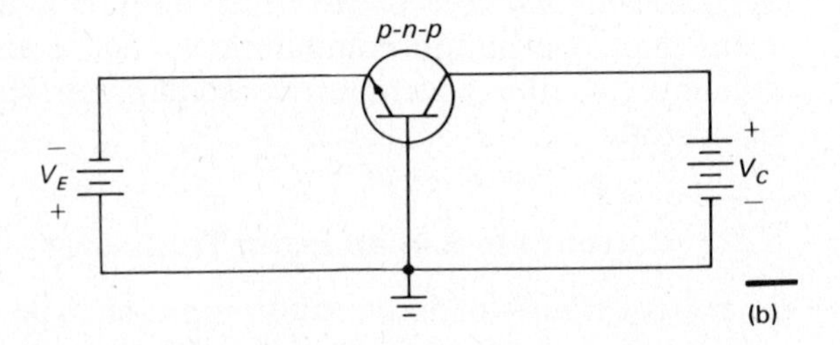

Figure 8.6 (a) Circuit symbols and typical voltage bias arrangement for an *n-p-n* transistor biased in the active region, in the common-base connection. (b) As (a) except the transistor is *p-n-p*.

three segments with three lead wires connected to the emitter, base, and collector regions. The currents conducted in each of these three leads are called the emitter, base, and collector currents and for purposes of analysis are conventionally assumed to flow *into* the device. Since one of Kirchhoff's laws requires that

$$I_E + I_B + I_C = 0, \tag{8.2a}$$

this indicates that only two of the three currents are independent of each other. The specification of any two currents fixes the third. The Kirchhoff law relating to voltages requires that

$$V_{EB} + V_{BC} + V_{CE} = 0, \tag{8.2b}$$

and again only two of the terminal voltages are independent. In later modeling of this transistor, it will be seen that the device operation is completely specified by a pair of input voltage and current values and a pair of output voltage and current values.

There are four basic modes of transistor operation:

1. Active mode—emitter junction forward biased, collector reverse biased.
2. Saturated mode—both emitter and collector junctions forward biased.
3. Cutoff mode—both emitter and collector junctions reverse biased.
4. Inverse mode—collector junction forward biased and emitter reverse biased.

The active mode is of primary interest when the transistor is utilized as an amplifier. The operation of the saturated and cutoff modes is of interest when the transistor is used as an electronic switch. In certain special applications the inverse mode is of interest, although it should be pointed out that the homojunction bipolar transistor is a completely symmetric device in principle. Hence the labeling of one junction as emitter and the other as collector is arbitrary if the device is designed symmetrically. Practically, however, the emitter region is most often more heavily doped with impurities than the collector region, and the collector junction cross sectional area is normally greater than the emitter junction area (see Fig. 8.2). From this viewpoint the transistor discussed here should properly be labeled an n^+-p-n transistor, with the emitter most heavily doped. This describes a device which may be fabricated by impurity diffusion or ion implantation techniques, both of which are used in microchip technology.[1] Here the base region is automatically more heavily doped than the collector region. Certain discrete devices are of an "epitaxial" variety where the collector region can be as heavily doped as the emitter. In this case the junction with larger cross sectional area is termed the collector.

Let us describe the bipolar transistor from the quantum mechanical energy band viewpoint. First consider three *separated* sections of a semiconductor crystal material, the first heavily doped n-type, the second part p-type, and the last section n-type, as shown in Fig. 8.7a. Now let us perform the mental experiment of bringing these three parts together. Majority holes from the p-section will spill over into the n-regions where they are few in number, and correspondingly the majority electrons in the n-sections will flow into the p-region, where electrons are in the minority. This charge transfer will result in the creation of two separate p-n junction space-charge regions containing ionized acceptors and donors and hence electric fields therein which tend to limit the charge flow. This identical process is described for an individual p-n junction in Section 6.1.2. Figure 8.7b shows the energy band diagram for this composite n^+-p-n structure after equilibrium has been established and the corresponding Fermi levels of the three sections are lined up. In equilibrium minority holes from the n^+-region can flow downhill (for holes) into the p-region but of course an equal number of majority holes from the p-region will mount the n^+-p potential barrier, so the net hole flow is indeed zero. A corresponding description applies for the detailed balance of the electron charges flowing across this junction and for electrons and holes flowing across the collector junction. Finally, Fig. 8.7c shows the energy band diagram for the n^+-p-n transistor biased in the active or amplifying mode as indicated in Fig. 8.6a. Note that the n^+-p emitter barrier is reduced in height by the forward bias, inducing a *net* flow of injected electrons into the p-type base region. Note too that the collec-

[1]The technology and processing for constructing these devices will be described in Chapter 11. Newer fabrication techniques utilizing vapor-phase deposition can yield quite different, useful transistor structures.

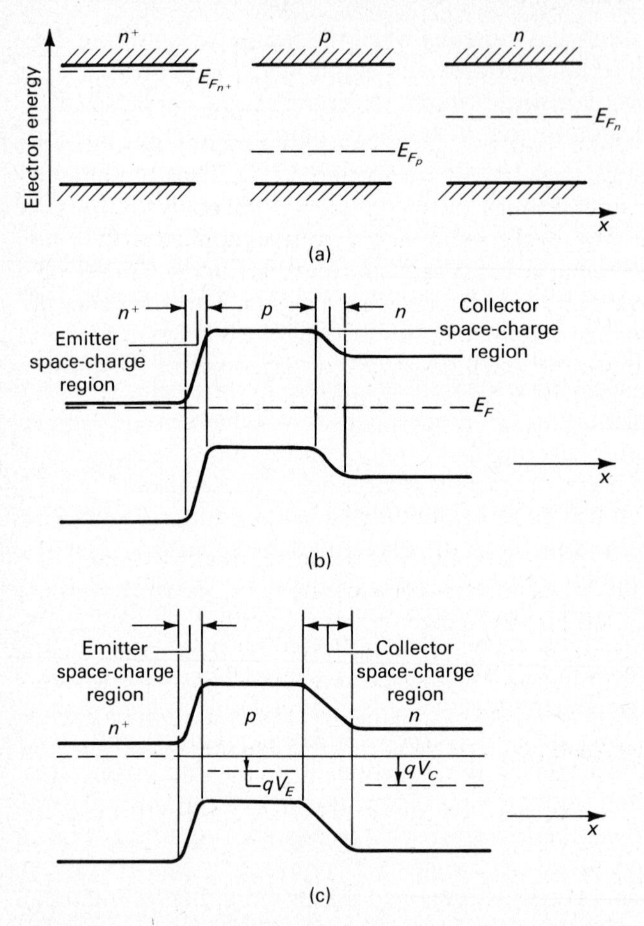

Figure 8.7 (a) Energy band diagram for three separate pieces of a semiconductor crystal material doped n^+, p, and n.
(b) Energy band diagram for an n^+-p-n transistor in thermal equilibrium. The Fermi energy E_F is constant throughout the crystal.
(c) Energy band diagram for an n^+-p-n transistor biased in the active mode. The emitter is forward biased with a voltage V_E and the collector is reverse biased with a voltage V_C. The emitter space-charge region is thinner and the collector space-charge region is wider than in the corresponding equilibrium case.

tor p-n barrier is substantially increased in height as a result of the reverse biasing of the collector junction. This barrier prevents holes from entering the n-type collector region but is very effective in collecting electrons from the p-type base region, which have mainly been injected into this region from the emitter. The efficient (without recombination) transfer of electrons from the emitting junction to the collecting junction is the basis of the transistor current gain as discussed in Section 8.1.2.

An analytic device description will now be pursued for an n^+-p-n transistor operating in the active mode under small signal, low current conditions. In this type of device and under these conditions, minority carrier flow is of primary importance. The analysis procedure will be to define the minority carrier concentrations at the edges of the emitter and collector *p-n* junction space-charge regions in terms of the applied voltages (see Section 6.2.2). Then the emitter-injected minority carrier flow through the base region to the collector junction will be considered. Under low level conditions the voltages applied to the emitter and collector junctions appear essentially across these junctions and very little electric field penetrates into the transistor base region. This simplifies the analysis since the minority carrier current flow through the base layer will now have only a diffusion component; the field-dependent drift current component is taken as negligible (see Section 6.2.1 and Appendix D).

8.1.4 Minority Carrier Flow in the Transistor

The junction transistor may be looked upon as a control device in which the potential applied across the input emitter–base junction, V_{EB}, controls the current I_C through the output collector junction. When the emitter junction is forward biased, excess carriers are produced at the edges of the emitter space-charge region [see Eq. (6.30)]. In the description of the p^+-n diode, the number of excess holes injected into the *n*-side of the junction is taken as much greater than the number of excess minority electrons injected into the p^+-side of the junction, this number being dependent on the minority carrier concentration in each respective region. In this case the minority carrier current through the n^+-p junction is composed primarily of electrons. Also here the *p*-region base of the transistor is assumed to be very thin, for good gain. This corresponds more exactly to the short diode treatment of Section 6.2.3. Specifically, the base region width W is taken as much less than the minority electron diffusion length L_n in the base. Consider this n^+-p-n transistor operating in the **active** mode with the emitter junction forward biased and the collector junction reverse biased. Under low level static or slowly varying (quasistatic) current conditions, the excess electron density at the base region edge of the emitter space-charge region, $x = 0$ (see Fig. 8.8), is given in analogy with Eq. (6.30) by

$$\Delta n_E(0) = n_{p0}(e^{qV_{BE}/kT} - 1) \approx n_{p0}e^{qV_{BE}/kT}, \tag{8.3}$$

where it is assumed that $V_{BE} \gg kT/q$, so that $e^{qV_{BE}/kT} \gg 1$. Here n_{p0} represents the minority electron density in the neutral *p*-type base region at equilibrium.

At the base region edge of the collector space-charge region, $x = W$, the excess electron density is

$$\Delta n_C(W) = n_{p0}(e^{qV_{BC}/kT} - 1) \approx -n_{p0}, \tag{8.4}$$

since here too it is assumed that $|V_{BC}| \gg kT/q$, but V_{BC} is negative. The negative sign indicates an excess electron concentration below the equilibrium value or a *total* electron concentration there of approximately zero. Since n_{p0}

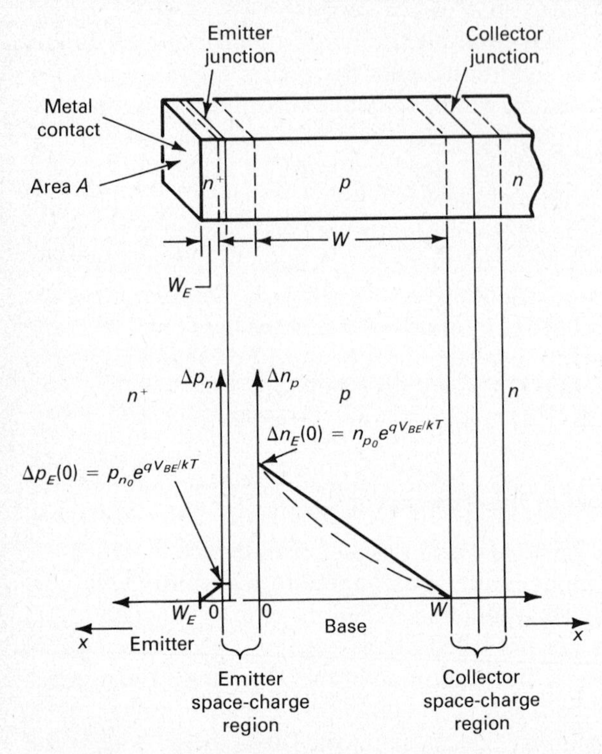

Figure 8.8 Linear distribution of minority electrons in the base region and holes injected into the emitter region of a typical n^+-p-n transistor operating in the active region, where the base current due to minority carrier recombination is negligible (solid line). Note that electron injection through the emitter–base junction predominates over holes owing to the much heavier n-doping of the emitter region, since $p_{no} \ll n_{po}$ [see Eq. (6.30)]. The dotted line indicates the distribution of electrons in the base region when the base current is not negligible.

is normally quite small, especially in silicon transistors, the excess electron density at the collector can be taken as essentially zero.

Hence to a good approximation the excess density of electrons in the base region varies from $n_{po}\, e^{qV_{BE}/kT}$ at the emitter to zero at the collector for a device biased in the active mode. To a first approximation this variation is linear, as shown in Fig. 8.8, and is more nearly so as the base current of holes required for minority electron recombination becomes more negligible compared with the emitter or collector currents. That this is a valid approximation will now be shown. The current through the base region, from emitter to collector, is determined essentially by an electron diffusion current which is constant throughout the base region since there is negligible electron recombination there; supporting this is a constant carrier gradient equal to $\Delta n_E(0)/W$. Then the magnitude of the emitter current essentially must equal the collector current, which is given in analogy with Eq. (6.22) as

$$I_E \approx -I_C = qAD_n \frac{d\,\Delta n_p}{dx} \approx -qAD_n \frac{\Delta n_E(0)}{W}. \tag{8.5}$$

Equation (8.2a) indicates that this is only strictly true if the base current due to recombination is negligible compared with I_E or I_C. Good current gain transistors have relatively small base currents and hence this is the case in most practical devices. Combining Eqs. (8.3) and (8.5) yields an expression for the collector current under the condition of negligible base recombination current:

$$I_C = \frac{qAD_n n_{p0}}{W}(e^{qV_{BE}/kT} - 1) \approx \frac{qAD_n n_{p0} e^{qV_{BE}/kT}}{W}, \tag{8.6}$$

where A represents both the emitter and collector cross sectional areas. In case electron recombination in the base is not negligible, the magnitude of the electron gradient at the emitter junction will be greater than the gradient at the collector junction, and hence the collector current will be less than the emitter current by the amount of the base current. This is indicated by the dotted line in Fig. 8.8.

A. BASE RECOMBINATION CURRENT

Although the base current normally is small, it nevertheless plays an essential role in transistor operation. For example, in the **grounded-emitter** circuit connection, an input current supplied to the transistor base lead is amplified by the transistor and appears as the output collector current. Physically, one function of the base current is to supply a majority hole for each minority electron which is lost by recombination in transit from emitter to collector. Hence the base lead must supply a current of majority holes equivalent to the minority carrier **recombination current.** In the steady state this represents the hole flow that must be supplied in order to maintain the linear minority carrier distribution as indicated in Fig. 8.8. This is for reasons of space-charge neutrality normally assumed. In order to maintain this charge distribution, holes must be supplied at a rate given in analogy with Eq. (7.21a) by

$$I_B = -Q_F/\tau_n, \tag{8.7}$$

where Q_F represents the steady-state *total* charge of excess electrons in the base region and τ_n is the time in which this charge tends to disappear (τ_n has previously been defined as the minority electron lifetime). An approximate expression for Q_F can be obtained by integrating the excess charge density distribution indicated in Fig. 8.8. By estimating the area under this curve one obtains, in the base region,

$$Q_F = -\tfrac{1}{2}qAW\,\Delta n_E(0), \tag{8.8}$$

and hence the base current that must be supplied is, from Eq. (8.7),

$$I_B = qAW\,\Delta n_E(0)/2\tau_n. \tag{8.9}$$

If in fact this base current is small compared to the current of Eq. (8.5), this would require that

$$\left|\frac{I_B}{I_C}\right| = \frac{qAW\,\Delta n_E(0)/2\tau_n}{qAD_n\,\Delta n_E(0)/W} \ll 1, \tag{8.10a}$$

or

$$\frac{W^2}{2D_n\tau_n} = \frac{W^2}{2L_n^2} \ll 1. \tag{8.10b}$$

Hence good current gain and a linear minority electron distribution in the base region require a very thin base region W compared with the minority carrier diffusion length L_n. In a typical silicon *n-p-n* transistor, as a result of the low electron recombination rate in the base region $W^2/2L_n^2 \ll 1$. The role played here by the base current in supplying charge to operate the device is part of a more complete **charge-control** model of the transistor to be discussed later.

B. EMITTER INJECTION EFFICIENCY

Actually the base current must also compensate for another source of inefficiency in the transfer of current from the emitter to the collector. This results from the fact that the n^+-p emitter current is not entirely made up of electron current: in practical devices about 1–5% of hole current is present.[2] This hole current is supplied through the base lead and is **injected back** through the emitter junction, representing a deficiency in emitter injection. To calculate this hole component of the emitter current a procedure identical to that used in calculating the electron current can be used, assuming a very thin emitter width W_E. Neglecting recombination we assume a linear distribution of holes in the n^+-type emitter region, as indicated in Fig. 8.8.[3] This **defect** current which is supplied through the base lead then becomes

$$I_{E\text{hole}} = I'_B \approx -\frac{qAD_p\,\Delta p_E(0)}{W_E} = -\frac{qAD_p p_{n0}(e^{qV_{BE}/kT}-1)}{W_E}, \tag{8.11}$$

in analogy with Eq. (8.5). Here $\Delta p_E(0)$ represents the density of injected holes at the n^+-edge of the emitter space-charge region. One can define an emitter **injection efficiency** γ, representing the ratio of the magnitude of the electron current injected by the emitter n^+-p junction to the total emitter current flowing, as

$$\gamma = \frac{I_{E\text{electron}}}{I_{E\text{electron}} + I_{E\text{hole}}}. \tag{8.12}$$

[2]No hole component of the emitter current may contribute to the collector current since the reverse-biased collector potential barrier rejects holes and collects electrons.

[3]The excess hole density at the metal emitter contact is taken as zero since an ohmic contact is considered to be a region of infinite minority carrier recombination.

With Eqs. (8.6) and (8.11) this gives

$$\gamma = \frac{1}{1 + D_p p_{n0} W / D_n n_{p0} W_E}. \tag{8.13}$$

Multiplying the second term in the denominator of Eq. (8.13) by $n_{n0}p_{p0}/n_{n0}p_{p0}$, noting that $n_{p0}p_{p0} = p_{n0}n_{n0} = n_i^2$, and using $D_n/D_p = \mu_n/\mu_p$ from the Einstein relation gives

$$\gamma = \frac{1}{1 + \sigma_p W / \sigma_n W_E}. \tag{8.14a}$$

Here σ_n and σ_p are correspondingly the conductivity of the emitter and base regions. Assuming that the base width W and the emitter width W_E are comparable in magnitude and that $\sigma_p/\sigma_n \ll 1$, we may approximate the injection efficiency from Eq. (8.14a) as

$$\gamma \approx 1 - \sigma_p W / \sigma_n W_E, \tag{8.14b}$$

and so the electron injection efficiency of an n^+-p junction emitter can be better than 0.99, or 99%.

C. EMITTER HEAVY DOPING EFFECTS AND BAND GAP NARROWING

Experimental results on bipolar transistors have indicated that the injection efficiency expression of Eq. (8.14a) gives too optimistic a prediction. The explanation for this result is attributed to a heavy doping effect in the emitter region called **band gap narrowing.**[4] The model considers the electrostatic energy of a minority carrier surrounded by a dense cloud of majority carrier charges (from impurities) that are attracted to it. This interaction screens the minority carrier and reduces the activation energy necessary to separate an electron from a hole and hence effectively acts like an energy gap reduction. The greater the concentration of majority carriers (the heavier the doping) in the semiconductor, the greater is the reduction in energy gap. For example, measurements show that the decrease in energy gap in n-type silicon is 7 meV for the case of $5 \times 10^{17}/\text{cm}^3$ donor doping and 159 meV for a $5 \times 10^{19}/\text{cm}^3$ doping concentration. An expression for calculating the band gap narrowing ΔE_g in this range in meV is[4]

$$\Delta E_g = 22.5(n/10^{18})^{1/2}, \tag{8.15}$$

where n is the density of majority carriers. The new effective intrinsic carrier density due to the reduced band gap now becomes, from Eq. (4.29),

$$\begin{aligned} n_{i\text{eff}}^2 &= N_c N_v e^{-(E_g - \Delta E_g)/kT} \\ &= n_i^2 e^{\Delta E_g/kT}. \end{aligned} \tag{8.16}$$

[4]H. P. D. Lanyon and R. A. Tuft, "Bandgap Narrowing in Heavily Doped Silicon," *IEEE Trans. Electron Devices,* **ED-26,** 1014 (1979).

The electron–hole density product in the emitter region under equilibrium conditions using Eq. (4.29) then becomes

$$n_{n0}p_{n0} = n_{i\text{eff}}^2 = n_i^2 e^{\Delta E_g/kT} \tag{8.17a}$$

or

$$p_{n0} = (n_i^2 e^{\Delta E_g/kT})/N_d, \tag{8.17b}$$

where $n_{n0} = N_d$ is the emitter donor doping density. Heavy emitter doping alters the band gap by ΔE_g, and hence increases the minority hole concentration there as indicated in Eq. (8.17b). Hence the emitter injection efficiency is reduced from the value predicted by Eq. (8.13), in which this effect is not included, and explains the discrepancy noted previously in the injection efficiency estimation. Equation (8.13) may be corrected to account for band gap narrowing by the introduction of an effective emitter doping

$$N_{d\text{eff}} = N_d(n_i^2/n_{i\text{eff}}^2). \tag{8.18}$$

*D. THE WIDE GAP EMITTER, HETEROJUNCTION BIPOLAR TRANSISTOR

The requirement for heavy emitter doping relative to the base region in order to obtain good emitter injection efficiency and hence transistor current gain can be relaxed with the aid of the use of a wide gap semiconductor material for the emitter and a smaller energy gap material for the base region. As was previously seen in the discussion of the heterojunction *p-n* diode in Section 7.6, a wide gap semiconductor *p*-region grown onto a narrower gap *n*-region can result in the presence of an increased potential barrier in the conduction band which tends to prevent electrons in the *n*-region from entering into the *p*-region (see Fig. 7.17b). Hence the current through this heterojunction tends to be exclusively hole current and is not sensitive to the doping in the *p*-region. Similar arguments can be given for a wide gap *n*-region grown onto a smaller gap *p*-region. This will result in an *n-p* junction with a strong dominance of electron current and can be used to form the emitter junction of an *n-p-n* transistor with high electron injection efficiency (i.e., holes from the base region will be blocked from entering the emitter). Since a lightly doped *p*-type base region (relative to the emitter region) is no longer necessary to yield good injection efficiency of electrons, a heavier doping there will decrease the transistor series base resistance. This resistance can reduce the bipolar transistor power gain, particularly at high frequencies, a fact which will be discussed later.

E. OTHER SOURCES OF EMITTER DEFECT CURRENT

There are two other sources of emitter defect current which must be supplied through the transistor base lead. These are particularly of consequence at low levels of emitter current and in fact tend to limit the current gain of a silicon transistor as the current is reduced to very low values. The first of these results from carrier generation and recombination in the emitter space-charge region. This effect was discussed in some detail for the *p-n* junction diode in Section

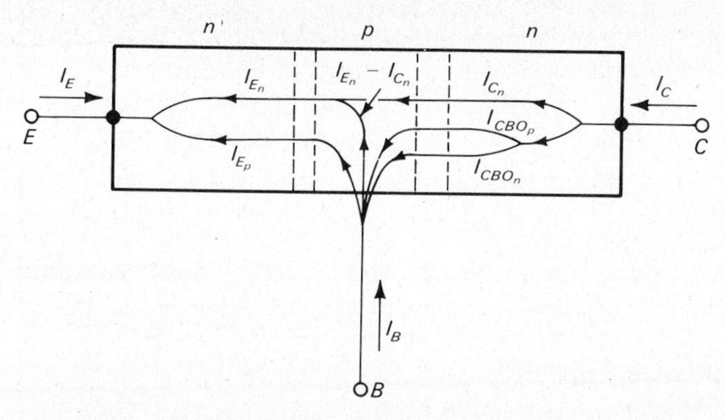

Figure 8.9 Schematic diagram identifying the various components of the base current of an n^+-p-n transistor, biased in the active region, as well as the emitter and collector currents; I_{E_p} represents the emitter hole defect current, $I_{E_n} - I_{C_n}$ represents the base electron recombination current, and I_{CBO} represents the collector "leakage" current, which includes both the diffusion component as well as the space-charge-generated current.

6.2.4. The recombination current was indicated to be significant compared to the diffusion current at low forward current levels for junctions fabricated from semiconductor materials with an energy gap equal to that of silicon or even greater. Hence the base lead must supply holes for recombination with electrons in the space-charge region of the emitter junction and in fact at the surface near the junction. The other source of low level emitter defect current results from surface effects of the type discussed in Sections 5.5 and 6.2.5.

Finally, since the collector junction is reverse biased with the transistor operating in the active region, a component (albeit small) of the collector current must be due to the normal diode leakage currents of Eqs. (6.32) and (6.39) and the surface leakage. This leakage current is also supplied through the base lead, but it is normally much smaller than the other defect currents, except at very high temperatures in silicon transistors. The various sources of the base current components are summarized in Fig. 8.9.

EXAMPLE 8.1

A silicon n^+-p-n transistor has a base width of 1.0×10^{-4} cm, a base resistivity of 0.10 Ω-cm, and an electron diffusion length in the base region of 10 μm. The emitter region thickness is 1.0×10^{-4} cm and has a resistivity of 0.005 Ω-cm. The emitter and collector areas are both 2.5×10^{-5} cm^2.

(a) Show that the component of the base current necessary for supplying electrons or recombining holes in the base region is small compared to the collector current.

(b) Calculate the injection efficiency of electrons for the emitter of this transistor. Neglect heavy doping effects in the emitter.
(c) Repeat (b) at 300°K, including emitter band gap narrowing. Use the graph in Appendix B to convert from resistivity to carrier density.

Solution

(a) According to Eq. (8.10b), to show that the hole current for recombining electrons is negligible, it is necessary to prove that $W^2/2L_n^2 \ll 1$, where $L_n^2 = D_n\tau_n$. Now

$$\frac{W^2}{2L_n^2} = \frac{(1.0 \times 10^{-4})^2 \text{ cm}^2}{2(10 \times 10^{-4})^2}$$

$$= \underline{5.0 \times 10^{-3} \ll 1.}$$

(b) From Eq. (8.14a)

$$\gamma = \frac{1}{1 + \dfrac{\sigma_p W}{\sigma_n W_E}} = \frac{1}{1 + \dfrac{[(1/0.10)\ (\Omega\text{-cm})^{-1}](1.0 \times 10^{-4}\text{ cm})}{[(1/0.005)\ (\Omega\text{-cm})^{-1}](1.0 \times 10^{-4}\text{ cm})}}$$

$$= \frac{1}{1 + (5.0 \times 10^{-2})} = \underline{0.95} \quad \text{or} \quad 95\%.$$

(c) From Appendix B

$\rho_n = 0.005\ \Omega\text{-cm}$, $N_d = 8 \times 10^{18}/\text{cm}^3$, $\mu_n = 150\ \text{cm}^2/\text{V-sec}$,

$\rho_p = 0.10\ \Omega\text{-cm}$, $N_a = 2 \times 10^{17}/\text{cm}^3$, $\mu_p = 250\ \text{cm}^2/\text{V-sec}$.

From Eq. (8.15)

$$\Delta E_g = 22.5\left[\left(\frac{n}{10^{18}\text{ cm}^{-3}}\right)^{1/2} \text{mV}\right]$$

$$= 22.5\left[\left(\frac{8 \times 10^{18}}{1 \times 10^{18}}\right)^{1/2} \text{mV}\right] = 64 \text{ mV}.$$

From Eqs. (8.18) and (8.16)

$$N_{d\text{eff}} = N_d(n_i^2/n_{i\text{eff}}^2) = N_d e^{-\Delta E_g/kT}$$

$$= (8 \times 10^{18}/\text{cm}^3)e^{-(64/26)}$$

$$= 6.8 \times 10^{17}/\text{cm}^3.$$

From revised Eq. (8.14a)

$$\gamma = \frac{1}{1 + \dfrac{N_a\mu_p W}{N_{\text{deff}}\mu_n W_E}} = \frac{1}{1 + \dfrac{(2 \times 10^{17}/\text{cm}^3)(250\ \text{cm}^2/\text{V-sec})(1.0\ \mu\text{m})}{(6.8 \times 10^{17})(150\ \text{cm}^2/\text{V-sec})(1.0\ \mu\text{m})}}$$

$$= \underline{0.68}.$$

8.2 THE EBERS AND MOLL TRANSISTOR EQUATIONS

In order to analyze electronic circuits containing the transistor it is convenient to model this device as a two-port circuit element. Two currents and two voltages are sufficient to specify the operation of this component modeled as a "black box." The **common-base** transistor circuit connection of an *n-p-n* transistor is given schematically in Fig. 8.10, showing the input current and voltage I_E and V_{EB} and the output current and voltage I_C and V_{CB}. By analogy with the junction diode current–voltage equations (6.32) and (6.39), one would expect similar types of current–voltage relationships for a transistor, which consists basically of two coupled diodes. These can in fact be written as

$$I_E = A_{11}(e^{qV_{BE}/kT} - 1) + A_{12}(e^{qV_{BC}/kT} - 1) \tag{8.19a}$$

and

$$I_C = A_{21}(e^{qV_{BE}/kT} - 1) + A_{22}(e^{qV_{BC}/kT} - 1), \tag{8.19b}$$

where A_{12} and A_{21} are coupling coefficients. Here the input current I_E and the output current I_C are specified in terms of the input and output voltages V_{BE} and V_{BC} as well as the A_{ij} coefficients which decribe the internal design of the particular transistor. These equations plus the two Kirchhoff laws of Eqs.(8.2a) and (8.2b) comprise four equations with six unknown currents and voltages, once the transistor design parameters A_{ij} are specified. Hence if two of these currents or voltages are given, all others are determined in principle. In contrast to the description in the last section, which was confined to the transistor

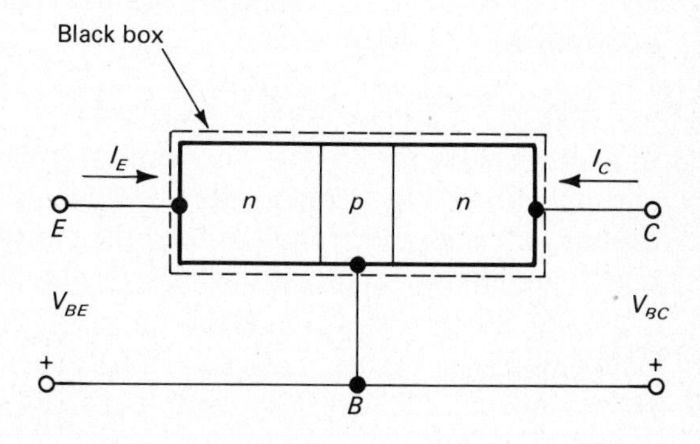

Figure 8.10 The common-base transistor circuit connection for an *n-p-n* transistor represented as a "black box," with two input ports and two output ports. Note that V_{BC} is positive when the base potential is positive relative to the collector. In addition, $V_{CB} = -V_{BC}$.

operating in the active mode, this formulation applies more generally to all four modes of transistor operation. Equations (8.19) are in a form which can be very useful for transistor modeling, particularly in computer-aided circuit analysis, and are not restricted to low level conditions. They are of a type commonly referred to as the **Ebers–Moll equations.**[5] By comparing Eqs. (8.3) and (8.4) and (8.19), Eqs. (8.19a) and (8.19b) may be rewritten as

$$I_E = a_{11}\,\Delta n_E(0) + a_{12}\,\Delta n_C(W) \tag{8.20a}$$

and

$$I_C = a_{21}\,\Delta n_E(0) + a_{22}\,\Delta n_C(W). \tag{8.20b}$$

In this form the transistor currents are seen to be *linearly* dependent on the excess minority carrier concentrations, respectively, at the emitter and collector edges of the base region. This is in contrast with the nonlinear relations in terms of the emitter and collector junction voltages as expressed in Eqs. (8.19). Since physical processes related to current flow in the transistor when expressed in terms of minority carriers are linear, the principle of superposition may be applied. This will be of considerable value later in the formulation of the charge-control model of the transistor.

8.2.1 Ebers–Moll Parameter Formulation in Terms of Physical Device Parameters

Expressions for the transistor emitter and collector currents expressed in terms of physical device parameters may be derived by considering the excess minority carrier flow in a manner similar to what has already been done for the *p-n* junction diode. Under low level steady-state conditions the excess electron density distribution in the *p*-type base region of the transistor, as shown in Fig. 8.11, may be determined from the continuity equation

$$D_n \frac{d^2\,\Delta n_p}{dx^2} = \frac{\Delta n_p}{\tau_n} \tag{8.21}$$

as in the treatment of the diode in Eq. (6.23b), except that the minority carrier here is the electron. The solution of this second-order linear differential equation may now be written as

$$\Delta n_p(x) = Be^{-x/L_n} + Ce^{x/L_n}, \qquad 0 \leq x \leq W, \tag{8.22}$$

in analogy with Eq. (6.24). The transistor base width W for reasons of good current gain will be taken as very narrow; i.e., $W/L_n \ll 1$. The boundary conditions for excess electron density at the emitter and collector edges of the base region are already expressed in Eqs. (8.3) and (8.4). Introducing these expres-

[5]J. J. Ebers and J. L. Moll, "Large-Signal Behavior of Junction Transistors," *Proc. IRE,* **42,** 1961 (1954).

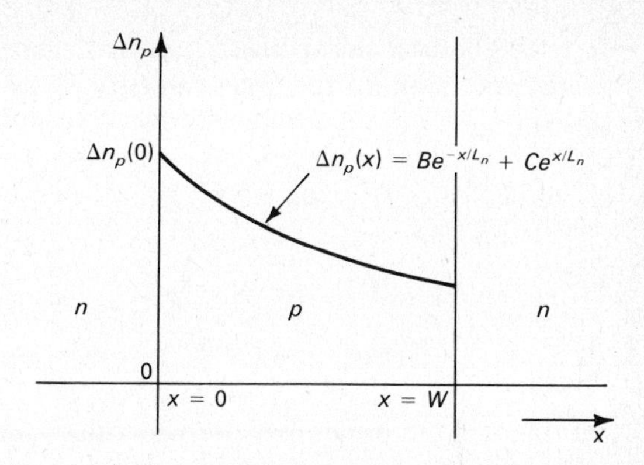

Figure 8.11 General type of low level excess electron density distribution in the *p*-type base region of an *n-p-n* transistor.

sions into Eq. (8.22) permits determination of the integration constants *B* and *C* as

$$B = \frac{\Delta n_E(0)e^{W/L_n} - \Delta n_C(W)}{e^{W/L_n} - e^{-W/L_n}} \tag{8.23}$$

and

$$C = \frac{\Delta n_C(W) - \Delta n_E(0)e^{-W/L_n}}{e^{W/L_n} - e^{-W/L_n}}.$$

The complicated general solution of Eq. (8.22) may be simplified by noting that in practical cases, $W/L_n \ll 1$. Then the exponentials may be series expanded and the solution for the excess electron density in the base region approximated as

$$\Delta n_p(x) \approx \frac{C - B}{L_n} x + (B + C), \qquad 0 \leq x \leq W. \tag{8.24}$$

This is an expression of the nearly linear distribution of excess electrons in the thin base region for **all** values of emitter and collector junction potential. This linear distribution of minority carriers has already been discussed for the special case of the transistor biased in the active region.

The minority hole diffusion current flow in the transistor base region can now be calculated, assuming the emitter current is exclusively electron current (the injection efficiency $\gamma = 1$), as

$$I_E = qAD_n \left. \frac{\partial \, \Delta n_p}{\partial x} \right|_{x=0} \tag{8.25a}$$

and

$$I_C = -qAD_n \left. \frac{\partial \, \Delta n_p}{\partial x} \right|_{x=W}. \tag{8.25b}$$

Here the emitter and collector areas are both taken as A. Differentiating the exact expression for the excess electron density of Eq. (8.22), we have for Eqs. (8.25)

$$I_E = \frac{qAD_n}{L_n}(C - B) \tag{8.26a}$$

and

$$I_C = \frac{qAD_n}{L_n}(Be^{W/L_n} - Ce^{-W/L_n}). \tag{8.26b}$$

Introducing the previously derived values for B and C given in Eq. (8.23), we may write Eqs. (8.26) in the Ebers–Moll form of Eqs. (8.20a) and (8.20b) as

$$I_E = a_{11}\,\Delta n_E(0) + a_{12}\,\Delta n_C(W)$$

and

$$I_C = a_{21}\,\Delta n_E(0) + a_{22}\,\Delta n_C(W),$$

where[6]

$$a_{11} = a_{22} = \frac{-qAD_n}{L_n}\coth\left(\frac{W}{L_n}\right), \qquad a_{12} = a_{21} = \frac{qAD_n}{L_n}\operatorname{csch}\left(\frac{W}{L_n}\right). \tag{8.27}$$

Again assuming $W/L_n \ll 1$, we can simplify the expression for the a_{ij} by the following approximations:

$$\coth\left(\frac{W}{L_n}\right) \approx \frac{L_n}{W} + \frac{W}{3L_n}, \quad \operatorname{csch}\left(\frac{W}{L_n}\right) \approx \frac{L_n}{W} - \frac{W}{6L_n} \tag{8.28}$$

to read

$$a_{11} = a_{22} = \frac{-qAD_n}{W} - \frac{qAW}{3\tau_n}, \quad a_{12} = a_{21} = \frac{qAD_n}{W} - \frac{qAW}{6\tau_n}. \tag{8.29}$$

Then the base current can be obtained by combining Eqs. (8.20) and the Kirchhoff law of Eq. (8.1) as

$$I_B = -[(a_{11} + a_{21})\,\Delta n_E(0) + (a_{12} + a_{22})\,\Delta n_C(W)]. \tag{8.30}$$

If values of the a_{ij} as given in Eq. (8.29) are introduced into Eq. (8.30) and the active region of transistor operation considered $[\Delta n_C(W) \approx 0]$, agreement with

[6]In this formulation, the emitter and collector junctions are not considered dissimilar. This accounts for the symmetry of the equations for I_C and I_E expressed by Eqs. (8.27). In real transistors this symmetry does not exist in general. The Ebers–Moll-type equations will nevertheless still apply although $a_{11} \neq a_{22}$ when the emitter and collector junctions are not identical. However, in general $a_{12} = a_{21}$, even in the case of lack of symmetry. This expresses **reciprocity** of the two-port transistor model.

Eq. (8.9) is obtained; note that the latter equation was derived beginning from a different viewpont.

EXAMPLE 8.2

Assuming the transistor design of Example 8.1, calculate the magnitude and dimensions of a_{11}, a_{12}, a_{21}, and a_{22} using

(a) the approximate expressions of Eq. (8.29) and
(b) the exact expressions of Eq. (8.27).

Solution

(a) From Eq. (8.29)

$$a_{11} = a_{22} = -\frac{qAD_n}{W} - \frac{qAW}{3\tau_n}$$

$$= -\frac{(1.6 \times 10^{-19}\ \text{C})(2.5 \times 10^{-5}\ \text{cm}^2)(17\ \text{cm}^2/\text{sec})}{1.0 \times 10^{-4}\ \text{cm}}$$

$$- \frac{(1.6 \times 10^{-19}\ \text{C})(2.5 \times 10^{-5}\ \text{cm}^2)(1.0 \times 10^{-4}\ \text{cm})}{3(5.9 \times 10^{-8}\ \text{sec})}$$

$$= -(6.8 \times 10^{-19}\ \text{A-cm}^3) - (2.3 \times 10^{-21}\ \text{A-cm}^3)$$

$$= \underline{-6.8 \times 10^{-19}\ \text{A-cm}^3}.$$

$$a_{12} = a_{21} = \frac{qAD_n}{W} - \frac{qAW}{6T_n} = 6.8 \times 10^{-19}\ \text{A-cm}^3.$$

(b) From Eq. (8.27)

$$a_{11} = a_{22} = \frac{-qAD_n}{L_n} \coth\left(\frac{W}{L_n}\right)$$

$$= \frac{-(1.6 \times 10^{-19}\ \text{C})(2.5 \times 10^{-5}\ \text{cm}^2)(17\ \text{cm}^2/\text{sec})}{10 \times 10^{-4}\ \text{cm}}$$

$$\times \coth\left(\frac{1.0 \times 10^{-4}\ \text{cm}}{10 \times 10^{-4}\ \text{cm}}\right)$$

$$= -(6.8 \times 10^{-20}\ \text{A-cm}^3) \coth(0.10)$$

$$= \underline{-6.8 \times 10^{-19}\ \text{A-cm}^3}$$ [within the approximation of (a)].

$$a_{12} = a_{21} = \frac{qAD_n}{L_n} \operatorname{csch}\left(\frac{W}{L_n}\right)$$

$$= (6.8 \times 10^{-20}\ \text{A-cm}^3) \operatorname{csch}\left(\frac{2.0 \times 10^{-4}\ \text{cm}}{10 \times 10^{-4}\ \text{cm}}\right)$$

$$= (6.8 \times 10^{-20}\ \text{A-cm}^3) \operatorname{csch}(0.10)$$

$$= \underline{6.8 \times 10^{-19}\ \text{A-cm}^3} \text{ [within the approximation of (a)].}$$

8.2.2 Ebers–Moll Parameter Determination by Measurements on Devices

Some simple measurements will now be described which can be made on a real transistor in the laboratory in order to determine the coefficients A_{ij} in Eqs. (8.19), in an attempt to represent the device by these equations:

1. Connect the base and collector leads together ($V_{BC} = 0$) and then measure the current–voltage (I-V) characteristic of the emitter–base diode. Equation (8.19a) then becomes

$$I_E|_{V_{BC}=0} = A_{11}(e^{qV_{BE}/kT} - 1), \tag{8.31}$$

which is the ordinary diode law of Eq. (6.32), where $-A_{11}$ is the emitter diode leakage current, usually denoted by I_{ES}.[7] Experimentally this is usually determined by fitting the measured forward-biased emitter diode characteristic to Eq. (8.31).

It is also instructive to measure the ratio of collector to emitter current under the shorted-collector condition.[8] In this situation Eqs. (8.19a) and (8.19b) can be combined to define a **short-circuit common-base current gain** as

$$\alpha_F \equiv -\left.\frac{I_C}{I_E}\right|_{V_{BC}=0} = -\frac{A_{21}}{A_{11}} = \frac{\operatorname{csch}(W/L_n)}{\coth(W/L_n)} \simeq 1 - \frac{1}{2}\left(\frac{W}{L_n}\right)^2. \tag{8.32}$$

Here Eqs. (8.27) are used to give an expression for α_F in terms of device parameters, assuming unity injection efficiency. Since the latter factor, γ, is less than

[7]The subscript S here refers to the electrical shorting of the collector junction. In small-signal transistors this quantity is of the order of picoamperes.

[8]Experimentally this condition can be approximated by placing a *very low resistance* ammeter between collector and base to measure the collector current. However, more commonly the common-base current gain α is measured at a specified collector–base potential and is defined as $|I_C/I_E|_{V_{BC}} =$ const.

one, the last terms in Eq. (8.32) should be multiplied by γ to predict the measured gain. Indeed, in modern bipolar transistors the injection efficiency controls the gain.

2. Now connect the base and emitter leads together ($V_{BE} = 0$) and then measure the collector–base diode I-V characteristic. Equation (8.19b) then becomes

$$I_C|_{V_{BE}=0} = A_{22}(e^{qV_{BC}/kT} - 1), \tag{8.33}$$

which is the ordinary diode law for the collector junction, where A_{22} is the collector diode leakage current with the emitter shorted, usually denoted by I_{CS}. Except in the special case in which the emitter and collector diodes are identical, in general $I_{CS} \neq I_{ES}$. Now the ratio of emitter to collector current under the shorted-emitter condition may be written as

$$\alpha_R \equiv -\left.\frac{I_E}{I_C}\right|_{V_{BE}=0} = -\frac{A_{12}}{A_{22}}, \tag{8.34}$$

where α_R is the short-circuit common-base current gain with the transistor operating in the reverse condition, i.e., the collector is forward biased and emitting and the emitter junction is collecting.

With the definitions described above, Eqs. (8.19a) and (8.19b) coupled with (8.1) may now be expressed for an n-p-n transistor as

$$I_E = -I_{ES}(e^{qV_{BE}/kT} - 1) + \alpha_R I_{CS}(e^{qV_{BC}/kT} - 1), \tag{8.35a}$$

$$I_C = \alpha_F I_{ES}(e^{qV_{BE}/kT} - 1) - I_{CS}(e^{qV_{BC}/kT} - 1), \tag{8.35b}$$

$$I_B = (1 - \alpha_F)I_{ES}(e^{qV_{BE}/kT} - 1) + (1 - \alpha_R)I_{CS}(e^{qV_{BC}/kT} - 1). \tag{8.35c}$$

For a p-n-p transistor the Ebers–Moll equations are

$$I_E = I_{ES}(e^{qV_{EB}/kT} - 1) - \alpha_R I_{CS}(e^{qV_{CB}/kT} - 1), \tag{8.35d}$$

$$I_C = -\alpha_F I_{ES}(e^{qV_{EB}/kT} - 1) + I_{CS}(e^{qV_{CB}/kT} - 1), \tag{8.35e}$$

$$I_B = -(1 - \alpha_F)I_{ES}(e^{qV_{EB}/kT} - 1) - (1 - \alpha_R)I_{CS}(e^{qV_{CB}/kT} - 1). \tag{8.35f}$$

These are the original forms of the equations as published by Ebers and Moll.[5] In general $I_{ES} \neq I_{CS}$ and $\alpha_F \neq \alpha_R$ but always

$$\alpha_R I_{CS} = \alpha_F I_{ES}. \tag{8.36}$$

This is a consequence of the fact that in general $a_{11} \neq a_{22}$ but always $a_{12} = a_{21}$ (see footnote 6, this chapter). Eliminating I_{ES} from Eqs. (8.35a) and (8.35b) yields the useful equation

$$I_C = -\alpha_F I_E - I_{CBO}(e^{qV_{BC}/kT} - 1), \tag{8.37a}$$

where by definition[9]

$$I_{CBO} \equiv (1 - \alpha_F \alpha_R) I_{CS}. \quad (8.38a)$$

Similarly,

$$I_E = -\alpha_R I_C - I_{EBO}(e^{qV_{BE}/kT} - 1), \quad (8.37b)$$

where

$$I_{EBO} \equiv (1 - \alpha_F \alpha_R) I_{ES}. \quad (8.38b)$$

Note that I_{CBO} may be determined by measuring the reverse collector leakage current ($V_{BC} \ll 0$) with the emitter unconnected ($I_E = 0$), according to Eq. (8.37a). Similarly, I_{EBO} may be determined by measuring the reverse emitter leakage current with the collector unconnected ($I_C = 0$), according to Eq. (8.37b). In the active region ($V_{BC} \ll 0$), Eq. (8.37a) states that the negative of the collector current is just the emitter current multiplied by the short-circuit common-base current gain less the collector leakage current as measured with the emitter open circuited, i.e.,

$$I_C = -(\alpha_F I_E - I_{CBO}). \quad (8.39)$$

Since α_F in practical devices is normally between 0.95 and 1.0, the common-base connection yields nearly unity current gain when the injection efficiency is also nearly unity. Since the collector junction is normally reverse biased with three or more volts and the emitter junction requires less than a volt for a silicon transistor to provide a practical emitter current, the device in this connection can have appreciable voltage and hence power gain (see Section 8.1.2). Figure 8.12 shows a plot of the measured **output** characteristics of a typical common-base-connected *n-p-n* bipolar transistor relating the collector current to collector–base voltage for different input values of emitter current.

In a corresponding manner the **input** characteristics of the transistor in the common-base connection may be derived. From Eq. (8.37b) the emitter–base *I-V* characteristic of the transistor operating in the active region becomes

$$I_E = -I_{EBO}(e^{qV_{BE}/kT} - 1) - \alpha_R I_C. \quad (8.40)$$

This is essentially the expression for the ordinary junction diode except for an additional coupling term $\alpha_R I_C$ which is small in devices in which α_R is small. The typical input characteristic of a common-base-connected *n-p-n* bipolar transistor relating the emitter current to the emitter–base voltage is shown in Fig. 8.13. Here the effect of higher collector reverse voltage is to widen the collector space-charge region, bringing the collector and emitter space-charge regions closer together. This will increase the values of the parameter I_{ES} and α_R [see Eqs. (8.19), (8.29), and (8.35)], and hence the absolute value of I_E for each V_{BC}.

[9]The subscript *O* refers to the open circuiting of the emitter. In small-signal transistors this quantity is of the order of picoamperes.

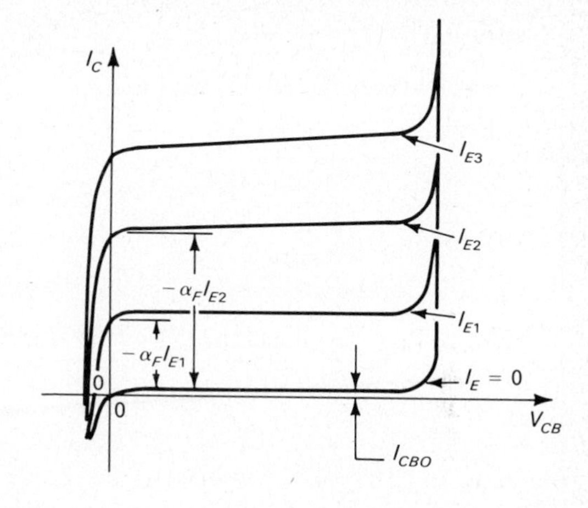

Figure 8.12 Measured common-base **output** current–voltage characteristic of an *n-p-n* bipolar transistor. The collector current versus collector–base voltage at a fixed emitter current is plotted for different values of emitter current $|I_{E3}| > |I_{E2}| > |I_{E1}| > 0$. The forward gain factor α_F is used to define the collector output current in terms of the emitter input current. The sudden increase in current at high $|V_{CB}|$ is due to avalanching of the collector junction. Note that in this text generally symbols with letter subscripts only (V_{CB}) refer to variable quantities, whereas those with a combination of letters and numbers refer to specific numerical values of the quantity. Note that I_{CBO} is normally small, of the order of nanoamperes or less, for a small-signal transistor. It is drawn here greatly expanded for illustrative purposes.

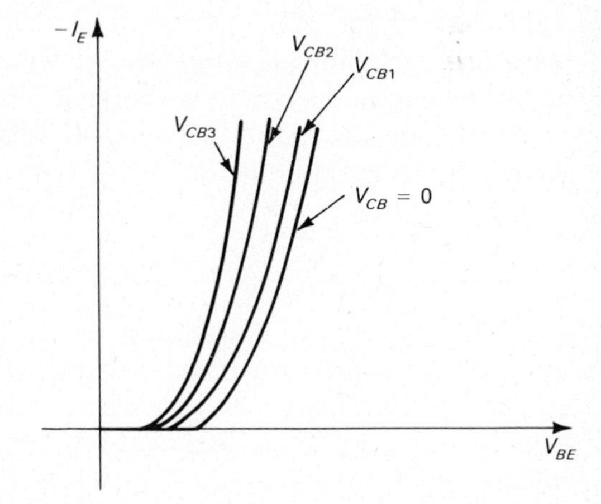

Figure 8.13 Measured common-base **input** current–voltage characteristic of an *n-p-n* bipolar transistor for different values of collector output voltage. As the collector potential is increased ($V_{CB3} > V_{CB2} > V_{CB1}$) space-charge widening reduces the effective transistor base width, causing additional current to flow through the emitter diode (I_{EBO} and α_R increase).

EXAMPLE 8.3

Assuming the transistor design of Example 8.1, calculate the short-circuit common-base current gain α_F assuming unity injection efficiency, using

(a) the approximate expressions of Eqs. (8.29) and (8.32) and
(b) the exact expressions of Eq. (8.27).

Solution

(a) Using Eq. (8.32) we have

$$\alpha_F \simeq 1 - \frac{1}{2}\left(\frac{W}{L_n}\right)^2 = 1 - \frac{1}{2}\left[\frac{1.0 \times 10^{-4}\ \text{cm}}{10 \times 10^{-4}}\right]^2$$

$$= \underline{0.95}.$$

(b) Combining Eqs. (8.27) with the definition of Eq. (8.32) gives

$$\alpha_F = -\frac{(qAD_n/L_n)\ \text{csch}(W/L_n)}{-(qAD_n/L_n)\ \text{coth}(W/L_n)} = \text{sech}\left(\frac{W}{L_n}\right)$$

$$= \text{sech}\left[\frac{1.0 \times 10^{-4}\ \text{cm}}{10 \times 10^{-4}}\right] = \text{sech}(0.10)$$

$$= \underline{0.95}.$$

Hence the two expressions give the same value within the accuracy of the given data.

8.2.3 The Ebers–Moll Equivalent Circuit

The Ebers and Moll equations may be expressed in terms of circuit models to aid in the solution of circuit network problems which involve transistors. One of these models is shown in Fig. 8.14. This is the model for the Ebers–Moll equations expressed in the form

$$I_C = -\alpha_F I_E - I_{CBO}(e^{qV_{BC}/kT} - 1), \tag{8.37a}$$

$$I_E = -\alpha_R I_C - I_{EBO}(e^{qV_{BE}/kT} - 1), \tag{8.37b}$$

where I_{EBO} and I_{CBO} are respectively the emitter and collector leakage currents, when the opposite terminal is open circuited. From reciprocity $\alpha_F I_{EBO} = \alpha_R I_{CBO}$, and hence only three measurements are needed to determine the four parameters α_F, α_R, I_{EBO}, and I_{CBO} and hence specify completely the *I-V* characteristics of a transistor.

The most common connection of the transistor in amplifier and switching applications is the common-emitter configuration, shown schematically in Fig. 8.15. Here the applied voltages V_{CE} and V_{BE} are specified relative to the common emitter terminal. Now the input current I_B is supplied to the transistor base, and the output current is the collector current I_C. A typical set of measured *I-V* characteristics for a transistor in this connection is shown in Fig. 8.16. When the collector current is expressed in terms of a base current input, one can write with the aid of Eqs. (8.37a), (8.37b), and (8.1)

$$I_C = \frac{\alpha_F}{1 - \alpha_F} I_B - \frac{I_{CBO}}{1 - \alpha_F}(e^{qV_{BC}/kT} - 1). \tag{8.41}$$

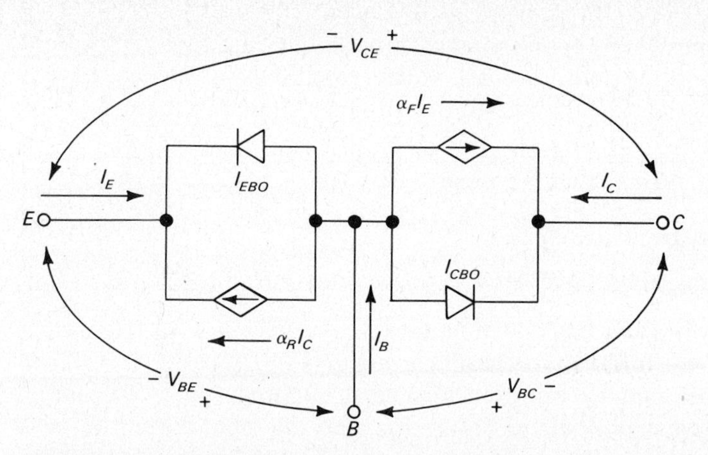

Figure 8.14 Ebers–Moll common-base *n-p-n* transistor model. The model consists of two current generators and two diodes coupled together. The diode symbol labeled I_{EBO} has a current–voltage characteristic given by $I = I_{EBO}(e^{qV_{BE}/kT} - 1)$; the symbol labeled I_{CBO} represents the *I-V* characteristic given by $I = I_{CBO}(e^{qV_{BC}/kT} - 1)$.

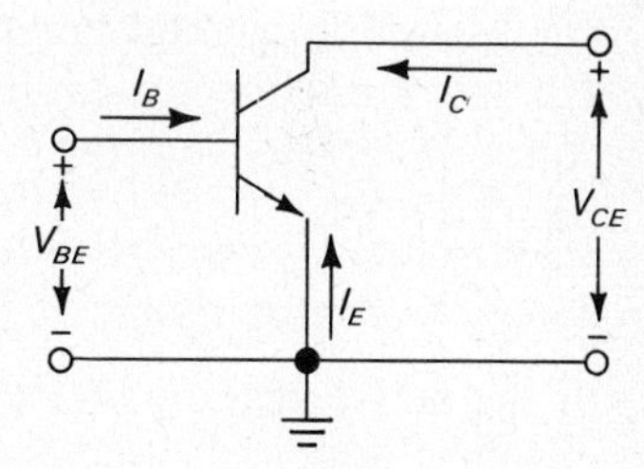

Figure 8.15 Common-emitter circuit configuration for an *n-p-n* transistor. This is sometimes referred to as the grounded-emitter connection.

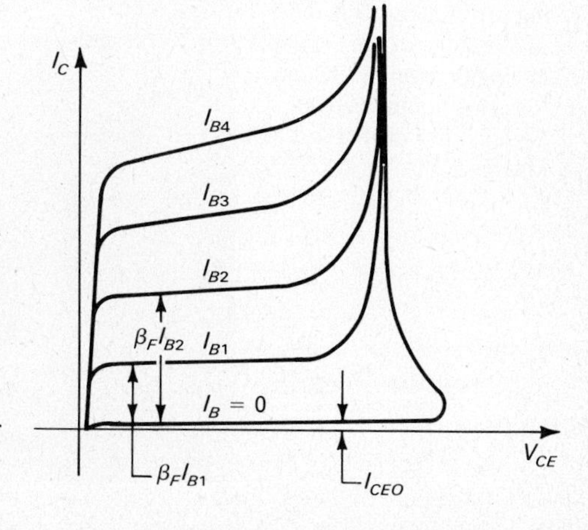

Figure 8.16 Measured common-emitter output current–voltage characteristic of an *n-p-n* transistor. The collector current versus collector-emitter voltage at a fixed base current is plotted for different values of base current $I_{B4} > I_{B3} > I_{B2} > I_{B1} > 0$. The common-emitter forward gain factor β_F is used to define the collector output current in terms of the base input current. A set of specification sheets for a typical commercial *n-p-n* silicon planar transistor is included in Appendix F1; Appendix F2 gives specifications for a *p-n-p* device.

This equation expresses the common-emitter output current I_C in terms of the input current I_B and the collector–base voltage V_{CB}. Since in this connection the applied voltage should be expressed in terms of V_{CE} (see Fig. 8.15), Eq. (8.2) may be used to give

$$V_{BC} = -(V_{CE} + V_{EB}) = -V_{CE} + V_{BE}$$

and Eq. (8.41) may be rewritten as

$$I_C = \beta_F I_B - I_{CEO}(e^{qV_{BE}/kT}e^{-qV_{CE}/kT} - 1), \tag{8.42}$$

where by definition $\beta_F \equiv \alpha_F/(1 - \alpha_F)$ and $(\beta_F + 1)I_{CBO} \equiv I_{CEO}$. Here β_F is termed the **common-emitter short-circuit current gain** (I_C/I_B with $V_{BC} = 0$) and has practical values ranging from ten to several hundred; I_{CEO} is the output current in the active region where $V_{CE} \gg V_{BE}$, and the input current $I_B = 0$; I_{CEO} can be a few orders of magnitude greater than I_{CBO} and is indicated in Fig. 8.16. From Eq. (8.32),

$$\beta_F \equiv \frac{\alpha_F}{1 - \alpha_F} = \frac{1 - \frac{1}{2}(W/L_n)^2}{1 - [1 - \frac{1}{2}(W/L_n)^2]} \simeq 2\left(\frac{L_n}{W}\right)^2 \tag{8.43}$$

for $W/L_n \ll 1$. The expression for β_F in terms of device parameters assumes unity injection efficiency. When $\gamma \neq 1$ then $\gamma\alpha_F$ should replace α_F in this expression. In general practice the common-emitter current gain is measured at a specified collector–emitter voltage and is defined as $\beta \equiv I_C/I_B|_{V_{CE}=\text{const}}$. The common-emitter input I-V characteristics of a transistor can be derived from Eq. (8.35c) as

$$I_B = (1 - \alpha_F)I_{ES}(e^{qV_{BE}/kT} - 1) - (1 - \alpha_R)I_{CS} \tag{8.44}$$

in the active region, where $V_{BC} \ll 0$. Here again this is the equation for the ordinary junction diode except for the additional term $(1 - \alpha_R)I_{CS}$, which is

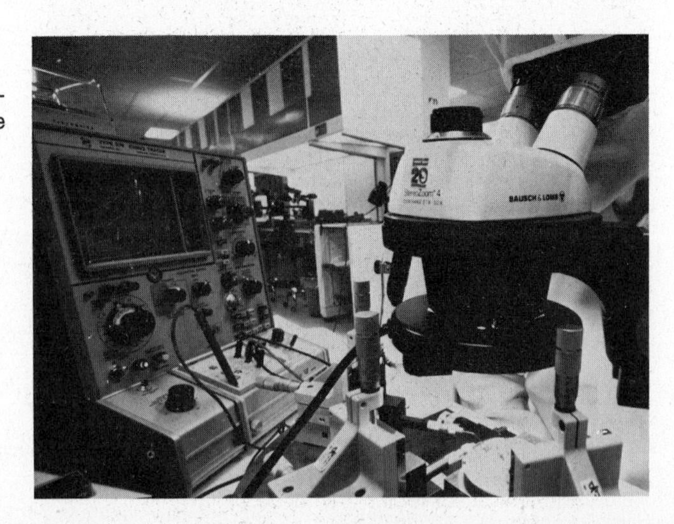

Operator performing in-line electrical testing during processing. This measurement is of the β value for the bipolar transistor. (Courtesy of the Digital Equipment Corporation, Maynard, MA.)

normally small. Some typical measured input characteristics of a common-emitter-connected *n-p-n* transistor relating the base current to the base–emitter voltage are shown in Fig. 8.17. Again the effect of higher collector voltage is to widen the collector space-charge region, increasing α_F but decreasing $(1 - \alpha_F)I_{ES}$. Hence the effect here is opposite to that for the common-base case (Fig. 8.13).

The Ebers–Moll circuit model for the transistor in the common-emitter connection is shown in Fig. 8.18. Note that the collector current source is specified in terms of the input base current through β_F.

8.3 THE CHARGE-CONTROL TRANSISTOR MODEL

Mention has already been made of the function of the base current in supplying charge to maintain a constant steady-state minority carrier distribution in the base region. It will now be pointed out how static or low frequency transistor operation may be conveniently described in terms of the charge distributed in the transistor base region by relating the terminal currents to this charge. The concept will prove to be useful in extending the description of low frequency transistor behavior to medium frequency transistor performance. This method also will be seen to be particularly powerful in explaining the operation of transistors in the switching mode. In this case use will be made of the linear relationship between the terminal currents and the charge in the transistor base region which permits the principle of superposition to be applied.

8.3.1 The Charge-Control Equations

Equation (8.7) is an expression relating the terminal base current to the *total* excess minority carrier charge Q_F in the transistor base region in the steady-state, active mode. By combining Eqs. (8.5) and (8.8) an equation relating the

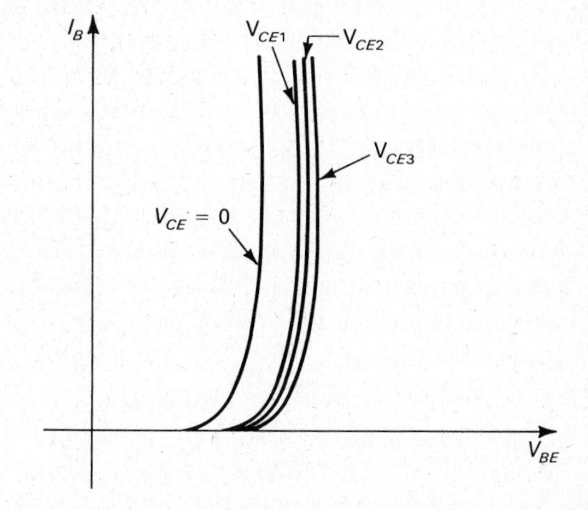

Figure 8.17 Common-emitter input current–voltage characteristic of an *n-p-n* transistor for different values of collector output voltage. As the collector potential is increased ($V_{CE3} > V_{CE2} > V_{CE1}$), space-charge widening reduces the effective transistor base width, which causes less current to flow through the emitter diode. This is in contrast to the common-base input characteristic of Fig. 8.13.

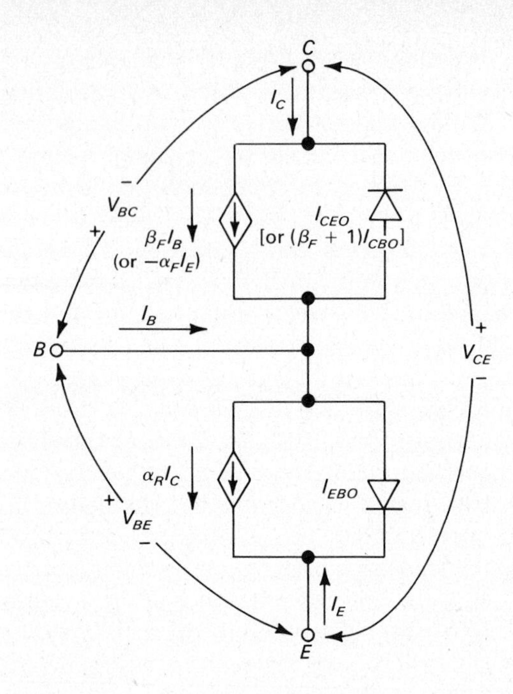

Figure 8.18 Ebers–Moll common-emitter transistor model. Here the collector current generator is specified in terms of the input current I_B. The diode symbol labeled I_{CEO} has an I-V characteristic given by $I_{CEO}(e^{qV_{BC}/kT} - 1)$; the diode symbol labeled I_{EBO} represents the I-V characteristic given by $I_{EBO}(e^{qV_{BE}/kT} - 1)$.

collector current to this total charge in the base region may be written for the transistor operating in the active mode as

$$I_C = -\frac{Q_F}{W^2/2D_n} \equiv -\frac{Q_F}{\tau_t}. \tag{8.45}$$

Here the quantity $W^2/2D_n$ has the dimensions of time and hence is defined as τ_t. This time is in fact essentially the time it takes minority electrons to traverse from the emitter junction to the collector junction where they are collected. This may be proved by solving the diffusion equation for minority carrier flow in the base region. Alternatively, this fact may be made plausible by verbal arguments as follows: The average time for electron recombination in the base region is τ_n. In order to maintain a steady-state charge distribution in the base the electrons thus lost must be reinjected through the emitter with an accompanying supply of holes through the base lead for reasons of space-charge neutrality. However, during the time for one electron recombination in the base, many more electrons will have successfully reached the collector space-charge region constituting the collector current. Hence the ratio of base current to collector current should reflect the ratio of the emitter–collector transit time to the recombination time. This is verified mathematically by combining Eqs. (8.7) and (8.45) to give

$$I_C/I_B = \tau_n/\tau_t, \tag{8.46}$$

where τ_t can be identified clearly as the transit time. From Eqs. (8.2), (8.7), and (8.45) the emitter current can now be expressed in the active mode as

$$I_E = Q_F(1/\tau_t + 1/\tau_n). \tag{8.47}$$

Finally, if we assume unity injection efficiency, the DC current gains of the transistor in the active region can be expressed in terms of the charge-control parameters τ_n and τ_t as

$$\beta_F = I_C/I_B = \tau_n/\tau_t \tag{8.48a}$$

and

$$\alpha_F = -\frac{I_C}{I_E} = \frac{1}{1 + \tau_t/\tau_n}. \tag{8.48b}$$

The formulation thus far has referred to the active region with the emitter forward biased and collector reverse biased. As explained in Section 8.1.3, in the saturated or inverse mode of operation the collector junction is injecting minority carriers into the base region. Hence for a more general formulation of the charge-control model this "reverse" operation should be considered and may be described by an analogous set of equations to those for "forward" operation, i.e.,

$$I_B = -Q_R/\tau_n, \tag{8.49a}$$

$$I_E = -Q_R/\tau_t, \tag{8.49b}$$

$$I_C = Q_R(1/\tau_n + 1/\tau_t). \tag{8.49c}$$

Here Q_R represents the total electron charge injected into the base by the collector junction and τ_n represents the minority electron lifetime in the base region in reverse operation. The latter is taken equal to the forward lifetime for simplicity, although this is not the case in general. As discussed before, because of the linear relationship between the terminal currents and the charge in the base region, the principle of superposition may be applied. Therefore the more general charge control equations for the transistor in the DC or static situation may be obtained from Eqs. (8.7), (8.45), (8.47), and (8.49) as

$$I_B = -\frac{Q_F + Q_R}{\tau_n}, \tag{8.50a}$$

$$I_C = -Q_F\frac{1}{\tau_t} + Q_R\left(\frac{1}{\tau_n} + \frac{1}{\tau_t}\right), \tag{8.50b}$$

$$I_E = Q_F\left(\frac{1}{\tau_n} + \frac{1}{\tau_t}\right) - Q_R\frac{1}{\tau_t}. \tag{8.50c}$$

Here the three terminal currents are expressed in terms of the charge-control parameters Q_F, Q_R, τ_n, and τ_t for the transistor in any of its four operating modes. If Q_F and Q_R are expressed in terms of the terminal voltages by using equations like (8.3) and (8.8), the analogy between these equations and the Ebers–Moll equations may be verified.

A plot of the minority charge distribution in an *n-p-n* transistor base region when both the emitter and collector junctions are in forward bias (as in the saturation mode) is shown in Fig. 8.19a. This linear distribution may be considered to be composed of the forward and reverse charge components Q_F and Q_R, shown in Figs. 8.19b and 8.19c, and may be constructed by addition or **superposition** of these distributions.

Thus far only the minority charges in the base region have been considered and the minority charges in the emitter and collector regions ignored. This is an excellent approximation as far as the emitter region is concerned because of the high doping and low injected-minority-carrier concentration there. However, bipolar transistors in integrated circuits usually have lightly doped collector regions and hence a more general treatment must take into account the

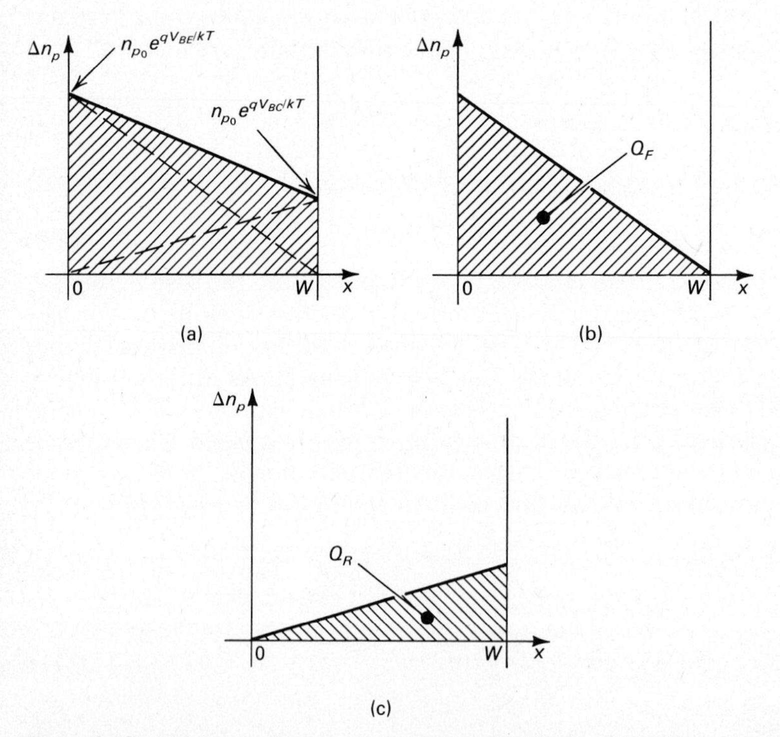

Figure 8.19 (a) Minority electron charge density distribution for a bipolar transistor in **saturation** (both emitter and collector junctions are forward biased). The solid line indicates the total electron distribution $Q_F + Q_R$.
(b) Forward charge density distribution whose integral is Q_F.
(c) Reverse charge density distribution whose integral is Q_R.

minority hole charges in the *n-p-n* transistor collector region. This is accomplished in the Gummel–Poon treatment described in the next section.

The power of the charge-control model will be seen in the case of transient or AC transistor operation to be described in the next chapter. Of interest then will be the *change* ΔQ_F of the charge in the base with a signal applied to the emitter junction. Under AC conditions an additional alteration of charge will occur in the space-charge region of the emitter and collector junctions due to charging of their space-charge capacitances. In the charge-control model this may easily be included by simply adding additional charge parameters.

*8.4 COMPUTER-AIDED MODELING OF BIPOLAR JUNCTION TRANSISTORS

Microchips can contain hundreds of thousands of transistors. Hence it is impractical to carry out the circuit analysis of transistor-containing circuits for a given chip design without the aid of a computer. The analysis of these circuits involves the application of the Kirchhoff current and voltage equations. The current–voltage relationship for the various circuit elements must be utilized. This is straightforward in the case of resistors, capacitors, and inductors. However, a circuit model is needed for transistors. The Ebers–Moll model or the charge-control model may be used for bipolar transistors in this type of analysis. However, the Ebers–Moll model uses superposition of collector and emitter diode currents but real transistors violate superposition owing to certain nonlinear effects. For example, the current gain is a function of collector voltage owing to the **Early** effect.[10] This effect is particularly important in narrow-base transistors where the increase of collector potential decreases the base width and hence increases the gain. Another effect which is not modeled by the simple Ebers–Moll equations already discussed is the **Kirk** effect.[11] This phenomenon is particularly important at high collector current densities where the minority charges injected by the emitter modify the fixed charges in the collector depletion region, which effectively increases the base width. This results in the falloff of current gain and high frequency performance at high currents.

8.4.1 The Gummel–Poon Transistor Circuit Model

The **Gummel–Poon** bipolar junction transistor (BJT) circuit model permits the incorporation of the effects mentioned above and other real effects into an Ebers–Moll-like circuit model.[12] This model is based on an **integral charge**

[10] J. M. Early, "Effects of Space-Charge Layer Widening in Junction Transistors," *Proc. IRE,* **40,** 1401–1406 (1952).

[11] E. T. Kirk, "A Theory of Transistor Cutoff Frequency, f_T, Falloff at High Current Densities," *IEEE Trans. Electron Devices,* **ED-9,** 164–174 (1962).

[12] H. K. Gummel and H. C. Poon, "An Integral Charge Control Model of Bipolar Transistors," *Bell Syst. Tech. J.,* **49,** 827 (1970).

relation that relates the electrical terminal characteristics to the base charge and hence bears a relationship to the charge control model. The Gummel–Poon model is very useful in that it takes into account accurately many physical phenomena such as the Early and Kirk effects and the transport of minority carriers both by diffusion and electric field in the base region. A large number of physical design parameters are needed for its accurate utilization. However, it is extensively used in computer-aided complex circuit analysis programs such as SPICE.[13]

The Gummel–Poon model is based on the integral charge relation for the current I_{CC} that would flow from emitter to collector *if* the transistor had unity current gain, where

$$I_{CC} = -\frac{(qn_iA)^2 D_B}{Q_B}(e^{qV_{BE}/kT} - e^{qV_{BC}/kT}) \tag{8.51}$$

and where the total majority hole base charge Q_B is given by

$$Q_B = qA\int_0^W p(x)\,dx = qA\int_0^W n(x)\,dx + qA\int_0^W N(x)\,dx; \tag{8.52}$$

here A is the cross sectional area of the emitting region and D_B is the base-region carrier diffusion coefficient. Equation (8.51) links junction voltages, collector current, and base charge. Traditionally, $N_g \equiv \int_0^W N(x)\,dx$ is defined as the **Gummel number;** it gives the total number of ionized impurities per square centimeter in the base region, and determines the value of voltage for the collector space-charge region to "punch through" to the emitter. The first term on the right-hand side of Eq. (8.52) represents the total minority electron charge in the base region. Increase of this quantity at high current levels has the effect of reducing the increase of I_C and hence the current gain [see Eq. (8.51)]. Now if the effects of the base recombination current and emitter defect current are included, the collector, emitter, and base currents can be written as

$$I_E = -(I_{CC} + I_{BE}) \tag{8.53a}$$

and

$$I_C = I_{CC} + I_{BC}, \tag{8.53b}$$

so

$$I_B = -(I_E + I_C) = I_{BE} - I_{BC}. \tag{8.54}$$

Here I_{BE} represents the emitter defect current due to recombination and injection inefficiency effects described in Section 8.1.4 and I_{BC} represents the collector junction leakage current of holes and electrons. Note that the dominant current I_{CC} cancels out in the base current expression, so the base current is

[13]SPICE—A Simulation Program for Integrated Circuit Equations. See W. Nagel, "SPICE 2, A Computer Program to Simulate Semiconductor Circuits," University of California, Berkeley, Memo No. ERL-M520 (1975).

determined by the defect and recombination currents as expected. The details of the manner in which the Early and Kirk effects as well as the high injection current and base electric field effects are included in this model are described in footnote 12 of this chapter.

8.4.2 Computer-Aided Modeling of Current Flow in the Transistor

Determination of the details of the transport of electron and hole current carriers in a semiconductor device structure is an exceedingly difficult problem when *both* diffusion and drift current effects are included and three-dimensional flow is considered. Hence computer-aided or "exact" analysis is required. In essence the problem is to solve the current continuity equations for electrons and holes consistent with the Poisson charge equation. This must be done in the emitter and collector regions of a BJT as well as the base region. An example of the approximate method for solution of this type of problem is given for the n-region of a p^+-n diode structure in Section 6.2. Here the hole continuity equation is solved in only *one dimension* by considering just carrier motion by *diffusion* [see Eq. (6.23b)]; the approximate junction law [Eq. (6.29)] is assumed in an ad hoc manner.

The first solution of the one-dimensional carrier flow problem considering both electron and hole flow in all regions of the bipolar transistor and including motion due to both diffusion and drift, without the a priori assumption of the junction law and charge neutrality, was accomplished with the aid of a digital computer in 1968.[14] The three semiconductor equations which were solved are

$$\frac{\partial^2\psi}{\partial x^2} = -\frac{q}{\epsilon}(p - n + N), \tag{8.55a}$$

$$-\frac{1}{q}\frac{\partial J_p}{\partial x} - (R - G) = \frac{\partial p}{\partial t}, \tag{8.55b}$$

$$\frac{1}{q}\frac{\partial J_n}{\partial x} - (R - G) = \frac{\partial n}{\partial t}. \tag{8.55c}$$

These equations constitute a set of three coupled differential equations which are nonlinear because of the form of the current density expressions for J_p and J_n [see Eq. (5.22)], and because of the dependence of the electrostatic potential ψ on the carrier densities p and n [see Eq. (4.31)]. The first is Poisson's charge equation in which N represents the net density of ionized impurities. It expresses the fact that the divergence of the electric field $\mathscr{E}$ $(= d\psi/dx)$ from a volume results from the net charge in that volume (Gauss's law). The next two equations are respectively the current continuity equations for holes and elec-

[14]H. K. Gummel, "A Self-Consistent Iterative Scheme for One-Dimensional Steady-State Transistor Calculations," *IEEE Trans. Electron Devices,* **ED-11,** 455–465 (1964).

trons where R represents the recombination rate and G the generation rate of electron–hole pairs. This expresses the fact that the volume rate of change of carrier density in time depends on the balance of the divergence of current flow from the volume and the net generation rate in that volume.

Equations (8.55) may be solved simultaneously for a particular two- or three-dimensional device structure for the three unknowns, p, n, and ψ, once the boundary conditions are assigned for the entire device region.[15] Included in these boundary conditions are the electric potential values which must be assigned to the emitter, base, and collector metal electrodes to represent the bias voltages applied to the device. The device structure is defined by its geometry plus the impurity distribution everywhere in the device. Semiconductor material parameters such as energy gap, carrier mobilities, and lifetime must be known.

The equations are solved by using numerical techniques, aided by a digital computer. First the differential equations are converted into a set of algebraic, nonlinear difference equations written at every point of intersection of a grid covering the device structure. Then the Newton iterative method is employed which results in a set of *linear* equations to be solved. In a two-dimensional analysis this requires the simultaneous solution of about 5000 linear equations; a set of three is written for each point of a grid formed by 40×40 lines typically, covering the 2D device structure. Three-dimensional analysis requires the solution of even more equations. Matrix techniques are used to solve this large set of equations simultaneously. Computer-aided iterative numerical techniques are often employed.

Once the electrostatic potential ψ and the carrier densities p and n are solved for at every grid line intersection in the device structure, the current density vector everywhere in the device can be obtained from Eq. (5.22). Now the terminal currents can be calculated. When this procedure is repeated for different applied terminal voltages, the device's current–voltage characteristics can be generated. Both steady-state and transient analyses can be carried out. *Any* semiconductor device can be modeled in this way once the structure is defined; *any* semiconductor device material may be assumed if its material parameters are known. This procedure is commonly used in connection with new device designs since very considerable expense is incurred in fabricating each new design.

☐ SUMMARY

An explanation of why the bipolar transistor provides power gain was given. The four basic modes of bipolar transistor operation—active, saturated, cutoff, and inverse—were described. Then the minority carrier emitter injection was modeled in terms of the junction bias voltages. The emitter current was considered to flow through the transistor base region at low injection levels only

[15]For an excellent review of the methods for solution of these semiconductor equations, see N. L. Engl, H. K. Dirks, and B. Meinerzhagen, "Device Modeling," *Proc. IEEE,* **71,** 10–33 (1983).

by the diffusion mechanism since the electric field is small there. An essentially linear distribution of electrons in an *n-p-n* transistor base results from the fact that the electron recombination is minimal in a good gain transistor. The emitter defect current limits the injection efficiency of electrons because of the back-injection of holes into the emitter region. This can be reduced by heavy doping of the emitter relative to the base region. However, this results in heavy doping effects, which limit the emitter injection efficiency; it also limits the base doping. The wide gap emitter heterojunction transistor uses different semiconductor materials for the emitter and base regions. This structure offers the possibility of both high gain even at high current levels and low base resistance by allowing high base doping. The Ebers–Moll bipolar transistor equations were introduced to provide a circuit model for the transistor for use in analyzing circuits containing these devices. These equations relate the emitter and collector currents to the emitter and base junction voltages. The Ebers–Moll circuit parameters were expressed in terms of the physical device design parameters. Both the common-base and common-emitter forms of these equations were presented. The charge-control bipolar transistor model was introduced in order to provide equations for the convenient solution of large signal switching problems. The basic parameters in this model are Q_F, the total injected base region charge; τ_n, the minority electron lifetime in the base; and τ_t, the transit time for these carriers to traverse the base region. The Gummel–Poon integral charge circuit model was described as a convenient way to include the Early and Kirk effects, as well as collector region minority injection. Finally, the more exact computer-aided modeling of the current flow in a transistor was presented, this being useful for new transistor physical design.

PROBLEMS

8.1 The transistor of Example 8.1 is operating in the active region, under steady-state conditions.

(a) Calculate the total charge of electrons stored in the base region when the transistor base current is 100 μA.

(b) Determine the collector current under the conditions of (a).

(c) Calculate the electron density just at the edge of the emitter space-charge region, in the *p*-type base region.

(d) Calculate the emitter–base voltage V_{BE} in this case.

8.2 Prove that Eq. (8.18) follows from Eqs. (8.17) and (8.16).

8.3 Repeat Example 8.1(b) if the emitter region doping is $N_d = 1.0 \times 10^{19}/\text{cm}^3$, including the heavy doping, band gap narrowing effect there. (*Hint:* Use the data of Appendix B.)

8.4 Determine the relationship between the coefficients a_{ij} and A_{ij} of Eqs. (8.19) and (8.20).

8.5 Show that the integration constants B and C of Eq. (8.22) may be expressed as given in Eqs. (8.23).

8.6 **(a)** Derive Eq. (8.24) from Eqs. (8.22) and (8.23).
(b) Under what conditions is the linear approximation of Eq. (8.24) valid?

8.7 Show that Eqs. (8.27) derive from Eqs. (8.26), (8.23), and (8.20).

8.8 Show how Eqs. (8.29) derive from Eqs. (8.27).

8.9 Using the a_{ij} expressions from Eqs. (8.29), show that Eq. (8.30) reduces to Eq. (8.9).

8.10 Introduce the expressions of Eqs. (8.29) into (8.20) and identify the physical significance of the terms in these equations for I_E and I_C.

8.11 A transistor with undetermined parameters is to be used in the circuit shown below. Data from which the transistor parameters can be computed are taken as follows: (i) V_{BC} is made zero and V_{BE} is made -10 V (reverse bias), then I_E is measured to be 50 nA and I_C to be -48 nA; (ii) V_{BE} is made zero and V_{BC} is made -10 V (reverse bias), then I_C is measured to be 64 nA and I_E -48 nA. Determine
(a) α_F;
(b) α_R;
(c) I_{CBO};
(d) I_{EBO};
(e) the forward emitter voltage needed to produce a collector current of 5 mA at 300°K, when used in the active mode.

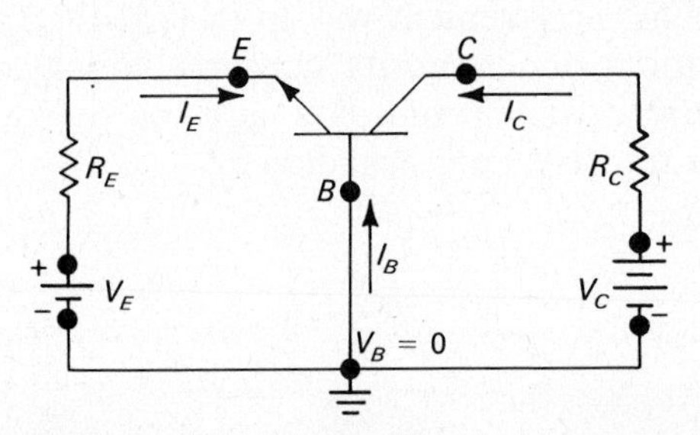

8.12 By combining Eqs. (8.35a) and (8.35b) under emitter open-circuit conditions, prove Eq. (8.37a) is correct.

8.13 **(a)** Calculate the common-base current gain α_F for the transistor of Problem 8.1.
(b) Calculate the common-emitter current gain β_F for the transistor of Problem 8.1.
Assume that the emitter injection efficiency is one.

8.14 Draw the approximate Ebers–Moll equivalent circuit for a transistor operating in the active region, where $\alpha_F \gg \alpha_R$ and $I_{CBO} \ll \alpha_F |I_E|$.

8.15 Derive Eq. (8.42) from Eqs. (8.37) and (8.1).

8.16 For a transistor with strongly reverse-biased collector junction and emitter–base open circuited, show that the floating emitter–base voltage is $kT/q[\ln(1 - \alpha_F)]$.

8.17 Derive an expression for the collector "offset voltage" V_{offset} for an *n-p-n* transistor in the grounded-emitter connection when $I_C = 0$, with the emitter junction forward biased. Show that this can be expressed as $V_{CE} = -(kT/q) \ln \alpha_R$, under certain conditions. [*Hint:* Use Eqs. (8.37) and (8.41) and assume that I_{EBO}, $I_{CBO} \ll I_B$.]

8.18 Calculate the transit time for electrons across the base region of the transistor of Problem 8.1.

8.19 The collector leakage current of a transistor in the grounded-emitter connection, with a resistor R connected between base and emitter, is called I_{CER}. Show that

$$I_{CER} = \frac{I_{CBO}(1 + qI_{EBO}R/kT)}{1 - \alpha_F\alpha_R + (qI_{EBO}R/kT)(1 - \alpha_F)}.$$

8.20 A silicon *n-p-n* switching transistor with a base width of 5.0 μm and a uniform base doping of 5.0×10^{14} acceptors/cm^3 has a heavily doped emitter and collector region. With a device operating in the saturated mode, the collector potential is 0.10 V and the base potential 0.70 V relative to the grounded emitter. Its emitter area is 6000 μm^2. The electron lifetime in the base is 1.0 μsec.
(a) Calculate the total electron charge stored in the base region of this transistor.
(b) If the base width is reduced to 1.0 μm, calculate the new value of the stored charge.
(c) What does (b) suggest about reducing the turnoff time of the transistor used as a switch?

8.21 Show the relationship between the charge-control equation (8.50b) and the Ebers–Moll equation (8.35b).

8.22 Find the Gummel number for the transistor of Example 8.1.

CHAPTER 9

Operation of the Bipolar Junction Transistor in Circuits

9.1 THE BEHAVIOR OF REAL BIPOLAR TRANSISTORS

The discussion in the last chapter was confined to what may be termed the **intrinsic** transistor. In modeling a semiconductor diode, factors other than the intrinsic *p-n* junction characteristic had to be considered for effective representation of the behavior of a real device (see Section 6.3). In the case of the BJT even more additional factors are important in describing the behavior of a real transistor. Which of these additional factors are particularly important depends greatly on the specific transistor design and operating conditions. Some of these additional effects are listed below:

1. The emitter, base, and collector regions have bulk series resistance and the ohmic voltage drop due to these resistances in the presence of current flow must be accounted for in the external *I-V* characteristics of the transistor.

2. The current gain, α or β, for a given transistor design is not a constant but is dependent on the collector voltage and the collector current.

3. The collector current will increase abruptly above a certain collector voltage as a result of avalanche multiplication in the collector junction space-charge region, in a manner similar to the diode (see Section 6.3.3).

4. The minority carrier transport across the base region of a transistor is not exclusively by diffusion but is electric field aided in some transistor designs.

Some of these effects can easily be incorporated into one of the transistor models. Sections 9.2 and 9.3, which describe the low frequency modeling of bipolar transistors, present material normally covered in a prior electronics course. This review is offered to aid in the understanding of the modeling of the high frequency operation of bipolar transistors.

9.1.1 Transistor Ohmic Voltage Drops

In order to accurately formulate the current–voltage characteristic of the transistor as illustrated for example in Fig. 8.12, possible voltage drops across resistances in the transistor bulk need to be considered. The voltage drop in the emitter bulk can normally be neglected in planar transistors, as illustrated in Figs. 8.2, owing to the heavy doping and hence high conductivity of the region. Also, this region is usually very thin.

Since the base doping is only moderate and since the base current which flows transversely (perpendicular to the emitter current) has a rather long path, the voltage drop in the **base resistance** R_B is of consequence. Since V_{BE} in the Ebers–Moll equations (8.35) represents the actual voltage drop across the emitter junction space-charge region, the relation between this voltage and the actual voltage V_{BL} applied externally between the emitter and base lead is given by $V_{BE} = V_{BL} - I_B R_B$. Since the transistor I-V characteristics are expressed in terms of terminal voltages, this equation should be introduced into Eqs. (8.35). Because this expression contains a terminal current and appears in the exponent of a term in the equations relating current and voltage, explicit relations between the terminal currents and voltages become very complicated. Fortunately, in a properly designed transistor, the base resistance is small and the base current is small if the device has good current gain and operates at low level. Hence for many purposes this voltage drop can be neglected in comparison with the voltage drop across the emitter junction. At high current levels, however, this voltage drop increases in proportion to the base current whereas the junction voltage hardly changes owing to the exponential nature of the junction I-V characteristic. Under these conditions the base drop cannot be neglected.

Finally, the importance of the series voltage drop in the transistor collector bulk should be considered. In planar transistors the collector bulk is more lightly doped than the base region, and since the collector current is often greater than the base current, this **collector series resistance** R_{CS} should be considered. This is particularly true in the case of microchip integrated-circuit (IC) bipolar transistor structures (see Fig. 8.2b), in which a long high-resistance path to the collector contact can be present. This is due to the requirement that all IC device contact interconnections be made on the chip surface. In the active region, the collector junction is reverse biased and hence represents a rather high impedance. Then the collector voltage drop $I_C R_{CS}$ is normally small compared to the collector junction resistance and can be neglected. However, in the saturation region, where the collector junction is forward biased, the collector bulk voltage drop generally even exceeds the potential drop across the junction. (The latter can be determined from the Ebers–Moll Eqs. (8.35).] The effect of this extra collector series resistance can be seen in Fig. 9.1, where the common-emitter characteristic is plotted for a transistor with different values of R_{CS}. The effect is most pronounced at low collector–emitter voltages where the transistor is in saturation and hence is collecting minority carriers inefficiently (the collector is back-injecting to the emitter). The collector–emitter potential drop represents the voltage across the transistor when operated as a switch. With the switch closed (the transistor ON), this voltage should *ideally* be zero and hence any drop across series collector resistance represents power loss or switching inefficiency.

9.1.2 Current Gain Variation with Collector Voltage and Current

The previously described expressions for transistor current gain, transport factor, and injection efficiency as given in terms of transistor parameters by Eqs. (8.32) and (8.14), involve the transistor base width W. This represents the distance the minority injected holes must travel before collection, or the distance between the edges of the emitter and collector space-charge regions. In Section 6.1.3 it was pointed out that the space-charge region of a p-n junction widens

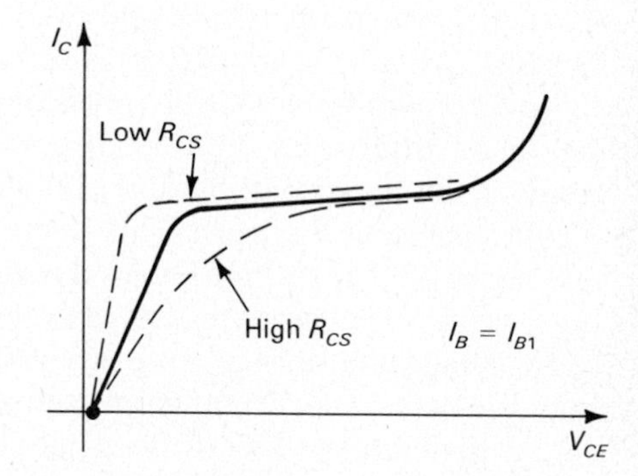

Figure 9.1 A common-emitter transistor current-voltage characteristic showing the effect of series collector resistance R_{CS}. The solid curve essentially represents a typical characteristic and the dashed lines represent the characteristics of the same device with either lower or higher R_{CS}.

as increased reverse bias is applied to the junction. Hence since the collector junction of a transistor in the active region is reverse biased, the edge of its space-charge layer should extend toward the emitter junction as the collector voltage is increased. Then the transistor base width, and correspondingly also the current gain, becomes voltage dependent. Hence for a given emitter current, the collector current will increase somewhat as the collector–base voltage V_{CB} is increased. This may be seen by examining Fig. 8.12 and in a similar sense Fig. 8.16 where I_C increases with V_{CE}. An extreme situation in this respect is called **punchthrough,** where the collector space-charge region reaches through to the emitter region—causing the collector current to increase drastically, limited only by the external circuit. The extent to which this effect occurs may be expressed by noting that the rate of increase of I_C with V_{BC} in the active region can be expressed by taking the differential of Eq. (8.37a) (neglecting the small I_{CBO} term) as

$$\left(\frac{\partial I_C}{\partial V_{BC}}\right)_{I_E=\text{const}} = -I_E\left(\frac{\partial \alpha_F}{\partial V_{BC}}\right)_{I_E=\text{const}} . \tag{9.1}$$

Therefore the slope of I_C versus V_C at constant emitter current is proportional to this current as well as the rate of change of α_F with V_{BC}. The latter effect results from the reduction in effective base width with collector space-charge widening due to increased collector potential and is known as the **Early effect.** This will be large when the base doping is small (see Section 6.1.3). The expression corresponding to Eq. (9.1) for the common-emitter transistor connection, obtained by differentiating Eq. (8.41), is

$$\left(\frac{\partial I_C}{\partial V_{CE}}\right)_{I_B=\text{const}} = I_B\left(\frac{\partial \beta_F}{\partial V_{CE}}\right)_{I_B=\text{const}} . \tag{9.2}$$

The increase in the I_C-V_{CE} slope at higher I_B can be seen in Fig. 8.16. In effect Eqs. (9.1) and (9.2) represent the transistor output conductance, which should be maintained low for good power gain.

The variation of current gain with collector current at a given collector voltage is a somewhat more complex situation. Equation (8.37a) as stated indicates a linear variation of collector current with emitter current. If a plot were made of I_C versus I_E, this should result in a straight line. A typical variation of I_C with i_E for silicon transistors is indicated in Fig. 9.2. Deviation from linearity occurs at both very small values of emitter current and at high values of I_E. In silicon transistors the loss of current gain at low emitter current values results from the recombination of injected carriers in the emitter space-charge region (see Section 8.1.4E). This fraction of the emitter current is not collected and hence effectively constitutes a reduction in emitter efficiency. As the emitter current is increased, though, this recombination current becomes negligible compared to the normal minority-carrier-injected diffusion current (see Section 8.1.4) and the current gain rises.

At high current levels the effective injection efficiency again falls off, but

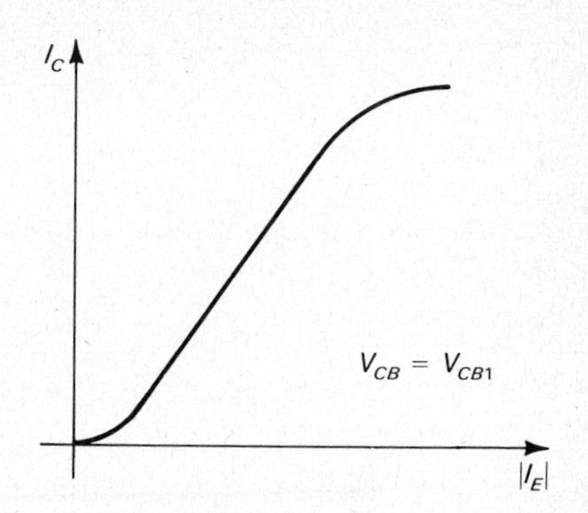

Figure 9.2 Variation of collector current with emitter current for a given collector potential in a typical silicon transistor. The linear portion of the curve represents constant current gain, whereas the curvature at both high and low currents indicates falloff of gain at these extremes.

for another reason. This can be understood by referring to Section 8.1.4. There it is pointed out that the doping on the *p*-side of the emitter *p-n* junction should be much greater than on the *n*-type base side of the junction for good injection efficiency. The expression derived as Eq. (8.14) assumes that the carriers present on each side of the junction derive from the impurity atoms there only, as under equilibrium conditions. This is a good approximation at low current levels, but at high levels of injection a significant concentration of electrons are injected into the *p*-type base region of an *n-p-n* transistor, comparable to or greater than the density of majority holes present initially. Hence the conductivity of the base region will be significantly altered or **modulated.** However, the n^+-type emitter side of the junction is not similarly modulated, relatively speaking, since there are an enormous number of carriers there to begin with as a result of the high doping of this n^+-p junction. Hence the ratio of the conductivity of the *n*-region to that of the *p*-region is reduced at high injection levels and so the injection efficiency as well as the current gain is reduced.

In bipolar transistors with lightly doped collector regions two other factors can cause the falloff of current gain at high current density. The first is caused by the collector **quasisaturation** effect caused by high collector series resistance. Since the transistor at its operating point may not quite be biased into the active region, the output collector current is lowered (see Fig. 9.1). This results in an apparent reduction in current gain. Also, the large density of injected electrons can alter the charge in the collector space-charge region, moving this high field region toward the collector contact, effectively widening the base region and reducing the current gain as well as the high frequency capability. This will be discussed in Section 9.1.5.

9.1.3 Avalanche Breakdown of the Collector Junction

Current multiplication in a *p-n* junction such as the collector junction of a transistor has been described in Section 6.3.3. This results from avalanche breakdown caused by charge carriers diffusing into the space-charge region and being accelerated to an energy high enough for impact ionization of some valence electrons of the semiconductor lattice atoms. However, in the case of the collector junction of an *n-p-n* transistor operating in the active region, the charge carriers entering the space-charge region may originate from injection by the emitter; this differentiates the transistor phenomenon from the diode effect. The *I-V* characteristics of the collector junction of an *n-p-n* transistor in avalanche breakdown can be seen by examining the transistor collector curves already drawn in Figs. 8.12 and 8.16. It is seen that the collector current increases rapidly above a certain collector voltage, for a given input current.

Note that the shape of the breakdown curve for a given transistor depends on the input current and the type of circuit connection, i.e., common base or common emitter. A few of these curves are repeated for clarity in Fig. 9.3. Curve 1 represents the normal collector–base *p-n* junction breakdown curve since it is measured with the emitter open circuited. The breakdown voltage here is defined as BV_{CBO}. If reverse potential is now applied between the collector and emitter leads, as in the common-emitter connection, and the base lead is left open circuited, curve 2 results. Note that the breakdown voltage $BV_{CEO} < BV_{CBO}$ occurs here and the voltage across the device becomes lower, with avalanching still maintained, as multiplication causes the collector current to increase. The lowest voltage to maintain the breakdown is called the sustaining voltage $V_{CE(\text{sus})}$ in analogy with the gaseous discharge that occurs for example in a fluorescent lamp. Curve 3 represents again the common-emitter connection, but in this case some external base current is applied and current

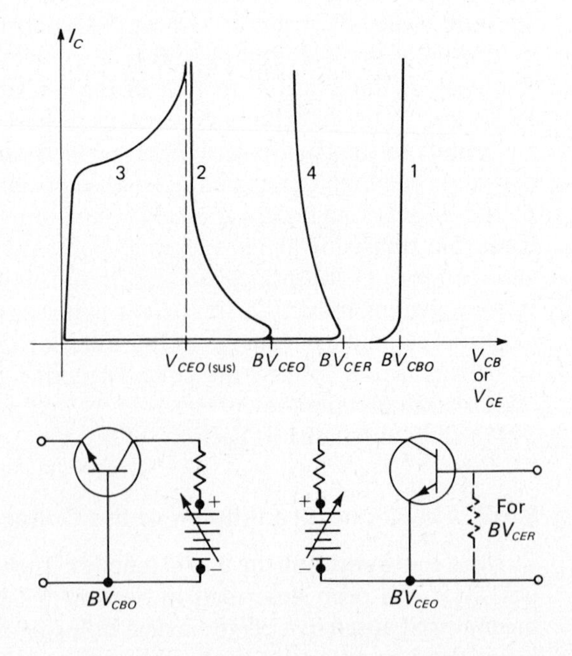

Figure 9.3 Common-emitter and common-base characteristics of an *n-p-n* transistor, emphasizing the voltage breakdown region; BV_{CBO} represents essentially the collector diode avalanche characteristic (curve 1) with the emitter lead open (the BV_{CBO} measuring circuit is indicated at bottom left); BV_{CEO} represents the peak voltage which can be supported, collector-to-emitter, in the common-emitter connection with the base lead open (curve 2) (the BV_{CEO} measuring circuit is indicated at bottom right); BV_{CER} is the peak supported collector–emitter voltage with a resistor between emitter and base (curve 4). The voltage supported at higher currents is less than at low currents, causing a negative resistance portion in the curve. Also indicated is a curve for the common-emitter connection when some base current is assumed to flow (curve 3).

gain is indicated. Carrier multiplication increases the collector current in the neighborhood of breakdown and the sustaining voltage is again approached asymptotically at high current values. This voltage, $V_{CE(sus)}$, typically is of the order of $\frac{1}{2}BV_{CBO}$ and can be even less if the transistor intrinsically has high gain; BV_{CER} is the peak supported collector–emitter voltage with a resistor connected between emitter and base. The breakdown curve in this case is marked 4 in Fig. 9.3 and lies between the curves for BV_{CBO} and BV_{CEO}, since the transistor gain is reduced from the open base case.

The relation between the sustaining voltage and the transistor current gain will now be formulated. The multiplication factor M of Section 6.3.3 is defined as the ratio of the current leaving the junction space-charge region relative to that entering this region. Using Eq. (8.37a) we express the transistor collector current in the active region, with the collector avalanching, by

$$I_C = M(-\alpha_F I_E + I_{CBO}). \tag{9.3}$$

For the common-emitter configuration it is convenient to express this output current in terms of the input base current. Using Eq. (8.1), we can write Eq. (9.3) as

$$I_C = \frac{M\alpha_F}{1 - M\alpha_F} I_B + \frac{M}{1 - M\alpha_F} I_{CBO}. \tag{9.4}$$

Hence it is clear that the collector current will increase, limited only by the circuit supplying the current, with or without base current, when

$$\alpha_F M = 1. \tag{9.5a}$$

In terms of the common-emitter gain β_F, this collector avalanche condition, using Eq. (8.42) becomes

$$\beta_F(M - 1) = 1. \tag{9.5b}$$

Introducing the expression for M from Eq. (6.50) gives an equation from which the transistor avalanche I-V characteristic can be computed in the common-emitter connection, if the current dependence of β_F, $\beta_F(I_C)$, is known. This is

$$\beta_F(I_C)\left(\frac{1}{1 - (V_{CB}/BV_{CBO})^n} - 1\right) = 1. \tag{9.6}$$

Since the transistor current gain at low currents increases with current (see Fig. 9.2), this expression predicts the reduction in collector blocking voltage V_{CB} with collector current. Hence the voltage capability of a transistor in the common-emitter connection is less than in the common-base connection, particularly for high gain devices.

In all of the above expressions the collector junction leakage current with the emitter open circuited, I_{CBO}, should be understood to be the leakage current characteristic of a p-n junction diode, as already described in Sections 6.2.2, 6.2.4, and 6.2.5. In addition, surface-limited voltage breakdown effects similar to those described in Section 6.3.3C apply here as well. In silicon planar transistor structures used in integrated circuits (see Fig. 8.2b), the collector region is usually more lightly doped than the emitter or base regions. Therefore the collector junction in these devices behaves like an n^+-p junction in a p-n-p transistor and a p^+-n junction in an n-p-n transistor. Hence positive charges in the oxide covering the collector junction will induce an n-type surface channel in a p-n-p transistor which can be limited by a p-type "guard ring" to avoid premature voltage breakdown at the surface; n-p-n transistors can suffer from low voltage breakdown due to the presence of a p^+-n^+ surface junction (see Section 6.3.3C). Lateral p-n-p structures used in integrated circuits have heavily doped collectors as well as emitters and thus suffer surface breakdown effects typical of p^+-n junctions.

9.1.4 The Built-In Field Due to Nonuniform Base Doping

Thus far it has been assumed that the three transistor regions are uniformly doped. In fact the planar transistors in integrated circuits have a substantially greater base doping near the emitter junction than near the collector junction. This is a natural consequence of the fact that these devices are fabricated by diffusing sequentially, from one surface, first p-type impurities and then n-type impurities into lightly doped n-type silicon.[1] The impurity profile of the transistor so obtained is shown in Fig. 9.4. The gradient of the impurity density in the transistor base region is of particular interest. This acceptor variation introduces an electric field in the base region which is in a direction to accelerate minority electrons through the p-type base region, reducing the transit time,

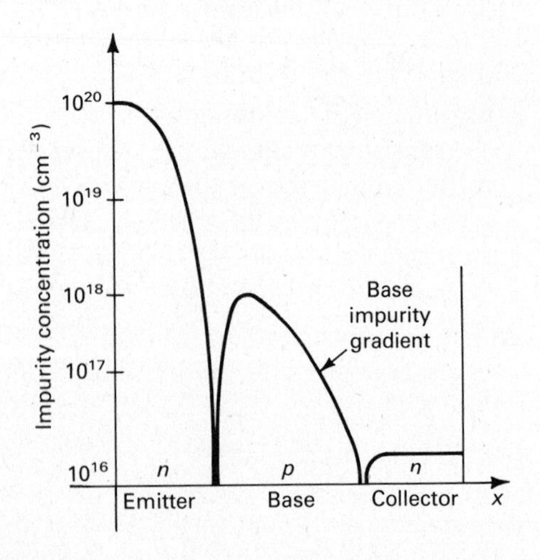

Figure 9.4 The impurity density distribution in a diffused n-p-n transistor. The emitter doping is n-type and about 10^{20} donors/cm^3 at the edge of the emitter. The base doping is p-type, reaching a maximum of about 10^{18} acceptors/cm^3 close to the base edge nearest the emitter, falling off toward the collector edge. This impurity grading produces a built-in electric field in the base region.

[1]In integrated circuits n-p-n transistors are most common, but similar arguments apply for p-n-p devices with the role of electrons and holes reversed.

emitter-to-collector, and hence increasing the current gain and transistor high-frequency capability. That is, these latter values are increased from the values obtained by assuming that carrier transport is solely by diffusion, as has been discussed thus far. However, the effect is moderate since the voltage drop across the graded base region is typically only about $5kT/q$ or 130 mV and hence the **built-in** field is small. This effect is only important at low current levels, for at high density the injected carriers similarly produce an aiding field which "swamps out" the built-in field due to impurities. Then the diffusion flow approximation is certainly no longer valid.

9.1.5 High Current Density Effects in Bipolar Transistors

Two additional effects need to be discussed in order to understand the operation of bipolar transistors operating at high current densities. One of these phenomena is **emitter current crowding** and the other the **Kirk effect.** These effects are not only of importance in power transistors designed for high current operation, but also in microchip IC transistors where the micrometer device dimensions cause the emitter current density to be high ($>10{,}000\ A/cm^2$) even at a few milliamperes of current.

Emitter current crowding refers to the concentration of emitter current at the edge(s) of the emitter closest to the base contact (see Figs. 8.2). In an *n-p-n* transistor the base current flows parallel to the emitter–base *p-n* junction and the $I_B R_B$ voltage drop across the base resistance under the emitter causes the electric potential at the emitter edge to rise above that at the center. This produces a greater voltage drop across the emitter–base junction at the emitter edge. The junction law of Eq. (6.32) dictates that the current increases exponentially with junction voltage and this causes the current to concentrate mainly near the edge(s) of the emitter nearest the base contact. This nonuniform emitter current distribution can result in reduced current gain due to the increased emitter current density (for a given emitter current) as pointed out in Section 9.1.2. It may also be responsible for raising the junction temperature at the edge of the emitter, which can give rise to a thermal instability known as **thermal second breakdown.**[2] The latter phenomenon results from the negative temperature coefficient of current of the emitter *p-n* junction (see Section 6.3.1). For, as the junction heats up, for a given emitter–base junction voltage, more emitter current flows, producing additional power dissipation there, which calls for even more current, etc. Local heating anywhere in the junction area will cause a relatively uniform current distribution to become concentrated in a small local region and hence burnout can occur there.

Also at high emitter current densities the large minority electron density injected into the *p*-type region enters the collector space-charge region. These mobile carriers can influence the charge density there previously made up of

[2]D. Navon and E. A. Miller, "Thermal Stability in Power Transistor Structures," *Solid-State Electron.*, **12,** 69 (1969).

High power silicon bipolar transistor in metal TO-3 packages which provide proper thermal design for the extraction of heat dissipated in operation. (Courtesy of Unitrode Corporation, Lexington, MA.)

fixed negatively charged ionized acceptors and positively charged ionized donors. These negatively charged electrons will tend to neutralize the positively charged donor ions there, so that the field can no longer be supported in the original (low current) collector depletion region. Hence the region of high electric field moves toward the collector contact and the base region will effectively widen. This will cause the current gain to fall off with increased current density; also it will increase the electron transit time through the base region [see Eq. (8.45)] and hence increase the transistor switching time and decrease its frequency capability, as will be discussed in Section 9.4.3. These effects were explained by Kirk.[3] A further increase in current density can cause a high electric field in a space-charge region dominated by the injected electron charges, and this space-charge there may no longer be capable of supporting the applied collector voltage. This will give rise to a large increase of collector current (termed **avalanche injection**) which may result in a device instability known as **electrical second breakdown** which can destroy the device.[4]

9.2 THE TRANSISTOR AS A CIRCUIT ELEMENT

In Section 8.1.2 the ability of a junction transistor in an appropriate circuit to provide power gain or amplification has been discussed. That is, when biased in the active region the transistor is capable of delivering more signal power to the load than the signal that is supplied to its input circuit. This input may be in the form of a current or voltage signal; it may be sinusoidal in time or have some other time dependence such as the voltage pulse input. In any case the usefulness of the device derives most often from its ability to take a small signal from a transducer or a communication channel and, after a number of stages of amplification, provide enough power to derive an output device such

[3]See footnote 11, Chapter 8.

[4]P. L. Hower and V. G. K. Reddi, "Avalanche Injection and Second Breakdown in Transistors," *IEEE Trans. Electron Devices,* **ED-17,** 320 (1970).

as a loudspeaker or other electromechanical driver. The manner in which this is accomplished will be described next with the aid of the load-line concept. This will be done for an *n-p-n* transistor connected to the common-emitter circuit configuration. Note that the emitter here is normally electrically connected to a metallic ground plane, as shown in Fig. 9.5a, and is taken as zero reference potential.

9.2.1 Common-Emitter Amplifier Load-Line Analysis

Because of the nonlinear character of the transistor *I-V* characteristics signal amplification is often studied by graphical methods. Consider the elementary common-emitter transistor amplifier circuit shown in Fig. 9.5a. The *I-V* characteristics of the transistor shown are given in Fig. 9.5b along with a load line representing the output load resistance R_L. The circuit DC operating point is established by the intersection of the load line and one of the *I-V* curves representing the transistor behavior, for a given base bias current. The intersection of this load line and the transistor characteristic for base current equal to I_{B2} is the DC **operating point,** labeled O in Fig. 9.5b.

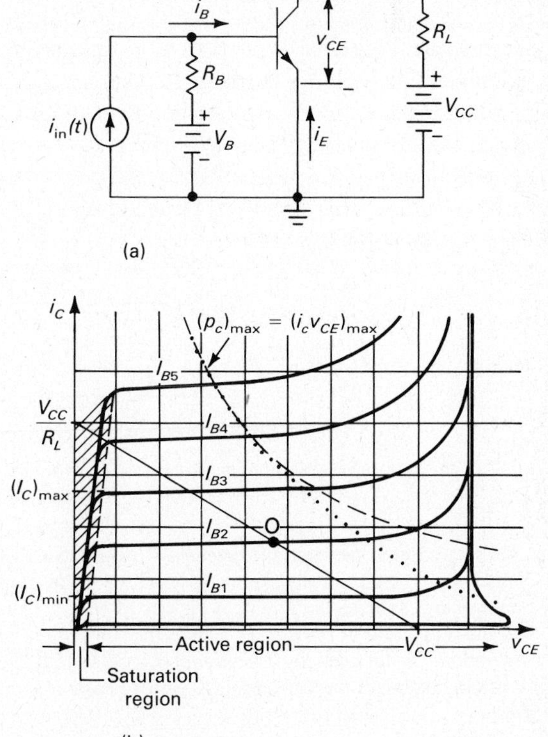

Figure 9.5 (a) Common-emitter transistor amplifier driven by a signal current. Note the grounding of the emitter in this common or grounded-emitter configuration. (b) The current–voltage collector characteristics of an *n-p-n* transistor for various values of base current. The load line for a battery voltage V_{CC} and a load resistor R_L is indicated. The operating point for a base current I_{B2} is marked by O. Also shown is the maximum power curve (dashed), representing the locus of constant $i_C V_{CE}$ product. A more realistic maximum power curve is indicated by the dotted curve.

These considerations have not included the effect of the signal current $i_{in}(t)$, which of course is the primary question. However, the effect of this input current can easily be ascertained with the aid of Fig. 9.5b. Consider that $i_{in}(t)$ is a small sinusoidal signal current alternating in time. This current superimposed on the DC bias base current I_{B2} represents the input current to the transistor. Assume that this total input base current oscillates between the extremes I_{B1} and I_{B3} as shown in Fig. 9.6. The corresponding extremes of collector current through the load resistor R_L are $(I_C)_{max}$ and $(I_C)_{min}$ as shown in Fig. 9.5b. Since the ratio of the incremental increase in collector current resulting from the incremental increase in base current due to the signal is much greater than one for a good transistor, the device exhibits current gain.[5] In fact this ratio is often referred to as h_{fe},[6] the common-emitter AC current gain, and is defined by

$$h_{fe} \equiv \left(\frac{\partial i_C}{\partial i_B}\right)_{V_{CE}\,=\,\text{const}} \tag{9.7}$$

If this quantity was constant and independent of base current a sinusoidal input signal current would yield an exact replica of output current, only increased in magnitude by the current gain. However, in general, this is not

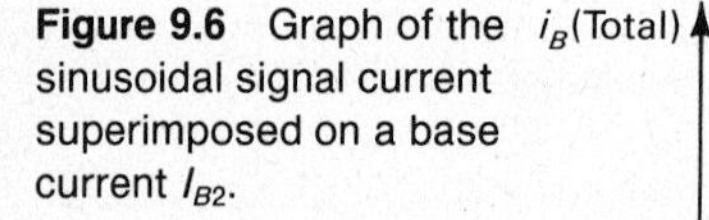

Figure 9.6 Graph of the sinusoidal signal current superimposed on a base current I_{B2}.

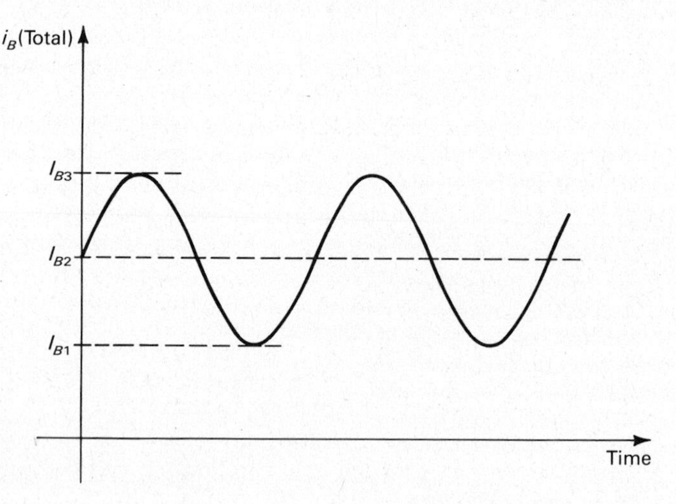

[5]The argument presented in Section 8.1.2 indicates why the device also gives power gain. Of course it can be shown that owing to a conservation-of-energy consideration, the power supplied by the energy sources (batteries) in the circuit exactly equals the power dissipated in the transistor and the remainder of the circuit. This is the subject of a problem at the end of this chapter. [See J. F. Gibbons, *Semiconductor Electronics* (McGraw-Hill, New York, 1966), p. 379.]

[6]In analogy with the DC common-emitter current gain, $|I_C/I_B|_{V_{CE}\,=\,\text{const}}$ is normally called h_{FE} or β.

the case except in a small current range. Hence in general a sinusoidal input signal will not result in an exactly sinusoidal amplified output signal. That is, the bipolar transistor as an amplifier introduces **distortion** in processing the signal. In small signal processing, where the current gain is reasonably constant over the range of current variation with signal, the device is said to be operating in the **linear** region. This is more true over an intermediate range of current values and less true at the high and low extremes of collector current (see Fig. 9.2).

This is no longer true in the case of large input signals; for example, if the peak value of the input current reaches I_{B4} and is increased still further to I_{B5} (see Fig. 9.5b), the value of the output current hardly increases. (The collector currents obtained for these base currents are determined by the intersection of the load line with the I-V characteristic, for I_{B4} and then I_{B5}.) Now the transistor is no longer operating totally in the active region, and the device is said to be driven into **saturation** or **bottoming.** This saturation region is indicated in Fig. 9.5b. The concept of current gain obviously has no meaning in this situation and can be applied only in the small signal or linear region of operation. The DC operating point is generally chosen depending on the nature of the input signal and extent of distortion allowable in amplification.

9.2.2 Temperature and Frequency Limitations of Transistors

The maximum power level at which the transistor can operate (or **safe operating area,** SOA), usually specified by the device manufacturer, also restricts the DC operating point. This power, handled by the transistor, causes internal heating, which raises the collector junction temperature above the package or case temperature. The maximum power dissipation $(p_C)_{\max}$ $[=(i_C v_{CE})_{\max}]$ graphs as a hyperbola and is indicated with the transistor output characteristics of Fig. 9.5b. Operating points below this curve are thermally stable, while those above are unstable. Here internal thermal dissipation limits the current and voltage that the transistor can handle.[7] The efficiency with which the heat generated in a given transistor is removed by mounting onto a metal heat sink can seriously affect the power handling capability of the device. A typical thermal mounting for a packaged transistor is shown in Fig. 9.7. The transistor package itself limits the efficiency with which heat can be removed from the transistor chip and transferred to the surrounding atmosphere. This limitation to heat removal is usually expressed as

$$\theta_{JC} \equiv \Delta T_C / p_C, \tag{9.8}$$

[7]In practice it is found that transistors can handle more power reliably at low voltage and high current, at a given power, than at high voltage and low current, so the hyperbolic power limitation curve is only approximate. This results from the transistor thermal failure mechanism known as "second breakdown." A more realistic maximum power curve is shown dotted in Fig. 9.5b. The stable region below this curve is called the "transistor safe operating area" (SOA).

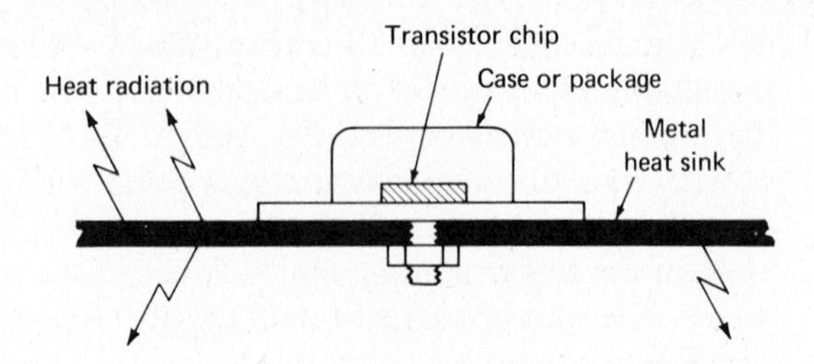

Figure 9.7 Sketch of a typical mounting of a transistor package onto a metal plate for heat sinking and thermal radiation into the surrounding atmosphere, resulting in cooling of the transistor chip.

where θ_{JC} is the transistor collector junction-to-case thermal resistance (usually specified by transistor manufacturers in degrees Celsius per watt) and ΔT_C is the rise in the transistor collector temperature above the package temperature when the transistor power p_C is dissipated in the device, mainly in the collector space-charge region. An additional case-to-ambient thermal resistance θ_{CA} must be added to θ_{JC} to account for the efficiency limitation of heat transfer from the case and heat sink to the ambient.

It should be understood that the ability of a transistor to function as an active device and produce power gain depends on it maintaining the nonlinear behavior characteristic of its *p-n* junctions. It is to be noted that when the device is raised to excessively high temperatures (greater than 300°C for silicon transistors), the large number of electron–hole pairs thermally generated in the semiconductor material tend to dominate current flow and the *p-n* junctions tend to conduct in an ohmic fashion, causing the power gain to fall off. Also,

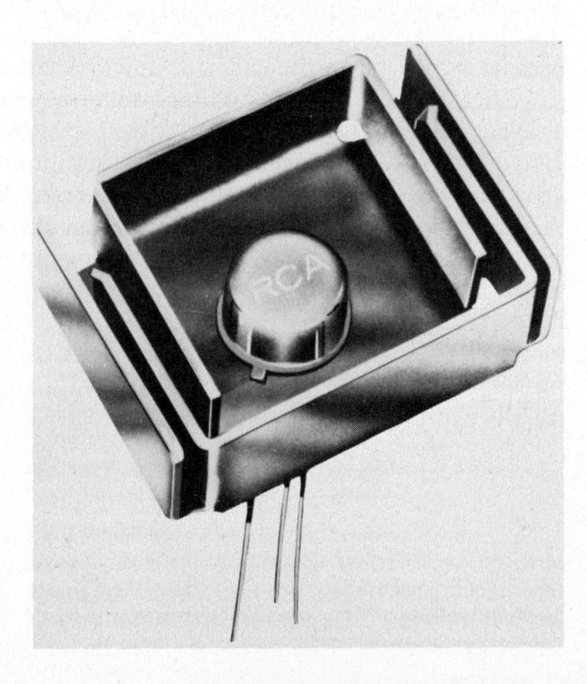

Power transistor package with heat sink. (Courtesy of RCA Solid State Division, Somerville, NJ.)

when the input signal to a transistor is of a sufficiently high frequency, the capacitive character of the *p-n* junctions of the transistor tends to dominate the behavior of the device. The high frequency current then passes through the emitter junction capacitance as a displacement current, which is a majority carrier current, and little minority carrier injection results. The junctions in this case become nearly linear, passive elements and hence the gain of the transistor falls off significantly with frequency. The hybrid-π model of the transistor will be presented later to explain this high frequency behavior of bipolar transistors.

9.3 SMALL-SIGNAL LINEAR TRANSISTOR MODELS

Useful AC circuit models for bipolar transistors will now be described. In Chapter 8 the Ebers–Moll and charge-control bipolar transistor models were discussed. The former is a rather general large-signal model intended primarily for DC and low-frequency analysis. The charge-control model is mainly useful for transistor switching applications. Consider now the transistor operating *strictly* in the active region with a small AC signal applied, in a typical linear amplification application. The device will be treated from a circuit viewpoint with the model parameters calculable from measured transistor characteristics. Initially a low frequency model will be developed but afterward the extension of the model to high frequency, small-signal operation will be made.

9.3.1 Two-Port Hybrid Transistor Model

The transistor may be considered as a black box with one input and one output port. The terminal characteristics are completely determined by the specification of two currents and two voltages. Two of the quantities may be specified as independent variables and the other two can then be determined from these plus certain device parameters. Such a transistor model is shown in Fig. 9.8, which shows the small-signal input current and voltage i_1 and v_1, respectively, and the output current and voltage i_2 and v_2, as well as some voltage polarities and current directions arbitrarily chosen.[8] If input current i_1 and output voltage v_2 are chosen as the independent variables and *linear, small-signal* conditions

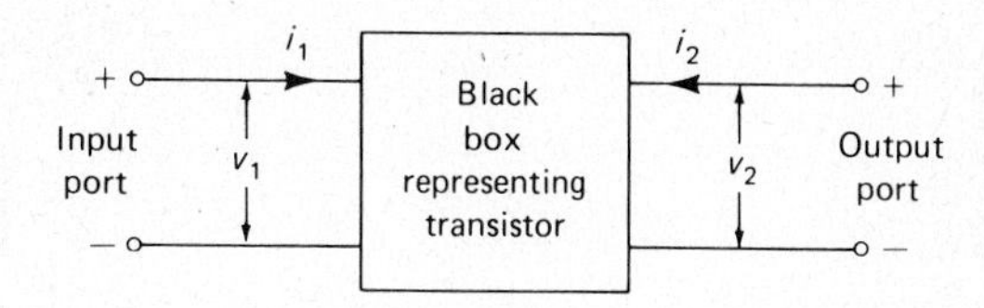

Figure 9.8 Two-port transistor model utilizing a black box transistor representation.

[8]Note that these small signals are superimposed on the DC current and voltage which bias the device into the active region.

are assumed, the remaining voltage and current can be written as

$$v_1 = h_{11}i_1 + h_{12}v_2, \tag{9.9a}$$

$$i_2 = h_{21}i_1 + h_{22}v_2. \tag{9.9b}$$

Here the quantities h_{11}, h_{12}, h_{21}, and h_{22} are parameters which describe the transistor characteristics at one particular DC operating point.

The letter h is chosen to describe these parameters since they are **hybrid** in the sense that they are not all dimensionally the same; one has the dimensions of ohms and another is dimensionally reciprocal ohms or **siemens** (S). They can be determined for a particular transistor by using a small sinusoidal input signal in a manner somewhat like that outlined in Section 8.2.2 for determination of the Ebers–Moll device parameters. For example, Eqs. (9.9) give $h_{11} = v_1/i_1|_{v2=0}$, which represents the transistor input resistance with the output short circuited; $h_{12} = v_1/v_2|_{i1=0}$, which is the reverse voltage gain with the input open circuited; $h_{21} = i_2/i_1|_{v2=0}$, which is the negative of the forward current gain with the output short circuited; and $h_{22} = i_2/v_2|_{i1=0}$, which is the output conductance with open-circuited input. At low frequency the h parameters are real numbers and the currents and voltages are in phase. However, in general they will be complex numbers and frequency dependent, so the currents and voltages will bear certain phase relationships to each other with respect to time. These device parameters are considered constant over the small-signal excursion about a DC operating point and are the ones normally specified on manufacturer specification sheets (see Appendixes F1 and F2).

A circuit model may now be constructed for the transistor for use in analyzing the performance of transistor amplifier circuits. This hybrid circuit model is shown in Fig. 9.9. Note that it contains a voltage source $h_{12}v_2$ (diamond shaped) as well as a current source $h_{21}i_1$ (diamond with arrow).[9] That this model can be derived from Eqs. (9.9) is the subject of a problem at the end of this chapter.

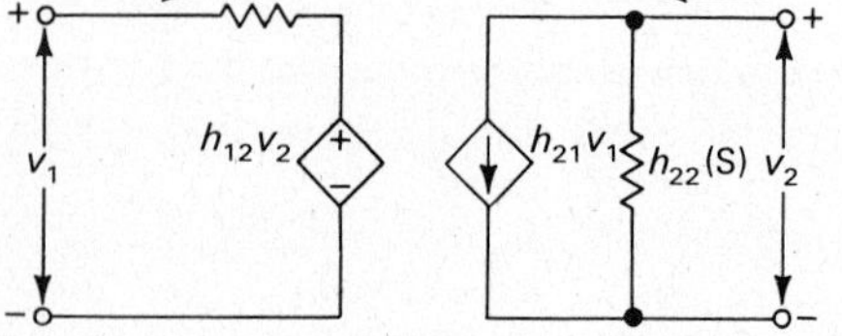

Figure 9.9 Hybrid circuit model for a transistor.

[9]See the Ebers–Moll equivalent circuit (Section 8.2.3) for an example of a current source.

9.3.2 The Transistor Hybrid Model for the Common-Emitter Connection

Consider the common-emitter transistor circuit of Fig. 9.10 containing an *n-p-n* transistor. It is desired to derive the hybrid model parameters for this device in this circuit configuration. If we assume that i_B and v_{CE} are chosen as the independent variables it follows that v_{BE} and i_C are functions of these variables and can be written, for the transistor operating in the active region, as

$$v_{BE} = f(i_B, v_{CE}), \tag{9.10a}$$

$$i_C = g(i_B, v_{CE}). \tag{9.10b}$$

Small-signal considerations permit a Taylor expansion of these equations about an operating point I_C and V_{CE}. This gives, considering only first-order terms,

$$\Delta v_{BE} = \left.\frac{\partial f}{\partial i_B}\right|_{VCE} \Delta i_B + \left.\frac{\partial f}{\partial v_{CE}}\right|_{IB} \Delta v_{CE} = \left.\frac{\partial v_{BE}}{\partial i_B}\right|_{VCE} \Delta i_B + \left.\frac{\partial v_{BE}}{\partial v_{CE}}\right|_{IB} \Delta v_{CE}, \tag{9.11a}$$

$$\Delta i_C = \left.\frac{\partial g}{\partial i_B}\right|_{VCE} \Delta i_B + \left.\frac{\partial g}{\partial v_{CE}}\right|_{IB} \Delta v_{CE} = \left.\frac{\partial i_C}{\partial i_B}\right|_{VCE} \Delta i_B + \left.\frac{\partial i_C}{\partial v_{CE}}\right|_{IB} \Delta v_{CE}. \tag{9.11b}$$

Since the symbols v_{be}, v_{ce}, i_b, and i_c represent actual incremental or small-signal quantities, Eqs. (9.11) may be rewritten as

$$v_{be} = h_{ie} i_b + h_{re} v_{ce} \tag{9.12a}$$

and

$$i_c = h_{fe} i_b + h_{oe} v_{ce}, \tag{9.12b}$$

where

$$h_{ie} \equiv \frac{\partial f}{\partial i_B} = \left.\frac{\partial v_{BE}}{\partial i_B}\right|_{VCE}, \qquad h_{re} \equiv \frac{\partial f}{\partial v_{CE}} = \left.\frac{\partial v_{BE}}{\partial v_{CE}}\right|_{IB},$$

$$h_{fe} \equiv \frac{\partial g}{\partial i_B} = \left.\frac{\partial i_C}{\partial i_B}\right|_{VCE}, \qquad h_{oe} \equiv \frac{\partial g}{\partial v_{CE}} = \left.\frac{\partial i_C}{\partial v_{CE}}\right|_{IB}.$$

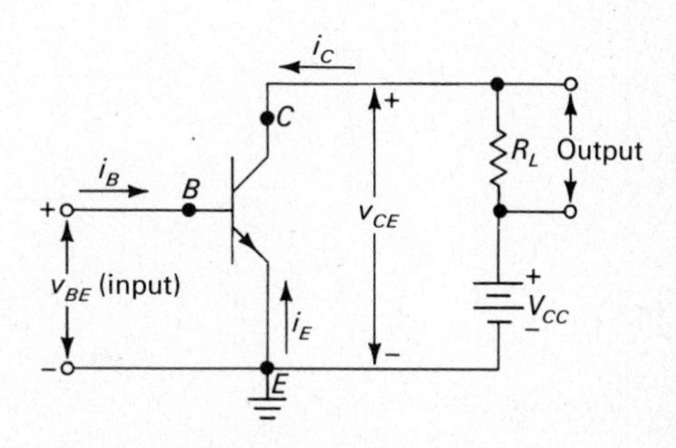

Figure 9.10 Common-emitter transistor circuit utilizing one battery. The input signal is v_{BE} and the output is taken across the load resistor R_L.

Equations (9.12) are of the same form as Eqs. (9.9) and hence they confirm that the equivalent circuit of Fig. 9.9 may be used to represent the common-emitter-connected transistor. Hence h_{ie} is the commonly accepted symbol for the transistor input resistance, h_{re} represents the feedback voltage ratio, h_{fe} denotes the small-signal current gain, and h_{oe} symbolizes the output conductance for the commit-emitter transistor connection. The definitions of the h parameters in Eqs. (9.12) suggest a method for finding their numerical values. A technique for determining these numbers by measurements on a given transistor is the subject of a problem at the end of this chapter. Typical values for these parameters are often indicated in the specification sheets supplied by the manufacturer for these devices (see Appendixes F1 and F2).

EXAMPLE 9.1

(a) Using Fig. 9.5b estimate the h_{fe} of the transistor represented by these characteristic curves at the operating point O indicated, assuming $I_{B1} = 0.10$ mA, $I_{B2} = 0.20$ mA, $I_{B3} = 0.30$ mA, $V_{CC} = 16$ V, and $R_L = 800\ \Omega$. Find also $h_{FE} = \beta \equiv (I_C/I_B)|_{V_{CE}}$.

(b) Using Fig. 9.5b estimate h_{oe} at the operating point O.

(c) Determine the power dissipated in the device operating in the quiescent condition at operating point O.

Solution

(a)
$$\frac{V_{CC}}{R_L} = \frac{16\ \text{V}}{800\ \Omega} = 20 \times 10^{-3}\ \text{A},$$

$$h_{fe} = \left.\frac{\partial i_c}{\partial i_B}\right|_{V_{CE}} = \frac{(11.0 - 6.0) \times 10^{-3}\ \text{A}}{(0.25 - 0.15) \times 10^{-3}\ \text{A}}$$

$$= \underline{50.}$$

$$\beta = h_{FE} = \left.\frac{I_C}{I_B}\right|_{V_{CE}} = \frac{8.7 \times 10^{-3}\ \text{A}}{0.20 \times 10^{-3}\ \text{A}} = \underline{44}\,.$$

(b)
$$h_{oe} = \left.\frac{\partial i_c}{\partial v_{CE}}\right|_{I_B} = \frac{(8.8 - 8.6) \times 10^{-3}\ \text{A}}{(11.2 - 7.2)\ \text{V}}$$

$$= \underline{5.0 \times 10^{-5}\ \text{S} = 50\ \mu\text{S}.}$$

(c)
$$P_{\text{diss}}|_O \approx I_C V_{CE}|_O = (8.7 \times 10^{-3}\ \text{A})(9.2\ \text{V})$$

$$= \underline{80 \times 10^{-3}\ \text{W} = 80\ \text{mW}.}$$

9.4 A HIGH FREQUENCY BIPOLAR TRANSISTOR MODEL

The two-port low frequency small-signal transistor model of Fig. 9.9 contains current and voltage sources as well as resistive elements. In this type of circuit the current and voltages are always in phase; there are no time delays involved. In an actual transistor the time required for the charge carriers injected by the emitter (input) junction to reach the collector (output) junction is often quite short owing to the narrow base width; in fact the criterion for low frequency operation is that this time constant be much smaller than the period of the input signal. However, when the period of the input signal becomes comparable to or less than the transport time for the injected carriers, provision must be made for the representation of this time delay in any valid high frequency transistor model. In addition, a finite time is required to charge the junction capacitance of the emitter junction before carrier injection can begin, and this time delay will also enter into the high frequency behavior of the transistor. It is the purpose of the next section to indicate how a high frequency bipolar transistor model can be derived which will accurately represent the behavior of the transistor as a function of frequency. One such simple equivalent circuit which has proved successful in representing the circuit performance of a common-emitter-connected transistor over a broad range of frequencies is the **hybrid-π** model. Here reactive circuit elements are introduced to provide proper phase relationships between the currents and voltages. A separate set of circuit parameter values must be known for each DC operating point in this AC small-signal model of the transistor operating in the active region.

9.4.1 The Hybrid-π Common-Emitter Transistor Model

The hybrid-π small-signal common-emitter transistor model is shown in Fig. 9.11a. A convenience of this model is that all passive circuit parameters are taken as independent of frequency. Also, the resistive components in the circuit can be derived from the low frequency h parameters already discussed. Additional parasitic resistances and capacitances can easily be added to this model to even more accurately represent the transistor circuit performance over a broad range of frequencies.

In this model, $r_{bb'}$, represents the small-signal ohmic transverse (parallel to the emitter junction) base resistance of the device in series with the base lead (see Fig. 9.11b). The resistor r_π represents the dynamic emitter resistance reflected[10] into the base input circuit configuration. That is,

$$r_\pi = (h_{fe} + 1)r_e, \tag{9.13}$$

where r_e is the dynamic emitter resistance at the set operating point and h_{fe} is the normal common-emitter AC current gain.

[10]This reflection may be understood by remembering that the base and emitter currents are related by $i_b = -i_e/(h_{fe} + 1)$.

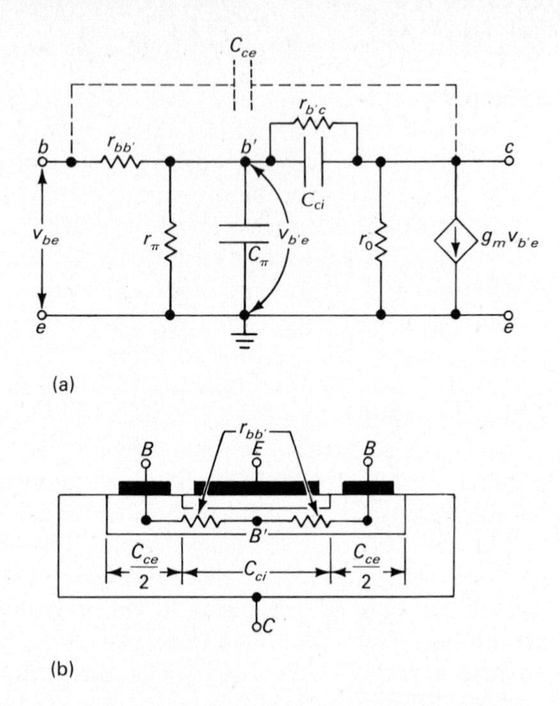

Figure 9.11 (a) A hybrid-π small-signal common-emitter transistor model useful in the megahertz frequency range. The various circuit parameters have a particular set of values for one value of bias point I_C, V_{CE}, and I_B. (b) Sketch of a typical planar transistor cross section showing the origin of the base resistance $r_{bb'}$, along with the intrinsic collector capacitance C_{ci}, and the extrinsic capacitance C_{ce}.

The capacitance C_π takes into account the incremental change of minority carrier charge stored in the transistor base region as a result of the AC input signal. This is essentially of the nature of the diffusion capacitance already introduced in Section 7.3.2 in connection with the high frequency behavior of the junction diode.[11] In parallel with the diffusion capacitance is the emitter junction depletion-layer capacitance, but this is often negligibly small compared to the diffusion capacitance and can be neglected when the emitter is forward biased (see Section 7.3.2).

For small voltage changes $v_{b'e}$ across the emitter junction, excess minority carriers are injected into the base region toward the collector junction, where they give rise to collector signal current. This component of the collector current is denoted by the current generator labeled $g_m v_{b'e}$, where g_m is the small-signal transistor **transconductance,** which is defined by

$$g_m = \left.\frac{\partial i_c}{\partial v_{b'e}}\right|_{V_{CE}}. \tag{9.14}$$

The **intrinsic** part of the collector junction transition or depletion-layer capacitance under the emitter junction is denoted by C_{ci} (see Fig. 9.11b) and this is shunted by the collector junction dynamic resistance, which derives from the variation in transistor base width (and hence current gain) with base-to-collector voltage variation. This has already been discussed in Section 9.1.2

[11]An expression for C_π in terms of transistor structure parameters is given in Section 9.5.3.

and an expression for this resistance can be derived from Eq. (9.1). The change in base width results in a change in the slope of the minority carrier distribution in the base, which determines the emitter diffusion current. Hence this feedback effect is taken into account by a resistor $r_{b'c}$ connected between the internal base point b' and the collector terminal. The collector-to-emitter resistance is denoted by r_0 and this together with $r_{b'c}$ reflects the slope of the output i_C-v_{CE} characteristic. The collector junction capacitance not under the emitter junction is termed **extrinsic** or **overlap** and is represented by C_{ce} in Fig. 9.11.

It can be shown that the high frequency hybrid-π circuit parameters can be expressed approximately in terms of the low frequency hybrid parameters[12] (see Section 9.3.2) at the transistor operating point I_C, V_{CE} as

$$g_m = \frac{q|I_C|}{kT}, \tag{9.15a}$$

$$r_\pi = h_{fe}/g_m, \tag{9.15b}$$

$$r_{bb'} = h_{ie} - r_\pi, \tag{9.15c}$$

$$r_{b'c} = r_\pi/h_{re}, \tag{9.15d}$$

$$1/r_0 = h_{oe} - (1 + h_{fe})/r_{b'c} \approx h_{oe} - g_m h_{re}. \tag{9.15e}$$

This model in practice is found to be useful in the hundreds of megahertz frequency range.

EXAMPLE 9.2

Given an *n-p-n* transistor operating at $I_C = 1.0$ mA, $V_{CE} = 5.0$ V at 300°K, with low frequency design parameters $h_{fe} = 60$, $h_{ie} = 1800\ \Omega$, $h_{re} = 2.0 \times 10^{-4}$, and $h_{oe} = 18 \times 10^{-6}$ S. Determine the hybrid-π circuit parameters

(a) g_m,
(b) r_π,
(c) $r_{bb'}$,
(d) $r_{b'c}$,
(e) r_o.

Solution

Using Eqs. (9.15a)–(9.15e), we have

(a) $g_m = \dfrac{|I_C|}{kT/q} = \dfrac{1.0 \times 10^{-3}\ \text{A}}{0.026\ \text{V}} = \underline{38\ \text{mS}}.$

(b) $r_\pi = h_{fe}/g_m = 60/0.038\ \text{S} = \underline{1600\ \Omega}.$

[12]See J. Millman and C. C. Halkias, *Integrated Circuits* (McGraw-Hill, New York, 1972). In some texts r_π is written as $r_{b'e}$, r_o as r_{ce}; C_π is represented by $C_{b'e}$, and C_{ci} by $C_{b'c}$.

(c) $r_{bb'} = h_{ie} - r_\pi = 1800 - 1600 = \underline{200\ \Omega}$.
(d) $r_{b'c} = r_\pi / h_{re} = 1600\ \Omega/(2.0 \times 10^{-4}) = \underline{8.0 \times 10^6\ \Omega = 8.0\ \text{M}\Omega}$.
(e) $r_o = 1/[h_{oe} - (1 + h_{fe})/r_{b'c}]$

$$= 1/[(18 \times 10^{-6}\ \text{S}) - (61)/(8.0 \times 10^6\ \Omega)]$$

$$= \underline{96{,}000\ \Omega = 96\ \text{k}\Omega}.$$

9.5 COMMON-EMITTER SHORT-CIRCUIT BJT CURRENT GAIN VERSUS FREQUENCY

It is of interest to determine the **intrinsic** device upper frequency limitation of a transistor operating in the common-emitter connection. This can be accomplished with the aid of the hybrid-π high frequency transistor model. The analysis will lead to the definition of the device parameters f_β and f_T, which are often given on transistor specification sheets supplied by the manufacturer. These parameters represent respectively the bandwidth of the device and the frequency at which the short-circuit common-emitter gain drops to unity. Of course a practical amplifier circuit containing this transistor will never achieve these frequency capabilities owing to limitations provided by circuit capacitances and inductances.

9.5.1 The Transistor Short-Circuit Bandwidth f_β

Consider the hybrid-π transistor equivalent circuit of Fig. 9.11 with the output short circuited. This revised circuit is shown in appropriate form in Fig. 9.12. In this version $r_{b'c}$, which should appear between terminals B' and C, is neglected since generally $r_{b'c} \gg r_\pi$. Also, r_o is eliminated since it is short circuited. In addition the output current supplied by C_{ci}, C_{ce}, and $r_{b'c}$ is assumed to be negligibly small compared with that supplied by the current generator, $g_m v_{b'e}$. This output current i_{out} is then simply $-g_m v_{b'e}$, where

$$v_{b'e} = \frac{i_{\text{in}}}{1/r_\pi + j\omega(C_\pi + C_c)}. \tag{9.16}$$

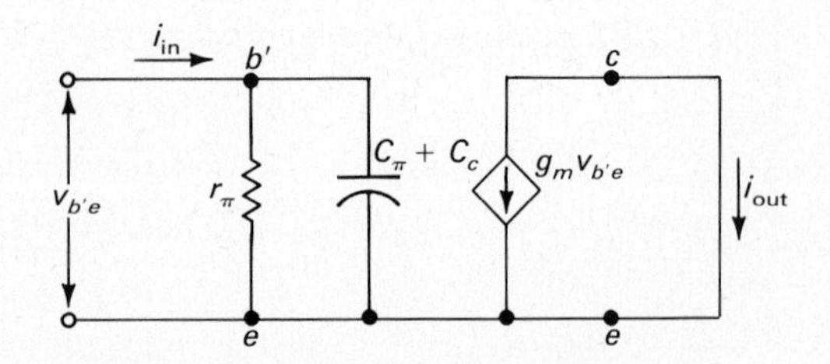

Figure 9.12 Hybrid-π transistor equivalent circuit with the output short circuited.

Here $\omega \equiv 2\pi f$ is the angular frequency of the input signal and $C_c = C_{ci} + C_{ce}$ the total collector junction capacitance. We can then write the short-circuit current gain A_i from Eq. (9.16) as

$$A_i \equiv \frac{i_{out}}{i_{in}} = -\frac{g_m}{1/r_\pi + j\omega(C_\pi + C_c)} \tag{9.17a}$$

or

$$A_i = -\frac{h_{fe}}{1 + j\omega[(C_\pi + C_c)h_{fe}/g_m]}, \tag{9.17b}$$

using the result of Eq. (9.15b).

This expression indicates correctly that the current gain numerically approaches h_{fe} as the signal frequency becomes vanishingly small. On the other hand, the current gain falls to $1/\sqrt{2}$ of its low frequency value (3 dB point) at a frequency f_β given by

$$f_\beta = \frac{1}{h_{fe}}\frac{g_m}{2\pi(C_\pi + C_c)}. \tag{9.18a}$$

The range of frequencies from zero to f_β is sometimes defined as the transistor **bandwidth,** which expresses the intrinsic ability of the transistor to maintain its current gain over this frequency range. Using Eq. (9.15b), we can rewrite Eq. (9.18a) as

$$t_{ch} \equiv 1/2\pi f_\beta = r_\pi(C_\pi + C_c). \tag{9.18b}$$

This can be simply interpreted as the time necessary to charge the diffusion and collector capacitances through the input resistor r_π, which accounts for the frequency response in this simplified model.

9.5.2 The Short-Circuit Common-Emitter Gain–Bandwidth Product

It is often convenient to specify the maximum useful high frequency capability of a transistor as the frequency at which the absolute value of the short-circuit common-emitter current gain reduces to one. This frequency is usually labeled f_T on transistor specification sheets (see Appendixes F1 and F2) supplied by manufacturers and is that operating frequency beyond which the transistor has little value as an amplifier. Equation (9.17b) can be used in order to derive an expression for f_T which, assuming $h_{fe} \gg 1$, is

$$f_T = g_m/2\pi(C_\pi + C_c). \tag{9.19}$$

Comparing Eq. (9.18a) with Eq. (9.19) gives

$$f_T = h_{fe}f_\beta. \tag{9.20}$$

This latter equation permits the interpretation of f_T as the **short-circuit current gain–bandwidth product.** Figure 9.13 is basically a log–log plot of Eq. (9.17), where the ordinate is expressed in decibels (dB), as is normally the practice. For values of frequency somewhat in excess of f_β this graph exhibits a linear falloff (in decibels) of the current gain, with a negative slope of 6 dB/octave (an octave drop in frequency corresponds to a falloff in frequency by a factor of 2).

The frequency cutoff f_T in some transistors extends into the hundreds of megahertz range and even into the gigahertz range. Special equipment is required to measure the transistor gain at these high frequencies and the measurement is also difficult to perform. Hence common practice is to measure the transistor current gain at some moderate frequency f_M, somewhat higher than f_β but much lower than f_T. Since f_M occurs along the straight-line falloff of 6 dB/octave, f_T can be calculated by making use of this linearity as

$$f_T = f_M(A_i)_M, \tag{9.21}$$

where $(A_i)_M$ is the current gain measured at f_M. This expression may be derived by first combining Eqs. (9.17b) and (9.18a) as $A_i = -h_{fe}/[1 + j(f/f_\beta)]$, and is offered as an exercise at the end of this chapter. Now it remains only to measure the transistor short-circuit current gain at some moderate frequency f_M. The experimentally determined value of the diffusion capacitance C_π can be calculated once f_T, C_c, and g_m are known by using Eq. (9.19).

9.5.3 The Transistor Diffusion Capacitance in Terms of Device Design Parameters

In Section 9.4.1 it was stated that the diffusion capacitance reflects the incremental change of minority carrier charge stored in the transistor base region as a result of the AC input signal v_{BE}. Now the total charge stored in the base

Figure 9.13 Log–log plot of transistor gain in dB versus operating frequency.

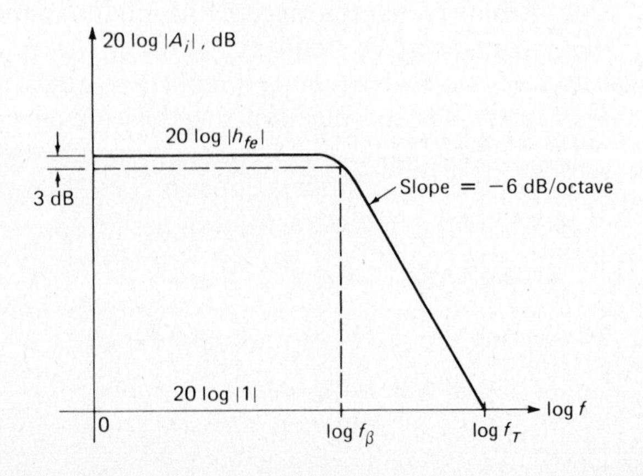

region when the transistor is normally biased in the forward direction, q_F, has been shown in Eq. (8.45) to be

$$q_F = -I_C W^2/2D_n, \tag{9.22}$$

where W is the transistor base width and D_n is the minority electron diffusion constant in the *n-p-n* transistor considered. The rate of change of this charge with respect to a voltage signal v_{BE} applied to the emitter junction is defined as C_{DE}, the **transistor diffusion capacitance;** by differentiating Eq. (9.22) this can be written as

$$C_{DE} \equiv -\frac{\partial q_F}{\partial v_{BE}} = \frac{W^2}{2D_n}\frac{dI_C}{dv_{BE}} \approx \frac{W^2}{2D_n}\frac{d|I_E|}{dv_{BE}}, \tag{9.23}$$

assuming negligible recombination of these carriers in transport from the emitter to the collector junction. The differential in the latter term of Eq. (9.23) can be evaluated starting with Eq. (8.6), the emitter junction I-V relationship, approximately as

$$\frac{d|I_E|}{dv_{BE}} = \frac{q|I_E|}{kT}. \tag{9.24}$$

This differential can be recognized as the reciprocal of the dynamic emitter resistance r_e. Introducing Eq. (9.24) into (9.23), we have for the diffusion capacitance

$$C_{DE} = qW^2|I_E|/2D_nkT. \tag{9.25}$$

At normal emitter currents for the transistor operating in the active region, this diffusion capacitance is much greater in value than the emitter transition-layer capacitance (but is in parallel with it) and hence $C_{DE} \approx C_\pi$ of Section 9.4.1 (this is the subject of a problem at the end of this chapter). Also $C_\pi \gg C_c$, the collector transition capacitance, so Eqs. (9.19) and (9.15a) can be used to yield an approximate expression for f_T in terms of transistor design parameters,

$$f_T \approx D_n/\pi W^2. \tag{9.26}$$

This confirms that the high frequency capability of a transistor requires that it be designed with narrow base width.

EXAMPLE 9.3

The *n-p-n* transistor of Example 9.2 has a bandwidth f_β = 1.0 MHz. The minority electron diffusion coefficient D_n = 20 cm^2/sec. Determine approximately the following design parameters for this device:

(a) C_π;
(b) W, the transistor base width;
(c) τ_n, the minority electron lifetime in the base region.

Solution

(a) Using Eq. (9.18a) and the results of Example 9.2, we have

$$C_\pi \approx g_m/2\pi h_{fe}f_\beta$$
$$= (0.038\text{ S})/(2\pi)(60)(10^6\text{ Hz}) = \underline{1.0 \times 10^{-10}\text{ F}}.$$

(b) From Eq. (9.20)

$$f_T = h_{fe}f_\beta = (60)(10^6\text{ Hz}) = 60 \times 10^6\text{ Hz} = 60\text{ MHz},$$

which from Eq. (9.26) gives

$$W^2 = \frac{D_n}{\pi f_T} = \frac{(20\text{ cm}^2/\text{sec})}{\pi(60 \times 10^6\text{ Hz})} = 1.1 \times 10^{-7}\text{ cm}^2$$

and

$$W = \underline{3.3 \times 10^{-4}\text{ cm} = 3.3\ \mu\text{m}}.$$

(c) Using Eq. (8.43) and noting that Eq. (8.43) with (9.7) gives $h_{fe} = \beta_F$, if β_F is independent of current, we get

$$h_{fe} = \beta_F = \frac{2D_n\tau_n}{W^2},$$

or

$$\tau_n \approx \frac{h_{fe}W^2}{2D_n} = \frac{(60)(1.1 \times 10^{-7}\text{ cm}^2)}{2(20)\text{ cm}^2/\text{sec}}$$
$$= \underline{1.7 \times 10^{-7}\text{ sec} = 0.17\ \mu\text{sec}}.$$

9.6 THE BIPOLAR TRANSISTOR AS A SWITCH

The bipolar transistor is used extensively as an electronic switch and gate to perform logic functions in digital computers. A schematic diagram of a typical common-emitter switching circuit is shown in Fig. 9.14, which shows the *I-V* characteristic of the transistor switch including a load-line representation of the resistive load R_L. Also shown is the circuit current response for a voltage input pulse to the base terminal. Before the input pulse comes on (goes positive) the device is cut off by a negative potential relative to ground; the transistor then conducts only a very small current and most of the battery voltage V_{CC} appears across the transistor collector junction. This is indicated as V_{OFF} in Fig. 9.14d. After the input pulse turns ON, the voltage and current of the device change along the load line until the steady-state collector current I_{C1} is reached. The voltage across the device then, marked V_{ON} in the figure, is often referred to as the **collector saturation voltage** and the device is then said to be

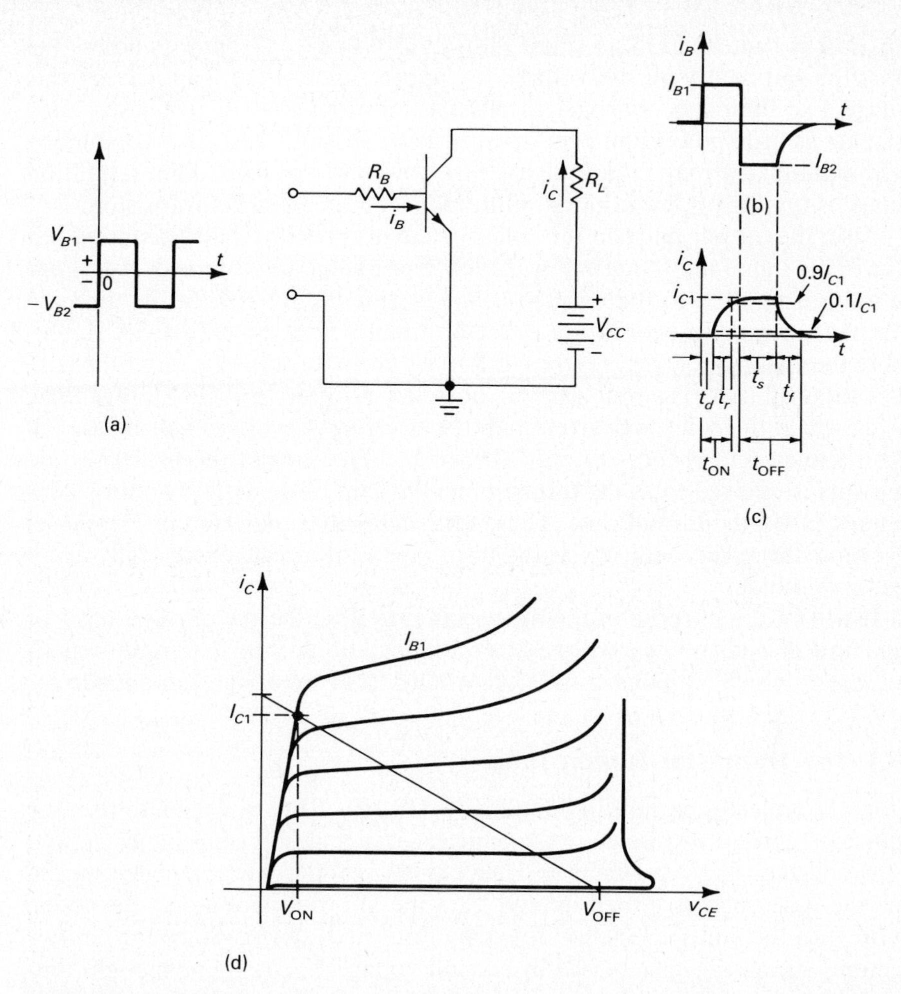

Figure 9.14 (a) Schematic diagram of a typical common-emitter switching circuit utilizing an *n-p-n* bipolar transistor.
(b, c) The base current and collector current versus time. The delay times for switching are indicated.
(d) Typical current–voltage characteristics of a transistor switch showing a load line. The voltage drops across the device when it is switched ON and OFF are indicated by V_{ON} and V_{OFF}, respectively.

in **saturation** (or bottoming). In this condition both the emitter and collector junctions are forward biased as illustrated in Fig. 8.19a. After a time the input voltage pulse goes negative, cutting off the transistor and causing the collector current to return eventually to zero. The base current and collector current are graphed versus time in Figs. 9.14b and 9.14c.

Hence when the switch is OFF practically all the voltage V_{CC} appears across the transistor and almost none appears across the load R_L. When the transistor is switched ON, a small saturation voltage V_{ON} appears across the transistor and nearly all the voltage V_{CC} appears across the load. Ideally this voltage V_{ON} should be as small as possible in order that the transistor may dissipate as little power and heat up as little as possible while it is conducting the maximum current I_{C1} and delivering power to the load. This saturation voltage is typically a few hundred millivolts for a small-signal transistor.

Also, the output pulse in the collector circuit is delayed and distorted relative to the signal input pulse (Fig. 9.14c). There is an initial time delay t_d after the pulse is applied to the emitter–base junction before 10% of the ultimate current I_{C1} begins to flow in the collector circuit. Then there is a rise time t_r before the collector current reaches 90% of its ultimate value I_{C1}. After the base pulse shuts off there is a time delay t_s, normally referred to as the **storage time,** during which the collector current reduces again to 90% of I_{C1}. During the fall time t_f the current reduces to only 10% of I_{C1}. The sum of the first two time constants is referred to as the **turnon time;** the sum of the latter two time constants is called the **turnoff time.** These time delays are indicated in Fig. 9.14c; they limit the speed with which the transistor can operate successfully as an electronic switch.

It will be the purpose of the next section to describe the physical basis of these time constants. This will permit the calculation of the maximum repetition rate at which a bipolar transistor switch of a given design can operate.

9.6.1 The Transistor Turnon Time

Before the collector current in the transistor of Fig. 9.14a can begin to rise, the input base current must charge the emitter–base junction capacitance as well as the collector–base capacitance. Hence there is a time delay t_d between the time the base pulse is turned on and the time the transistor enters the active region with the emitter junction injecting minority carriers. Since the emitter junction, initially reverse biased by a small voltage V_{B2} has to be brought into forward bias to initiate injection, its depletion-layer capacitance C_e under these bias conditions is relatively high compared to the reverse-biased collector junction. This is true in spite of the fact that the collector area is generally larger than the emitter junction area, since in planar transistors the doping in the base region near the emitter junction is much greater than that near the collector junction. Therefore the time delay can be approximated by calculating the time necessary to alter the charge on the emitter junction capacitance, to bring it to the edge of the active region at about V_{B1} volts. This delay time is then

$$t_d = \frac{C_e \, \Delta V}{\Delta I} \approx \frac{C_e V_{B1}}{I_{B1}}, \tag{9.27}$$

and represents the approximate time for the collector current to begin injecting or reach 10% of its ultimate value I_{C1} when driven by a base current I_{B1}.

In order to calculate the rise time t_r for the collector current to reach approximately 90% of I_{C1}, the charge-control model discussed in Section 7.4.1

in connection with the *p-n* junction diode will be utilized. The transistor charge-control model of Section 8.3.1 will also be useful. It has already been shown in Section 8.1.4A that, in order to maintain a steady-state electron distribution in the transistor base region, a majority hole base current $I_B = Q_B/\tau_n$ must be supplied in order to ensure charge neutrality in the face of minority electron recombination. Here Q_B is equal in magnitude to the total excess minority carrier charge stored in the base region. Under transient conditions it must be true that at any time t

$$i_B(t) = \frac{Q_B(t)}{\tau_n} + \frac{dQ_B(t)}{dt}. \tag{9.28}$$

The second term on the right-hand side represents rate of deviation of base charge from the steady state. During the initial phase of rise of collector current, the forward charge distribution of electrons is established in the base region by emitter injection, with the collector junction still in reverse bias, not injecting carriers into the base region. In this case the total excess base charge $Q_B(t) = -Q_F(t)$, where Q_F is the total excess minority-electron charge. During this time interval the base current pulse value is constant at I_{B1} and Eq. (9.28) becomes

$$-I_{B1} = \frac{Q_F(t)}{\tau_n} + \frac{dQ_F(t)}{dt}. \tag{9.29}$$

The solution of this differential equation, assuming approximately that Q_F at $t = 0$ is zero, is

$$Q_F(t) = -I_{B1}\tau_n(1 - e^{-t/\tau_n}) \tag{9.30}$$

for $0 < t < t_r$. The charge-control model of Section 8.3.1 for the transistor in forward operation at any time t is given in Eq. (8.45) as

$$i_C(t) = -Q_F(t)/\tau_t. \tag{9.31}$$

Just at time t_r, when the collector junction is at the edge of saturation (forward bias), the collector current is approximately

$$i_C(t_r) \approx I_{C1} \approx V_{CC}/R_L. \tag{9.32}$$

Introducing Eq. (9.32) into (9.31) and combining this result with Eqs. (9.30) and (8.48) gives finally

$$t_r = -\tau_n \ln(1 - I_{C1}/\beta_F I_{B1}). \tag{9.33}$$

The time variation of collector current obtained by combining Eqs. (9.31) and (9.30) is shown in Fig. 9.15. The collector current rises exponentially toward the value $\beta_F I_{B1}$ since the base input current is I_{B1}. However, the switching circuit of Fig. 9.14 constrains the collector current to rise no higher than the value I_{C1} given by Eq. (9.32), at approximately t_r, at which time the transistor current saturates and so the device is no longer in the active region of operation. Since normally $I_{C1}/I_{B1}\beta_F \ll 1$ the logarithmic term of Eq. (9.33) can

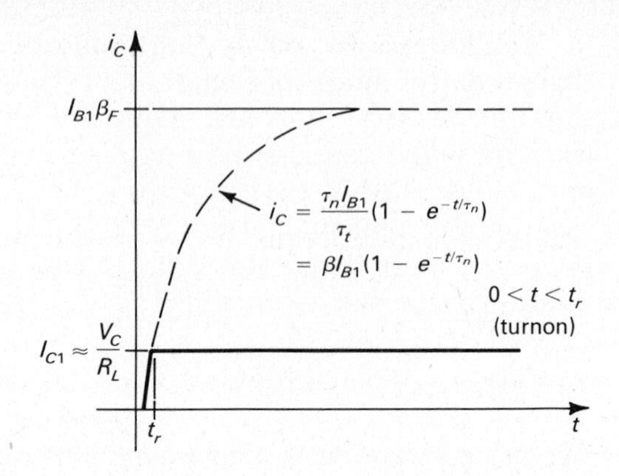

Figure 9.15 The time variation of the collector current of a transistor switch; I_{C1} represents the collector current of an *n-p-n* transistor driven into saturation by a base current I_{B1}. The dotted line represents the time variation of the collector current of the unsaturated transistor.

be approximated and a simple expression for the transistor rise time is seen to be

$$t_r \approx \tau_n(I_{C1}/\beta_F I_{B1}). \tag{9.34}$$

This indicates that the transistor may be caused to rise more quickly to collector current I_{C1} by driving it "harder" with a higher pulse base current I_{B1}. In the next section it will be shown that this is not necessarily a way to "speed up" the transistor since a long storage time may then result.

9.6.2 The Transistor Turnoff Time

In Fig. 9.14c it is seen that the collector current persists at a constant value I_{C1} for a time interval marked t_s called the storage time, after the base drive pulse shuts off and reverses polarity. During this period the collector junction is operating in saturation; that is, it is forward biased and injecting minority electrons into the transistor base region. This condition occurs when the collector potential drops below the base potential. Since both emitter and collector junctions are now forward biased, the device has only a few hundred millivolts across it typically, is in the low impedance state, and no longer limits the current, so I_{C1} is circuit determined as essentially V_{CC}/R_L. This low impedance state will be maintained after the base drive is removed until most of the excess minority carriers in the base undergo recombination or are extracted by a reverse base current supplied by the negative base pulse voltage $-V_{B2}$ in Fig. 9.14a and the device reenters the active region.[13]

[13] In the planar transistor structures illustrated in Fig. 8.2, the collector region is more lightly doped than the base region. Hence with the device operating in saturation, a significant density of holes is injected from the *p*-type base region into the *n*-type collector region, constituting a major component of the total minority charge stored. These charges must also recombine or be extracted through the base lead to turn the device off. For simplicity of formulation here, these charges are not included, although the time for their removal typically is a substantial portion of the planar bipolar transistor storage time.

To calculate the storage time consider again Eq. (9.28) under the condition that the base current is suddenly reduced to zero, which can be written as

$$0 = \frac{Q_B}{\tau_n} + \frac{dQ_B}{dt'}. \tag{9.35}$$

The solution of this equation, noting that at the beginning of turnoff ($t' = 0$), $Q_B = I_{B1}\tau_n$, is given by

$$Q_B(t) = I_{B1}\tau_n e^{-t'/\tau_n}. \tag{9.36}$$

This exponential decay of excess base charge is graphed in Fig. 9.16. Between $t' = 0$ and $t = t_s$ the collector current remains constant at a value I_{C1} until the base charge $Q_B(t)$ reaches the value

$$Q_B(t_s) = I_{C1}\tau_t, \tag{9.37}$$

at which time, $t' = t_s$, the transistor enters the active region (collector just reverse biased) and Eq. (9.31) applies again. Evaluating Eq. (9.36) at $t' = t_s$ and using Eq. (9.37) gives for the storage time

$$t_s \approx \tau_n \ln(\beta_F I_{B1}/I_{C1}). \tag{9.38}$$

This equation indicates that if I_{B1} is increased to reduce the current rise time, the transistor will be driven "deeper" into saturation and the storage time will increase correspondingly. The storage time is directly proportional to the minority carrier lifetime in the base region since, with the base current reduced to zero in this calculation, the transistor will recover depending on the carrier recombination rate in the base region. If reverse base current I_{B2} is drawn on turnoff, the storage time is reduced below that given by Eq. (9.38) since then carriers can be removed by this current as well.

Let us assume that the minority carrier charge stored in the transistor base region while the device is in saturation ($0 < t' < t_s$) is entirely removed by a reverse base current I_{B2}. Then this base current can be written approximately as

$$I_{B2} \approx \frac{\Delta Q_B}{\Delta t} = \frac{Q_B(0) - Q_B(t_s)}{t_s}. \tag{9.39}$$

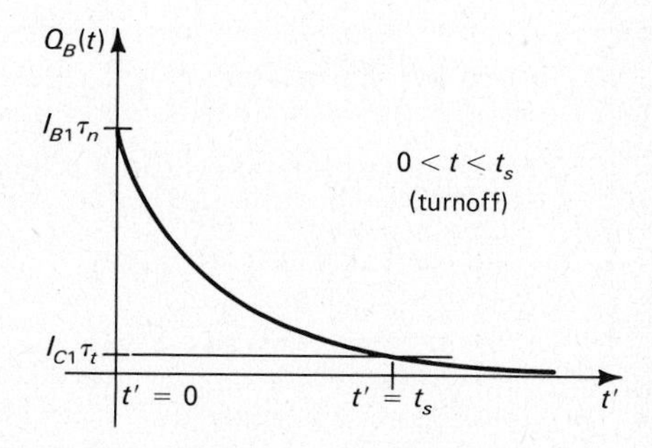

Figure 9.16 The exponential decay of the total excess charge in the base region of a transistor in time, after the base drive is turned OFF at time $t' = 0$; t_s represents the transistor storage time.

Using Eqs. (9.36) and (9.37), we can express this as

$$I_{B2} \approx (I_{B1}\tau_n - I_{C1}\tau_t)/t_s. \tag{9.40}$$

From Eq. (8.48), $\beta_F = \tau_n/\tau_t$, and introducing this into Eq. (9.40) gives for the storage time

$$t_s \approx \tau_n\left(\frac{I_{B1}}{I_{B2}} - \frac{I_{C1}}{\beta_F I_{B2}}\right) = \tau_n\frac{I_{B1}}{I_{B2}}\left(1 - \frac{I_{C1}}{I_{B1}\beta_F}\right). \tag{9.41}$$

Hence, if the reverse base current $I_{B2} \gg I_{B1}$, then the storage time can be reduced substantially below τ_n, but at the expense of additional energy. In this case fast rise time and short storage time can be achieved simultaneously.

The fall time t_f can be estimated for turnoff with $I_B = 0$ by noting that the transistor returns again to the active region and the collector current decays after the storage time according to

$$i_C(t) = \beta_F I_{B1} e^{-(t'-t_S)/\tau_n}, \tag{9.42}$$

which can be derived by combining Eq. (9.36) with (9.37). From this equation it can be shown that the fall time $t_f \approx \tau_n I_{C1}/\beta_F I_{B1}$ is numerically of the same order as the value for the rise time given by Eq. (9.34). Of course, this decay time can be greatly reduced if a base current $|I_{B2}| \gg I_{B1}$ is extracted during turnoff.

EXAMPLE 9.4

A switching transistor of the design described in Example 9.3 driven into saturation delivers a collector current I_{C1} when driven by a base current I_{B1}. The device is turned off by reducing the base current to zero. Assuming $\beta_F > I_{C1}/I_{B1}$, determine the change in rise time and storage time for this transistor if the base drive is increased by a factor of 4. Originally $t_r = 53.3$ nsec.

Solution

From Eq. (9.34)

$$\frac{t_{ro}}{t_{rf}} \approx \frac{\tau_n I_{C1o}/\beta_{Fo} I_{B1o}}{\tau_n I_{C1f}/\beta_{Ff} I_{B1f}} = \frac{I_{B1f}}{I_{B1o}},$$

where the subscripts o and f refer to original and final conditions if we assume that $\beta_{Fo} = \beta_{Ff}$ and $I_{C1o} = I_{C1f}$ (the collector current is determined by the load resistance). Then

$$t_{rf} \approx \frac{I_{B1o}}{I_{B1f}} t_{ro} = \tfrac{1}{4}(53.3\text{ nsec}) = 13.3\text{ nsec}$$

and

$$\Delta t_r = t_{rf} - t_{ro} \approx (13.3 - 53.3 \text{ nsec}) = \underline{-40.0 \text{ nsec}}.$$

Now from Eq. (9.38)

$$\Delta t_s = t_{sf} - t_{so} = \tau_n \ln(\beta_{Ff} I_{B1f}/I_{C1f}) - \tau_n \ln(\beta_{Fo} I_{B1o}/I_{C1o})$$

$$= \tau_n \ln(I_{B1f}/I_{B1o}) = (0.17 \times 10^{-6} \text{ sec}) \ln(4)$$

$$= \underline{0.24 \times 10^{-6} \text{ sec} = 240 \text{ nsec}}.$$

Hence increasing the base drive *reduces* the rise time by 40 nsec but *increases* the storage time by 240 nsec.

☐ SUMMARY

Several factors in addition to the intrinsic parameters governing the *I-V* characteristics of bipolar transistors were discussed. The ohmic voltage drops across the bulk base and collector resistances were of primary importance. The reduced current gain at both low and high current levels was explained as well as the increase of gain at higher collector voltages. At low current, carrier recombination in the space-charge region reduces the gain; at high current, loss of injection efficiency and current-induced base region widening causes the gain to fall off. At increased voltage collector space-charge widening can cause base width narrowing, enhancing the gain. The BJT voltage limitation was attributed to transistor action as well as avalanche carrier multiplication. The origin of the electric field which aids minority carrier transport through the base region was explained as being due to a doping gradient or a carrier gradient there. The circuit operation of the transistor as an amplifier and as an electronic switch was discussed. First the frequency and temperature behaviors were described. The two-port low frequency transistor circuit model was introduced followed by the hybrid-π circuit model useful at higher frequencies. This led to the derivation of an expression for the transistor frequency bandwidth in terms of physical device parameters. Finally the design factors governing the switching speed of the transistor were analyzed. The minority carrier storage time was seen to be a primary factor in limiting the turnoff time of the BJT.

PROBLEMS

9.1 The common-base current gain α_F of a transistor increases from 0.90 at low currents to 0.98 at higher currents. The collector breakdown voltage BV_{CBO} is 100 V. Determine the collector sustaining voltage at higher currents for the transistor in the common-emitter connection. Assume that the exponent $n = 3$ in Eq. (9.6).

9.2 For the transistor of Problem 8.20 operating in the active region, determine
(a) The collector-to-emitter "punchthrough" voltage.
(b) The output conductance $(\partial I_C/\partial V_{CE})_{IB=\text{const}}$ for the transistor operating at a base current $I_B = 0.10$ mA. [*Hint:* Use Eq. (8.43) for β_F and Eqs. (6.18) for the variation of base width with collector voltage.]

9.3 Show that in the circuit of Fig. 9.5a the power supplied by the battery exactly equals the power dissipated in the transistor and the other circuit elements.

9.4 A transistor operating at a case temperature of 350°K has a junction-to-case thermal resistance θ_{JC} of 1.0°C/W. What is the maximum power which can be dissipated in this transistor if the collector junction temperature may never exceed 200°C?

9.5 Derive the transistor circuit model of Fig. 9.9 starting with Eqs. (9.9).

9.6 **(a)** Draw a hybrid-π *low frequency* transistor model starting with Fig. 9.11.
(b) Estimate below what frequency the capacitor C_π across r_π may be neglected in deriving the low frequency version of the hybrid-π transistor model of (a).

9.7 Derive the hybrid-π circuit resistive components from the low frequency h parameters.

9.8 Prove the validity of the approximation of Eq. (9.15e). Under what conditions is this approximation good?

9.9 Consider a planar *n-p-n* transistor as shown in Fig. 8.2a with emitter region doping 5.0×10^{19} donors/cm^3, base region doping 1.0×10^{17} acceptors/cm^3, and collector doping 5.0×10^{15} donors/cm^3. The emitter junction area is 2.0×10^{-8} cm^2 and the transistor base width is 2.0×10^{-4} cm. For a transistor operating in the active region at 300°K with an emitter current of 10 mA and an emitter–base potential of 0.60 V, find the emitter depletion-layer capacitance and the emitter diffusion capacitance; compare these values.

9.10 Given an *n-p-n* transistor with an f_T of 50 MHz, low frequency $h_{fe} = 50$, and collector capacitance of 5.0 pF operating at a collector current of 2.0 mA, determine its diffusion capacitance and f_β at 300°K.

9.11 Show, beginning with Eq. (9.17b), that $f_\beta = g_m/2\pi h_{fe}(C_\pi + C_c)$.

9.12 Show that $f_T = f_M(A_i)_M$, as in Eq. (9.21), after proving that $A_i = h_{fe}/[1 + j(f/f_\beta)]$.

9.13 Prove that Eq. (9.24) follows from Eqs. (8.5) and (8.6).

9.14 Show that the falloff of gain is 6 dB/octave (or 20 dB/decade) from Fig. 9.13 by determining the slope of the curve in the falloff region.

9.15 Prove that $C_\pi r_\pi \approx \tau_p$.

9.16 Consider an *n-p-n* silicon switching transistor with $C_c = 3.0$ pF, $C_e = 6.0$ pF, $\tau_n = 1.0\ \mu$s, $I_{C1} = 10$ mA, $I_{B1} = 1$ mA, $\beta_F = 50$. Calculate approximately
(a) the time delay t_d before the rise of the collector current after a positive current pulse I_{B1} is applied to the transistor base;
(b) the rise time t_r;
(c) the storage time t_s after the base current pulse is turned off;
(d) the storage time t_s if a reverse base current pulse $I_{B2} = 2.0$ mA is applied to the transistor base.

9.17 Show that Eq. (9.30) is a solution of Eq. (9.29).

9.18 Prove that Eq. (9.33) follows from Eqs. (9.30)–(9.32) and (8.48).

9.19 Show that Eq. (9.38) follows from Eqs. (9.36) and (9.37).

9.20 Show that Eq. (9.40) follows from Eqs. (9.39), (9.36), and (9.37); then derive Eq. (9.41).

CHAPTER 10

The Unipolar Field-Effect Transistor

10.1 THE SEMICONDUCTOR FIELD-EFFECT TRANSISTOR

The **unipolar** or **field-effect transistor** is another type of three-terminal device, capable of power gain and functioning as an electronic switch like the bipolar transistor, but it operates on a basically different physical principle. It is the purpose of this chapter to discuss the principle of operation of this device, which is used extensively, particularly in digital integrated circuits and electronic memories, as well as in amplifier circuits. Comparisons will be drawn with the bipolar transistor.

First the operation of the **junction field-effect transistor (JFET)** will be described; here the input voltage is applied to a *p-n* junction whose space-charge widening constricts current flow in the channel of the device. A modification of the JFET called the **MESFET** uses instead a Schottky diode as the input electrode. The **MOSFET** device is introduced by discussing the charge accumulation, depletion, and inversion modes of operation of its input MOS capacitor. An expression is derived for the MOSFET current-initiating threshold voltage. Both the enhancement- and depletion-type silicon MOSFETs are described. Next the *I-V* characteristics of the enhancement-type MOSFET are derived. The device's output conductance is analyzed. Short-channel effects which modify the characteristics of the device are discussed. Finally, an expression for the frequency cutoff of the field-effect transistor is derived and a comparison is made with the junction bipolar transistor in this regard. An equivalent circuit for analyzing the high frequency small-signal performance of the FET is presented.

The term "unipolar" derives from the fact that only one type of current carrier is required to basically describe this transistor's mode of operation. In contrast, the consideration of both electron and hole flow must be taken into account to analyze the operation of the bipolar transistor as an amplifier or switch. Hence minority carrier storage effects do not occur to slow down the operation of unipolar devices. The term "field-effect" describes the manner in which control of current flow is achieved in this device, namely by adjusting the electric field in the junction or capacitorlike "gate" input electrode. This permits modulation of charge flow in the transistor channel without drawing any appreciable DC input current. Hence the field-effect transistor is voltage driven rather than current driven as is the bipolar transistor.

There are basically two versions of the unipolar field-effect transistor. Both **insulated-gate** and **junction** field-effect transistors are used in electronic circuits. However, the insulated-gate device has achieved numerically much more extensive use than the junction device because of its simplicity of fabrication. It is the insulated-gate field-effect transistor (IGFET) which will be described later in more detail. The sandwichlike metal–oxide–semiconductor (MOS) input structure of this transistor is typical of the most popular version of the device. This gives rise to a DC input resistance often in excess of $10^{14}\ \Omega$. Schematic drawings of a typical MOS-IGFET device and a junction field-effect transistor (JFET) are shown in Figs. 10.1 and 10.2, respectively.

The principle of operation of the JFET simply involves the control of the source-to-drain channel current by the narrowing of this current path by p^+-n junction space-charge widening into the n-type channel. MOS device operation is a bit more difficult to describe. The electrical characteristics of the JFET shown in Fig. 10.3 are very similar to those for the MOSFET shown in Fig. 10.12. A cursory view of the device structures tends to indicate strong similarities between the construction of field-effect devices and the planar bipolar transistor of Fig. 8.2. For example, both devices contain two *p-n* junctions and some associated metal contacts. However, when the fabrication of each of these devices is described in detail in the next chapter in connection with integrated-circuit technology, it will be seen that the processing of MOS devices is significantly simpler and the steps less critical than for the comparable bipolar transistor or junction field-effect device. In addition, MOS-IGFETs can be made several times smaller in surface area than bipolar devices; hence their extensive use in large electronic memory circuits which employ thousands of devices in a small-area silicon chip.

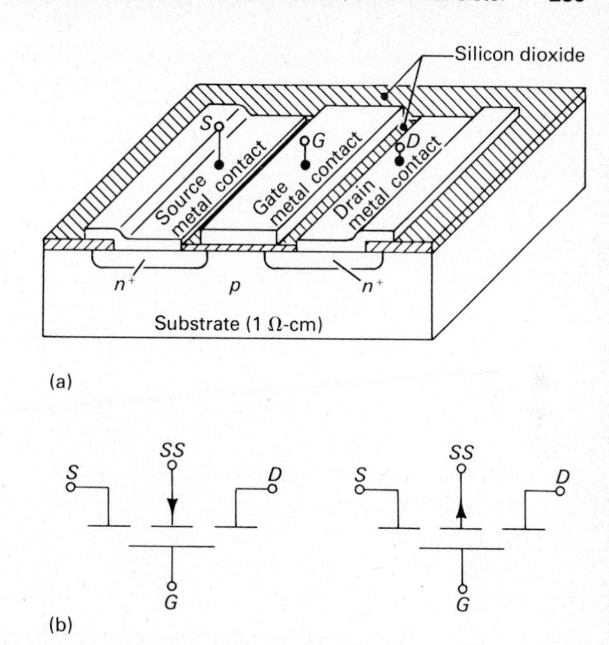

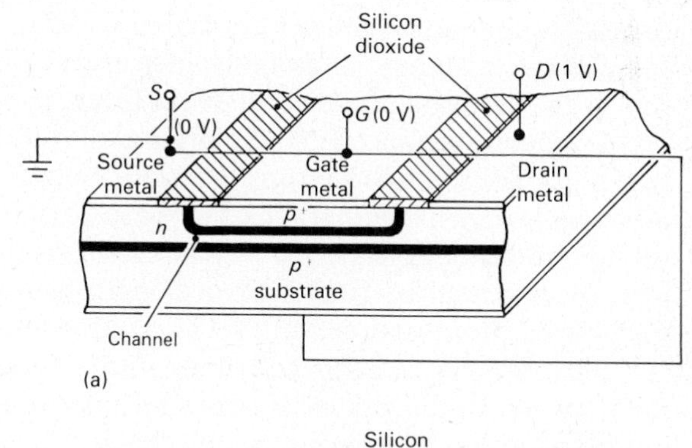

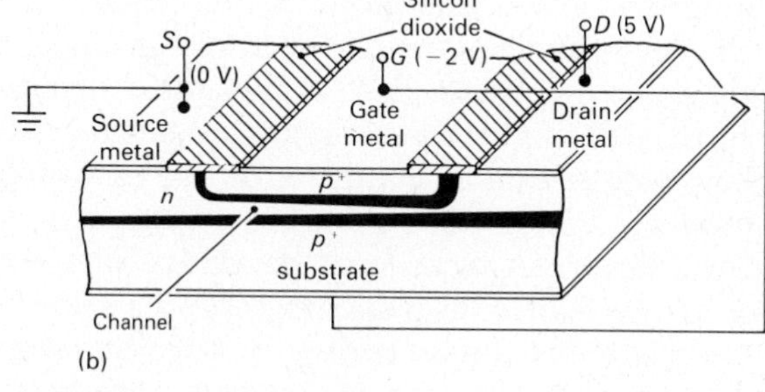

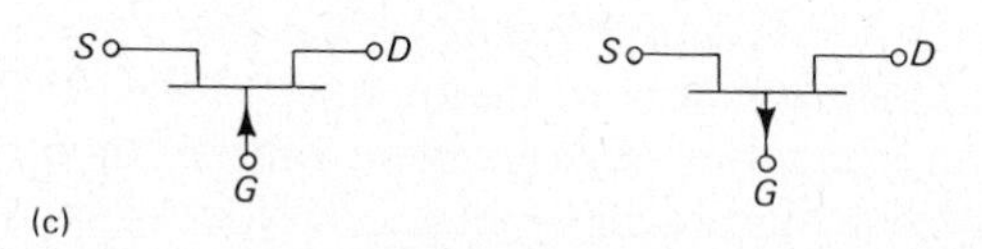

Figure 10.1 (a) Cross sectional view of an *n*-channel metal–oxide–semiconductor enhancement-type insulated-gate field-effect transistor showing the basic device structure.
(b) Circuit symbols for an *n*-channel (left) and a *p*-channel (right) enhancement-type MOSFET with source electrode (*S*), drain (*D*), and gate (*G*). The symbol to the left has an arrow indicating a *p*-type substrate (*SS*). The symbol to the right has an arrow indicating an *n*-type substrate. The arrows are often left off since the battery polarity indicates whether *n*- or *p*-channel devices are used.

Figure 10.2 (a) Cross sectional view of an *n*-channel junction field-effect transistor with gate shorted to source and a small applied drain voltage. The darkened regions shown represent the junction space-charge regions.
(b) Cross sectional view of an *n*-channel junction field-effect transistor biased just at the point of channel pinch-off. The darkened regions shown represent the junction space-charge regions.
(c) Circuit symbols for the junction field-effect transistor with source electrode (*S*) and drain (*D*). The symbol to the left has an arrow indicating a *p*-type gate region (*G*) for this *n*-channel device. The symbol to the right has an arrow indicating an *n*-type gate for this *p*-channel device.

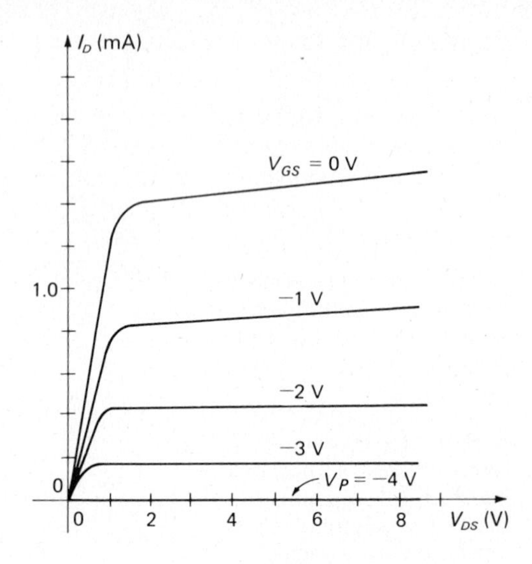

Figure 10.3 Typical set of output characteristics for an n-channel junction field-effect transistor with a pinch-off voltage of −4 V.

10.2 THE JUNCTION FIELD-EFFECT TRANSISTOR (JFET)

Consider first the junction field-effect structure of Fig. 10.2a. The two p^+-regions connected together form the JFET gate. Take the gate initially at the same electrical potential as the source metal electrode; this metal contact as well as the drain metal electrode make low resistance ohmic connections to the n-type silicon region. If now the drain electrode is raised to a potential above that of the source, an electron current will flow from drain to source along the ***n*-channel** sandwiched between the space-charge regions of the two p^+-n junctions. The current does not flow into these space-charge regions because of the electron potential barriers provided by the p^+-n junctions. Initially a further increase in drain–source voltage V_{DS} will result in a linear increase in drain current I_D since the channel conducts current like a linear resistor. In addition, the increased drain potential will tend to reverse-bias the gate p^+-n junction more and more near the drain electrode. This will cause additional space-charge widening of the depletion regions there, tending to **pinch off** the n-type channel at the drain end, as illustrated in Fig. 10.2b. Now the rate of drain current increase with voltage can no longer remain linear owing to this current flow constriction and the rate of rise of I_D becomes considerably less as V_{DS} increases.

Still further increase of the drain voltage results in a **saturation** of drain current, which tends to increase only very slowly with increased voltage. Figure 10.3 shows the output current–voltage characteristics of an n-channel JFET in the common-source connection with zero gate-to-source potential V_{GS}. The linear $I_D - V_{DS}$ region is shown at low values of V_{DS}, along with the sharp bend in this curve due to the tendency toward drain pinchoff, followed by the current saturation region. Note that as the drain voltage continues to increase beyond the pinchoff point, the additional voltage cannot be absorbed by the drop along the channel but instead appears across a space-charge region which forms near the drain, in a sense somewhat similar to the case of the increase of reverse bias across the depletion region of a *p-n* junction diode. Eventually avalanche breakdown occurs just as in the case of the diode.

Detailed analysis shows that complete pinchoff or the coming together of the gate space-charge regions does not always occur to yield current saturation at higher V_{DS}. In certain cases there can remain a very narrow, long channel which supports a large voltage drop and hence includes a very high field.[1] Then carriers accelerated in this high field may achieve a **scattering-limited** maximum velocity (**saturation** velocity, see Section 5.2.4) accounting for a limiting (or saturation) of current in spite of increased voltage. Since the current density J can always be written as $J = qnv$, for a given channel carrier density n, constant velocity v means constant current density. This is particularly important in short channel (of order 1 μm) devices. Eventually avalanching can occur as a result of this high electric field.

The same type of discussion as was given above for the situation with zero potential difference between gate and source can be repeated for the case of the gate at a negative potential relative to the source. Because of this reverse bias across the gate junctions, the channel width is now narrower to begin with, so pinchoff near the positive potential drain occurs at a lower drain-to-source voltage, causing the drain current to saturate at a lower value of drain current. This is illustrated in Fig. 10.3. Repeating this procedure again by increasing the reverse gate potential still further until the two gate–junction space-charge regions essentially touch reduces the channel width to zero; this prevents any appreciable increase of current flowing from drain to source regardless of the value of the drain potential. The gate–source voltage required for this pinchoff along the total channel length is called the **pinchoff voltage** and is labeled V_P on the graph of Fig. 10.3.

Note the similarity between the general shape of the curves of Fig. 10.3 and the common-emitter output characteristics of the bipolar transistor given in Fig. 8.12. There are a number of differences of course. The unipolar transistor is voltage driven rather than current driven as is the case for the bipolar device. The FET often exhibits better current saturation and hence higher dynamic output resistance. JFETs are generally fabricated by using the semiconductor silicon.

The above discussion indicates that the saturation value of the drain current, $I_{D\text{sat}}$, is a function of the gate–source voltage V_{GS}. A simple relationship has been found to predict the variation of $I_{D\text{sat}}$ with V_{GS} rather accurately[2]:

$$I_{D\text{sat}} = I_{DSS}(1 - V_{GS}/V_P)^2, \tag{10.1}$$

[1]D. P. Kennedy and R. R. O'Brien, "Computer-Aided Two-Dimensional Analysis of the JFET," *IBM J. Res. Dev.*, **14,** 95 (1970).

[2]See S. M. Sze, *Physics of Semiconductor Devices,* 2nd ed. (Wiley, New York, 1981), Chap. 6. In specification sheets g_m is sometimes referred to as g_{fs}, the forward transconductance with output shorted (see Appendix G1), in units of siemens or amperes/volt.

where I_{DSS} is the saturated drain current with the gate shorted to the source ($V_{GS} = 0$) and V_P is the gate–source voltage required to pinch off the drain-to-source channel. This **square-law** variation of drain saturation current with gate voltage is typical of field-effect devices and illustrates how the input gate voltage is used to modulate the output drain saturation current. An indication of the gain of this transistor may be obtained by specifying the rate of change of output current with input voltage by defining a **tranconductance,** $g_m \equiv (\partial I_{\text{sat}}/\partial V_{GS})_{V_{DS=\text{const}}}$. Performing the appropriate differentiation on Eq. (10.1), one may write g_m as[2]

$$g_m = -\frac{2I_{DSS}}{V_P}\left(1 - \frac{V_{GS}}{V_P}\right) = -\frac{2}{V_P}(I_{DSS}I_{D\text{sat}})^{1/2}. \tag{10.2}$$

Since V_{GS} and V_P are always at a negative potential relative to the source for an n-channel device, g_m is always a positive quantity.

The input to the JFET is in the form of a reverse-biased p^+-n junction, which has been described in Section 6.2 as exhibiting a very small leakage current of the order of several picoamperes. This small input current ensures the high power gain of the device at low frequencies. The leakage current of the input electrode of the MOSFET (a capacitor with a pure SiO_2 dielectric layer), however, is typically several orders of magnitude even lower than that of the JFET and hence is useful when extremely high input impedance is required for such a device. In contrast, the very low noise properties typical of JFETs dictate the use of this type of device when the circuit application requires that a very minimum of noise be contributed by the transistor. Since the FET is a majority carrier device, no noise results from minority carrier fluctuations that occur in bipolar transistors.

10.3 THE METAL–SEMICONDUCTOR FIELD-EFFECT TRANSISTOR (MESFET)

The active device with the highest frequency capability manufactured today is the gallium arsenide metal–semiconductor field-effect transistor or MESFET. The principle of operation of this device is in many respects similar to that of the JFET just described. The primary difference lies in the fact that the controlling gate is not a p-n junction but instead is a Schottky metal–semiconductor diode of the type described in Section 7.5.1. Space-charge widening due to a reverse voltage applied to a Schottky barrier is described there. This effect can be used to pinch off a thin, conducting semiconductor layer as was the case for the JFET. Any semiconductor material can be used to form the conducting channel; however, n-type GaAs is utilized presently because of its five times higher electron mobility and two times higher peak electron-scattering-limited velocity compared with silicon (see Section 5.2.4).

A sketch of a typical GaAs MESFET is shown in Fig. 10.4. The device is fabricated by depositing a thin (a few micrometers thick), lightly doped n-GaAs layer epitaxially onto a semi-insulating GaAs substrate. The substrate material

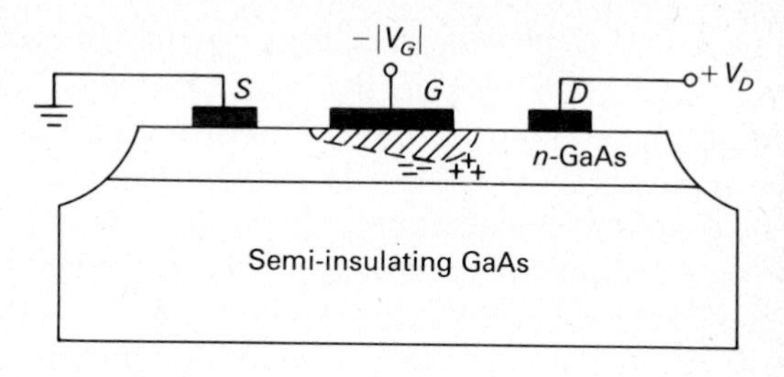

Figure 10.4 Sketch of a gallium arsenide MESFET. The thin conducting *n*-GaAs layer is formed by epitaxial deposition onto the semi-insulating GaAs substrate. A negative gate potential relative to the source (*S*) contact produces a depletion region under the gate electrode (*G*). This depletion region (cross-hatched) constricts the current flow from drain to source near the drain region, causing an electric dipole to form there.

is cut from a GaAs crystal usually doped with chromium which serves to trap out mobile carriers, causing the material to become virtually nonconducting. Hence when the Schottky metal (typically aluminum) gate is negatively biased relative to the source ohmic contact (typically a gold–germanium alloy), an electron depletion region is formed under the gate electrode, tending to constrict or pinch off the electron current flow in the conducting *n*-GaAs channel between drain and source. This gives rise to a set of current–voltage characteristics very similar to those shown in Fig. 10.3 for the JFET.

Since the interest in this type of device is for high frequency, fast switching circuit applications, the source-to-drain electron transit time must be made as short as possible by reducing the channel length typically to 0.5 μm. Velocity saturation tends to produce the current saturation observed in the current-saturating *I-V* characteristics for this device which are similar to those sketched in Fig. 10.3. This current saturation comes about as follows: for a given gate–source negative bias voltage, an electron depletion region is formed under the gate electrode. As the drain potential is raised relative to the source, the conducting *n*-GaAs channel narrows most near the drain side of the channel, as shown in Fig. 10.4.

Since the electron current I must be constant everywhere along the channel (this current can be written as $I = Aqnv$, where A is the channel cross sectional area and v is the carrier velocity), a constriction or decrease in A under the condition of velocity saturation must result in an increase in the electron concentration n there. The latter effect is very likely to occur even at small drain voltages as a result of the high field caused by the shortness of the channel. This increase in electron concentration above the number contributed by the *n*-type donor impurity atoms in the channel gives rise to a negatively charged electron **accumulation** layer on the source side of the channel constriction. On the drain side, the constriction widens, the electron concentration decreases there even below that due to the donor atoms in the channel, and a positively charged space-charge region is formed, completing the electric dipole in the constricted channel region. Additional drain voltage applied in the current saturation region appears mainly across this dipole (see Fig. 10.4). This dipole formation is predicted to occur in short-channel MESFETs fabricated from GaAs, the velocity–field characteristic of which is as shown in Fig. 12.4.

The search for even faster switching transistors has led to the invention of the HEMT[3] (high electron mobility transistor) device. This device is produced presently by successively depositing several epitaxial intermetallic compound semiconductor materials forming heterojunctions onto a GaAs substrate by a technique called molecular beam epitaxy (MBE). A highly conducting *n*-type layer supplies a large concentration of electrons which are injected into a lightly doped, relatively pure GaAs layer where they can conduct with a high electron mobility.[4] In both structure and operational principle this device is similar to the MESFET and MOSFET; however, the separation of the functions of electron carrier supply and a region where these electrons can move with very high electron velocity accounts for the higher speed or frequency possibilities of this device. The more usual method of obtaining high channel conductance for high frequency performance involves heavy doping to yield a higher carrier density, but this reduces the carrier mobility.

10.4 THE METAL–INSULATOR–SEMICONDUCTOR OR INSULATED-GATE FIELD-EFFECT TRANSISTOR (IGFET)

In the MOS-IGFET device illustrated in Fig. 10.1, the metal–oxide–semiconductor gate is used to control the concentration of electron carriers which flow from source to drain, and hence the drain current. In order to understand the principle of operation of this device it is essential to have a detailed knowledge of this input **MOS capacitor.** The importance of charges in the oxide in controlling the behavior of silicon devices has already been discussed in Section 5.5. Charges in the insulator and at the semiconductor–insulator interface are of primary importance in determining the electrical behavior of IGFETs since they can induce charges in the semiconductor underneath the insulator, which in turn can conduct current. The MOS capacitor structure can be used to evaluate the type and quantity of charge in the insulator. This same structure is the basic element of the charge-coupled device (CCD) to be described in Chapter 12, arrays of which are used to construct high density shift registers and image photodetectors. The electrical properties of the MOS capacitor will now be discussed.

10.4.1 The MOS Capacitor: Charge Accumulation, Depletion, and Inversion

The MOS structure of Fig. 10.5a is basically a capacitor with one metal plate electrode and the *p*-type semiconductor substrate as the other electrode, with

[3]H. Morkoc and P. M. Solomon, "The HEMT: A Superfast Transistor," *IEEE Spectrum,* **21,** 28 (1984).

[4]R. Dingle, J. L. Störmer, A. C. Gossard, and W. Wiegmann, "Electron Mobilities in Modulation-Doped Heterojunction Superlattices," *Appl. Phys. Lett.,* **33,** 665 (1978).

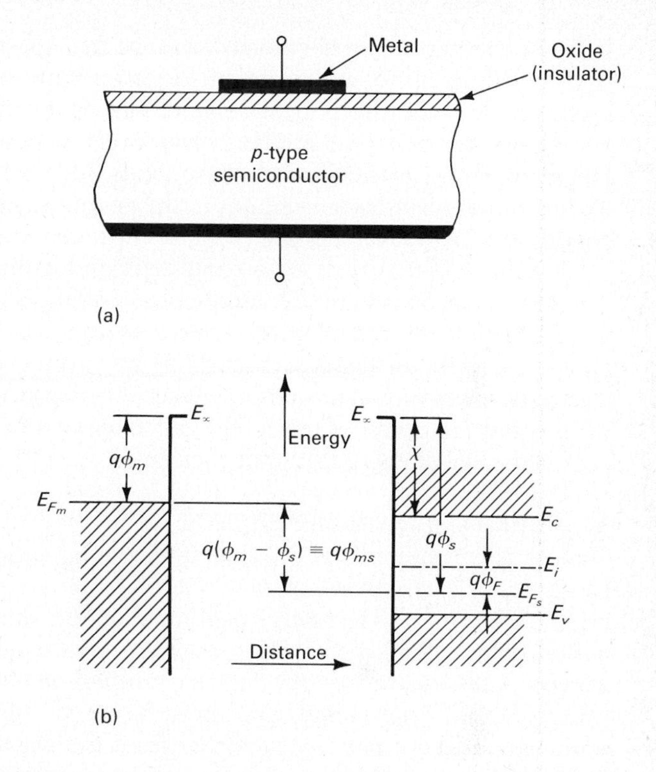

Figure 10.5 (a) The MOS capacitor structure. (b) Electron energy band diagrams of a metal (left) and *p*-type semiconductor (right) separated remotely from each other; $q\phi_m$ is the metal work function and $q\phi_s$ is the semiconductor work function; $q\phi_F$ represents separation of the semiconductor Fermi level from the energy gap center E_i.

an insulating oxide sandwiched in between. A negative potential applied to the metal gate relative to the substrate will attract holes to the oxide–semiconductor interface, and so the majority carrier concentration there exceeds the hole concentration coming from the local ionized acceptors. This is referred to as charge carrier **accumulation.** Reversing the gate potential to positive relative to the substrate tends to reject holes from the interface which are associated with the acceptor impurity atoms there. This is known as charge carrier **depletion** and is similar in nature to the formation of the depletion layer that occurs when a *p-n* junction or a Schottky contact is reverse biased (see Sections 6.1.3 and 7.5.1). Still higher positive biasing of this MOS capacitor will cause the collection at the oxide–semiconductor interface of additional minority electrons from the *p*-type semiconductor substrate. When the density of these electrons exceeds the density of the majority holes associated with the substrate acceptor impurity atoms there, the semiconductor surface tends toward "*n*-type." This is referred to as semiconductor surface **inversion.**

In preparation for the quantitative description of the MOS capacitor as the gate input element of the MOSFET, it is useful to describe semiconductor surface charge accumulation, depletion, and inversion in terms of the energy band theory. Figure 10.5b shows the energy band structures of a metal and a *p*-type semiconductor in thermal equilibrium at 300°K, when located very far from each other so that no electrical interaction occurs. Here E_∞ is the reference energy of an electron far enough away in free space that the metal and the semiconductor no longer electrically interact with it. Hence $q\phi_m$ and $q\phi_s$ represent the **work function** for these materials, or the energy required to remove an electron from their respective Fermi levels to a point where the electron no longer interacts electrically with these solids. Also, χ is called the **electron affinity** for the semiconductor and $q\phi_F$ is the **energy separation** between the semiconductor Fermi level and the intrinsic (near center gap) energy E_i.

The band structure of metals and semiconductors was discussed in Section 3.5. The energy levels in the conduction band of a monovalent metal is half filled and all levels are filled with electrons up to the Fermi energy E_{F_m}. In a *p*-type semiconductor the energy levels in the valence band are nearly filled with electrons, except for a few empty states or holes at the top; the Fermi energy E_{F_s} is below the center of the energy gap. Note that in this illustration the semiconductor Fermi energy is lower than that of the metal.

A. WORK-FUNCTION DIFFERENCE

Now consider the process of bringing the metal and semiconductor into close proximity of each other, separated only by a thin oxide (or insulator), to form an MOS capacitor. The energy band diagram for this thermal equilibrium situation requires that the Fermi levels of the metal and semiconductor line up. The resultant electron energy band structure is shown in Fig. 10.6. The downward bending of the energy bands near the surface of the *p*-type semiconductor pushes the valence band edge farther from the Fermi level and the surface is depleted of some holes and *tends toward n*-type. Since this is an electron energy diagram, the lower electron energy near the surface and the positive slope of the band edges (and hence negative slope of electric potential) there call for an electric field directed in the positive *x*-direction, which forces holes from the surface. Note that the positive, constant energy slope in the oxide means that an electric field also exists in the oxide, uniform in magnitude and directed in the positive *x*-direction. These fields are **built-in** under the condition of zero external applied potential, similar to the field built into a *p*-*n* junction (see Section 6.1.2), and result from the difference in work function between the metal and semiconductor.

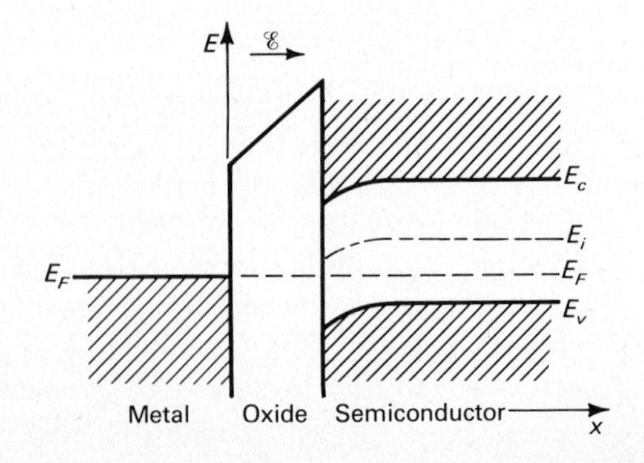

Figure 10.6 Electron energy band diagram for the metal and *p*-type semiconductor, illustrated in Fig. 10.5b, in close proximity, separated only by a thin oxide. The band bending at the semiconductor surface is caused by the difference of the metal and semiconductor work functions, and results in a depletion layer there.

B. OXIDE CHARGES

A further bending of the bands in this same direction will occur if the oxide contains positive charges. For example, it is found that ionized, positively charged alkali atoms are found even in carefully grown and relatively pure oxides grown onto silicon by exposure to oxygen or water vapor at high temperatures. This will be discussed further in Chapter 11. Also, positive charges probably due to partially oxidized silicon atoms occur near the silicon–oxide interface. When these charges, usually referred to as q_{ss}, are sufficiently numerous, the surface of the p-type semiconductor may even become inverted to n-type owing to the collection of these minority electrons there by the x-directed field. The band bending will bring the conduction band edge near the surface closer to the Fermi level than the valence band edge. Hence the surface of the semiconductor can be "inverted" to n-type even under thermal equilibrium conditions (zero applied voltage bias).

C. APPLIED BIAS

If a battery is connected across this MOS structure, the semiconductor surface may become more inverted, less depleted, or accumulate holes as discussed in the previous section. That is, if the metal plate is made positive with respect to the p-type semiconductor an electric field will be induced which will repel the holes associated with the acceptors at the surface, causing even more band bending than that shown in Fig. 10.6 and eventually inverting the surface to n-type. This is illustrated in the energy band diagram of Fig. 10.7a. If the metal electrode is now made negative relative to the substrate, majority holes will be attracted to the surface, causing them to accumulate there in excess of the thermal equilibrium density. Figure 10.7b illustrates this condition of hole accumulation.

10.4.2 The MOSFET Threshold Voltage

A. THE IDEAL STRUCTURE

The MOS capacitor as the input gate element of a MOSFET will now be described in a quantitative sense. Initially the absence of charges in the oxide and the work-function difference between metal and semiconductor will be assumed. Poisson's charge equation is applied to the MOS structure for this purpose.[5] The one-dimensional solution of this equation subject to appropriate boundary conditions including a potential V_G applied to the metal gate will yield the electric potential $\phi(x)$ at a distance from the oxide interface into the p-type semiconductor substrate, as illustrated in Fig. 10.8. Commonly the zero of potential is taken as the electric potential corresponding to the intrinsic energy level E_i, i.e., $\phi_i = E_i/q = 0$; the quantity $q\phi$ is referenced to the equilib-

[5]Gauss's law applied to the p-n junction in Section 6.1.3 is another way of stating Poisson's equation and the solution presented here parallels the one given in that section. That the two equations are analogous may be seen by noting that the electric field $\mathcal{E} = -\partial\phi/\partial x$.

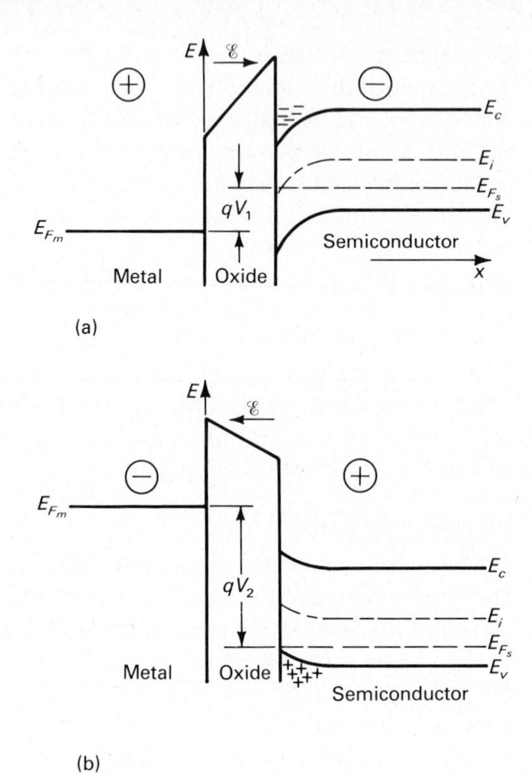

Figure 10.7 (a) A positive voltage applied to the metal relative to the semiconductor rejects holes and attracts electrons, inverting the semiconductor surface to *n*-type. Band bending causes the conduction band edge to be nearer the Fermi level than the valence band edge at the surface.
(b) A negative voltage applied to the metal relative to the semiconductor accumulates holes at the surface of the semiconductor. Band bending causes the valence band edge to approach the Fermi energy.

rium value of E_i and is taken positive for energy values below E_i and negative for values above E_i. Poisson's equation may be written as

$$\frac{\partial^2 \phi}{\partial x^2} = -\frac{q[N_d^+ - N_a^- + p(x) - n(x)]}{\epsilon_0 \epsilon_s}, \tag{10.3}$$

where N_d^+ and N_a^- represent the ionized donor and acceptor densities in the semiconductor, respectively; $p(x)$ and $n(x)$ correspond to the mobile hole and

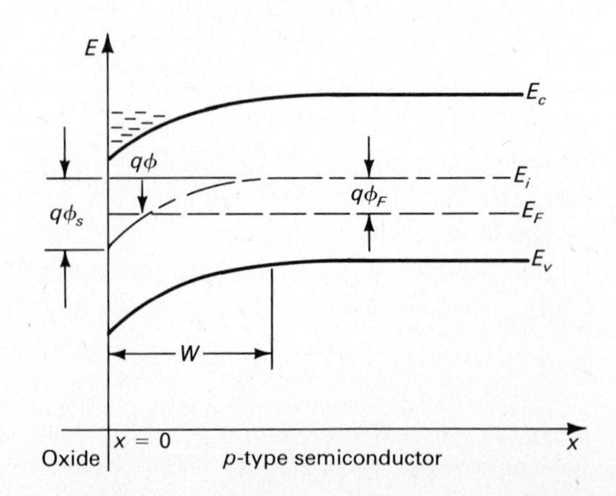

Figure 10.8 Energy band diagram at the oxide–semiconductor interface in an MOS capacitor structure under the condition of strong surface inversion. This is defined as the condition under which the band bending at the surface causes the surface potential ϕ_s to be just twice the Fermi potential ϕ_F deep in the bulk of the semiconductor, where no electric field is present.

electron carrier densities at a distance x from the oxide–semiconductor interface, respectively; and ϵ_0 and ϵ_s are the respective dielectric constants of free space and the semiconductor. The thermal equilibrium carrier densities n_{p0} and p_{p0} in the p-type substrate may be expressed in terms of the electric potential by referring back to Eqs. (4.31) of Section 4.4.3, which are

$$n_{p0} = n_i e^{(E_F - E_i)/kT} \tag{4.31a}$$

and

$$p_{p0} = N_a = n_i e^{(E_i - E_F)/kT}. \tag{4.31b}$$

In terms of the Fermi electric potential these equations can be written as

$$n_{p0} = n_i e^{-q\phi_F/kT} \tag{10.4a}$$

and

$$p_{p0} = N_a = n_i e^{q\phi_F/kT}. \tag{10.4b}$$

It follows that under nonequilibrium conditions, if a voltage is applied between the metal gate and the p-type semiconductor substrate, the carrier densities $n(x)$ and $p(x)$ can be written in terms of the electric potential $\phi(x)$ as

$$n(x) = n_i e^{-q(\phi_F - \phi)/kT} = n_{p0} e^{q\phi/kT} \tag{10.5a}$$

and

$$p(x) = n_i e^{-q(\phi - \phi_F)/kT} = p_{p0} e^{-q\phi/kT}. \tag{10.5b}$$

Note that in the case of strong inversion, illustrated in Fig. 10.8, where $\phi_s = 2\phi_F$, the surface concentration of electrons n_s is just equal to the bulk hole concentration p_{p0}. This is the subject of a problem at the end of this chapter.

In principle Eq. (10.3) can now be solved for the potential $\phi(x)$ once the carrier density expressions of Eqs. (10.5) are introduced since $\phi(x)$ is the only remaining unknown. The solution is somewhat complex.[6] Instead much information can be obtained about the operation of the MOS capacitor and MOSFET by making a depletion approximation similar to that in Section 6.1.3 for the p-n junction; i.e., the region where a strong electric field occurs near the oxide–semiconductor interface is assumed to be depleted of carriers [$n(x) = p(x) = 0$]. Then Eq. (10.3) can be readily integrated. The results are analogous to those obtained in the case of an n^+-p junction.

In the depletion approximation the total charge of ionized acceptors *per unit area* located in the depletion region is

$$q_d = -qN_a W, \tag{10.6}$$

where W is the depletion width as indicated in Fig. 10.8. In the semiconductor, at the interface, minority electrons of charge q_n per unit area can be collected by means of a positive charge q_G applied to the metal gate. The thickness of

[6]For an exact solution of this problem see S. M. Sze, *Physics of Semiconductor Devices,* 2nd ed. (Wiley, New York, 1981), Section 7.2.1.

this electron inversion layer is typically only a few tens of angstroms compared to the depletion width, which is typically of the order of a micrometer; hence just at the strong inversion condition shown in Fig. 10.8, the total depletion region charge of ionized acceptors greatly exceeds the electron inversion charge.

A positive potential V_G applied to the metal gate appears partly across the oxide and partly across the semiconductor depletion region, or

$$V_G = V_{ox} + \phi_s, \tag{10.7}$$

where V_{ox} is the voltage drop across the oxide and ϕ_s is the potential drop across the semiconductor depletion region (see Fig. 10.8). Now, assuming that the electric field is confined to the oxide and the semiconductor depletion region, Gauss's law requires that

$$\epsilon_0 \epsilon_{ox} \mathscr{E}_{ox} = q_G = -q_s = -(q_d + q_n) = \epsilon_0 \epsilon_s \mathscr{E}_s, \tag{10.8}$$

where ϵ_{ox} and ϵ_s are the respective dielectric constants for the oxide and the semiconductor, q_s is the total unit area charge in the semiconductor, and $\mathscr{E}_{ox}$ and $\mathscr{E}_s$ are the electric field intensities in the oxide and semiconductor, respectively. Since in this treatment it is assumed that the oxide of thickness t_{ox} is charge free,

$$\mathscr{E}_{ox} = V_{ox}/t_{ox}. \tag{10.9}$$

Now combining Eqs. (10.8) and (10.9) gives

$$V_{ox} = -\frac{t_{ox}(q_d + q_n)}{\epsilon_0 \epsilon_{ox}} = -\frac{q_d + q_n}{C_{ox}}, \tag{10.10}$$

where C_{ox} is the conventional parallel-plate oxide-layer capacitance *per unit area.*

In the absence of oxide charges and metal–semiconductor work-function difference the gate voltage just under the condition of strong inversion of Fig. 10.8, now termed the **threshold voltage** V_T, is obtained by combining Eqs. (10.6), (10.7), and (10.10) as

$$V_T = -\frac{q_d + q_n}{C_{ox}} + \phi_s \approx \frac{qN_a W}{C_{ox}} + 2\phi_F \tag{10.11}$$

since $q_d \gg q_n$ and $\phi_s = 2\phi_F$ just at the point of strong inversion. Using the depletion approximation, in analogy with Eq. (6.18a), we have

$$W = \left(\frac{2\epsilon_0 \epsilon_s \phi_s}{qN_a}\right)^{1/2} = \left(\frac{2\epsilon_0 \epsilon_s (2\phi_F)}{qN_a}\right)^{1/2}. \tag{10.12}$$

Increasing the gate voltage slowly beyond this point mainly serves to collect more minority electrons at the silicon–oxide interface rather than expand the depletion-layer width, and hence this maximum layer width is given by

$$W_{\max} = 2\left(\frac{\epsilon_0\epsilon_s(kT/q)\ln(N_a/n_i)}{qN_a}\right)^{1/2}, \tag{10.13}$$

where Eq. (10.4b) has been utilized.

EXAMPLE 10.1

Determine the maximum depletion-layer width at 300°K for an MOS capacitor with p-type substrate doping of 2.0×10^{16} acceptors/cm^3, assuming the absence of oxide charges and oxide–semiconductor work-function difference. The dielectric constant of silicon is 11.9.

Solution

Using Eq. (10.13) we have

$$W_{\max} = 2\left(\frac{(8.85 \times 10^{-14}\ \text{F/cm})(11.9)(0.026\ \text{V})\ln\left(\dfrac{2.0 \times 10^{16}}{1.5 \times 10^{10}}\right)}{(1.6 \times 10^{-19}\ \text{C})(2.0 \times 10^{16}\ \text{cm}^{-3})}\right)^{1/2}$$

$$= \underline{2.2 \times 10^{-5}\ \text{cm} = 0.22\ \mu\text{m}}.$$

B. THE REAL DEVICE

Generally in a practical device there is a difference in work function between the gate metal and the semiconductor as illustrated in Fig. 10.5b which gives rise to a built-in field even under thermal equilibrium conditions, as shown in Fig. 10.6. To *undo* this band bending a negative bias voltage must be applied to the metal gate relative to the semiconductor substrate so that no field exists in the latter. The applied voltage necessary to accomplish this is just $\phi_m - \phi_s \equiv \phi_{ms}$, the work-function potential difference, as may be seen by studying Figs. 10.5b and 10.6. This latter quantity varies with semiconductor doping since the semiconductor Fermi energy varies with impurity concentration. Figure 10.9 gives the variation of ϕ_{ms} with semiconductor silicon doping when the gate metal is aluminum. Similar curves could be plotted for other gate metals or polysilicon gates.

When positive charges are included in the gate oxide or at the oxide–semiconductor interface, a further equilibrium band bending occurs which can even invert the semiconductor surface to n-type. To undo the band bending due to these positive charges, again a negative potential must be applied to the metal gate in order to repel the electrons near the interface and flatten the energy band edges there. Quantitatively these charges are often taken as an

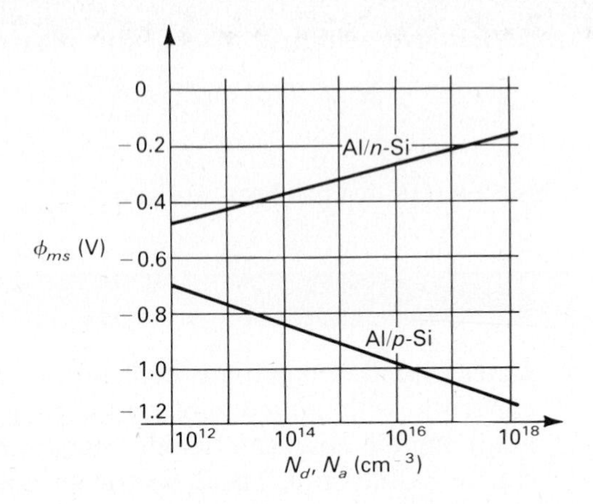

Figure 10.9 Aluminum gate/silicon work-function potential difference ϕ_{ms} versus substrate doping concentration.

equivalent unit area charge located at the oxide–silicon interface which is termed q_{ss}. Then this required voltage is simply $-q_{ss}/C_{ox}$. Now the gate voltage necessary to undo the band bending caused both by these charges plus the work-function difference, the **flat-band voltage,** is given by

$$V_{FB} = \phi_{ms} - q_{ss}/C_{ox}. \tag{10.14}$$

Finally the total gate potential necessary to bring the surface of a semiconductor just to the condition of strong inversion, the threshold voltage, as shown in Fig. 10.8, is given by

$$V_T = \phi_{ms} - \frac{q_{ss}}{C_{ox}} - \frac{q_d}{C_{ox}} + 2\phi_F, \tag{10.15}$$

where use is made of Eq. (10.11). For an Al–p-type silicon MOS capacitor, ϕ_{ms} is negative, q_{ss} is positive, q_d is negative, and ϕ_F is positive. For an Al–n-type silicon capacitor, ϕ_{ms} is negative, q_{ss} is positive, q_d is positive, and ϕ_F is negative.

EXAMPLE 10.2

Calculate the threshold voltage V_T for an Al–p-Si MOS capacitor with an oxide thickness of 0.10 μm and a substrate doping of $2.0 \times 10^{16}/cm^3$. The effective oxide unit-surface-area charge number density q_{ss}/q is $5.0 \times 10^{10}/cm^2$.

Solution

From Fig. 10.9, $\phi_{ms} = -0.97$.

$$C_{ox} = \epsilon_0\epsilon_{ox}/t_{ox} = (8.85 \times 10^{-14}\text{ F/cm})(3.9)/(0.10 \times 10^{-4}\text{ cm})$$
$$= 3.5 \times 10^{-8}\text{ F/cm}^2.$$

Hence

$$q_{ss} = (1.6 \times 10^{-19}\ \text{C})(5.0 \times 10^{10}\ \text{cm}^{-2})$$
$$= 8.0 \times 10^{-9}\ \text{C/cm}^2$$

and

$$q_{ss}/C_{ox} = (8.0 \times 10^{-9}\ \text{C/cm}^2)/(3.5 \times 10^{-8}\ \text{F/cm}^2)$$
$$= 0.23\ \text{V}.$$

From Eq. (10.6) and Example 10.1

$$q_{d\max} = -qN_aW_{d\max}$$
$$= -(1.6 \times 10^{-19}\ \text{C})(2.0 \times 10^{16}\ \text{cm}^{-3})(2.2 \times 10^{-5}\ \text{cm})$$
$$= -7.0 \times 10^{-8}\ \text{C/cm}^2.$$

Hence

$$q_{d\max}/C_{ox} = -2.0\ \text{V}.$$

From Eq. (10.4b),

$$\phi_F = \frac{kT}{q}\ln\frac{N_a}{n_i}.$$

Hence

$$\phi_F = (0.026\ \text{V})\ln(2.0 \times 10^{16}/1.5 \times 10^{10})$$
$$= 0.37\ \text{V}.$$

Finally, Eq. (10.15) gives

$$V_T = -0.97 - 0.23 - (-2.0) + 2(0.37\ \text{V})$$
$$= \underline{1.5\ \text{V}}.$$

10.4.3 Measurement of q_{ss}

The measurement of the small-signal capacitance of an MOS capacitor structure versus DC bias provides a useful technique for determining the effective oxide charge density q_{ss}. If q_G is the charge per unit area on the metal gate, the small-signal ideal MOS capacitance per unit area *in the absence of oxide charges and metal–semiconductor work-function difference* is given by

$$C_{\text{MOS}} \equiv \frac{dq_G}{dV_G} = -\frac{dq_s}{dV_G} = -\frac{dq_s}{-(dq_s/C_{ox}) + d\phi_s} \tag{10.16}$$

from Eqs. (10.7) and (10.10). Dividing numerator and denominator by dq_s gives

$$C_{\text{MOS}} = \frac{1}{1/C_{\text{ox}} + 1/C_s}, \tag{10.17}$$

where by definition

$$C_s \equiv -\frac{dq_s}{d\phi_s} \approx \frac{\epsilon_0\epsilon_s}{W} \tag{10.18}$$

is the unit area capacitance of the semiconductor surface space-region while the surface depletion width is increasing because of a rising positive voltage applied to the metal gate. An interpretation of Eq. (10.17) is that the MOS capacitance results from a series combination of the oxide and the semiconductor depletion-layer capacitances. Introducing Eqs. (10.12) and (10.7) into Eq. (10.17) gives, at a DC bias voltage V_G, assuming that flat band occurs at $V_G = 0$,

$$\frac{C_{\text{MOS}}}{C_{\text{ox}}} = \left[1 + \left(\frac{(2\epsilon_0\epsilon_{\text{ox}})^2}{qN_a\epsilon_0\epsilon_s t_{\text{ox}}^2} V_G\right)^{1/2}\right]^{-1}, \tag{10.19}$$

predicting that this ratio will decrease while the p-type semiconductor surface is being depleted. The derivation of Eq. (10.19) is the subject of a problem at the end of this chapter.

A plot of the ratio $C_{\text{MOS}}/C_{\text{ox}}$ versus gate voltage for an MOS capacitor with a p-silicon substrate is given in Fig. 10.10, indicating the decrease of this quantity as V_G is increased in a positive sense. This is caused by the increase of the depletion-layer width in the silicon just under the oxide as a result of the rejection of holes from the Si/SiO_2 interface, leaving behind negatively charged acceptor ions. This effectively reduces the silicon capacitance C_s, and hence

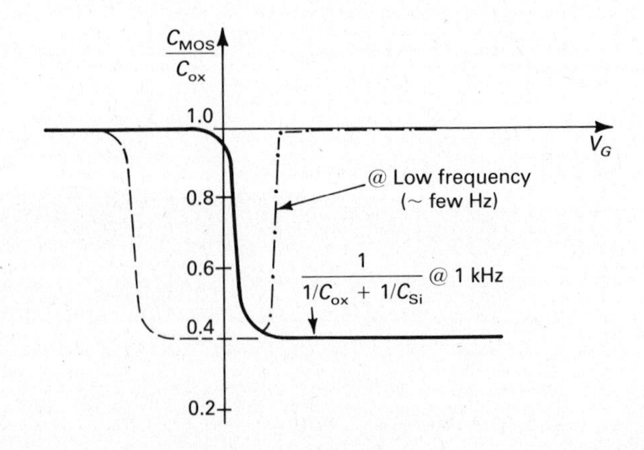

Figure 10.10 A plot of $C_{\text{MOS}}/C_{\text{ox}}$ versus gate voltage for an ideal MOS capacitor at a high measuring frequency of a few kilohertz (solid line). The dot-dashed curve represents values at a low measuring frequency of a few hertz. The dashed curve indicates a shift of the solid curve due to an oxide charge q_{ss} plus the metal–semiconductor work-function difference.

reduces the C_{MOS}/C_{ox} ratio according to Eq. (10.17). The minimum value of this ratio occurs when the depletion width increases to its maximum value W_{max}. Additional applied positive gate charge then results in the collection of mobile minority electrons in the *p*-silicon at the Si/SiO_2 interface.

Eventually any electric flux lines generating from additional positive charges placed on the metal gate will terminate on these electron charges and the MOS capacitance will revert to the value controlled by the oxide capacitance only, C_{ox}. This is indicated in Fig. 10.10 by the dot-dashed line and will be observed only if the capacitance measuring signal is a very low frequency sinusoid of the order of a few hertz. The minority electrons respond very slowly to the measuring signal and hence at a typical measuring frequency of 1 kH there is not sufficient time for these electrons to accumulate and the capacitance increase to C_{ox} does not take place. This is indicated by the solid line of Fig. 10.10, which shows that C_{MOS}/C_{ox} remains constant at a low value typical of the maximum-width depleted *p*-region as the gate voltage is increased positively. A negative gate potential relative to the *p*-substrate will accumulate additional holes at the Si/SiO_2 interface, so the MOS capacitance value is simply that of the oxide layer, C_{ox}, for these negative voltages.

The solid-line C_{MOS} versus gate voltage plot of Fig. 10.10 represents the ideal case when there is no metal–semiconductor work-function difference and the oxide charge density q_{ss} is zero. A positive charge density when present in the oxide may be treated as an effective charge density in the oxide just at the Si/SiO_2 interface. This positive charge will serve effectively as an extra potential applied to the metal plate and is given by

$$V_{G\text{extra}} = q_{ss}/C_{ox}. \tag{10.20}$$

Hence the ideal C_{MOS}/C_{ox} versus V_G solid line curve of Fig. 10.10 is shifted to the left (or in the negative gate voltage direction) by an amount given by Eq. (10.20), corresponding to the need for negative metal gate charge to terminate the electric flux lines created by the q_{ss} charge. This is shown as the dashed curve of Fig. 10.10. The shift of this curve can be used to determine the oxide charge density q_{ss} if the metal–semiconductor work-function difference is known.

EXAMPLE 10.3

An MOS capacitance versus voltage curve similar to Fig. 10.10 is shifted from the calculated curve which includes the metal–semiconductor work-function difference by -2.0 V for a structure whose SiO_2 thickness is 0.10 μm. Find the effective oxide interface surface charge density q_{ss} and the number of charges per cm^2, N_{ss}.

Solution

The gate voltage shift from Eq. (10.20) is

$$V_{G\text{shift}} = -q_{ss}/C_{\text{ox}},$$

where

$$C_{\text{ox}} = \epsilon_0\epsilon_{\text{ox}}/t_{\text{ox}} = (8.85 \times 10^{-14}\ \text{F/cm}(3.9)/(0.10 \times 10^{-4}\ \text{cm})$$

$$= 3.5 \times 10^{-8}\ \text{F/cm}^2.$$

Hence

$$q_{ss} = -V_{G\text{shift}}C_{\text{ox}} = -(-2.0\ \text{V})(3.5 \times 10^{-8}\ \text{F/cm}^2)$$

$$= \underline{7.0 \times 10^{-8}\ \text{C/cm}^2},$$

and

$$N_{ss} = q_{ss}/q = (7.0 \times 10^{-8}\ \text{C/cm}^2)/(1.6 \times 10^{-19}\ \text{C})$$

$$N_{ss} = \underline{4.4 \times 10^{11}/\text{cm}^2}.$$

10.4.4 The Silicon MOS Enhancement-Type Transistor

The metal–oxide–semiconductor field-effect transistor (MOSFET) is normally fabricated with silicon as the semiconductor, its oxide SiO_2 as the insulator, and with a metal such as aluminum to provide ohmic contact to each of the three transistor regions. These electrodes contact the three main sections of the device: the source, the gate, and the drain, as shown in Fig. 10.1. (Sometimes a fourth electrical contact is made to the chip substrate.) If a rough analogy is drawn with the bipolar transistor, the source would correspond to the emitter, the gate to the base, and the drain to the collector. The current which flows from drain to source is modulated by a potential of the order of a few volts applied to the gate electrode, insulated from the silicon chip by a thin silicon dioxide layer 100–1000 Å thick. In some of these devices the structure is completely symmetric, so either of the two n^+-p junctions may be labeled the source while the other then becomes the drain.

To understand the current flow in this device, assume that the gate electrode of the MOS device pictured in Fig. 10.1 is biased a few volts positive with respect to the p-type silicon substrate beneath. Because of the capacitor-like gate structure of this device consisting of an insulating oxide layer sandwiched between the conducting metal gate electrode and the conducting (about 1 Ω-cm) p-type silicon substrate, a positively charged gate will induce a negative charge at the oxide–substrate interface.[7] This negative charge in the *p*-type silicon of course must consist of minority carrier electrons, which are collected at the oxide-silicon interface, just under the gate electrode, as well as some negatively charged, ionized acceptors there. When a sufficiently large gate potential is applied, a significant number of these electrons, collected at the interface, will outnumber the holes there and the original *p*-type region will essentially convert to *n*-type. This *n*-region will then produce a conducting path between the *n*-type source and the *n*-type drain, permitting majority electrons to flow between these latter regions, should a potential be applied between them. This gives rise to the designation of this transistor as a majority carrier device.

This situation is sketched in Fig. 10.11. Here for convenience the source is electrically connected to the substrate and both are assumed at a reference potential of zero volts or ground. Consider the gate electrode also connected for the moment to ground potential and the drain at +5 V relative to ground. Then the situation is simply one of 5 V connected across two back-to-back *p-n* junction diodes, with the drain junction reverse biased and conducting perhaps only a few picoamperes of current. However, when the gate potential is made +2 V, the induced *n*-layer so produced will provide an *n*-type bridge or **channel** between source and drain and typically milliamperes of current will flow. This "enhancement" of drain-to-source current by about nine orders of magnitude due to the charging of the gate electrode accounts for the designation of this device as an **enhancement-type** MOS transistor. In addition it is referred to as an ***n*-channel** device. It is also possible to construct a device similar in every respect except that the source and drain will be p^+-regions and the substrate low conductivity *n*-type silicon. In this case a sufficiently large negative potential applied to the gate electrode will collect holes under the gate, forming a *p*-type region at the oxide–substrate interface, providing a conducting bridge between source and drain. This device is referred to as a ***p*-channel enhancement-type** device. Typical sets of output characteristics for both *n*- and *p*-channel MOSFETs are shown in Fig. 10.12. In the *n*-channel device, while

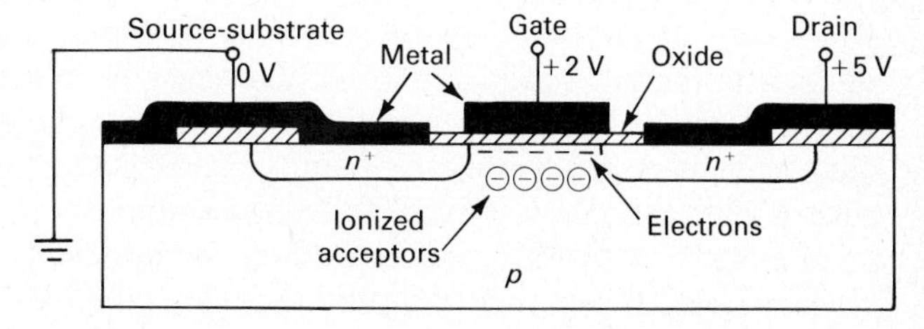

Figure 10.11 An *n*-channel enhancement-type MOSFET with gate bias showing the negative electrons which form the *n*-type channel as well as the ionized acceptors, which are depleted of holes.

[7]This discussion will not make specific reference to charges which appear near the oxide-substrate interface owing to charges in the oxide. This topic has already been considered in Section 10.4.2.

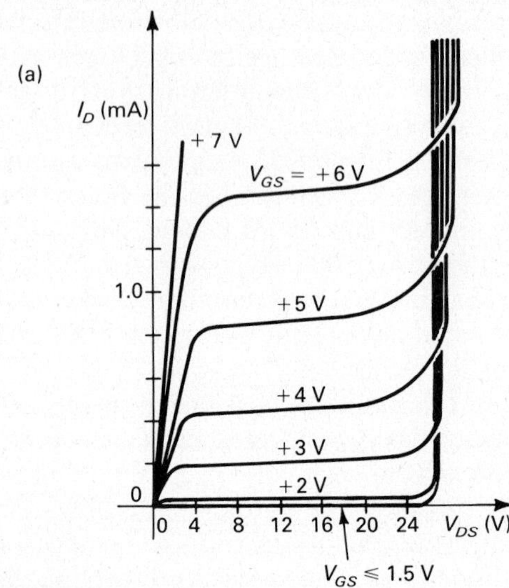

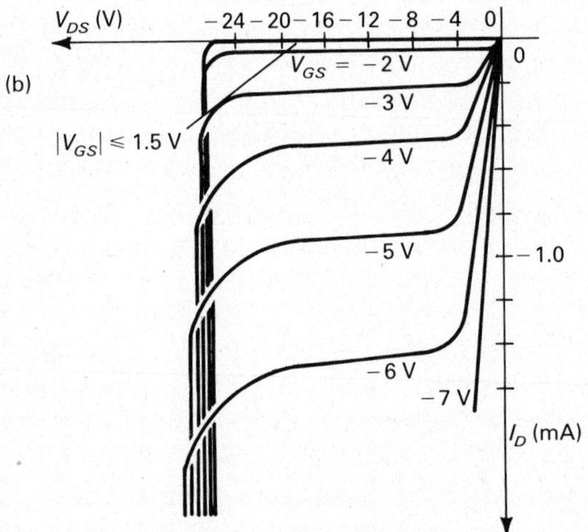

Figure 10.12 (a) Typical set of output characteristics for an *n*-channel enhancement-type MOSFET. (b) Typical set of output characteristics for a *p*-channel enhancement-type MOSFET.

the gate potential is raised in a positive sense from zero to 1 V, the drain current remains in the picoamperes range. An insufficient number of electrons have been attracted to the oxide–substrate interface to convert the *p*-type silicon beneath the gate to *n*-type, forming an *n*-type conducting channel. The few picoamperes of current then represents essentially the leakage current of the reverse-biased drain junction. When the gate potential is made somewhat more positive than 1 V, a drain current of the order of a milliampere begins to flow and the device becomes active in the sense that increasing the positive gate potential results in higher output drain current. This critical voltage of +1 V is called the gate threshold voltage and is normally of the order of a volt or two in a well-designed MOS silicon device, which has few charges in the oxide.

Scrupulous care must be exercised in the preparation of the gate oxide and gate metallization for a *p*-channel enhancement-type device to avoid a high surface density of negative charges at the silicon surface induced by some positive charges q_{ss} in the oxide and at the silicon–oxide interface (see Section 10.4.2). This raises the threshold voltage. Proprietory cleaning procedures, introducing the chlorine ion into the oxidation furnace, annealing steps, and aluminum gate depositions using electron-beam heating yield *p*-channel MOSFETs with low threshold voltage. These steps tend to reduce sodium contamination and minimize q_{ss}. The replacement of the aluminum gate metalization with amorphous silicon can result in threshold voltages of a volt or less. Positive oxide charges will induce an *n*-type skin on the surface of an intended *n*-channel enhancement MOSFET, giving rise to drain current even without applied gate voltage and thus limiting the usefulness of this device as an enhancement-type transistor.

Note that the output characteristics of this unipolar transistor, in analogy with the bipolar transistor, include an active region as well as a bottoming (or linear) region and a drain junction avalanche region. Note also that the drain current in the active region at a given drain voltage increases at an increasing rate as the gate potential is made more positive for an *n*-channel device. In fact it will be shown in the next section that the drain current I_D increases essentially as the *square* of the gate voltage increase, giving rise to the term **square-law** device to describe this type of transistor. No such simple analytic type of description has been made for the bipolar transistor.

10.4.5 The Silicon MOS Depletion-Type Transistor

A **depletion-type** MOSFET is pictured in Fig. 10.13. There are two basic differences between the structure of this type of device and the enhancement-type MOSFET: (i) the *n*-channel, a lightly doped layer created by introducing donor impurities, is already present without the application of gate potential; (ii) generally the gate electrode covers only a small fraction of the channel length (see Fig. 10.13a). Hence there is no gate threshold voltage necessary to begin source-to-drain current conduction. Source-to-drain current is conducted even at zero gate potential. In fact the device operates by supplying negative gate potential which when sufficiently high tends to **cut off** drain current. Typical output characteristics of this *n*-channel depletion type of device are shown in Fig. 10.14a. Characteristics for a *p*-channel device are shown in Fig. 10.14b. These curves are very similar in shape to those shown for a JFET in Fig. 10.3.

In the case of the *n*-channel device a negatively applied gate potential with respect to the substrate will tend to repel majority carrier electrons from the channel underneath the gate, leaving behind positively charged, ionized donor impurities, as shown in Fig. 10.13b. This "depletion" of electrons from the channel is analogous to the formation of the depletion layer of a reverse-biased *p-n* junction. Since the mobile carriers tend to be driven from this type of region, the semiconductor *n*-layer becomes devoid of mobile carriers (nearly

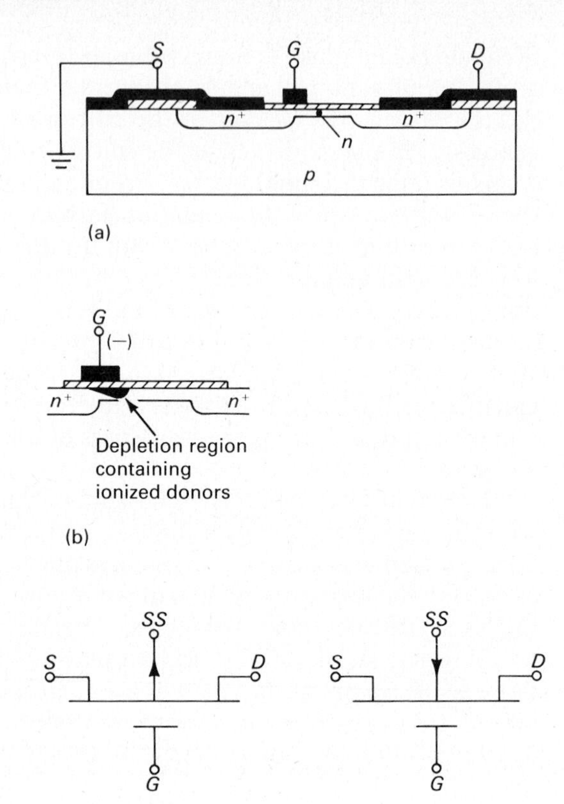

Figure 10.13 (a) Cross section of an *n*-channel depletion-type MOSFET. The crosshatched region represents the silicon dioxide insulator.
(b) Sketch of a cross section of the *n*-type channel of a depletion-type MOSFET showing the manner in which a negative gate potential creates a depleted region which tends to pinch off the channel.
(c) Circuit symbols for the MOS depletion-type field-effect transistor with source (*S*) and drain (*D*). The symbol to the left has an arrow indicating an *n*-type substrate (*SS*) region for this *p*-channel device. The symbol to the right has an arrow indicating a *p*-type substrate region for this *n*-channel device. The arrows are often left off since the battery polarity indicates whether *p*- or *n*-channel devices are used.

intrinsic) and the source-to-drain current is reduced nearly to zero. This resembles the pinchoff mechanism already described for JFETs and accounts for the similarity of the *I-V* characteristics of this MOSFET and the JFET. A gate potential more negative than -3 V is required for the device whose characteristics are shown in Fig. 10.14a to achieve this condition known as current cutoff or channel **pinchoff.** Since the pinchoff of any portion of the channel will interrupt source-to-drain current, the gate electrode need only cover a small fraction of the channel length.

The depletion type of device is often used in high-frequency amplifier applications. This fact results directly from the small-area gate electrode required in this transistor compared to the full size gate needed for the enhancement-type device. The input gate electrode capacitance can hence be significantly less in value in the depletion-type MOSFET. It will be shown later that this input capacitance limits the high frequency performance of the IGFET in a sense similar to the frequency limitation of the bipolar transistor described in Section 9.4. It is common practice to fabricate the depletion device with the gate electrode only near the source region. Any overlap of the

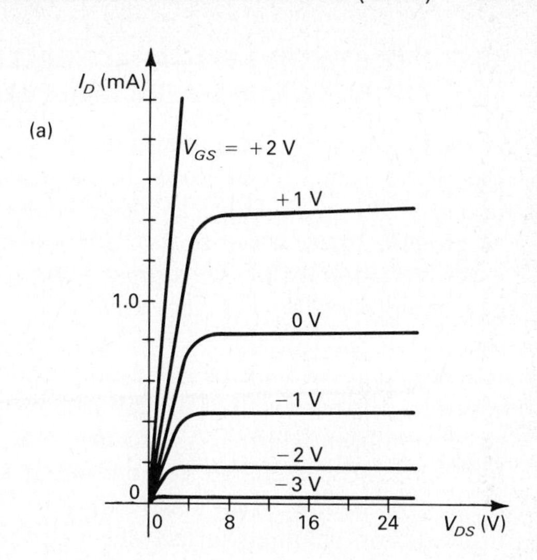

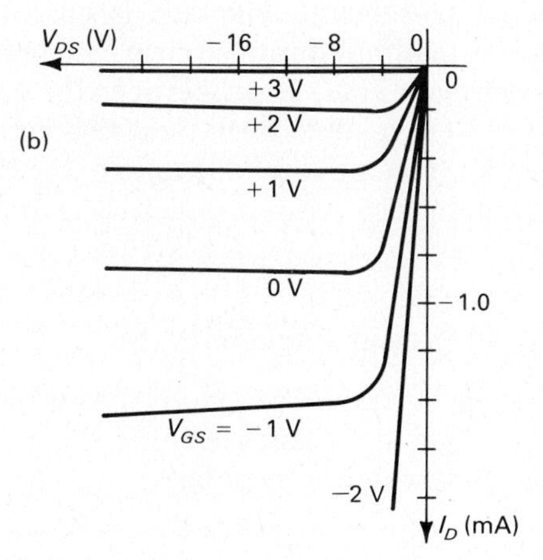

Figure 10.14 (a) Typical set of output characteristics for an *n*-channel depletion-type MOSFET.
(b) Typical set of output characteristics for a *p*-channel depletion-type MOSFET.

gate and drain regions will tend to introduce direct capacitive coupling between input and output. This capacitance will substantially reduce the high frequency gain of the device. The depletion type of MOSFET is also used as the load device of a digital gate circuit in which the enhancement-type MOSFET is the driver (see Section 11.6.1E).

As is the case with the enhancement-type MOSFET, the depletion-type device can be fabricated with either a *p*-channel or an *n*-channel. Since the *n*-channel device conducts by electrons and the *p*-channel MOSFET operates by hole conduction the former type tends to have a higher frequency capability. This is true since the mobility of electrons exceeds that of holes in silicon and so the source-to-drain transit time will be shorter in the *n*-channel device. It will be shown later that this carrier transit time limits the high frequency performance of the IGFET.

10.5 THE STATIC *I-V* CHARACTERISTIC OF THE ENHANCEMENT-TYPE MOSFET

Consider the n-channel enhancement-type MOSFET drawn in cross section in Fig. 10.15. For convenience the source and substrate are taken at ground potential ($V_S = 0$). The calculation of the device I-V characteristic which will be presented here assumes the "gradual channel" approximation. That is, the exact computation of the source-to-drain characteristic involves the solution of a two-dimensional current flow problem even when only a cross section of the device is considered.[8] However, this complex analysis is approximated by reducing it to two simpler one-dimensional problems. In the first place it is assumed that nearly all of the gate control voltage appears across the oxide and so the control field in the y-direction, normal to the oxide–substrate interface, induces a sheet of charge in the semiconductor at the interface. In the second place it is assumed that source-to-drain current conduction takes place in a thin (shallow channel) surface sheet and that the electric field driving this current in the x-direction, parallel to the oxide–substrate interface, is small compared with the gate field. The approach is that the control electric field induces the collection of mobile charges at the surface of the silicon, between the source and drain regions, which serves to conduct current between these regions. The more positive the gate control voltage, the greater the conductance of the channel and the larger the drain current conductecd when a potential is applied between drain and source.

10.5.1 Derivation of the MOSFET *I-V* Output Characteristic

Consider the n-channel enhancement-type MOSFET structure shown in Fig. 10.16. If a gate control voltage V_{GS} is applied to the gate electrode, the electric field in the oxide in the y-direction at any point x is given by

$$\mathscr{E}_{\text{ox}}(x) = [V_{GS} - V(x)]/t_{\text{ox}}, \tag{10.21}$$

where $V(x)$ is the voltage in the channel (referenced to the source–substrate

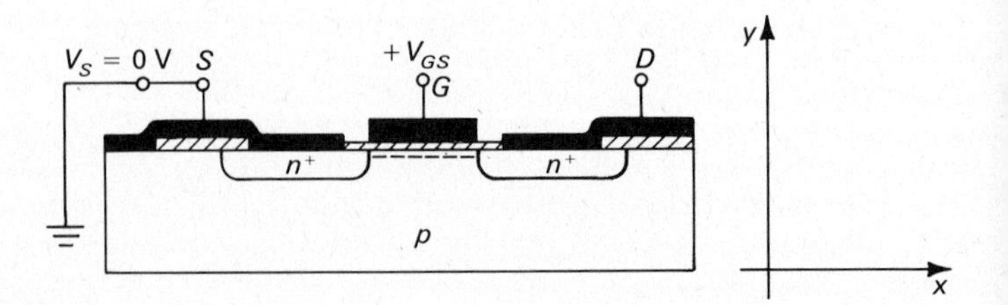

Figure 10.15 Cross sectional view of an n-channel enhancement-type MOSFET showing the negative electron charges forming the conducting channel between drain and source.

[8]This reduces the basically three-dimensional problem to a two-dimensional problem. See for example D. Frohman-Bentchkowsky and A. S. Grove, "Conductance of MOS Transistors in Saturation," *IEEE Trans. Electron Devices,* **ED-16,** 108 (1969).

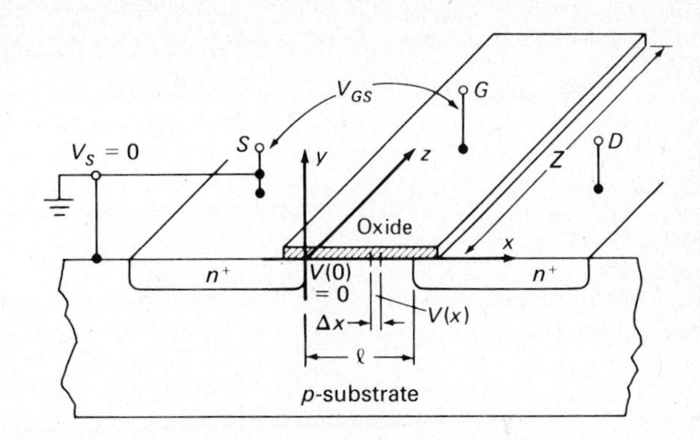

Figure 10.16 Sketch of an *n*-channel enhancement-type MOSFET showing a cross section used in the derivation of the *I-V* output characteristic of the device. The *x*-direction is taken along the channel and the *y*-direction is normal to the oxide layer.

taken as zero potential) at a distance of x from the source and t_{ox} is the oxide thickness. The surface charge density induced under the gate q_I (sheet charge in C/cm^2), is given by Gauss's law as

$$q_I(x) = \epsilon_0\epsilon_{ox}\mathscr{E}_{ox}(x) = (\epsilon_0\epsilon_{ox}/t_{ox})[V_{GS} - V(x)], \tag{10.22}$$

where ϵ_{ox} is the dielectric permittivity of the insulating oxide. However, a channel containing mobile charge carriers does not exist if $V_{GS} - V(x) \leqq V_T$ for any value of x between source and drain. Here V_T, the threshold voltage, is the minimum potential required to cause sufficient mobile surface charge to be induced that appreciable channel current can begin to flow. The **mobile** surface charge density is then given by[9]

$$q_m(x) = (\epsilon_0\epsilon_{ox}/t_{ox})[V_{GS} - V(x) - V_T] \tag{10.23a}$$

for $V_{GS} - V(x) > V_T$, and

$$q_m(x) = 0 \tag{10.23b}$$

for $V_{GS} - V(x) \leqq V_T$.

The incremental conductance $\Delta G(x)$ of an infinitesimally thin channel of width Z (z-direction) and length Δx is given by

$$\Delta G(x) = q_m(x)\bar{\mu}Z/\Delta x, \tag{10.24}$$

where $\bar{\mu}$ is the average electron carrier mobility[10] in the channel at the silicon surface just under the oxide. Applying Ohm's law to this incremental portion of the channel Δx, we obtain for the source-to-drain current there, using Eq. (10.24),

$$I_D = \Delta G(x)\,\Delta V(x) = q_m(x)\bar{\mu}Z\,\Delta V/\Delta x. \tag{10.25}$$

[9]This derivation neglects any variation with x of the induced ionized impurity charge density in the substrate and incorporates its effect in V_T. It also ignores oxide–semiconductor interface charges (see Section 10.4.2), except as included in the gate threshold voltage V_T.

[10]The surface mobility in general is less than the bulk carrier mobility (see Section 5.2.1) due to scattering at the Si/SiO$_2$ interface; it decreases as the gate field which drives the carriers to the interface increases.

Now this drain current must obviously be the same at every point x along the channel. Then integrating Eq. (10.25), taking into account this constancy of I_D, we have

$$I_D \int_0^l dx = \frac{\epsilon_0 \epsilon_{ox} \bar{\mu} Z}{t_{ox}} \int_0^{V_{DS}} [V_{GS} - V(x) - V_T]\, dV, \tag{10.26}$$

where the increment Δx has been assumed arbitrarily small (approaching zero), l is the channel length, and V_{DS} is the channel voltage at $x = l$ which is characteristic of the drain–source potential. Carrying out the integration gives

$$\begin{aligned} I_D &= (\epsilon_0 \epsilon_{ox} \bar{\mu} Z / l t_{ox})[(V_{GS} - V_T)V_{DS} - \tfrac{1}{2}V_{DS}^2] \\ &\equiv (C_{ox} \bar{\mu} Z / l)[(V_{GS} - V_T)V_{DS} - \tfrac{1}{2}V_{DS}^2], \end{aligned} \tag{10.27}$$

where C_{ox} is the gate oxide capacitance *per unit area*.

For $V_{DS} = 0$, $I_D = 0$. As V_{DS} increases in value, with the gate positively biased at some value V_{GS1}, I_D will correspondingly increase. However, when V_{DS} reaches the value given by $V_{DS} = (V_{GS1} - V_T)$, the current must tend to stop increasing. This is true since for this latter value of V_{DS}, termed $V_{DS_{sat}}$, there is zero electric field in the gate oxide just at the drain, and hence there is no induced mobile charge there [see Eq. (10.23a)]. This might seem to indicate that the current will cease to flow because the end of the channel is nonconducting. However, this conclusion is absurd since it is the IR voltage drop in the channel that permits the drain end of the channel to rise to a value $V_{DS_{sat}}$. Hence the drain current $I_{D_{sat}}$ at this drain voltage must be maintained, and for values of $V_{DS} > V_{DS_{sat}}$ the drain current remains essentially constant at a value $I_{D_{sat}}$ called the **saturation current.**[11]

The output characteristics of an n-channel MOSFET indicating the locus of the voltages $V_{DS_{sat}}$ and $I_{D_{sat}}$ for various values of V_{GS} are shown in Fig. 10.17. The value of this saturation current can be calculated by setting $\partial I_D/\partial V_{DS}|_{V_{GS}} = 0$, yielding $V_{DS} = (V_{GS} - V_T)$ and obtaining from Eq. (10.27)

$$I_{D_{sat}} = \frac{C_{ox} \bar{\mu} Z (V_{DS})_{sat}^2}{2l} = \frac{C_{ox} \bar{\mu} Z}{2l}(V_{GS} - V_T)^2. \tag{10.28}$$

Actually the drain current rises somewhat with drain voltage, so the device output conductance $\partial I_D/\partial V_{DS}$ is not zero but has a finite, if small, value. The "square-law" nature of the MOSFET characteristics is indicated in Eq. (10.28).[12] This is reminiscent of the square law of Eq. (10.1) which applies to

[11]This saturation region should not be confused with the bipolar transistor saturation region. As V_{DS} is increased beyond $V_{DS_{sat}}$ excess voltage can no longer contribute to this ohmic drop along the channel but must appear across the space-charge region formed at the drain. Eventually the avalanche field is reached and voltage breakdown occurs (see Fig. 10.12).

[12]This square law approaches a linear relationship for channel lengths less than 1 μm. This is due to gate–drain charge sharing (see Section 10.5.3).

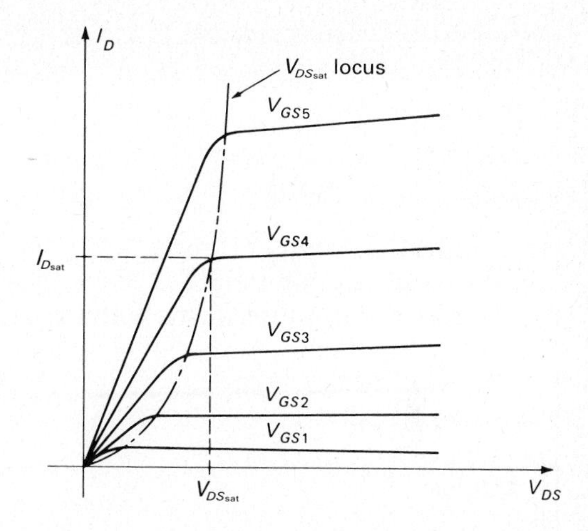

Figure 10.17 The output characteristics of an n-channel enhancement-type MOSFET showing the locus of drain–source voltage saturation points for different values of gate–source voltage.

the JFET. No such simple relationship between output current and voltage has been written for the bipolar transistor.

A measure of the field-effect transistor gain is given by the transconductance g_m, which is obtained by differentiating Eq. (10.28) with respect to gate voltage as

$$g_m \equiv \left(\frac{\partial I_{D\text{sat}}}{\partial V_{GS}}\right)_{V_{DS}} = \frac{C_{\text{ox}}\bar{\mu}Z}{l}(V_{GS} - V_T) = \frac{C_{\text{ox}}\bar{\mu}Z}{l} V_{DS\text{sat}}. \tag{10.29}$$

This represents the change in output drain current per unit change in input gate potential, at a fixed value of drain voltage, in the active region. This definition emphasizes the voltage-control nature of the IGFET. Note the direct dependence of the gain on C_{ox} and $\bar{\mu}$ and the inverse relationship to the channel length.

The device characteristic in the transistor "linear" region ($V_{DS} < V_{DS\text{sat}}$] can be studied by approximating the drain current I_D from Eq. (10.27) for $V_{DS} \to 0$ as

$$I_D = (C_{\text{ox}}\bar{\mu}Z/l)(V_{GS} - V_T)V_{DS}. \tag{10.30}$$

The channel conductance in this region where I_D increases linearly with V_{DS} (the so-called **ON conductance** of interest in digital switching circuits) is given by

$$g_{ds}|_{V_{DS}\to 0} = \left(\frac{\partial I_D}{\partial V_{DS}}\right)_{V_{GS}} = \frac{C_{\text{ox}}\bar{\mu}Z}{l}(V_{GS} - V_T). \tag{10.31}$$

Hence the **linear conductance** increases with increasing gate voltage as well as the inverse of the channel length. The reciprocal of this quantity is usually called r_{ds} (see Appendix G1).

EXAMPLE 10.4

Calculate the electric field in the gate oxide of a silicon n-channel enhancement-type MOSFET with gate voltage $V_{GS} = 5.0$ V and drain voltage $V_{DS} = 4.0$ V

(a) at the source ($x = 0$),
(b) at the drain ($x = l$).
(c) Compare these fields with the avalanche field in silicon.

The oxide thickness $t_{ox} = 8.0 \times 10^{-6}$ cm and the channel length $l = 2.0 \times 10^{-4}$ cm.

Solution

From Eq. (10.21)

$$\mathscr{E}(x) = [V_{GS} - V(x)]/t_{ox}.$$

(a) At $x = 0$,

$$\mathscr{E}(0) = (5.0 - 0 \text{ V})/(8.0 \times 10^{-6} \text{ m}) = \underline{6.3 \times 10^5 \text{ V/cm}}.$$

(b) At $x = l$,

$$\mathscr{E}(l) = (5.0 - 4.0 \text{ V})/(8.0 \times 10^{-6} \text{ m}) = \underline{1.3 \times 10^5 \text{ V/cm}}.$$

(c) These fields are of the same order as the electric field at which avalanching occurs in silicon. However, the breakdown field of SiO_2 is higher, about 6×10^6 V/cm.

EXAMPLE 10.5

(a) Derive an expression for the potential $V(x)$ at any point x along the conducting channel of the MOSFET of Example 10.4, at the edge of saturation.
(b) Then write an expression for the electric field along the channel $\mathscr{E}(x)$ and
(c) calculate the field parallel to the channel halfway between source and drain, if $V_{DS_{sat}} = 5.0$ V.

Solution

(a) Integrating Eq. (10.26), we have

$$I_D \int_0^x dx = \frac{\epsilon_0 \epsilon_{ox} \bar{\mu} Z}{t_{ox}} \int_0^{V(x)} [(V_{GS} - V_T) - V(x)]\, dV,$$

which gives

$$I_D x = B\{(V_{GS} - V_T)V(x) - \tfrac{1}{2}[V(x)]^2\},$$

where $B = \epsilon_0 \epsilon_{ox} \bar{\mu} Z / t_{ox}$ and which satisfies the boundary condition $V(0) = 0$. Now at the point of saturation, since $V_{GS} - V_T = V_{DS_{sat}}$

$$[V(x)]^2 - 2V_{DS_{sat}}V(x) + 2I_{D_{sat}}x/B = 0.$$

Solving this quadratic equation for $V(x)$ yields

$$V(x) = V_{DS_{sat}}\left[1 \pm \left(1 - \frac{2I_{D_{sat}}x}{BV_{DS_{sat}}^2}\right)^{1/2}\right].$$

Introducing Eq. (10.28) for $I_{DS_{sat}}$ then gives

$$\underline{V(x) = V_{DS_{sat}}[1 - (1 - x/l)^{1/2}],}$$

where the negative sign of the radical is chosen to ensure that $V(0) = 0$.

(b) The electric field is obtained by differentiation as

$$\underline{\mathcal{E}(x) = -\frac{dV(x)}{dx} = \frac{V_{DS_{sat}}}{2l[1 - (x/l)]^{1/2}}}.$$

(c) $\mathcal{E}(\tfrac{1}{2}) = 5.0\ \text{V}/2(2.0 \times 10^{-4}\ \text{cm})(1 - \tfrac{1}{2})^{1/2}$

$= \underline{1.25\sqrt{2} \times 10^4\ \text{V/cm} = 1.8 \times 10^4\ \text{V/cm},}$

which is large but much smaller than the field in the oxide.

10.5.2 Calculation of the MOSFET Output Conductance in the Active Region

For drain voltages beyond the knee in the *I-V* curve (see Fig. 10.17), the output conductance is determined by the change of channel length with drain voltage. In this current saturation region, the channel may be considered to be made up of two parts as shown in Fig. 10.18. The normal channel ohmic conduction region discussed thus far adjoining the source is labeled l_s. The portion of the channel near the drain, l_d, is a small section which represents the space-charge depletion region of the reverse-biased drain junction. This region begins to form as the drain voltage is increased to $V_{DS_{sat}}$ and then continues to expand as this voltage continues to increase. In fact the drain voltage in excess of $V_{DS_{sat}}$ appears across this depletion region in a manner very similar to the way in which the space-charge layer supports the reverse voltage applied to a simple *p-n* junction diode. Again space-charge widening occurs as a result of increasing the applied reverse voltage. In the case of the MOSFET, this increases l_d at the expense of decreased l_s, thereby reducing the effective conducting channel

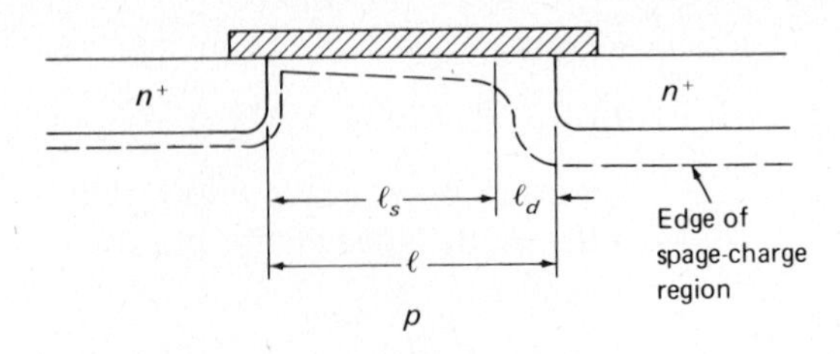

Figure 10.18 Sketch of the cross section of an n-channel enhancement-type MOSFET showing the two sections of the channel. Ohmic conduction takes place in the section marked ℓ_s. The section marked ℓ_d is the region occupied by the drain space-charge region. The dashed line indicates the boundary of the space-charge layer.

length. The output conductance g_{os} for some gate bias V_{GS} can then be computed by calculating the slope of the drain–source output characteristics beyond $V_{DS\text{sat}}$ as

$$g_{os} = \left(\frac{\partial I_{D\text{sat}}}{\partial V_{DS}}\right)_{V_{GS}} = \left(\frac{\partial I_{D\text{sat}}}{\partial l_s}\right)_{V_{GS}} \left(\frac{\partial l_s}{\partial V_{DS}}\right)_{V_{GS}}$$

$$= -\left(\frac{\partial I_{D\text{sat}}}{\partial l_s}\right)_{V_{GS}} \left(\frac{\partial l_d}{\partial V_{DS}}\right)_{V_{GS}}, \tag{10.32}$$

since $l_d = l - l_s$ (see Fig. 10.18). Using Eq. (10.28) we obtain

$$\left(\frac{\partial I_{D\text{sat}}}{\partial l_s}\right)_{V_{GS}} \approx \left(\frac{\partial I_{D\text{sat}}}{\partial l}\right)_{V_{GS}} = -\frac{I_{D\text{sat}}}{l}, \tag{10.33}$$

which together with Eq. (10.32) gives the output conductance

$$g_{os} = \left(\frac{(\partial l_d)_{\text{sat}}}{\partial V_{DS}}\right)_{V_{GS}} \frac{I_{D\text{sat}}}{l} \propto \frac{I_{D\text{sat}}}{l}. \tag{10.34}$$

Hence the output conductance is inversely proportional to the channel length and increases with increasing drain current. This is found to be approximately true in practice (see Fig. 10.12).

A set of specification sheets for a typical commercial silicon n-channel enhancement-type MOSFET is included in Appendix G1; Appendix G2 gives specifications for a p-channel device.

*10.5.3 Short-Channel MOSFET Effects

In order to achieve improved high frequency and fast switching MOSFET performance, the device channel length is shortened to reduce the carrier source-to-drain transit time. In this case the previously derived device characteristics must be reviewed.

A. MORE ACCURATE DERIVATION OF THE I_D-V_{DS} CHARACTERISTIC

For simplicity, in the derivation of Eq. (10.27), the variation along the channel of the induced ionized-impurity charge density in the semiconductor substrate depletion layer was neglected (see footnote 9). Because of the relatively high doping in the substrate used in the fabrication of short channel MOSFETs to avoid punchthrough, this approximation should be eliminated. The depletion-layer area charge density at a point x along the channel (see Fig. 10.16), where the potential is $V(x)$, is given by Eqs. (10.6) and (10.12) as

$$q_d = \{2q\epsilon_0\epsilon_s N_a[2\phi_F + V(x)]\}^{1/2}. \tag{10.35}$$

Noting that the total induced charge density $q_I(x) = q_d(x) + q_m'(x)$, solving for the new mobile charge density $q_m'(x)$, introducing this into Eq. (10.25), and integrating as was done in Eq. (10.26) yields for the corrected drain current

$$I_D = \frac{C_{ox}\bar{\mu}Z}{l}\left[\left(V_{GS} - V_{FB} - 2\phi_F - \frac{V_{DS}}{2}\right)V_{DS} - \frac{2}{3}\frac{(2\epsilon_0\epsilon_s qN_a)^{1/2}}{C_{ox}}[(V_{DS} + 2\phi_F)^{3/2} - (2\phi_F)^{3/2}],\right. \tag{10.36}$$

where V_{FB} is given by Eq. (10.14). This equation should be compared with the simpler but less exact Eq. (10.27).

It was assumed that channel pinchoff was the cause of drain current saturation in the above derivations. However, in very short channel MOSFETs with channel lengths in the micrometer and submicrometer range, the current limitation is at least partly provided by the velocity saturation of the mobile carriers in the inversion layer. This effect was described in Section 5.2.4 and occurs when the driving electric field exceeds about 10^4 V/cm in silicon and 3×10^3 V/cm in GaAs. In very short channel MOSFETs this field can be obtained over an appreciable portion of the channel even at relatively low drain voltages. At these high fields in silicon the velocity tends to saturate at approximately 10^7 cm/sec. If the electron velocity achieves this saturation value over the entire channel length, the drain current density, given in general by $J = qnv_{sat}$, will reach a constant maximum or saturation value, independent of drain voltage, since the carrier concentration n and the velocity v_{sat} are fixed. Under this condition the drain saturation current becomes

$$I_{D\text{sat}} = C_{ox}Z(V_{GS} - V_T)v_{sat}, \tag{10.37}$$

and the transconductance by definition becomes

$$g_m \equiv \frac{\partial I_{D\text{sat}}}{\partial V_{GS}} = C_{ox}Zv_{sat}. \tag{10.38}$$

This defines the maximum gain possible for a MOSFET.

B. THRESHOLD VOLTAGE AND SUBTHRESHOLD CURRENT EFFECTS

Another approximation used in the derivation of Eq. (10.27) must be reconsidered in applying this equation to short channel MOSFETs. In Eq. (10.22) it was assumed for simplicity that all the electric flux lines generated by the charge on the MOSFET gate electrode terminated on the induced mobile and depletion-layer charge just under the gate electrode. The depletion widths of the source and drain *p-n* junctions were considered to occupy a negligible portion of the channel length. Hence the gate threshold voltage was found to be a constant, depending on the substrate doping and independent of channel length.

For short channel MOSFETs this is no longer the case, as is shown in Fig. 10.19. Now some of the field lines originating from the source and drain electrodes terminate on charges in the channel region. Hence less gate charge and voltage is required to cause inversion of the channel region. And so the gate threshold voltage of a MOSFET with a given substrate doping begins to decrease as the channel length becomes a few micrometers or less. In addition, at these short channel lengths increased drain potential lowers the n^+-source barrier height, releases electrons, and gives rise to an increased drain current even at low gate voltages. This **subthreshold current** tends to draw current in an enhancement-type MOSFET even when the current in the device is intended to be shut off. Still further increase in the drain voltage can lead to drain-to-source punchthrough and hence to a voltage limitation.

C. ADJUSTMENT OF THE THRESHOLD VOLTAGE

To regulate the tendency of short channel devices to exhibit lower threshold voltages, two techniques are commonly used to adjust this voltage. In an *n*-channel device the implantation of charged boron ions by an ion accelerator

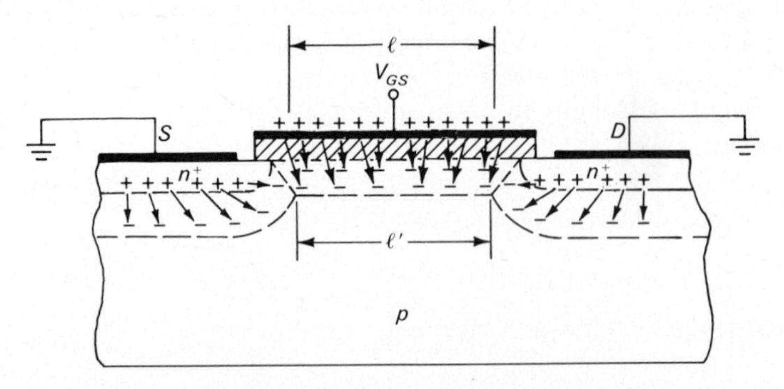

Figure 10.19 Illustration of charge sharing of some channel charge by the source and drain charges. According to the elementary treatment a rectangular channel region of side ℓ is needed to terminate the gate charge electric flux lines. In the short channel device illustrated here only a trapezoidal region with sides ℓ and ℓ' is needed. For simplicity of illustration the drain is taken here at zero potential relative to the source.

will increase the p-type substrate doping just beneath the Si/SiO_2 interface, thereby raising the threshold voltage [see Eq. (10.15)]. Ion implantation is also used to compensate for spurious oxide charges which affect the threshold voltage.

Finally, the use of a negative substrate bias relative to a grounded (zero potential) source will also tend to raise the gate threshold voltage V_T. In the derivation of Eq. (10.27) for the drain current, the substrate was assumed to be at the same potential as the source. However, if a reverse bias V_B is applied between the substrate and the grounded source, the channel depletion region is increased in width and so contains more negatively charged, ionized acceptors q_d (for an n-channel device). Then Eq. (10.12) for the depletion region width becomes

$$W = [2\epsilon_0\epsilon_s(2\phi_F - V_B)/qN_a]^{1/2} \tag{10.39}$$

and q_d correspondingly increases in magnitude according to Eq. (10.6), since V_B is negative. This will require extra positive charge on the gate electrode to provide electric flux lines to terminate on this additional charge. The increment in threshold voltage due to this negative substrate bias is obtained from Eq. (10.11) as

$$\Delta V_T = [(2\epsilon_0\epsilon_s qN_a)^{1/2}/C_{\text{ox}}][(2\phi_F - V_B)^{1/2} - (2\phi_F)^{1/2}]. \tag{10.40}$$

This technique may be used to adjust the threshold voltage for both n- and p-channel MOSFETs.

10.6 THE FREQUENCY LIMITATION OF THE FIELD-EFFECT TRANSISTOR

In order to estimate the gain–bandwidth capability of the FET, refer to the previous calculation of this quantity for the bipolar transistor given in Section 9.5.2. In analogy with that discussion the gain–bandwidth product for the MOSFET can be written as

$$G \times BW = \frac{g_m}{2\pi C_{\text{in}}} \approx \frac{g_m}{2\pi C_{\text{ox}} Zl}, \tag{10.41}$$

where C_{in} is the input or essentially the gate capacitance of the MOSFET, which is roughly the total oxide capacitance $C_{\text{ox}}Zl$. Combining Eq. (10.41) with Eq. (10.29) for g_m yields

$$G \times BW \approx \frac{\bar{\mu}}{2\pi l^2}(V_{GS} - V_T) = \frac{\bar{\mu}}{2\pi l^2} V_{DS\text{sat}}. \tag{10.42}$$

Hence the gain–bandwidth product may be increased four times by halving the channel length. However, this will increase the device's output conductance [see Eq. (10.34)], as well as limit the voltage capability of the transistor. For if the channel length is short enough, increased drain voltage will expand the drain space-charge region l_d, and the maximum drain voltage capability will be limited to that voltage which causes the drain depletion region to reach through to the source depletion region. At that point a rapid increase of drain current will occur, somewhat like that observed when the drain junction breaks down by avalanching. This "punchthrough" phenomenon also occurs in bipolar transistors, particularly with heavily doped emitter and collector regions. However, the planar-type bipolar transistor has heavy base doping near the emitter junction, which reduces the rate of space-charge depletion-layer increase as the emitter is approached. An FET fabricated by diffusion in a similar way, termed the DMOSFET, has a higher punchthrough voltage capability, but also generally higher threshold voltage.

Returning to Eq. (10.42), we may obtain a physical interpretation of this expression by rewriting this equation as

$$G \times BW \approx \frac{1}{2\pi}\frac{\bar{\mu}}{l}\frac{V_{DS\text{sat}}}{l} = \frac{1}{2\pi}\frac{\bar{\mu}\mathcal{E}_{D\text{sat}}}{l}, \tag{10.43}$$

where $\mathcal{E}_{D\text{sat}}$ is approximately the average electric field *along* the channel, just at saturation. Since the average channel mobility $\bar{\mu}$ is defined as the carrier velocity per unit electric field, $\bar{\mu}\mathcal{E}_{D\text{sat}}/l$ is the average channel electron carrier velocity divided by the channel length, or the reciprocal of the carrier source-to-drain **transit time.** Another way of arriving at the same result is to consider the time it takes to charge the gate capacitance through half of the channel resistance. This is the subject of a problem at the end of this chapter. The gain–bandwidth product also supplies an estimate for the switching speed of a MOSFET, since this is limited by the gate charging time and the carrier transit time.

EXAMPLE 10.6

Determine approximately

(a) The gain–bandwidth product for a MOSFET with a gate capacitance $C_{\text{ox}}Zl = 0.02$ pF and a transconductance $g_m = 1.2 \times 10^{-4}$ S $= 0.12$ mS.

(b) The carrier transit time for this device.

Solution

(a) Using Eq. (10.41) we have

$$(G \times BW) \approx g_m/2\pi C_{\text{ox}}Zl$$

$$= (0.12 \times 10^{-3}\ \text{S})/(2\pi)(0.02 \times 10^{-12}\ \text{F})$$

$$= \underline{0.95 \times 10^9\ \text{Hz} = 0.95\ \text{GHz}}.$$

(b) The carrier transit time is

$$\tau_t = 1/2\pi(G \times BW)$$
$$= 1/2\pi(0.95 \times 10^9 \text{ Hz})$$
$$= \underline{1.7 \times 10^{-10} \text{ sec} = 170 \text{ psec.}}$$

10.6.1 Comparison of the Bipolar and Unipolar Transistor Gain–Bandwidth Products

The gain–bandwidth product for bipolar and unipolar transistors of similar geometry may be compared by estimating the expression given in Eq. (10.41) for each type of device. The g_m of the bipolar transistor is given by Eq. (9.15a). The g_m of a MOSFET may be estimated by using Eq. (10.29). If we assume that these devices are similar in size, then their input capacitances are comparable. Hence the $G \times BW$ of these devices may be compared by taking the ratio of their transconductances.

EXAMPLE 10.7

Determine approximately the ratio of the gain–bandwidth product of a bipolar transistor operating at a collector current of 1.0 mA to that of a MOSFET of comparable size. The MOSFET has the following design and operating parameters: $C_{ox}Zl = 0.02$ pF, $\bar{\mu} = 600$ cm^2/V-sec, $l = 2.5 \times 10^{-4}$ cm, and $V_{GS} - V_T = 3.0$ V.

Solution

Using Eq. (9.15a) we have

$$g_{m\text{bipolar}} = \frac{I_C}{kT/q} = \frac{1.0 \times 10^{-3} \text{ A}}{0.026 \text{ V}}$$
$$= \underline{0.038 \text{ S.}}$$

From Eq. (10.29)

$$g_{m\text{MOSFET}} = \frac{(C_{ox}Zl)\bar{\mu}}{l^2}(V_{GS} - V_T)$$
$$= \frac{(0.02 \times 10^{-12} \text{ F})(600 \text{ cm}^2/\text{V-sec})(3.0 \text{ V})}{(2.5 \times 10^{-4})^2 \text{ cm}^2}$$
$$= \underline{0.58 \times 10^{-3} \text{ S} = 0.58 \text{ mS.}}$$

Using Eq. (10.41) we can write

$$\frac{(G \times BW)_{\text{bipolar}}}{(G \times BW)_{\text{MOSFET}}} = \frac{(g_m/C_{\text{in}})_{\text{bipolar}}}{(g_m/C_{\text{in}})_{\text{MOSFET}}}$$

$$\approx g_{m\text{bipolar}}/g_{m\text{MOSFET}},$$

if the input capacitances of these devices are comparable. Hence

$$\frac{(G \times BW)_{\text{bipolar}}}{(G \times BW)_{\text{MOSFET}}} \approx \frac{0.038}{0.58 \times 10^{-3}} \approx \underline{66}.$$

This example indicates the inherent higher frequency capability of the bipolar transistor compared with the MOSFET. However, because of the geometric simplicity of the MOSFET structure we can assume that its input (gate) capacitance can be made five times smaller than a bipolar device input (emitter) capacitance, by reducing the area of the MOSFET gate. Example 10.7 then predicts that the bipolar transistor will still have a gain–bandwidth product about 13 times greater than that of a comparable unipolar MOSFET device. Hence, intrinsically, the bipolar transistor has higher frequency and hence faster switching capability if we consider devices of comparable geometry, made of the same material. However, very small MESFETs have been developed recently which have gallium arsenide as their starting material with frequency capability greater than the best bipolar transistors of silicon. Of course the five times higher electron mobility and higher peak saturation velocity in GaAs compared with Si partially account for this result [see Eq. (10.29)].

It has already been mentioned that a primary application of the MOSFET is in high density electronic memories, where small size and simplicity of device structure are its primary advantages. High speed capability is not the most significant requisite here, rather device manufacturing yield.

10.6.2 Small-Signal Equivalent Circuit for the Field-Effect Transistor

A small-signal model of the FET useful for analyzing linear amplifier circuits in the active mode will now be presented. This will be developed along the lines already taken in the derivation of the two-port hybrid circuit model of the bipolar transistor given in Section 9.3.1. Since the drain current i_D is dependent on the gate–source voltage v_{GS}, as well as the drain–source voltage v_{DS}, an approximate (first-order) Taylor expansion for an incremental change in drain current Δi_D about a DC operating point in the saturation (active) region set by V_{DS} and V_{GS} is

$$\Delta i_D = \left.\frac{\partial i_D}{\partial v_{GS}}\right|_{V_{DS}} \Delta v_{GS} + \left.\frac{\partial i_D}{\partial v_{DS}}\right|_{V_{GS}} \Delta v_{DS}. \tag{10.44}$$

Using small-signal notation we can write this as

$$i_d = g_m v_{gs} + g_{ds} v_{ds}, \tag{10.45}$$

where the lower-case subscripts refer to incremental quantities and the following definitions are used:

$$g_m \equiv \left.\frac{\partial i_D}{\partial v_{GS}}\right|_{V_{DS}} \tag{10.46}$$

and

$$g_{ds} \equiv \left.\frac{\partial i_D}{\partial v_{DS}}\right|_{V_{GS}}. \tag{10.47}$$

A circuit which represents Eq. (10.45) is shown in Fig. 10.20a. This common-source circuit is useful both for JFETs and MOSFETs with substrate joined to source. Note that the open circuit between gate and source as well as gate and drain indicates the negligibly small current leakage through the gate electrode. However, the finite conductance between drain and source is indicated by g_{ds}. This quantity may be determined by the slope of the FET I_D-V_{DS} characteristic in the saturation or active region where the device is biased for amplifier operation, and is generally greater for JFETs than for MOSFETs.

The model just described is intended for low frequency circuit analysis. A higher frequency version of this equivalent circuit is given in Fig. 10.20b. Here the capacitor C_{gs} represents the capacitance between gate and source, which can

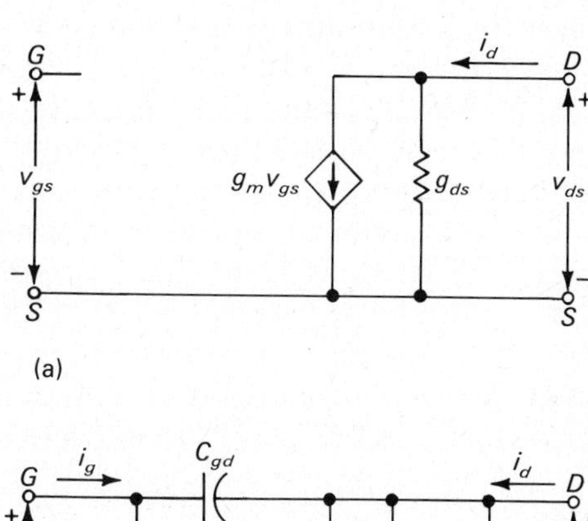

Figure 10.20 (a) Low frequency small-signal common-source equivalent circuit for a field-effect transistor; $g_m v_{gs}$ indicates a current source and g_{ds} is the dynamic drain–source conductance of the device biased in the saturation or active region.
(b) High frequency small-signal equivalent circuit for a field-effect transistor; C_{gs} and C_{ds} represent the gate–source and drain–source interelectrode capacitances; C_{gd} is the gate–drain capacitance.

be shown to be about two-thirds of the gate oxide capacitance C_{ox}[13] in the case of the MOSFET. It is less than the oxide capacitance since the charge induced in the semiconductor is nonuniformly distributed in the conducting channel just under the oxide in the current saturation region. The gate-to-drain capacitance C_{gd} is normally less than C_{gs}. Finally, a small drain-to-source capacitance C_{ds} completes the simple high frequency FET model of Fig. 10.20b.

Since the displacement current through a capacitor is given by $i = C\, dv/dt$, the gate and drain currents of the FET in this simple circuit model are given by

$$i_g = C_{gs}\frac{dv_{gs}}{dt} + C_{gd}\frac{dv_{gd}}{dt} \tag{10.48a}$$

and

$$i_d = g_m v_{gs} + g_{ds} v_{ds} + C_{ds}\frac{dv_{ds}}{dt} - C_{gd}\frac{dv_{gd}}{dt}. \tag{10.48b}$$

Here an incremental gate signal current i_g must be supplied to charge C_{gs} and C_{gd}. The equation for i_g is consistent with the summing of the currents at the gate node *G* of Fig. 10.20b. Similarly Eq. (10.48b) can be obtained by summing the currents at the drain node, marked *D*.

These interelectrode capacitances can be obtained by measurements which are next described. First the output drain-to-source reverse transfer capacitance C_{rss} is measured with the gate-to-source input terminals short circuited. This yields a value for C_{gd} in parallel with C_{ds}, which is essentially C_{gd} since normally $C_{gd} \gg C_{ds}$. Now with the output drain-to-source short circuited, a measurement of the input gate-to-source capacitance C_{iss} gives a value for C_{gd} and C_{gs} in parallel, or $C_{gd} + C_{gs}$. When C_{gd} from the first measurement is subtracted from this latter measurement, C_{gs} is obtained. Since these capacitances in the case of the JFET are voltage dependent, care must be taken to take all the measurements at the same value of voltage.

It is sometimes convenient to model the FET for small-signal high frequency calculations by utilizing a two-port black box equivalent circuit similar to that introduced for the bipolar transistor in Section 9.3.1 and shown in Fig. 9.8. In the present case, however, *y*-parameter admittances are commonly used. This circuit model is described by the following equations, where the subscripts 1 and 2 refer respectively to the input and output:

$$i_1 = y_{is}v_1 + y_{rs}v_2 \tag{10.49a}$$

and

$$i_2 = y_{fs}v_1 + y_{os}v_2. \tag{10.49b}$$

[13]P. E. Gray and C. L. Searle, *Electronic Principles—Physics, Models and Circuits* (Wiley, New York, 1969), p. 339.

The values of the y parameters can be determined by measurements at the device leads. Here y_{is} is the input admittance and y_{fs} is the forward transadmittance with the output short circuited; y_{rs} is the reverse transadmittance and y_{os} is the output admittance with the input short circuited. Note that the real part of y_{fs} is g_{fs}, the forward transconductance, and equals g_m when the FET output is short circuited. (See the device parameter specification sheets in Appendices G1 and G2.) These y parameters are frequency dependent.

In the common-source circuit configuration, i_1 and i_2 in Eqs. (10.49) are respectively i_g and i_d; the voltages v_1 and v_2 are respectively v_{gs} and v_{ds}. This common-source small-signal y-parameter equivalent is sketched in Fig. 10.21.

□ *SUMMARY*

The mode of operation of the JFET with electrical conduction modulated by a *p-n* junction input gate is described. Its current-saturating *I-V* characteristic is due to pinchoff of its conducting channel. The square-law variation of drain saturation current with gate input voltage is pointed out. A description is next given of the related MESFET, which utilizes a Schottky diode as the input gate. The GaAs MESFET is presently the highest frequency cutoff transistor in manufacture. A further increase in frequency capability may be achieved by utilizing a related high electron mobility transistor (HEMT). Next the most commonly employed field-effect device, the MOSFET, is described; it uses an MOS capacitor as an input gate. The charge accumulation, depletion, and inversion modes of potential bias of the MOS capacitor are discussed. An expression is derived for the current-initiating threshold voltage of the MOSFET, including the effect of positive charges in the gate oxide. An experimental method for determining the oxide charge density is indicated. The enhancement and depletion types of MOSFETs are described. The static *I-V* characteristic of the enhancement-type MOSFET is derived by using the "gradual channel" approximation. For a long channel device the drain current is quadratically related to the gate voltage. The MOSFET gain is expressed in terms of a transconductance g_m which is inversely related to the channel length. The elementary derivation predicts zero output conductance, but a modification of this theory yields a better expression for this conductance; an inverse relationship to channel length and a direct correspondence to drain current are predicted. A more accurate *I-V* characteristic is given which is important in modeling short channel (1 μm or less) MOSFETs. A short channel correction for the

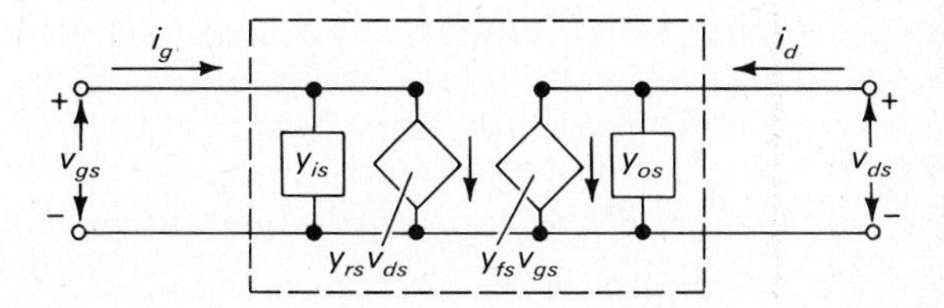

Figure 10.21 Common-source small-signal y parameter equivalent circuit of a field-effect transistor. In general the y parameters are frequency dependent.

threshold voltage is discussed as well as a modification of the threshold voltage formula when a substrate bias relative to source is applied. Finally the frequency limitation of the field-effect transistor is considered. An expression is derived for the gain–bandwidth product. The source-to-drain carrier transit time is seen to limit the switching speed of the FET. A comparison of the FET and bipolar transistor frequency capabilities indicates an advantage for the latter due to its intrinsically higher g_m. A small-signal, high frequency circuit model for the FET is presented.

PROBLEMS

10.1 Explain the output characteristic of the *p*-channel depletion-type MOSFET and the JFET for positive values of gate potential relative to the substrate.

10.2 Explain the primary reason why the GaAs MESFET frequency capability exceeds that of a silicon transistor.

10.3 Discuss the effect of increased substrate doping on the gate threshold voltage of an *n*-channel enhancement-type MOSFET. Use appropriate formulas in the discussion.

10.4 Show that for the case of strong inversion of a *p*-type semiconductor surface under an oxide in an MOS capacitor structure where $\phi_s = 2\phi_F$, the electron inversion layer concentration $n_s = p_{p0}$.

10.5 Show that Eq. (10.12) follows from Eq. (6.18a).

10.6 Derive Eq. (10.18) in terms of such physical device design parameters as substrate doping.

10.7 Calculate the charge per unit area at 300°K in the depletion layer of an aluminum gate MOS capacitor on a *p*-type silicon substrate doped with 1×10^{15} acceptors/cm^3, under the bias condition of strong inversion.

10.8 Find the flat-band voltage for the device of Problem 10.7 if the oxide surface number density is 8×10^{10}/cm^2 and the oxide thickness is 8.0×10^{-6} cm.

10.9 **(a)** Calculate the minimum capacitance value under positive bias of the MOS capacitor of Problems 10.7 and 10.8.
(b) At about what value of bias voltage will this minimum occur?

10.10 Derive Eq. (10.19).

10.11 Given an *n*-channel enhancement-type MOSFET made of silicon with source connected to substrate and $\epsilon_{ox} = 3.9$, $\bar{\mu} = 600$ cm^2/V-S, $l = 2.5 \times 10^{-4}$ cm, $t_{ox} = 8.0 \times 10^{-6}$ cm, $Z = 1.8 \times 10^{-3}$ cm, $V_T = 1.5$ V, and substrate acceptor concentration $N_a = 1 \times 10^{16}$/cm^3,
(a) compute the gate oxide capacitance;
(b) calculate I_{Dsat} for values of V_{GS} ranging from 0 to 5 V, in 1-V intervals;
(c) determine the linear conductance of this device near $V_{DS} = 0$ for $V_{GS} = 2, 3,$ 4, and 5 V.

10.12 Starting with Eq. (10.27), derive Eq. (10.28).

10.13 For the device described in Problem 10.11,
(a) Find the device g_m at V_{GS} = 2, 3, 4, and 5 V.
(b) Repeat (a) if the channel length is reduced in length by a factor of 2.

10.14 Assume that the drain space-charge width of the MOSFET of Problem 10.11 increases by 0.10 μm/V of drain–source voltage.
(a) Calculate the active-region output conductance g_{os} for gate voltages of 2 and 5 V.
(b) Repeat (a) if the channel is reduced in length by a factor of 2.

10.15 Determine approximately the maximum drain voltage capability of the device of Problem 10.11 if this is limited by punchthrough of the drain space-charge region to the source space-charge region. How is this value affected if the channel length is halved? How is this affected by the substrate doping?

10.16 Derive Eq. (10.36) starting with Eq. (10.25) and using Eq. (10.35).

10.17 Calculate the maximum possible g_m for the MOSFET of Problem 10.11.

10.18 Derive Eq. (10.40) for the shift in MOSFET gate threshold voltage with substrate bias.

10.19 Determine the shift in threshold voltage for the MOSFET of Problem 10.11 if the device's *p*-type substrate is biased 3 V negative relative to the source. Will the threshold voltage be raised or lowered?

10.20 For the device of Problem 10.11 at V_{GS} = 2 and 5 V,
(a) Determine the gain–bandwidth product of the device.
(b) Repeat (a) if the channel length is halved.

10.21 Derive an expression for the gain–bandwidth product of a MOSFET by calculating the time necessary to charge the gate capacitance through one-half the channel resistance.
(a) Compare this result with the expression in Eq. (10.43).
(b) Calculate the ratio of the gain–bandwidth product for a *p*-channel to $G \times BW$ for an *n*-channel silicon MOSFET of identical design.

CHAPTER

11

Microelectronics and Integrated Circuits

11.1 THE INTEGRATED CIRCUIT (IC): PHILOSOPHY AND TECHNOLOGY

The basic building block of most electronic equipment and computers today is the **integrated-circuit microchip.** This was described in Section 1.3, where the integrated-circuit chip was defined as "one piece of solid into which have been incorporated several components, passive as well as active, without external connection between these devices." Electronic components such as planar transistors (see Fig. 8.2b) and MOS transistors (see Fig. 10.1) are primary elements of these chips which "perform a complete circuit function." Vast families of these circuit blocks have been developed and the incentive for doing so has been fourfold:

1. Cost.
2. Size.

3. Reliability.
4. Performance.

The development of sophisticated semiconductor technology resulting in the fabrication of thousands and even hundreds of thousands of transistors onto a chip of single-crystal silicon, about 5 mm on each side and only 0.25 mm thick, has had low cost as a primary impetus. The planar technology of integrated circuits utilizes **batch processing,** which permits hundreds of electronic components to be incorporated onto a single chip of silicon at one time at nearly the same cost as producing a single discrete transistor chip. Hence the unit cost of a transistor has been reduced to a small fraction of a cent in an integrated-circuit microchip. Recently the gallium arsenide digital IC microchip has been introduced; this chip offers faster switching-speed capability than a corresponding silicon chip.

The use of photolithographic processing to delineate the many devices on the microchip not only provides a *cheap* process for simultaneously producing these components, but also permits them to be made extremely *small* in size. The theory of optics dictates that the minimum size of structures which may be resolved is limited only by the wavelength of the radiation used to expose the photographic emulsion utilized in processing. This radiation ranges from visible light with a wavelength of several thousand angstroms to electrons with a wavelength of a few angstroms.

The technique utilized for interconnecting the various components on the single-crystal chip involves the deposition of a thin film of metallization in a pattern delineated again by a photolithographic process. This method provides a cheap and highly *reliable* scheme for interconnection, so that the primary source of unreliability results from the external joining of these packaged circuits into a circuit board by soldering or welding, to create a complete piece of electronic equipment. The most complex circuitry will require several layers of interconnection metallization electrically insulated from each other by thin dielectric films. Reliability and cost dictate the largest assemblage of components possible on a single IC chip. Intensive development is under way to permit tens of these ICs, each performing a complete electronic function, to be joined together by thin-film metallization to constitute an entire unit of electronic equipment such as a computer plus its auxiliary circuits. However, it has been found recently that the resistance of these narrow metal interconnects limits the switching frequency of these circuits.

Finally the batch process aspect of the integrated-circuit technology permits the economic use of a large number of devices to perform a single circuit function. This provides for sophistication of circuit design not possible before the advent of these integrated devices and hence improvement in circuit *performance.* The IC format also permits the fabrication of new **merged devices** with very small geometry and very low power dissipation in digital applications.

A 64K packaged DRAM (dynamic response access memory) chip placed on a silicon slice which contains thousands of such chips. (Supplied by Texas Instruments Incorporated, Dallas, TX.)

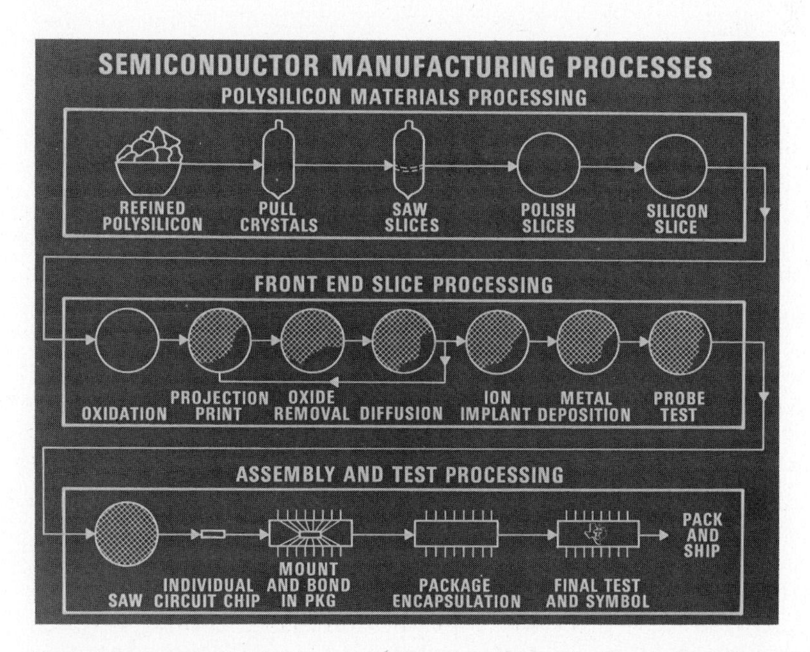

Typical steps in the fabrication of an integrated circuit chip. (Courtesy of Texas Instruments Incorporated, Dallas, TX.)

It is the purpose of this chapter to consider the different semiconductor materials that are used in the fabrication of integrated circuits (ICs) and the device technology employed in their manufacture. Methods for selectively introducing impurities into the single-crystal semiconductor wafer to produce devices, such as diffusion and ion implantation, are explained. A description is given of the various devices used for the construction of ICs and how they differ in form from the discrete component versions of these devices. The types of silicon and gallium arsenide digital ICs in manufacture presently are described. The basic elements of silicon analog IC design will be discussed. An explanation will be given of the fundamental limitations on microminiaturization which will restrict the maximum device density on microchips and guidelines will be developed to predict the maximum operating speed and frequency performance of these circuits. Limitations provided by size definition, materials, manufacturing chip yield, maximum power dissipation, and reliability considerations will be analyzed. Finally some computer-aided design methods in use for developing new IC chips will be presented.

11.2 MATERIALS FOR INTEGRATED CIRCUITS

Most integrated-circuit blocks in use today employ the semiconductor silicon as a starting material. This is accounted for by the following unique combination of favorable properties of this semiconductor material:

1. It is an element, of low atomic weight, and in plentiful supply.
2. Pure, single crystals are readily grown.
3. It has a relatively large energy gap.
4. In it *p-n* junctions can be easily formed by impurity introduction.
5. It is a good thermal conductor.
6. It readily forms an excellent native, electrically insulating, stable oxide.
7. Electron mobility in it is high.

The compound semiconductor gallium arsenide has an electron mobility five times greater than Si. However, GaAs is inferior to Si in five of the respects listed above. Nevertheless, GaAs digital integrated-circuit microchips have recently become available which have superior switching speed capability compared with silicon chips. The basic switching element of these new chips is the MESFET, which has a structure which is relatively easy to integrate. Germanium is an elemental semiconductor with an electron mobility more than twice that of silicon. However, Ge does not form a stable oxide and has a smaller energy gap which gives higher *p-n* junction leakage current and lower temperature capability. Hence in the near future silicon will remain the dominant starting material for IC chips.

The fact that silicon is an element (and not a compound) abundantly available in nature[1] makes its preparation for use relatively simple and economical. The *elemental* nature of silicon makes its purification and single-crystal growth straightforward. Its low atomic weight means it has a *low mass density* (about equal to that of aluminum). This is particularly avantageous in aerospace applications, where the lightweight nature of the integrated circuit is of importance.

Techniques for purifying silicon are readily available and the material is obtainable economically at a maximum impurity level of less than *one part per billion.* Single crystals can be grown from the material without much difficulty in the manner shown in Fig. 2.2. The need for the use of single crystals of silicon as the form of the starting material for the fabrication of integrated circuits, as stated in Section 2.1.2, stems from the requirement for perfection and controllability of the material. This is particularly true in the case of relatively large-area circuits, in which a minute imperfection can interfere with the proper functioning of the entire circuit chip.

By sequentially exposing the material to a vapor of donor and acceptor impurities at an elevated temperature in excess of 900°C or by **ion implantation,** *p-n* junctions are readily produced in single-crystal silicon. The technique of **solid-state diffusion** is not always successful in compound materials but is easily accomplished, very controllably, in the element silicon.

All electronic devices are limited in their performance by internally generated heat. Semiconductor devices in particular are temperature sensitive, as has already been discussed in some detail. An important requisite for an integrated-circuit starting material is that it have **good thermal conductivity,** particularly because of the minute size of these devices, so that the high heat density generated therein can be efficiently conducted to a heat radiator. Proper circuit functioning is often made possible by using appropriate circuit design at somewhat elevated temperatures as long as all the elements of the circuit are at a uniform temperature. Operation at high power density is made feasible by using silicon, which has a thermal conductivity at room temperature of about one-third that of the metal copper. This is particularly high for a semiconductor substance. For example, germanium has only about one-half the thermal conductivity of silicon and gallium arsenide somewhat less. In addition, the maximum operating temperature of a germanium device is about 100°C compared to 200°C for silicon devices because of germanium's smaller energy gap and the *p-n* junction leakage current due to thermal electron–hole pair generation. Because of the larger energy gap of GaAs, devices fabricated with this semiconductor material in principle can operate to temperatures in excess of 300°C.

In a final comparison of silicon with other semiconductors such as germanium and GaAs it is important to point out that silicon forms an **excellent insulating, stable oxide** which plays an important role in integrated-circuit technology. This is in contrast to germanium, which forms two types of oxides:

[1]About 20% of the Earth's crust is made up of silicon.

Operator running a Micro Automation Company automatic wafer saw. This machine uses a diamond-edged blade to saw along the scribelines to divide a wafer into individual dies. (Supplied by the Digital Equipment Corporation, Maynard, MA.)

one is water soluble and the other is unstable at high temperatures. The oxides of GaAs are unstable. In contrast, silicon dioxide (SiO_2) is an electrical insulator with a resistivity in excess of 10^{14} Ω-cm. The common name for this material is quartz, which is known to be very stable at high temperatures and has a melting point in excess of 2000°C. Silicon dioxide is used for two distinct purposes in the fabrication of integrated circuits. Its electrical insulating property is utilized in **coating and protecting** *p-n* junctions from contamination by unwanted impurities in the surrounding ambient. In addition, it is used as a stable, high temperature **masking** material which blocks donor or acceptor impurities from selectively diffusing into the silicon where they are not desired. The protective property of silicon dioxide has already been discussed in Section 5.5 and may often be enhanced by the additional coating of a layer of silicon nitride, Si_3N_4, over the oxide. SiO_2 and Si_3N_4 deposited onto GaAs are used for surface protection and masking functions.

11.2.1 Material Preparation

The basic form of the starting material for the silicon **monolithic** integrated circuit[2] is a slice of silicon cut from a single crystal of the type pictured in Figs. 2.2a and 11.1a. The diameter of crystals presently in use is 5–6 in. or about 12–15 cm. These are then cut with a diamond-primed saw,[3] like baloney, into slices less than half a millimeter thick. Thousands of individual circuit chips 25 mm^2 or more in area are eventually produced by dicing this slice by means

[2]**Monolithic** refers to its being of a uniform nature. Hence the monolithic circuit is one in which all devices, both active and passive, are fabricated basically from one material and in one block or chip of the substance.

[3]Diamond must be used since silicon is a very hard material.

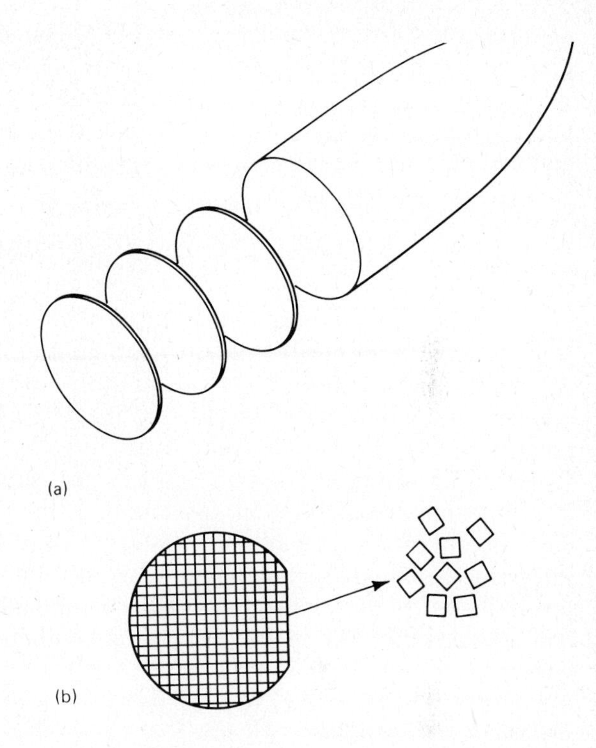

Figure 11.1 (a) Single crystal of silicon about 12 cm in diameter cut into slices about 0.5 mm thick.
(b) Single wafer of silicon diced into chips about 25 mm on each side.

of a tungsten carbide scribe which is used to cut silicon in a manner identical to the way in which glass is cut. This is shown in Fig. 11.1b. However, the silicon slice is kept intact during most of the fabrication so that batch processing may be carried out, resulting in a simultaneous fabrication of thousands of identical circuits at once. The dicing operation is one of the final steps in integrated-circuit fabrication and occurs just prior to circuit packaging.

After being sawed, the silicon slices are lapped and polished to a mirrorlike finish. The final step in the polishing is a chemical treatment in a mixture of nitric and hydrofluoric acids which dissolves away the roughness in the silicon crystal surface and results in a single-crystal surface with an extremely flat and smooth finish.

The slices so treated are typically cut from 1–10 Ω-cm *p*-type silicon single crystal. In the fabrication of bipolar ICs these are then placed in a reactor apparatus for the deposition of a thin (about 1–5 μm) layer of 0.1–1.0 Ω-cm *n*-type silicon single crystal on one side of the silicon slice by **epitaxial** growth. Epitaxial growth refers to the deposition of a thin single-crystal layer onto a single-crystal substrate of a basically similar material from the vapor phase. This chemical vapor-phase growth is in contrast to the growth from the liquid phase (illustrated in Fig. 2.2c) in that the seed in epitaxial growth is a flat silicon substrate slice which is exposed to the silicon-containing vapor at an elevated temperature *below* the silicon melting point. In this way a thin *n*- (or *p*-) layer of extremely uniform thickness may be prepared. The apparatus for accomplishing this deposition is sketched in Fig. 11.2a. The silicon source material is typically a few percent silicon tetrachloride or trichlorsilane in hydrogen gas, and a typical chemical reaction that takes place at temperatures between 1000 and 1200°C and results in silicon deposition is given by

$$SiCl_4 + 2H_2 \rightarrow Si\downarrow + 4HCl. \tag{11.1}$$

Hydrochloric acid (HCl) can be added prior to deposition to prepare the substrate surface.

In addition to the accurate control of the thickness of this deposited *n*-silicon layer, its impurity content must be accurately controlled to yield the proper resistivity material. For this purpose an extremely small, but accurate, quantity of donor material is added to the $SiCl_4$. This is usually in the form of phosphine, PH_3. (A *p*-type layer can be grown by introducing diborane, B_2H_6.) Figure 11.2b shows in cross section a typical starting slice for the fabrication of a bipolar integrated circuit, including a thin *n*-type silicon epitaxial layer deposited onto a *p*-type silicon substrate. All of the electronic devices in the integrated circuit are included in the *n*-type epitaxial film. The *p*-type substrate primarily serves as a base on which the epitaxial layer is deposited, for ease in handling this thin film. A *p*-type substrate without an epitaxial deposit is used

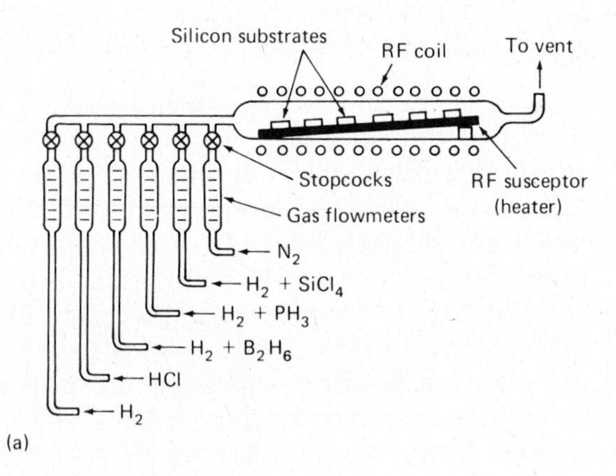

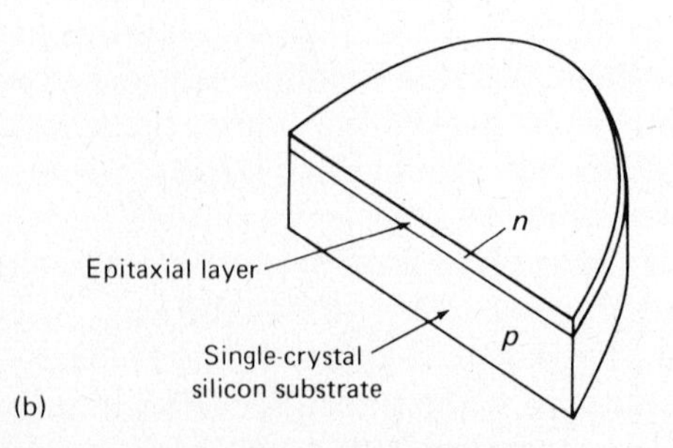

Figure 11.2 (a) Apparatus for depositing thin, single-crystal layers of silicon onto single-crystal silicon substrates; phosphine gas (PH_3) is introduced to yield *n*-type layers, while diborane gas (B_2H_6) is used for *p*-type layers.
(b) Cross section of a typical silicon epitaxial starting slice for the fabrication of integrated circuits.

for fabricating some MOSFET memory and digital ICs; however, an epitaxial deposit is used for complementary MOS circuits containing both *n*- and *p*-channel devices.

11.3 METHODS FOR THE CONTROLLED INTRODUCTION OF IMPURITIES INTO SEMICONDUCTORS

One method for the introduction of donor and acceptor impurities into a semiconductor is **solid-state diffusion.** This involves exposing the silicon to a vapor of an appropriate impurity at an elevated temperature in excess of 900°C. Another technique in use is the **ion implantation** of donors or acceptors into the semiconductor crystal. Here ions of doping atoms are accelerated to an energy of a few hundred thousand electron volts and "shot" (implanted) into the semiconductor. The simplicity of the method makes the diffusion technique popular for impurity introduction. All that is required in the way of equipment is a furnace capable of raising the semiconductor crystal to temperatures in excess of 900°C. However, the oxide masking technique must be employed in order to ensure that the impurity is only introduced into the crystal wafer in certain areas selected by design. In contrast, ion implantation requires a more complex and expensive ion accelerator and the slice throughput is slower. Nevertheless, the accurate control of the density of impurity atoms introduced by implantation, particularly at reduced concentrations, is in use extensively. Thermal diffusion is a more gentle process which causes much less damage to the semiconductor crystal than the bombardment required for ion implantation. Heat treatment is required after implantation to anneal out the structural defects induced in the crystal.

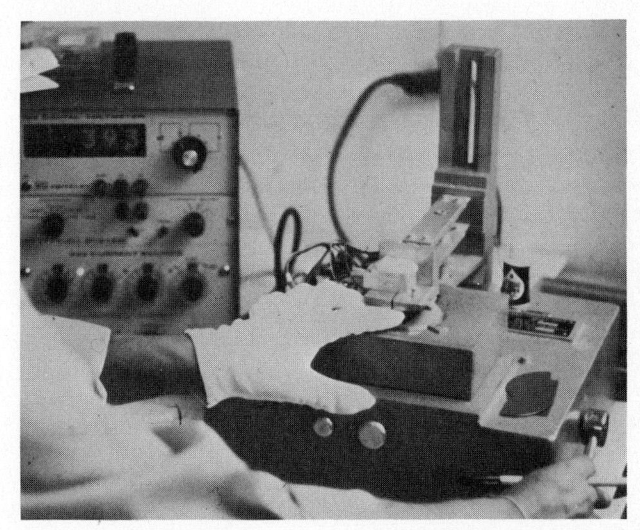

The testing of the sheet resistance of a thin epitaxial layer grown onto a silicon substrate using a 4-probe electrical measurement. (Supplied by Analog Devices Semiconductor, Wilmington, MA.)

11.3.1 The Solid-State Diffusion Technique

The process of solid-state diffusion is comparable in a sense to the diffusion of minority electrical carriers in the silicon lattice described in Section 5.3.1. That is, the atoms in the vapor undergo random motion but there is a statistical probability that the impurities will enter the pure silicon crystal lattice where few such doping atoms are present compared to the impurity-rich vapor. The formulation of the problem of the flow of these impurity atoms from the vapor into the solid results in an equation similar to Eq. (7.8) for the diffusion of minority carriers in a crystal. However, no recombination term is needed for the case of the diffusion of impurity atoms, so the relevant equation can be written as

$$D\frac{\partial^2 N}{\partial x^2} = \frac{\partial N}{\partial t}, \tag{11.2}$$

where $N(x,t)$ represents the density of impurity atoms at a distance x from the surface of the crystal after a diffusion time t at an elevated temperature.

Here D is the **diffusion constant** which represents the rate of penetration of the doping impurity into the crystal. This parameter may be expected to be temperature dependent, the penetration rate increasing with increasing temperature as the random motion of the impurity atoms becomes more vigorous. The variation of the diffusion constant with absolute temperature T is given by

$$D = D_0 e^{-E_D/kT}, \tag{11.3}$$

where E_D is the **activation energy for diffusion,** which for silicon has the value of about 3–4 eV for nearly all acceptor and donor substitutional impurities, and D_0 is a temperature-independent constant. The activation energy for interstitial diffusants such as gold and copper is only about 1 eV. A semilogarithmic plot of D versus $1/T$ results in a straight line the slope of which is determined by E_D. A graph of the diffusion constants of the primary donor and acceptor impurities in silicon at low concentrations versus temperature is given in Fig. 11.3.

If the silicon crystal is in contact at all times at an elevated temperature with a constant concentration of donor or acceptor atoms in a vapor, the impurity atom distribution in the crystal is given by the solution of Eq. (11.2) subject to appropriate boundary conditions as

$$N(x,t) = N_0[1 - \mathrm{erf}(x/2\sqrt{Dt})] = N_0\,\mathrm{erfc}(x/2\sqrt{Dt}). \tag{11.4}$$

Here N_0 represents the concentration of the impurity at the surface ($x = 0$) of the semiconductor, assumed independent of time, and erf is the same mathematical error function that previously occurred in Eq. (7.27); erfc is by definition the complementary error function ($\mathrm{erfc}\,Z = 1 - \mathrm{erf}\,Z$) which is a tabulated function that is graphed in Fig. 11.4a. Using Eq. (11.4) and Fig. 11.4a, and knowing N_0 one may predict the distribution of impurity atoms which have penetrated into the surface of a semiconductor crystal. This is represented

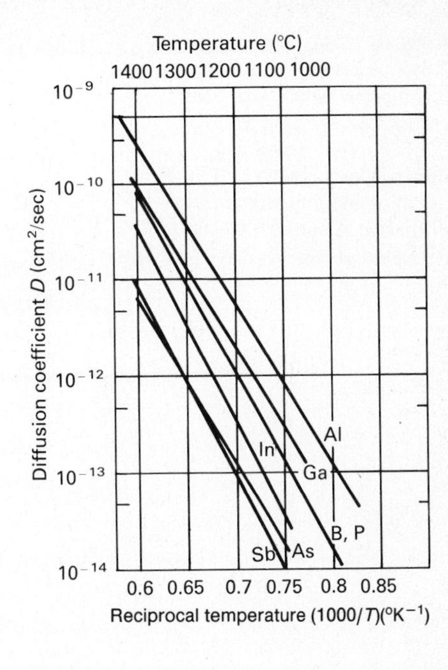

Figure 11.3 Low concentration diffusion coefficients of the primary substitutional donor and acceptor impurities in silicon versus 1000/absolute temperature in °K^{-1}. Data are shown for the *p*-type impurities aluminum, gallium, boron, and indium and the *n*-type impurities phosphorus, arsenic, and antimony.

Operator introducing a quartz vessel containing 50 wafers into a furnace at a temperature in excess of 1000°C for impurity diffusion. In the foreground are silicon integrated circuit wafers whose fabrication has been completed. (Supplied by the Digital Equipment Corporation, Maynard, MA.)

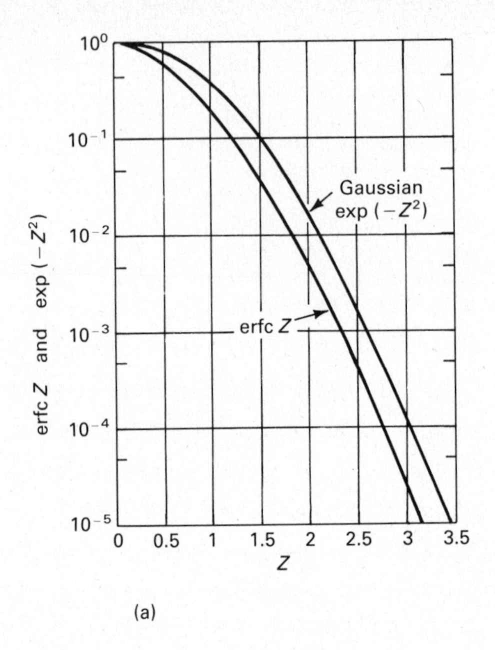

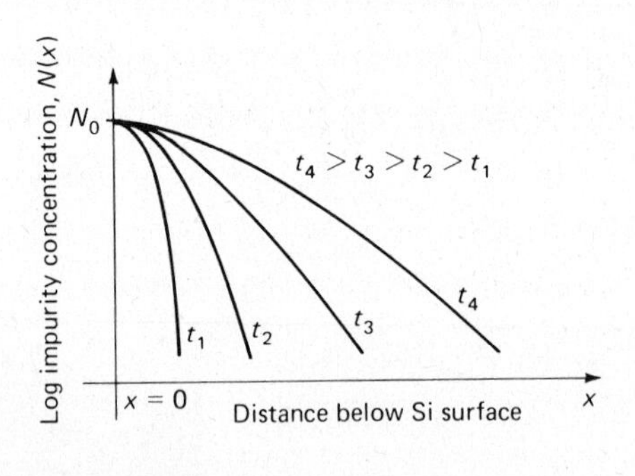

Figure 11.4 (a) Complementary error function and Gaussian function of Z versus Z, where $Z \equiv x/2\sqrt{Dt}$. (b) Graph of increasing penetration versus time at $T = T_1$ (°K) of impurity atoms below the silicon surface, which is located at $x = 0$.

in the graph of Fig. 11.4b, which shows the increasing penetration of the impurity atoms beyond the surface as time progresses with the crystal held at some temperature T_1. Since the crystal is normally exposed to a dense vapor of donors or acceptors, the crystal surface is saturated with this impurity and hence N_0 in fact represents the **maximum solubility** of the impurity atom which can be present in the semiconductor without disrupting the crystal lattice. Values for the solid solubility of various donor and acceptor atoms in silicon in the normal diffusion temperature range of 900–1300°C are graphed in Fig. 11.5.

The complementary error function type of diffusion is typically used to form the emitter region of a bipolar transistor and the source and drain regions of the MOSFET. However, in designing the diffusion profile $N(x)$ for the base

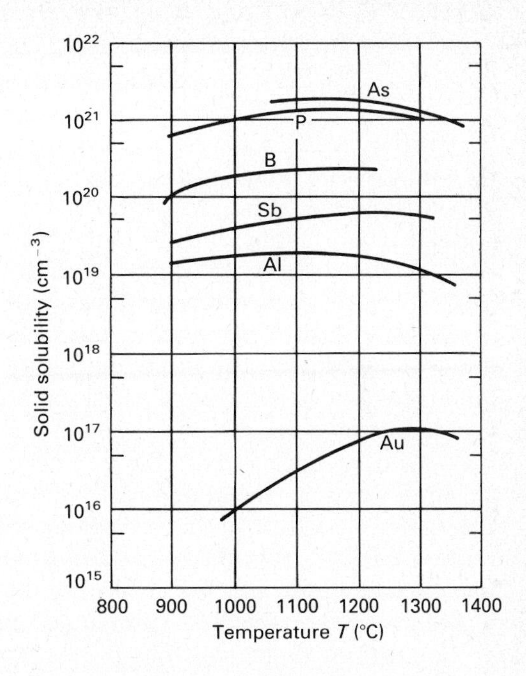

Figure 11.5 The solid solubility limits of some typical impurities in silicon versus temperature. [From R. A. Colclaser, *Microelectronics Processing and Device Design* (Wiley, New York, 1980).]

region of a bipolar transistor it is important for the surface concentration of the doping impurity to be significantly below that of the maximum solubility limit to obtain high emitter injection efficiency. This is accomplished by a two-step diffusion process, a short, low temperature erfc **deposition** followed by a high temperature **drive-in** carried out in an inert gas environment as free as possible of any impurity contaminant. The impurity diffusant profile in this case, obtained by solving Eq. (11.2), is said to be Gaussian and can be represented by

$$N(x,t) = [Q_0/(\pi Dt)^{1/2}]\exp(-x^2/4Dt), \tag{11.5}$$

where Q_0 is the delta function concentration of impurity atoms per unit area of surface present at $t = 0$; Q_0 can be approximated in the case of a shallow complementary error function predeposition by integrating Eq. (11.4) over x for the deposition time t.

A. DIFFUSION MASKING

To form *n-p-n*-type structures as for planar bipolar transistors (see Fig. 8.2), a sequential diffusion of *p*-type and then *n*-type impurities must take place into a crystal containing *n*-type impurities. In addition, these diffusions must selectively introduce the doping impurities only into certain areas of the surface of the crystal. The presence of silicon dioxide in specific regions on the silicon surface tends to retard the introduction of these impurities there. This results from the fact that certain impurities such as boron, arsenic, and phosphorus diffuse more slowly in silicon dioxide than in silicon. These impurities are said to be effectively **masked** by this appropriately thick, thermally stable oxide, which is used to delineate the required device structures on the silicon crystal surface.

Patterns in the oxide for masking can be created in turn by using photolithographic techniques. The oxidized silicon wafer is coated with a liquid polymeric photographic material which forms a very thin, solid film on the oxide surface on evaporation of a solvent. A light and dark pattern is now projected onto the photographic film by using ultraviolet (short wavelength) light. This will crosslink the polymer film where light is absorbed and a subsequent "developing" procedure in an organic solvent will wash away the unexposed film, leaving behind a negative impression of the projected light pattern; i.e., film is present where the light was absorbed and vice versa. This film will now be used to protect the silicon dioxide in areas where it is present from exposure to a wet etchant such as hydrofluoric acid or a dry, fluorine-containing gas etchant. In this manner the **photoresist** film is used to delineate an oxide masking pattern on the silicon wafer, according to design. Silicon dioxide or silicon nitride deposited from the vapor phase onto other semiconductor crystal surfaces such as GaAs and InP is also used to selectively mask the diffusion of doping impurities into these compound semiconductor crystals.

Boron from sources such as B_2O_3, BBr_3, and B_2H_6 is commonly used as a *p*-type impurity diffusant in silicon. Phosphorus from $POCl_3$ and PH_3 is used as an *n*-type impurity diffusant, and AsH_3 and As_2O_3 are sources of arsenic. Gold is normally introduced into silicon by diffusion from a metal source for the control of minority carrier lifetime, as discussed in Section 7.4.2. The diffusion coefficient of Au in Si is quite rapid, about 10^{-5} cm^2/sec, so this diffusion can be accomplished at temperatures of 800°C or less.

Zinc and cadmium are common *p*-type dopants in gallium arsenide; tin and selenium are common *n*-type dopants.

EXAMPLE 11.1

A p^+-n junction is formed by diffusing boron from an impurity-saturated vapor into *n*-type silicon substrate which contains 2.2×10^{15} donors/cm^3. The diffusion is performed at 1090°C for 1.0 h. Find the junction depth.

Solution

From Eq. (11.4)

$$N(x,t) = N_0 \operatorname{erfc}(x/2\sqrt{Dt}) = N_0 \operatorname{erfc} Z.$$

From Fig. 11.5, the solubility N_0 of boron at 1090°C is 2.2×10^{20}/cm^3 and from Fig. 11.3, D at 1090°C is 2.0×10^{-13} cm^2/sec. Since the junction is formed where the concentration of the diffusing boron atoms

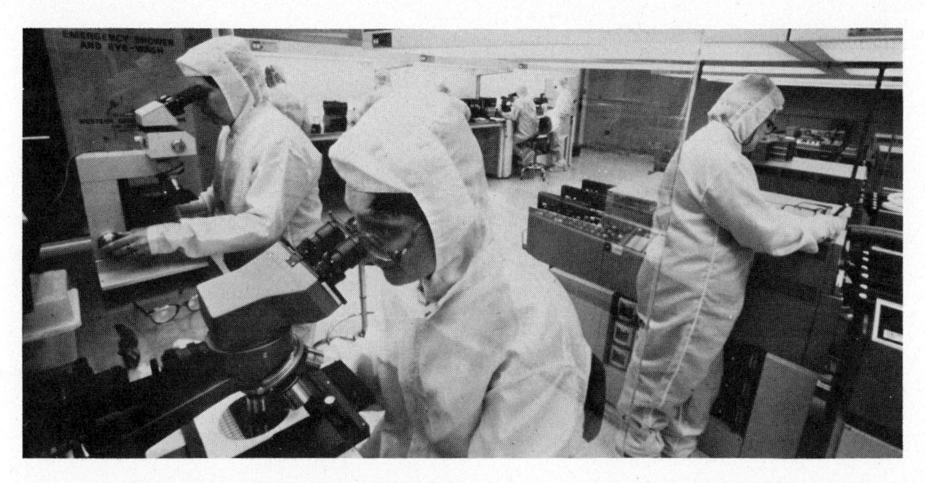

Operators conducting inspection during the photoresist processing steps. (Supplied by the Digital Equipment Corporation, Maynard, MA.)

just equals the donor substrate doping, $N(x_j,t) = 2.0 \times 10^{15}/\text{cm}^3$. Hence

$$N(x_j,t)/N_0 = 2.2 \times 10^{15}/2.2 \times 10^{20} = 1 \times 10^{-5}.$$

So erfc $Z = 1 \times 10^{-5}$, which from Fig. 11.4a yields $Z = 3.15$. From Eq. (11.4) given above then

$$x_j/[2(2 \times 10^{-13}\ \text{cm}^2/\text{sec})(3600\ \text{sec})]^{1/2} = 3.15$$

and

$$x_j = \underline{1.2 \times 10^{-4}\ \text{cm} = 1.2\ \mu\text{m}}.$$

11.3.2 Ion Implantation

The process of ion implantation of boron atoms into silicon requires the use of a boron ion source as well as an accelerator to raise the kinetic energy of the ions to a few hundred kilovolts. Positively charged boron ions can be produced by exposing a boron gas to a radiation source. These ions are then caused to fall through a large DC potential difference between a capacitorlike metal plate and the silicon slice. These high energy boron ions penetrate a distance of micrometers or fractions thereof into the negatively charged silicon crystal, becoming acceptor doping ions there. Also, *n*-type impurity atoms can be ion implanted into silicon. An apparatus for performing this implantation is

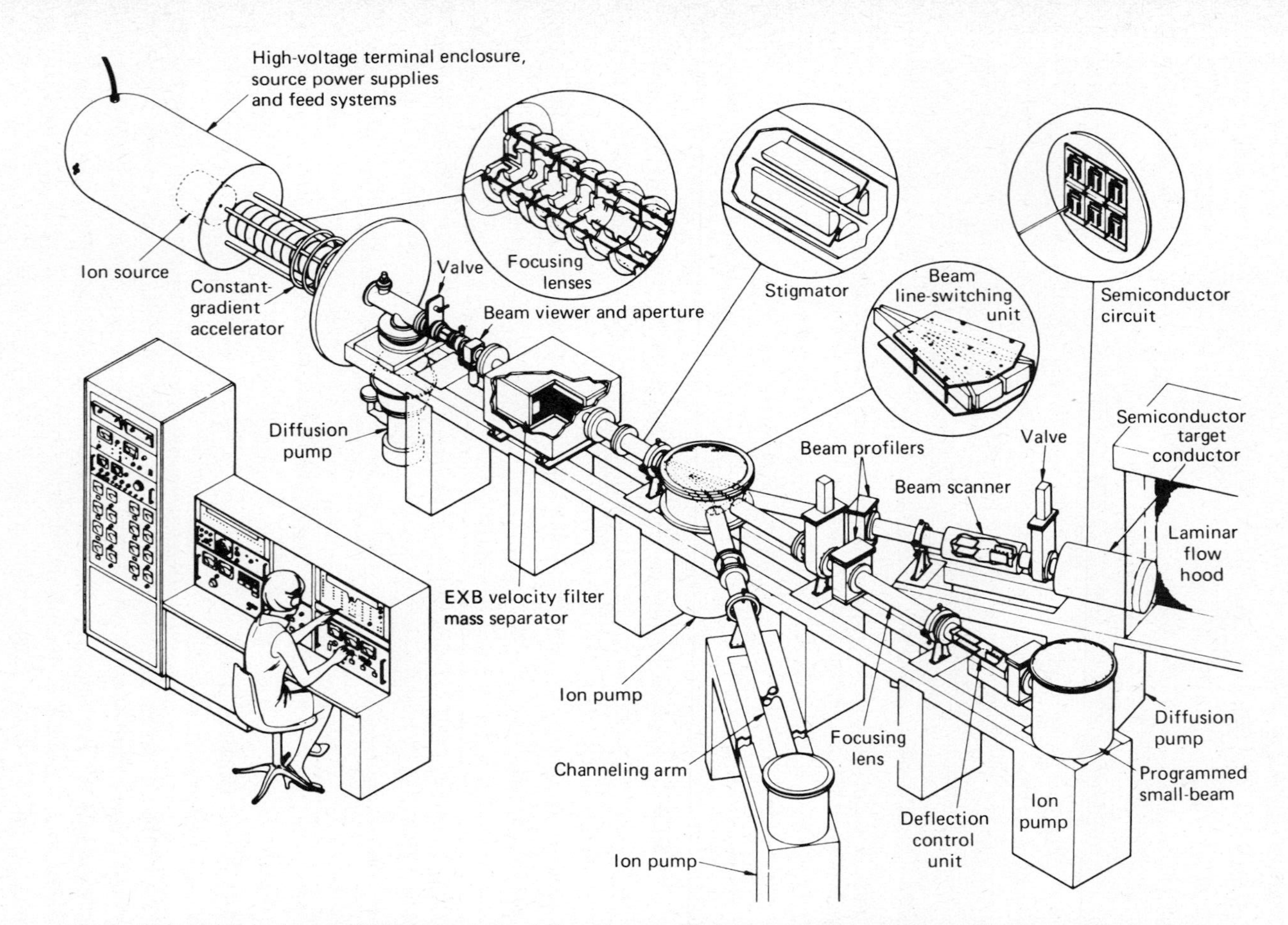

Figure 11.6 (a) Drawing of a 300-kV ion implantation system with three arms for separate applications of the ion beam.

sketched in Fig. 11.6. Ion implantation is also an important technique for introducing doping impurities into GaAs and InP.

Some advantages of the implantation method of impurity introduction compared with diffusion are

1. Accurate control of impurity level, particularly for low impurity concentrations.
2. Good shallow layer depth control.
3. Very low temperature processing.
4. Creation of unusual impurity profiles.
5. Possibility of implanting ions insoluble in the semiconductor.

Some disadvantages are

1. Lattice damage created by high energy implants must be postannealed, thermally or by laser annealing.
2. High concentration doping is limited.
3. Limited depth of penetration without severe crystal lattice damage.
4. Ion penetration is not isotropic (not the same in all crystal directions).
5. Throughput is limited.
6. Equipment is complex and expensive.

The distribution of dopant atoms introduced beneath the semiconductor surface (at $x = 0$) by the implantation technique is given by the Gaussian formula

$$N(x) = (\Phi/\sqrt{2\pi}\,\Delta R_p)\exp[-(x - R_p)^2/2(\Delta R_p)^2]. \tag{11.6}$$

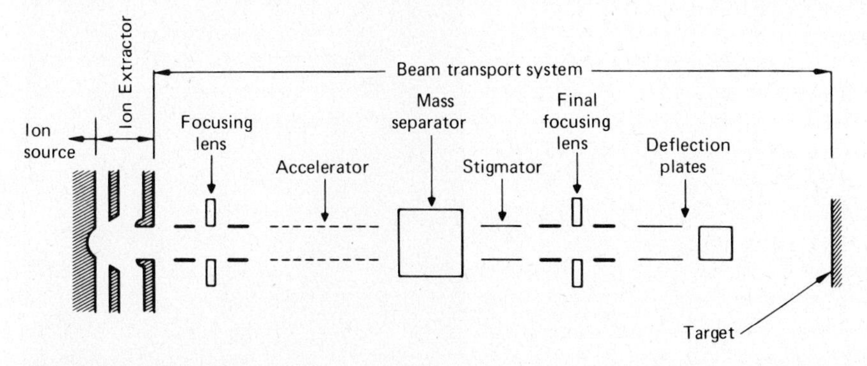

Figure 11.6 (continued)
(b) Sketch of an ion implantation system including an ion source, focusing electrodes, acceleration tube, mass analyzer, neutral trap, beam deflector, and wafer chamber. Since some ions are neutralized in transit these must be eliminated before reaching the target. [Both from R. G. Wilson and G. R. Brewer, *Ion Beams* (Wiley, New York, 1973).]

Here Φ is the ion flux or dose in ions per square centimeter, R_p is the **range** of the impurity ion in the semiconductor material which is a function of the ion mass and its implant energy, and ΔR_p is the **straggle** of the impurity and is also related to the ion mass and energy.

Implantation of impurity ions is typically carried out by accelerating the ions to energies from 50 to 200 keV and at ion currents ranging from the tens of microamperes to the many milliamperes level. Masking of implanted ions is accomplished using silicon dioxide, silicon nitride, or photoresist materials, which are also commonly used for diffusion pattern delineation. For high energy ion implantation it is sometimes necessary to deposit a pattern of a thin metal film such as gold in order to provide for effective ion masking.

An important application of ion implantation is to sensitively adjust the threshold voltage of a MOSFET by a shallow and accurate low concentration implant of impurity atoms in the channel region. Another implantation application is in the creation of a lightly doped *n*-region in a *p*-type semiconductor substrate (or vice versa) for the production of a *p*-channel MOSFET in complementary MOSFET fabrication, soon to be discussed.

11.4 DEVICE ISOLATION TECHNIQUES

A monolithic integrated circuit may include many transistors in a single block of silicon. If a number of bipolar *n*-*p*-*n* planar devices of the type illustrated in Fig. 8.2b are included in the epitaxial *n*-silicon layer, it is apparent that their collector regions will all be electrically connected together by the conducting

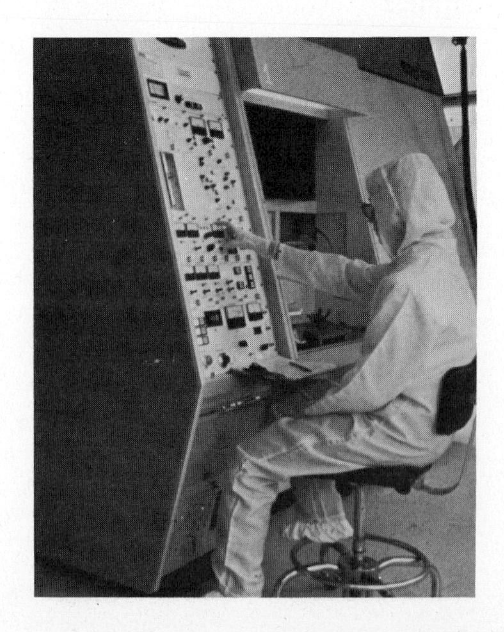

Operator at the console of the Extrion ion implanter. (Supplied by the Digital Equipment Corporation, Maynard, MA.)

silicon. Of course, this in general does not satisfy circuit connection requirements. Hence some method must be available to **electrically isolate** many of the devices on the silicon chip from one another. There are two basic techniques in general use for accomplishing this task:

1. Isolation by introducing a high impedance *p-n* junction between each of the components.

2. Isolation by producing an electrical nonconducting insulating material between the various components.

The first method involves diffusing *p*-type regions through openings in an oxide mask on the *n*-type epitaxial layer so as to produce *p*-type isolated ***n*-pockets,** as shown in Fig. 11.7a. Other somewhat different diffusion procedures have been proposed for accomplishing the same purpose and achieving more economy in the surface area devoted to this isolation step and hence permitting higher component density circuits to be integrated. In all these methods, isolation is accomplished by a very high resistance, low leakage current, reverse-biased *p-n* junction. The second method involves the use of a thin (1 μm)

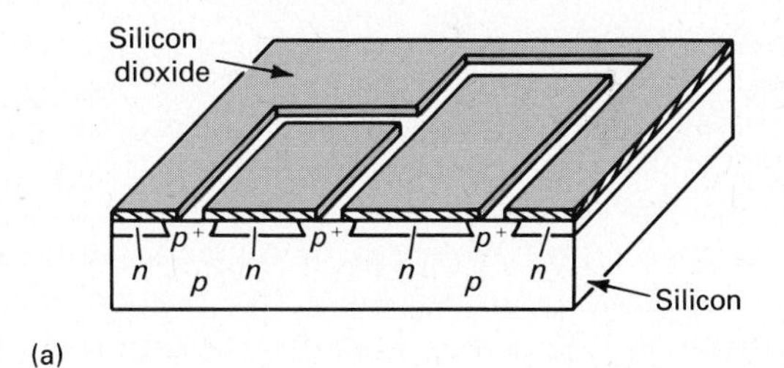

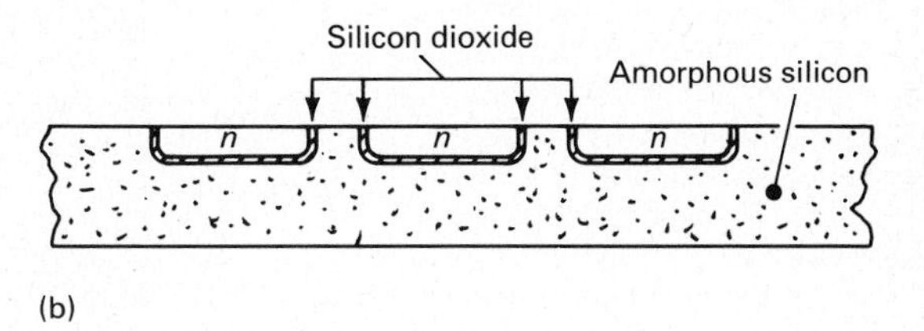

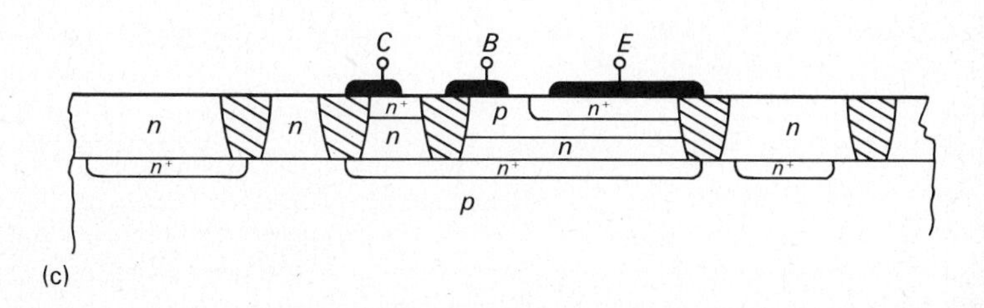

Figure 11.7 (a) Epitaxial wafer after *p*-type impurity diffusion resulting in isolated "*n*-pockets" in *p*-type silicon. (b) Silicon dioxide thin-film isolation of single-crystal *n*-type pockets of silicon embedded in an amorphous silicon substrate. (c) Another example of a dielectric isolation technique showing silicon dioxide (crosshatched) areas and n^+-regions which electrically insulate the *n-p-n* transistor from other *n*-type pockets. The solid block regions represent the emitter, base, and collector metallizations.

dielectric film of silicon dioxide to surround each *n*-pocket and isolate it from the amorphous silicon substrate. This is shown in Fig. 11.7b.

A third method of isolation uses both oxide regions and *p-n* junctions for this purpose. Silicon dioxide insulation is preferred wherever possible since it has a dielectric constant approximately three times smaller than silicon. This provides for lower coupling capacitances between isolated devices and hence higher possible operating frequencies. Such an isolation technique is sketched in Fig. 11.7c. The crosshatched oxide isolation regions pictured not only provide lateral insulation but also reduce the surface area occupied by a single device. This results from closer possible spacing of interconnection metallizations to the emitter, base, and collector regions. If *p-n* junction isolation were used instead, more space on the surface would have to be provided to prevent overlapping and electrically shorting out of the device *p-n* junctions.

Because of the confinement of the current flow in the N-MOSFET to the region between source and drain, no special isolation methods are required for ICs containing these devices. Only a single type of impurity-doped substrate is required. However, in the case of complementary MOSFET circuitry, a *p*-substrate is needed for *n*-channel devices and an *n*-substance is required for the *p*-channel devices. Hence *p-n* junction isolation is generally used to permit the fabrication of *n*- and *p*-channel MOSFETs on a single silicon substrate. This is sketched in Fig. 11.8a. Another method of isolation which is used for complementary field-effect transistor isolation requires the epitaxial deposition of single-crystal silicon from the vapor phase onto crystalline sapphire. The latter material is a good insulator and has a lattice constant which matches very closely that of silicon. Isolated islands of silicon are produced by chemically etching or removing the silicon material between devices. Subsequent diffusion of *n*- and *p*-type impurities into the thin-film deposited silicon permits the fabrication of isolated devices as shown in Fig. 11.8b.

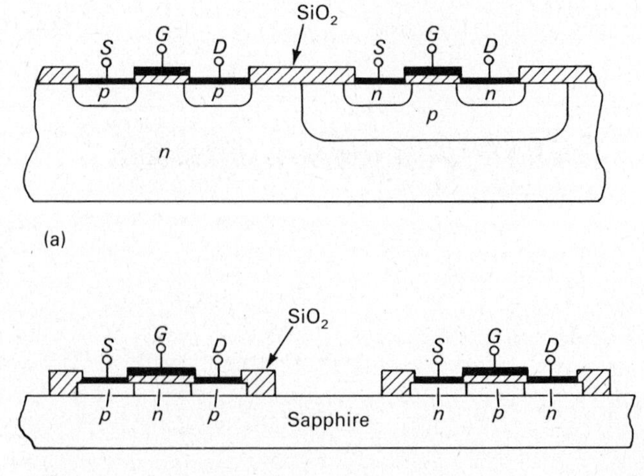

Figure 11.8 (a) Cross section of a complementary MOSFET circuit element including NMOS and PMOS field-effect transistors insulated from each other by a *p-n* junction. (b) Cross section of a complementary MOSFET circuit element made by depositing thin-film silicon onto sapphire and subsequently introducing *p*- and *n*-impurities by diffusion or ion implantation.

11.5 DEVICES FOR INTEGRATED CIRCUITS

The minute device structures of integrated circuits are delineated on the semiconductor wafer surface with the aid of photolithographic techniques. The first step in the device fabrication procedure for silicon ICs is to coat an oxidized epitaxial slice of silicon with a liquid photosensitive material called **photoresist,** which is often a poly(vinyl alcohol) in a solvent such as xylene. The solvent is then baked out to harden the coating. Then the whole coated slice is placed under a photographic negative glass plate that contains opaque and transparent areas which comprise device patterns that are to be transferred to the oxide. Now the oxidized silicon slice, covered by the masking plate, is exposed to ultraviolet radiation. The photoresist polymer becomes crosslinked and hence solvent insoluble in the areas exposed to ultraviolet light, whereas it is soluble in a solvent such as xylene where it has been protected from light by the mask. Now the coated slice is washed with a solvent (developed), leaving the desired photoresist pattern on the oxide. The oxide is then selectively removed in certain areas and allowed to remain in others by subjecting the slice to a buffered solution of hydrofluoric acid which *dissolves* silicon dioxide where it is *not* protected by the photoresist. In this way an oxide **mask** is produced which will later inhibit selectively the subsequent diffusion of such doping impurities as boron and phosphorus when the slice is subjected to a vapor of these impurities at a temperature in excess of 1000°C. A metallic interconnection pattern is also delineated like the oxide pattern by depositing the metal everywhere on the slice, using photoresist for protection and an appropriate acid to dissolve the metal where it is unwanted. Metals commonly used are aluminum and refractory metals. Polycrystalline silicon and metal silicides are also used for device interconnections.

The basic procedures outlined above are repeated in sequence to produce the final intergrated-circuit device structure. A typical set of steps to fabricate a monolithic silicon integrated device slice is outlined in Fig. 11.9.

The device structure resolution is limited by the wavelength λ of light used to project the mask pattern. An optical rule of thumb is that the finest linewidth obtainable by photographic procedures is approximately equal to the wavelength of radiation used to expose the photoresist. Deep ultraviolet light (0.2–0.3 μm) is being used to produce patterns with linewidths of 0.5 μm. For device dimensions much below this, electron irradiation is used since the intrinsic electron wavelength can be a few angstrom units.

The photolithographic process inherently permits the basic microcircuit pattern to be repeated on a semiconductor wafer five or six inches or more in diameter thousands of times and more. All of the processing steps outlined in Fig. 11.9 are of a **batch** fabrication type. This provides for the significant economic advantage of the microchip integrated-circuit technology. Literally thousands of circuits can be produced on each wafer and many wafers are processed at once; each circuit may contain hundreds of thousands of devices such as transistors.

After the semiconductor slice is completed, it is scribed and broken into

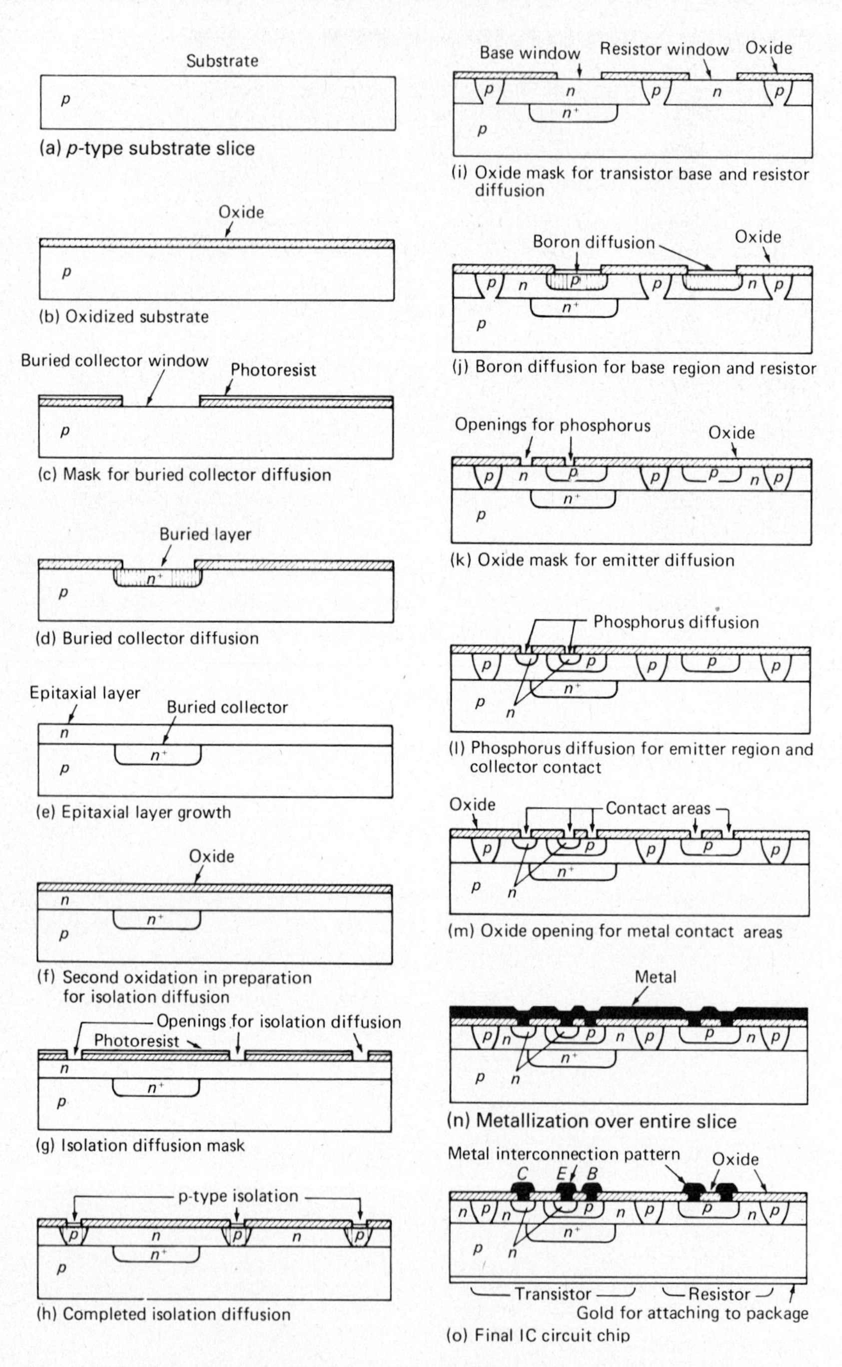

Figure 11.9 Process steps to fabricate a monolithic silicon integrated device slice.

Technician at the console of an electron-beam machine. This equipment is a Perkin-Elmer Manufacturing Electron Beam Exposure System (MEBES). In this installation it is used for making masks for optical and x-ray applications. (Supplied by the Digital Equipment Corporation, Maynard, MA.)

separate circuit chips, some millimeters or more on a side. Then the chips are individually mounted into packages and lead wires bonded by operators using semiautomatic machinery no longer in a batch procedure. Hence the latter steps add considerably to the cost of manufacture and some techniques have been developed to attach the leads all at once, in a batch process. The beam-lead technique is one such process which involves the simultaneous electroplating of all the leads, from 14 to over a hundred in number, at one time.

The above description of microchip processing applies to integrated circuits fabricated with any semiconductor starting material. A specific process for silicon bipolar transistor IC microchip fabrication is illustrated in Fig. 11.9. A simpler but in many respects similar procedure is used in the manufacture of silicon MOSFET ICs. Gallium arsenide microchip digital IC fabrication has more recently been implemented. The basic building block of these circuits is the MESFET. Optically coupled ICs using GaAs and other compound semiconductors are in development. Although the individual technologies differ, lithography is generally used for microstructure delineation as well as several of the other techniques already described for junction formation. Because of the absence of a natural, stable, insulating oxide and the high density of surfaces states present on GaAs surfaces, MOSFET technology does not appear to be feasible for this material. However, the significant interest in GaAs ICs stems from the higher electron mobility and peak velocity in this semiconductor compared with silicon.

Having described the procedures for producing the integrated-circuit devices, we will next discuss the various components in these circuits. The differences between the components included in these monolithic integrated circuits and comparable discrete components, which have heretofore been described, will be emphasized.

11.5.1 Silicon Integrated Transistors and Diodes

Figure 11.10a shows a cross section of a typical integrated-circuit version of a silicon *n-p-n* bipolar transistor. Note how this device is electrically isolated from the *p*-type substrate by a *p-n* isolation junction. Note too that the collector contact as well as the emitter and base metal contacts are made on the upper surface of the silicon slice, since all interconnections must be made on the top surface of this monolithic integrated-circuit chip (see Fig. 11.11). This is in contrast to the discrete planar transistor structure of Fig. 8.2a, in which the collector contact is made at the bottom of the chip. The relatively long path through the collector region to the collector contact in the integrated transistor compared to the discrete transistor accounts for the relatively high inherent series collector resistance of the integrated component. The n^+**-buried-collector,** low resistivity region shown in the figure is introduced in order to effectively "short-out" this high resistance collector current path.

The *n-p-n* transistor structure is generally used in preference to the *p-n-p* design because high mobility electrons transport the injected current from emitter to collector. This yields higher frequency capability and faster switching speed for the component. In certain types of circuits, however, there is a need for *p-n-p* bipolar transistors to complement the *n-p-n* components. The technology to produce these devices is not compatible with producing complementary *n-p-n* and *p-n-p* devices of a "vertical" structure at once. However, the *p*-type diffusion for the *n-p-n* base region may be used at the same time to produce the emitter and collector regions of a "lateral" *p-n-p* transistor, as shown in Fig. 11.10b. This lateral *p-n-p* transistor normally has necessarily rather wide spacing between emitter and collector and hence the component has lower current gain, lower frequency, and slower switching speed capability.

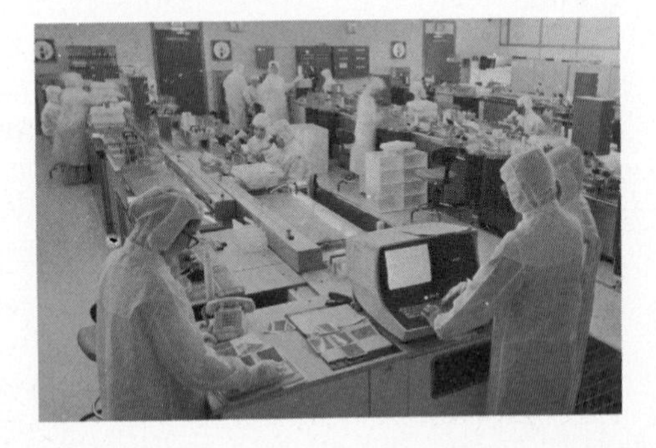

Typical clean room area for processing integrated circuit microchips. (Courtesy of the Texas Instruments Incorporated, Dallas, TX.)

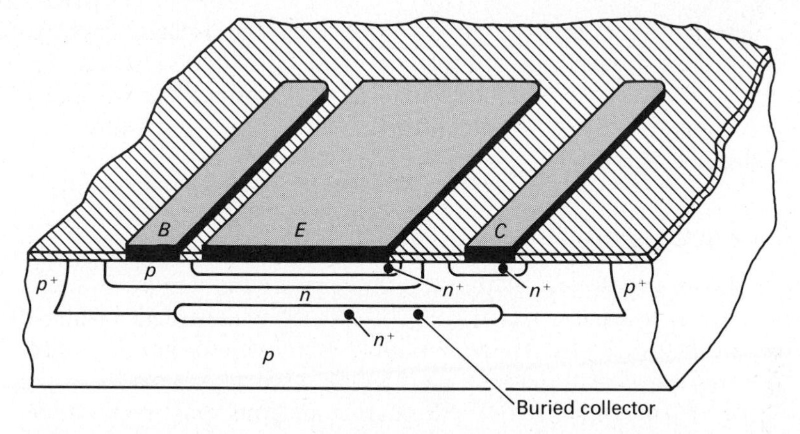

Figure 11.10 (a) Cross section and top view of a typical integrated version of an *n-p-n* bipolar silicon transistor. Note the utilization of a "buried" n^+-layer in the collector region to reduce series collector resistance. The emitter, base, and collector metallizations are marked *E*, *B*, and *C*. The crosshatched regions represent the insulating oxide.

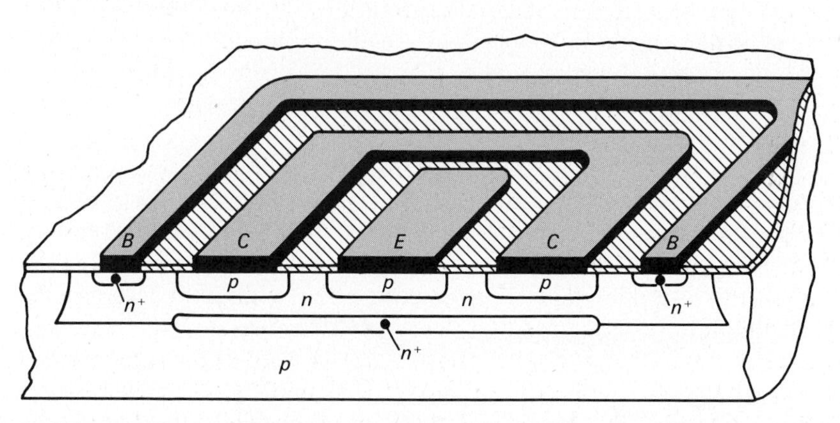

(b) Cross section of a typical "lateral" *p-n-p* bipolar transistor often utilized in integrated circuits. The "buried" n^+-collector layer is used here to improve the current gain. This gain is normally low relative to the gain of "vertical" transistor structures [as in (a)] owing to typically 5–10 times larger emitter-to-collector spacings. The base and collector metallizations are usually in the form of a U surrounding the emitter.

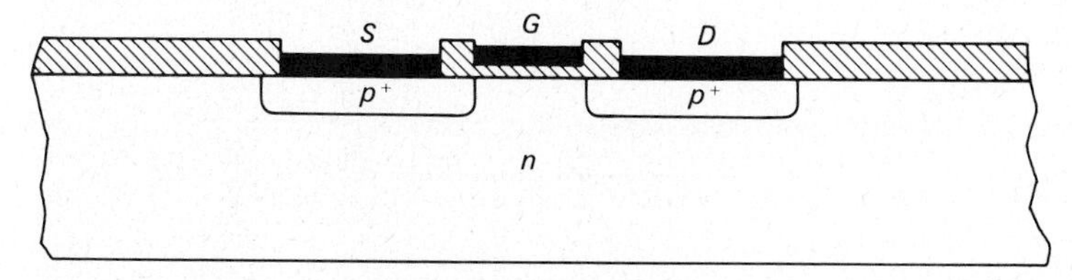

(c) Cross section of a *p*-channel MOS transistor. The source, gate, and drain metallizations are marked *S*, *G*, and *D*, respectively. The crosshatched regions represent the insulating oxide which is thinnest under the gate electrode to provide high transconductance.

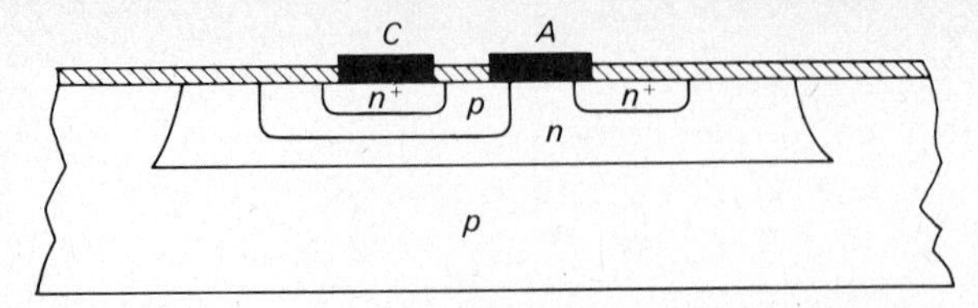

(d) Cross section of an n^+-p integrated form of a diode. In this version the p- and n-regions (base and collector) are electrically shorted to form the anode (A). An emitter region is the cathode (C). This results in a fast switching but low avalanche breakdown voltage (6–8 V) diode.

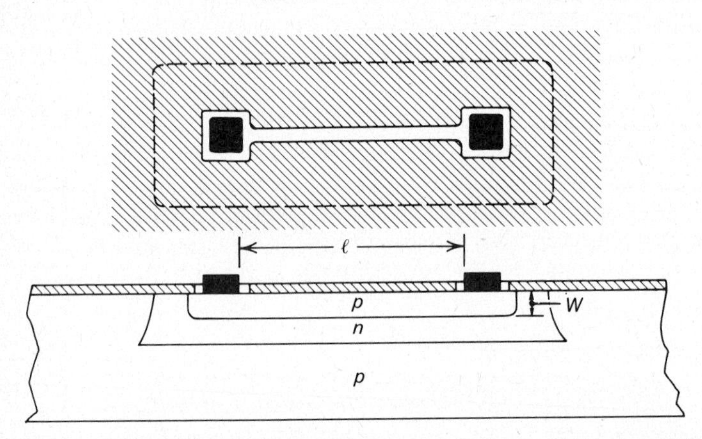

(e) Cross section of a diffused semiconductor resistor used in monolithic integrated circuits. The current flows from one metallized area (darkened area), through the thin p-region to the other metallized area. The high impedance p-n junction isolates the current in the upper p-region and prevents it from entering the n-region or the substrate. The crosshatched areas represent insulating or protective coating.

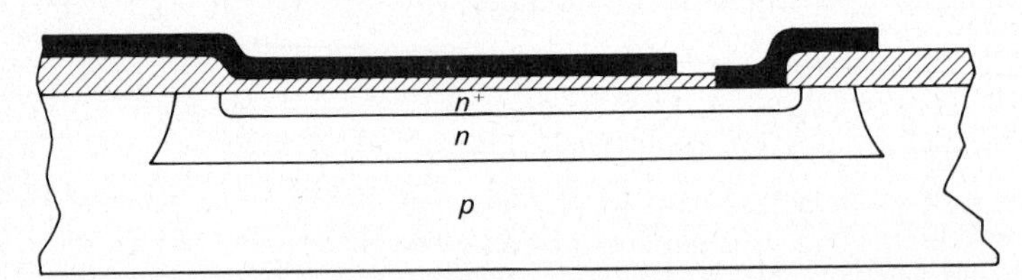

(f) Cross section of a silicon monolithic integrated capacitor. Here the thin insulating layer (shown crosshatched) is usually silicon dioxide thermally grown onto the silicon substrate by exposing it to oxygen or water vapor at a temperature exceeding 1000°C.

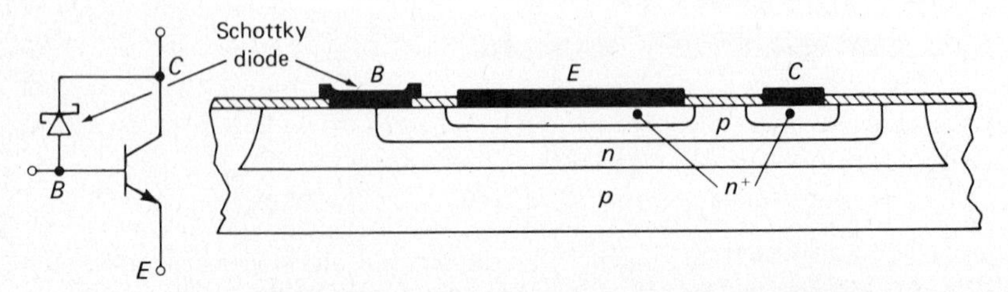

(g) Cross section of an integrated Schottky diode connected in parallel with the collector–base junction of an integrated n-p-n transistor. The metallization (B) is both the anode of the Schottky diode and the transistor base contact. The cathode of the Schottky diode is the n-type collector region of the transistor. A schematic diagram of this device is shown to the left.

A number of linear amplifier integrated circuits use this device for polarity reversal, level shifting, and current sources where gain and frequency capability is of no consequence. A vertical *p-n-p* transistor is used in applications in which the *p*-type collector can be connected to the substrate (see Fig. 11.10a).

A single *p*-type diffusion will produce a *p*-channel MOSFET component as shown in Fig. 11.10c. Comparing this structure with the bipolar structure of Fig. 11.10a confirms the earlier claim of simplicity of MOSFET fabrication technology. This is particularly evident since no isolation pockets are needed for MOSFET circuits. They are **self-isolated** since the current is automatically confined to the small channel region just at the silicon surface, between source and drain, as discussed in Chapter 10.

The diode in integrated circuits is actually a transistor structure with two regions shorted together by aluminum metallization. For example, the base and collector regions may be connected together to form one electrode of the diode while the emitter electrode forms the other connection. Any other two regions may be connected together to form a different diode structure; the electrical behavior of the component is determined by which two are interconnected.[4] A form of integrated diode is shown in Fig. 11.10d.

11.5.2 Integrated Diffused Resistors

The form that discrete resistors often take is that of thin metallic films composed of high resistivity materials such as Nichrome. These films could also be deposited onto the oxide or insulator of the integrated-circuit chip and delineated into the form of thin-film resistors using photoresist. This would require the introduction of still another material and step in the processing of integrated circuits. Instead **diffused** resistors are normally used for the purpose of limiting current and establishing biases in silicon bipolar integrated circuits. The integrated resistor is shown in its usual form in Fig. 11.10e. It consists of a long, narrow diffused *p*-type region embedded in an *n*-type region and hence electrically isolated from it by a *p-n* junction. This *p*-type region is diffused into the silicon at the same time as the bipolar integrated transistor base region. It is this compatibility with the standard diffusion technology that accounts for its general use in bipolar ICs. The resistance value R of a diffused resistor is usually specified in terms of the sheet resistance ρ_s of the diffused region of depth W, as $R = \rho l/A = \rho_s/W$ (see Fig. 11.10e). Here ρ_s is expressed in terms of resistance per square of surface area.

Since the basic material of the diffused resistor is a semiconductor, its resistance value is subject to rather large variations with temperature. In fact the silicon integrated resistor increases in value by about 3000 parts per million per degree centigrade temperature increase, in contrast to only 50 ppm/°C for a thin-film Nichrome resistor. However, this is of little consequence in digital

[4]For a more complete discussion of integrated circuits see for example D. J. Hamilton and W. G. Howard, *Basic Integrated Circuit Engineering* (McGraw-Hill, New York, 1975).

switching circuits which do not require accurate resistor values since it is only necessary to maintain two separate circuit states, ON and OFF. Linear amplifier circuits require much more strict control of resistor values, however. In the commonly used differential amplifier integrated circuits it is only the matching and tracking with temperature of a pair of resistors which is required. This is automatically accomplished in the integrated-circuit technology since both resistors in the pair are produced at the same time and in close proximity to each other on the silicon chip. Resistor values ranging from about 50 to 30,000 Ω are conveniently made by this method. Larger value resistors occupy too much chip area ("real estate") and are hence uneconomical.

11.5.3 Integrated Capacitors

It is quite natural to expect the silicon dioxide layer on the silicon integrated-circuit chip to be used as the dielectric material in a parallel-plate type of capacitor structure. This is in fact the form of the capacitor commonly used in silicon integrated-circuit chips, as is illustrated in Fig. 11.10f. The upper plate of this sandwichlike structure is a thin-film metallization; the middle section is the insulator silicon dioxide. The lower plate is basically a heavily doped *n*-type silicon layer which is contacted with a metallization. The relative dielectric constant of silicon dioxide is unfortunately only 3.9, but in other respects it is an excellent insulating material. Since the thinnest pinhole free oxide that can be produced with good yield at this time is only a few hundred angstroms, it can be shown that the capacitance per unit area achieved in this manner is only about 3500 pF/mm^2. However, a square millimeter constitutes a substantial part of an integrated-circuit chip. In fact, several thousand transistors can be integrated per square millimeter. Hence it becomes clear that the use of integrated capacitors in integrated circuits must be restricted. This accounts for the general use of directly coupled amplifier circuits in the integrated form which do not require isolating or bypass capacitors.

11.5.4 Integrated Schottky Diodes

The important use of the Schottky diode for speeding up digital switching circuits was mentioned in Section 7.5.1. This majority carrier device in its integrated form is shown in Fig. 11.10g shunting the collector junction of an *n-p-n* bipolar transistor. The structural simplicity of this metal–semiconductor diode accounts for its frequent use in transistor–transistor logic (TTL) digital integrated circuits (see Section 11.6) to reduce storage time in this saturating form of logic circuitry. For this purpose the anode of the diode is connected to the transistor base and the cathode to the collector region. The Schottky diode metallization can be done at the same time as the interconnection metallization for the integrated circuit and hence is compatible with standard integrated-circuit technology. The etching of a hole in the silicon dioxide exposing bare semiconductor for the metal–silicon Schottky contact is accomplished in the same way that openings are made in the oxide for impurity diffusion.

The low barrier height and starting voltage of the Schottky diode make it useful as a coupling element in integrated injection logic (I^2L) ICs. It is also used to form the control gate of the MESFET used in GaAs digital logic circuits. Figure 11.11 shows an integrated inverter circuit in order to illustrate a very simple version of a monolithic silicon integrated circuit. The interconnection metallization which is confined to the silicon wafer upper surface is shown. The metallization paths eventually terminate at square bonding "pads" at the edges of the integrated-circuit chip of silicon so that fine (20 μm diameter) gold or aluminum/1% silicon lead wires may be attached. These wires are then connected to the various metal terminals which are part of the device package. The common methods of lead attachment are by thermal compression bonding or ultrasonic bonding as shown in Figs. 11.12a and 11.12b.

In the more complex integrated circuits several layers of metallization are required in order to provide the necessary electrical interconnections without any metal path crossovers. For this to be accomplished the different metal layers need to be insulated from each other by appropriate dielectric thin films. The most common materials used for device interconnection and conducting electric current are aluminum and doped polycrystalline silicon. However, because of the light weight of Al atoms they can be *moved* by the electrons conducting the electric current by a process termed **electromigration.** This effect can give rise to interconnection voids and hence IC failure, particularly for high density circuits in which narrow interconnection metallizations are used and the current densities are high. Adding a fraction of a percent copper to aluminum will substantially reduce the mean time-to-failure for this type of metallization and can be carried out by simple evaporation. However, a more direct solution of this problem is the use of the heavy, high melting point metals such as tungsten, titanium, tantalum, molybdenum, etc., to minimize electromigration. These can be deposited by a somewhat more complex process

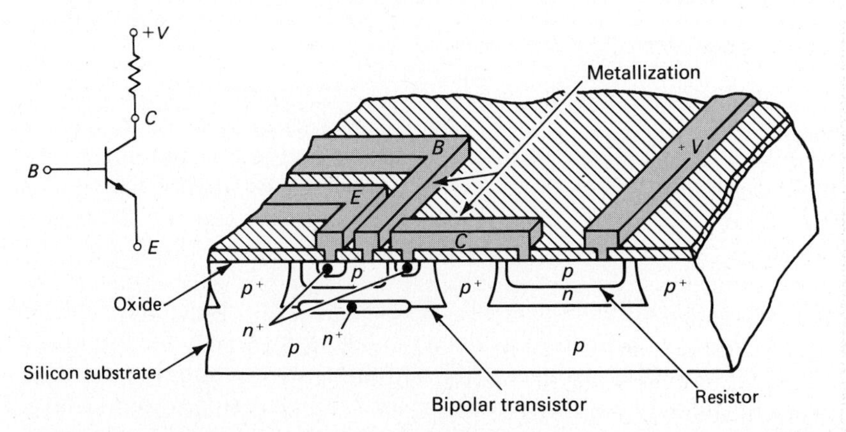

Figure 11.11 Simple silicon monolithic integrated inverter circuit showing the surface interconnection metallization. A schematic circuit diagram of this structure is shown to the left.

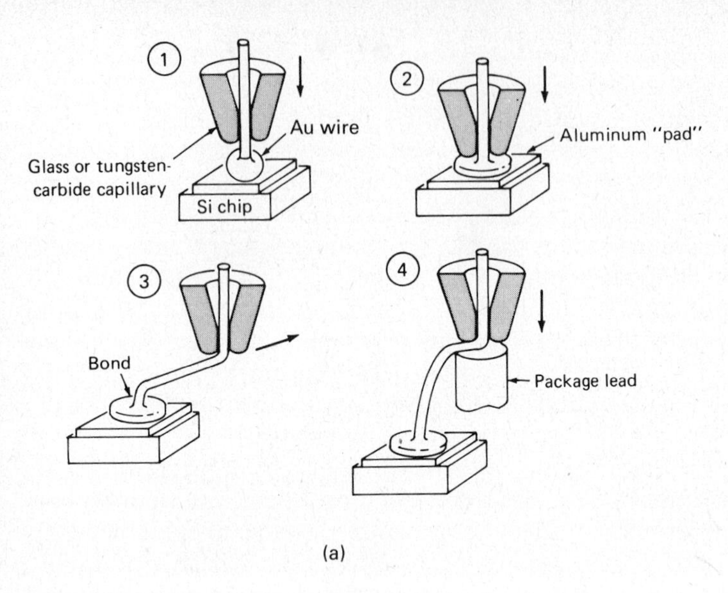

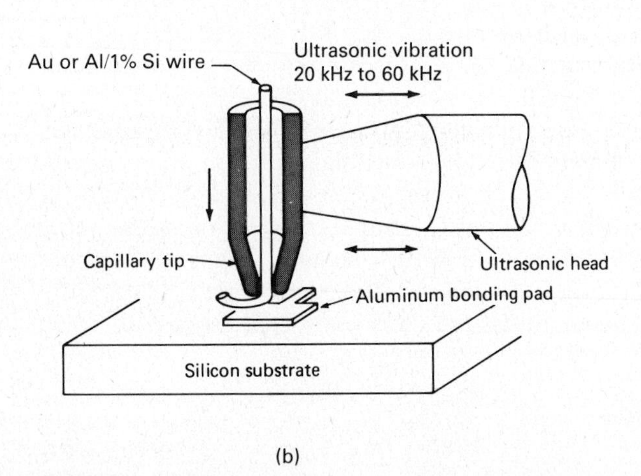

Figure 11.12 (a) Sequence of steps used in the "thermocompression" bonding of gold wire to a metal pad on a silicon circuit chip. The chip is heated to about 150°C in this operation. (b) Ultrasonic bonding of a fine gold or aluminum/1% silicon wire to a metal pad on a chip. The ultrasonic vibrations break through any metal oxide. No heating of the wafer is required in this process.

called **sputtering.** A family of metal silicides such as $TaSi_2$, WSi_2, and $TiSi_2$, are also in use. Doped polysilicon films have too high a resistivity for use in high speed circuits. For insulating layers SiO_2 and Si_3N_4 are commonly used. Polysilicon films doped with oxygen also can be used for insulating adjoining layers of conducting materials.

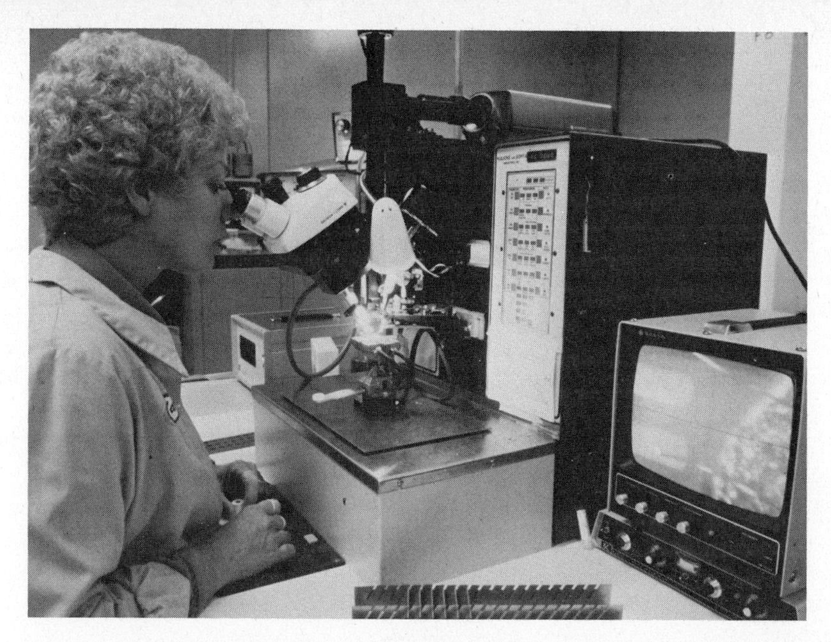

Operator running a Kulicke and Soffa #1419 automatic wafer bonder. This machine connects gold wires between a die and the package leads at a rate of 4 wires per second. (Supplied by the Digital Equipment Corporation, Maynard, MA.)

A vacuum chamber used for vapor deposition of interconnection metallization on the round integrated circuit wafers. (Courtesy of Analog Devices Semiconductor, Wilmington, MA.)

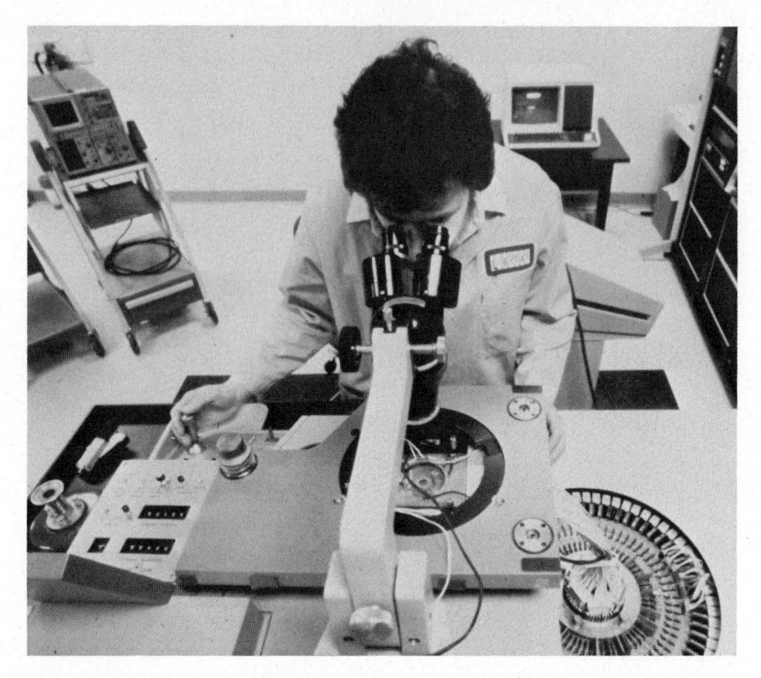

Technician operating an Electroglass wafer prober, interfaced to a Tektronix 5-3270 LSI tester to locate the good dies on a newly-fabricated wafer. (Courtesy of the Digital Equipment Corporation, Maynard, MA.)

11.6 INTEGRATED DIGITAL CIRCUITS

The integrated-circuit microchip is the basic building block of electronic computers. The chip contains a multitude of semiconductor devices of the types described in the last section, interconnected so as to perform an entire electronic function. Some chips perform digital or computational operations; others perform analog or amplifying functions. Most microchips manufactured today use silicon as a starting semiconductor material. More recently gallium-arsenide-based integrated circuits have become available and have exhibited

Steps in the packaging of an IC chip in a plastic package, beginning with the chip at the far right and ending with the labeled package on the far left.

higher speed performance than chips made of silicon. However, at present silicon chips are cheaper to produce.

The electronic logic gate is the basic building element of most digital integrated circuits. The logic gate is a basic binary circuit which represents the number 1 when it is ON and conducting and 0 when it is OFF and nonconducting (or vice versa). Generally one gate drives the next gate and is driven by a previous gate. A combination of these gates are interconnected to perform a complete computer function such as an arithmetic calculation or to store information as in an electronic memory. A combination of these computational functions are combined in a microprocessor or **computer-on-chip.** It is not the function of this book to discuss the architecture of a computer. Instead the basic classes of logic gates in general use today in the design of digital microchips will be described next.

11.6.1 Silicon Digital ICs

A variety of types of logic gates are used in the design of silicon digital circuitry. Some of these circuits use bipolar transistor switches, while others use MOSFET switching devices. Bipolar circuits are used where the ultimate in switching speed is desired. The layout simplicity of MOSFET circuits allows for the high packing density possible with these devices and hence accounts for their general use in fabricating large memory chips as well as logic chips.

The two most used bipolar logic gates are those employing transistor–transistor logic (T^2L) or emitter-coupled logic (ECL).

A. TRANSISTOR–TRANSISTOR LOGIC (T^2L)

A basic two-input T^2L gate is shown in Fig. 11.13a. Transistor Q_1 is the input device. Input voltages V_1 and V_2 are applied to the two emitters of the *n-p-n* transistor Q_1. Push–pull output is provided by the **totem pole** output devices Q_3 and Q_4, with the output voltage V_o taken from the collector of Q_3. The **phase-splitting** transistor Q_2 provides drive for the two output transistors Q_3 and Q_4. Clamping diodes D prevent negative overshoot at the input.

Consider that both inputs V_1 and V_2 are initially at ground potential taken at a reference value of zero volts. Current will be conducted by resistor R_1, through the transistor Q_1 base region, and via the two emitters to ground. Hence the potential at V_{B1} will be approximately 0.9 V relative to ground since 0.7 V is about the voltage drop across a conducting silicon emitter *p-n* junction and the output voltage of the previous driving stage is 0.2 V. In tracing a path to ground through the collector junction of transistor Q_1 and the emitter of Q_2, this 0.9 V will be shared between these two junctions and so very little current will take this path to ground. Hence transistor Q_2 will be OFF or nonconducting and only a small collector leakage current will pass through resistor R_{C2}. So the collector potential of Q_2 and V_{B4}, will be about the same as V_{CC}, or 5.0 V. Correspondingly the emitter potential of Q_2, V_{BE3}, will be near ground potential, or zero volts. This will keep output transistor Q_3 OFF; however, transistor

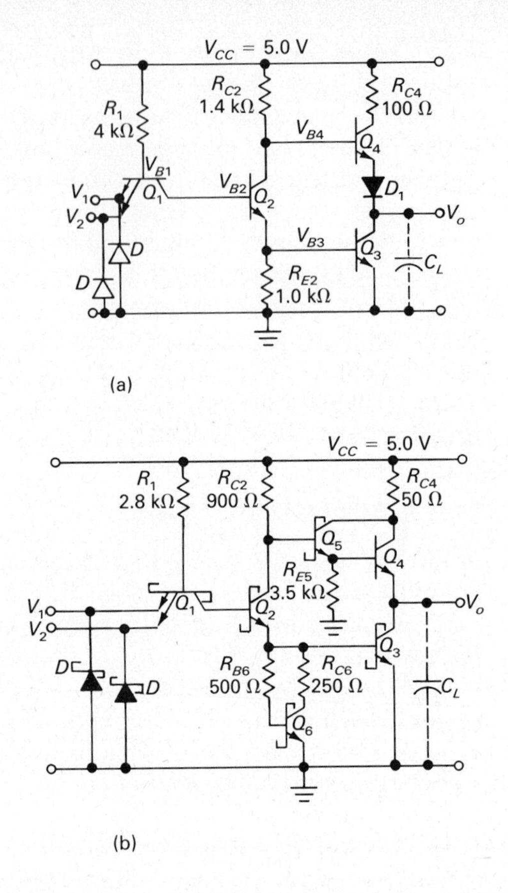

Figure 11.13 (a) Standard transistor–transistor logic (T^2L) NAND gate driving a capacitive load.
(b) Schottky-clamped T^2L NAND gate.

Q_4 will perform as a transistor operating in the active mode, since its input is high, causing the output potential V_o to be two diode drops (emitter junction of Q_4 and diode D) below V_{B4} or 3.6 V. Hence the logic gate output voltage V_o is high (3.6 V) when *either* input V_1 or V_2 is low (0.2 V).

Now consider that both inputs V_1 and V_2 are raised to 3.6 V. The potential of the base of input transistor Q_1 cannot rise above 2.1 V since this constitutes the voltage drop across three diodes to ground; the three diodes are the collector of Q_1, the emitter of Q_2, and the emitter junction of Q_3. Note that input transistor Q_1 is operating in the *inverse* active mode since the two emitter junctions of Q_1 are reverse biased and the collector is forward biased. The reverse gain of transistor Q_1 is kept low by design in order to ensure a negligible drain of current by the previous OFF stage.

Since Q_3 is driven ON, the output voltage V_o drops to about 0.2 V above ground, this being about the saturation voltage of a conducting bipolar transistor. Since transistor Q_2 is ON, this brings its collector potential V_{B4} to about 0.9 V, the diode voltage V_{BE3} plus the saturation voltage of transistor Q_2. Since the output voltage V_o is 0.2 V, only 0.7 V is dropped across two diodes, the emitter junction of transistor Q_4 plus diode D_1. This ensures that Q_4 is OFF. Hence the logic gate output is low (0.2 V) when the *two* inputs V_1 *and* V_2 are high (3.6 V). Also, the gate output is high (3.6 V) when *either* of the inputs V_1 or V_2 are low (0.2 V). This constitutes the negative (input high, output low, and vice versa) AND function and hence this logic circuit is termed a two-input **NAND** gate.

Consider the case in which the inputs are high so that the output V_o is low and input transistor Q_1 is operating in the inverse active mode. If one of the inputs is now lowered in voltage, Q_1 is placed in the inverse saturated mode and hence provides a low impedance path to discharge the charge previously stored in transistor Q_2, which had been in saturation. This speeds the turnoff of this logic gate circuit. Hence transistor Q_1 acts not only as a coupling element but also to facilitate the turnoff of Q_2, which then turns off output transistor Q_3. The totem pole output driving circuit also provides a speed advantage in driving the next stage. When transistor Q_4 is in the forward active region it has a low output impedance due to emitter follower action that speeds up the charging of the capacitive load C_L, which represents the input to the next stage.

Typically such manufactured T^2L circuits can drive ten similar circuits or provide a maximum **fanout** of 10. The average **power dissipation per gate** is typically about 10 mW and the time it takes for a digital signal applied to the input to reach the output or the **propagation delay** is typically about 6 nsec. The supply voltage required is between 4.75 and 5.25 V and the specified temperature range for operation is from −55 to 125°C.

B. SCHOTTKY-CLAMPED T^2L LOGIC

There are a variety of T^2L circuit chips manufactured today, e.g., standard, low power, high power, and the Schottky T^2L chips. These circuits represent a tradeoff between power dissipation and propagation delay. If the resistor values in the circuit of Fig. 11.13a are reduced, the current available to drive the switching transistors and the next logic stage will be increased, resulting in a more rapid charging of capacitances and hence faster switching. However, T^2L is a saturating form of logic circuitry which is limited in speed by minority carrier charge storage in the base and collector regions of the transistor (see Section 9.6.2). In Section 11.5.4 it was shown that a Schottky diode, shunting the base–collector junction of a transistor, can effectively prevent the device from being driven into deep saturation by the voltage limiting effect of the low forward voltage drop of a Schottky diode (0.3–0.4 V). All the transistors in the standard T^2L circuit of Fig. 11.13a may be thus effectively Schottky clamped, avoiding charge storage effects for the most part.

A Schottky-clamped T^2L NAND gate circuit is illustrated in Fig. 11.13b. Here all the transistor base–collector junctions are shunted by Schottky diodes except transistor Q_4, which always operates in the active region and never saturates. In this Schottky T^2L circuit an additional driving transistor Q_5 is introduced to speed the switching of transistor Q_4, and hence speed the pullup of the output voltage V_o toward the battery potential. This device also introduces an additional diode voltage drop, hence removing the need for diode D_1 in the standard T^2L logic circuit of Fig. 11.13a. Transistor Q_6 acts as an active pull-down device, replacing the resistor R_{E2} of Fig. 11.13a, providing a low resistance path to ground, and turning off transistor Q_3 by extracting its stored charge. Finally, note that the resistor values in the Schottky gate are lowered relative to the standard T^2L gate, improving the switching speed even more. For the same power dissipation the Schottky gate has a propagation delay two or three times shorter than the standard T^2L circuit.

EXAMPLE 11.2

Calculate the output voltage V_o of the T^2L gate shown in Fig. 11.13a when one of the inputs V_1 is low (0.2 V).

Solution

The output voltage is given by

$$V_o = V_{CC} - I_{B4}R_{C2} - V_{BE4} - V_{D1},$$

where V_{D1} is the voltage drop across the diode $D1$ and V_{BE4} is the base–emitter diode drop of transistor Q_4. When an input V_1 or V_2 is low the output is high, the following logic gate stage is OFF, transistor Q_2 is OFF, and transistors Q_3 and Q_4 draw only the leakage current of the following input device; hence I_{B4} is negligibly small. So

$$V_o = 5.0 - 0 - 0.7 - 0.7 = \underline{3.6\ \text{V}},$$

where the diode drops V_{BE4} and V_{D1} are taken as 0.7 V when drawing a small (not zero) current from the succeeding stage.

C. EMITTER-COUPLED LOGIC (ECL)

Emitter-coupled logic or **current-mode logic** is a nonsaturating form of circuitry where the transistors are constrained to always operate in the active region; no minority carrier charge storage occurs and hence it is the fastest operating circuitry in the family of logic circuits. Figure 11.14a is a schematic diagram of a standard ECL OR-NOR gate.

The basic input stage of this logic circuit consists of the differential transistor pair Q_2 and Q_3. A fixed reference voltage V_R is applied to the base of Q_3 and the input to the gate is via Q_2 or, alternatively, Q_1 in this two-input ECL gate. The relatively large resistor R_E serves as a current source for the input stage. A simplified schematic diagram simulating the input circuit is shown in Fig. 11.14b. Transistors Q_2 and Q_3 are accurately matched since they are fabricated at the same time and are located quite close to each other on the micro-

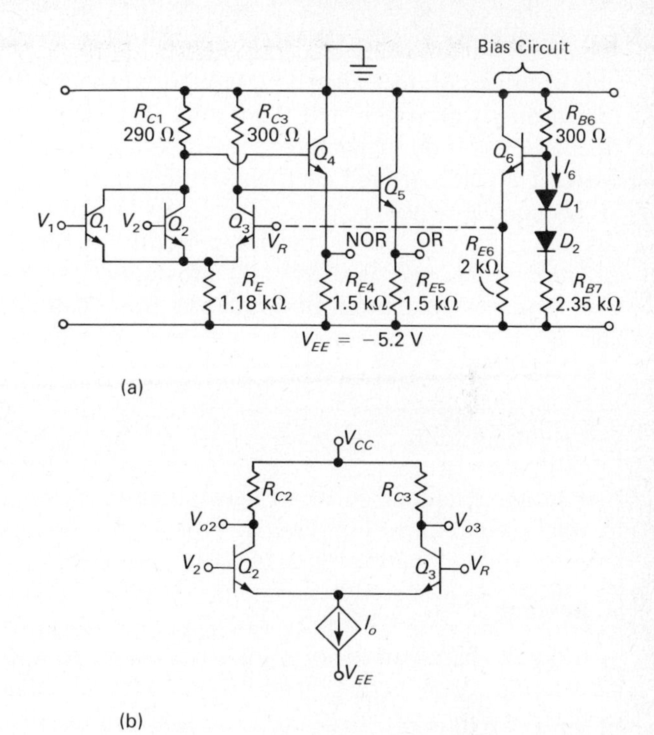

Figure 11.14 (a) Standard ECL OR-NOR logic gate. The reference potential V_R is provided by the bias circuit shown.
(b) Basic input circuit schematic for ECL gate.

chip. Hence with input voltage V_2 equal to V_R, the input circuit is balanced and the current drawn by transistor Q_2 is equal to that drawn by Q_3. The sum of the currents drawn by these transistors is supplied by the current source I_o. When, however, V_2 is raised in potential a little above V_R (about 0.1 V) nearly all of I_o will be conducted by transistor Q_2, bringing the output potential V_{o2} down low; with V_2 below V_R, nearly all the current I_o will be conducted by transistor Q_3, so V_{o2} will go high. This constitutes a NOR function. If the output voltage is taken as V_{o3}, a high input voltage will result in a high output voltage and vice versa, constituting an OR function. The circuit components are chosen so the transistors operate only in the active and cutoff regions, *never in the saturation region,* so that minority carrier charge storage effects are avoided. This accounts for the ability of carefully designed ECL circuitry to operate with propagation delays in the subnanosecond range.

In the actual ECL circuit of Fig. 11.14a the current source I_o of Fig. 11.14b is provided by a large resistor R_E. The two input transistors Q_1 and Q_2 share the current through R_E when both input voltage V_1 and V_2 are high. The NOR and OR outputs are taken from emitter followers Q_4 and Q_5, both of which operate in the active region since their collectors are at the highest available potential, zero, and their base voltages always operate below ground potential. These emitter followers provide a low output impedance to provide more current drive for the next logic stage. The voltage drop of the emitter junction of these emitter followers also provides a diode drop which brings the output quiescent voltage to a level equal to that of the input voltage so that one stage may properly drive the next. The small input voltage excursion (0.1 V) needed to charge the input capacitance reduces the charging time for a given driving current, which helps reduce the switching time.

Fanout for this type of logic circuitry is typically 20 or more owing to the emitter follower drive. Typical propagation delay is 1 nsec, and when oxide isolation technology is used, subnanosecond performance is obtained. However, a power dissipation of 20 mW or more is incurred to achieve these fast speeds.

EXAMPLE 11.3

Find the reference voltage V_R generated by the circuit of transistor Q_6 for the ECL circuit of Fig. 11.14a.

Solution

Take the base current of transistor I_6 to be small compared with I_6 and the diode drops across D_1 and D_2 to each be V_D. Then Kirchhoff's voltage law gives

$$-V_{EE} = I_6(R_{B6} + R_{B7}) + 2V_D.$$

Also $V_{B6} = -I_6R_{B6}$, and the base–emitter voltage drop of transistor Q_6, $V_{BE6} = V_{B6} - V_R$. Hence

$$V_R = V_{B6} - V_{BE6} = -I_6R_{B6} - V_{BE6}$$

or

$$V_R = [R_{B6}/(R_{B6} + R_{B7})](V_{EE} + 2V_D) - V_{BE6}.$$

Since the base–emitter voltage drop for transistors operating in the active region in ECL circuitry is about 0.75 V,

$$V_R = (0.3\ \text{k}\Omega/2.65\ \text{k}\Omega)(-5.2 + 1.5\ \text{V}) - 0.75\ \text{V} = \underline{-1.17\ \text{V}}.$$

D. INTEGRATED-INJECTION LOGIC (I^2L)

Integrated-injection logic or **merged transistor logic** uses bipolar transistors, integrated in such a way as to provide a very high packing density of gates per chip. No isolation *p-n* junctions are needed for integration and the power dissipation per gate is low. The basic logic gate of this type of circuitry consists of a lateral *p-n-p* bipolar transistor for a current source and a vertical *n-p-n* transistor with multiple collectors for outputs, as shown in the schematic circuit diagram of Fig. 11.15a. The merged nature of the devices is shown in Figs. 11.15b and 11.15c, where a standard chip layout of two I²L basic units is illustrated along with the cross section of one unit. Note the lateral bipolar *p-n-p* transistor and the vertical *n-p-n*. All the *n-p-n* emitters are common to the substrate. All *p-n-p* emitters (or injectors) are common to a *p*-type injector rail. The base region of the *p-n-p* transistor is the same as the emitter of the *n-p-n*;

Figure 11.15 (a) Schematic circuit diagram of a basic I²L unit.
(b) Chip layout for two I²L units.
(c) Cross section of a basic I²L unit.

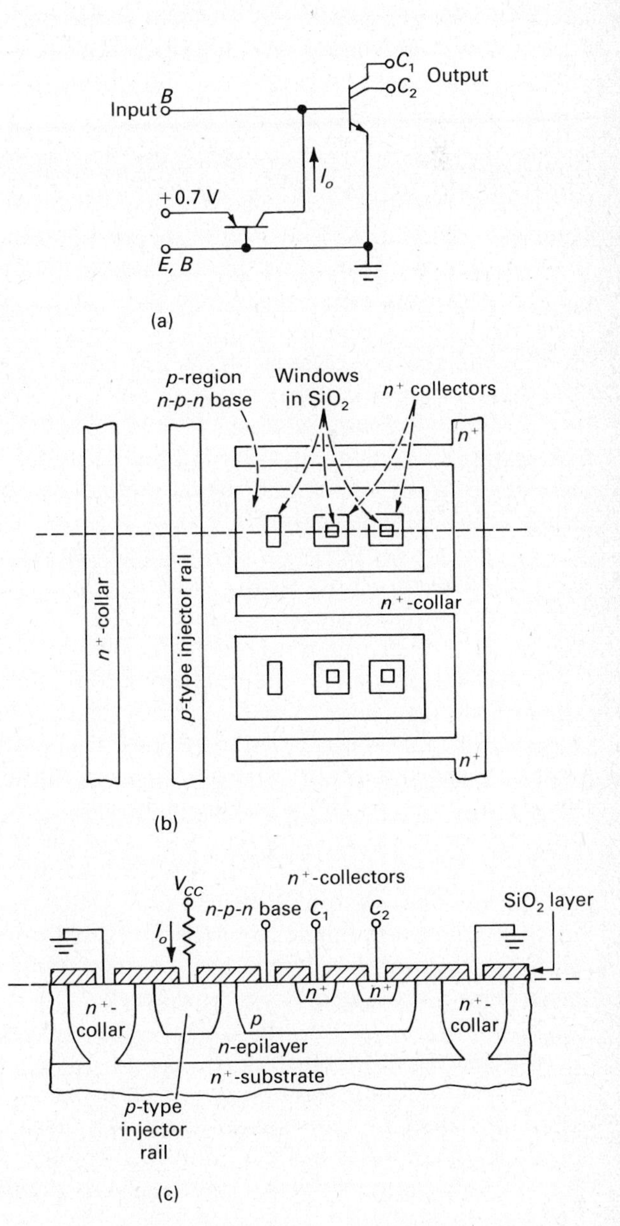

the collector of the *p-n-p* is the base of the *n-p-n*. Hence the term "merged transistor."

The switching operation of this basic unit may be described as follows: the *p-n-p* acts as a current source or supply. If the base input B of Fig. 11.15a is high (0.7 V), the *n-p-n* transistor gate conducts current, in saturation, and the collector (C_1 and C_2) potentials are low (0.1 V), representing the saturation voltage of this transistor. Since in this case a high input voltage means that the previous driving stage is OFF (Fig. 11.16), the current is supplied by the *p-n-p* source. When the base input B is low (0.1 V, the saturation voltage of the previous stage), the *p-n-p* supplies current for the previous ON conducting stage. Hence the logic function for I^2L circuitry consists of current steering of the *p-n-p* current source. It supplies current to the *n-p-n* transistor of the logic gate when it is ON and current to the driving circuitry when the logic gate is OFF. The fanout of this logic circuitry is provided by the multiple collector junctions. A typical I^2L logic circuit configuration is shown in Fig. 11.16a.

The primary features of I^2L circuits are the moderately small power dissipation and propagation delay as well as the high packing density on a chip. The high packing density results from the merged nature of the transistors as well as the absence of resistors, which normally occupy substantial areas on the chip. An important advantage of this type of logic circuitry in competition with even higher packing density MOS digital logic is the compatibility of its fabrication technology with linear (analog) bipolar transistor processing. Hence combinations of analog and digital circuits can be efficiently integrated on a single semiconductor chip.

A typical logic delay for an I^2L gate is 10 nsec with a power dissipation of 50 μW. This constitutes slower switching speed compared with ECL and T^2L, but lower power dissipation, which is also consistent with high chip packing density. However, the power–delay product of 0.5 pJ per gate surpasses both ECL and T^2L circuitry. Attempts at speeding up I^2L circuitry have led to the use of Schottky diodes in each collector contact or by shunting the collector–base junction of the *n-p-n* transistor by a Schottky device to limit the logic voltage swing and minority carrier storage in this type of circuitry. An example of Schottky I^2L circuitry is shown schematically in Fig. 11.16b. A type of digital circuitry known as **ISL** combines some of the speed advantages of T^2L with the high packing density of I^2L. A schematic diagram of this type of circuitry plus a chip cross section of the ISL gate is shown in Fig. 11.16c. Here the basic input *n-p-n* device inherently adds a vertical *p-n-p*. Also, a lateral *p-n-p* parallels the vertical *n-p-n*. The *p-n-p* device limits the saturation of the *n-p-n* transistor, improving the switching speed.

E. NMOS TRANSISTOR LOGIC

The silicon digital integrated-circuit chips most in use today employ NMOS (*n*-channel, metal–oxide–semiconductor) transistors. However, inroads on this prime market position are rapidly being made by CMOS (complementary metal–oxide–semiconductor) circuitry. The popularity of NMOS logic chips

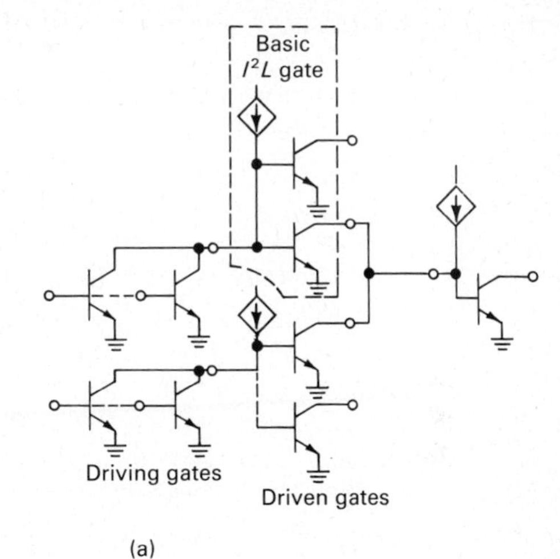

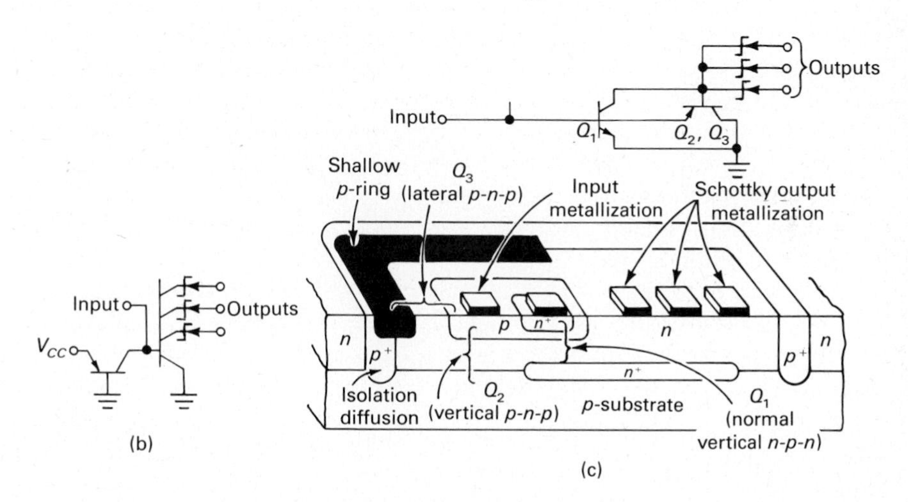

Figure 11.16 (a) An I^2L circuit configuration showing the basic gate driven by a similar gate and driving similar gates.
(b) A type of Schottky I^2L circuitry.
(c) ISL digital circuit which occupies 40% greater chip area than a conventional I^2L gate, but which switches in about one-quarter of the time of the latter.

stems from the simplicity of the MOS logic gate structure and hence its high packing density, as well as moderate switching speed and power dissipation. No *p-n* junction isolation is needed for this circuitry, which is self-isolating. The attraction of CMOS circuitry results from its low power dissipation due to the negligible power consumption of these types of logic circuits in the static state between switching operations. In this section the standard digital NMOS logic circuit will be described; the next section will provide a description of the CMOS logic gate.

Figure 11.17a shows a schematic diagram of a standard NMOS logic circuit. Note that the driver (or inverter) transistor is of the enhancement-mode type; the load device is of the depletion type. The operation of this logic gate may be described as follows: assume that the input voltage V_i is just a few tenths of a volt above ground potential, taken as zero. Since this is below the

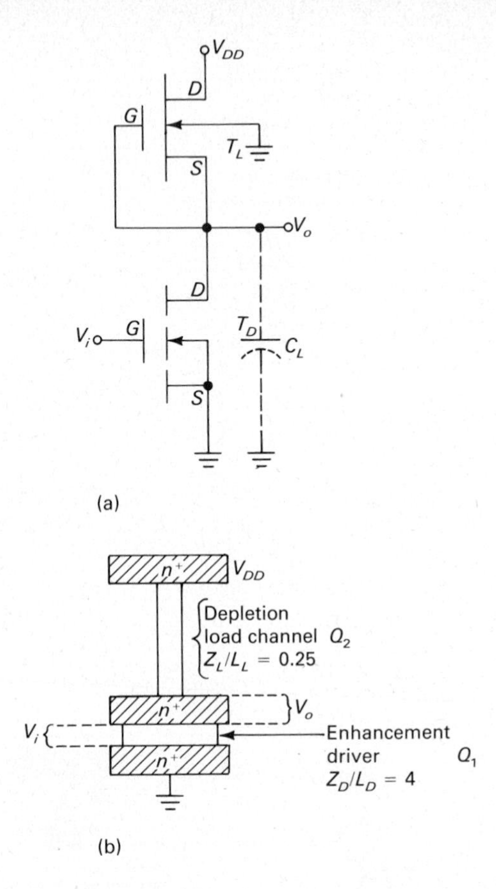

Figure 11.17 (a) NMOS gate with an enhancement-type driver and a depletion load driving the next stage represented by a capacitive load C_L.
(b) Chip layout of the NMOS gate showing the short channel driver transistor and the long channel load transistor.

threshold voltage of the driver enhancement transistor T_D this device conducts only a small leakage current characteristic of its reverse-biased drain junction. The output voltage of this logic gate will then tend toward the battery supply voltage V_{DD}. This high potential will turn ON the next logic stage driven by this gate, but very little current will be drawn in the steady state because of the capacitive gate input of the next stage. Next assume that the logic gate input is essentially at the supply potential V_{DD}. This will cause the driver transistor to conduct and this current will also be conducted by the depletion load transistor T_L. The low ON output voltage of the enhancement driver transistor will keep the next logic stage OFF. A typical chip layout for this gate is shown in Fig. 11.17b.

Typical current–voltage (I-V) characteristics for both these enhancement and depletion NMOS devices are shown in Fig. 11.18. Initially, with low input voltage, the gate is OFF and the output voltage is high (enhancement transistor drain potential is indicated by the state labeled A in the figure), and there is negligible current conduction. When the input voltage is raised to 5 V, the quasistatic switching states are represented by the intersection of the enhance-

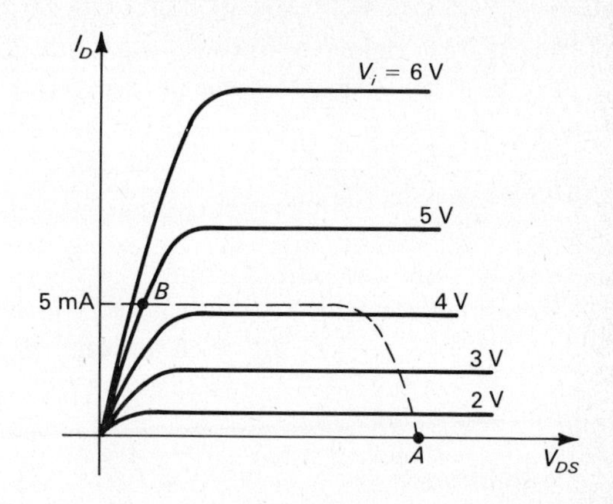

Figure 11.18 Current–voltage characteristics for the enhancement driver transistor T_D (solid curve) and the depletion transistor T_L load characteristic (dashed curve) of the logic gate of Fig. 11.17.

ment transistor characteristics (solid lines) and the depletion device load characteristic (dashed curve). Finally, the steady-state ON condition marked *B* in Fig. 11.18 is reached with a current of 5 mA conducted by both these devices. An analytic treatment of these states *A* and *B* will be presented next to determine the gate output voltage as a function of the device design parameters.

The *I-V* equations for NMOS transistors, as developed in Section 10.5.1, are for the driver and load devices

$$I_D^{(\text{driver})} = \beta_D[(V_i - V_{TD})V_o - \tfrac{1}{2}V_o^2] \qquad \text{(linear region)}, \tag{11.7a}$$

$$I_D^{(\text{driver})} = \tfrac{1}{2}\beta_D(V_i - V_{TD})^2 \qquad \text{(saturation region)}, \tag{11.7b}$$

$$I_D^{(\text{load})} = \beta_L[(-V_{TL})(V_{DD} - V_o) - \tfrac{1}{2}(V_{DD} - V_o)^2] \qquad \text{(linear region)}, \tag{11.7c}$$

$$I_D^{(\text{load})} = \tfrac{1}{2}\beta_L V_{TL}^2 \qquad \text{(saturation region)}, \tag{11.7d}$$

where $\beta_{D,L} = (C_{\text{ox}}\,\bar{\mu}Z/L)_{D,L}$ for the driver and load transistors, respectively; V_{TD} is the threshold voltage for the driver transistor; and V_{TL} is the negative pinch-off voltage needed to turn off the depletion-type load device. In the gate OFF state transistor T_D is operating in the saturation region and T_L in the linear region (point *A* of Fig. 11.18). In the gate ON state transistor T_D is operating in the linear region (point *B* of Fig. 11.18). In an intermediate state both T_D and T_L may be operating in their saturation regions. By setting the drain current of transistor T_D equal to that of T_L, the output voltage V_o can be obtained. The output voltage for state *A* can be shown to be

$$V_{oA} = V_{DD} + V_{TL} + [V_{TL}^2 - (\beta_D/\beta_L)(V_i - V_{TD})^2]^{1/2}, \tag{11.8a}$$

and for state *B*

$$V_{oB} = (V_i - V_{TD}) - [(V_i - V_{TD})^2 - (\beta_L/\beta_D)V_{TL}^2]^{1/2}, \tag{11.8b}$$

where of course the quantities whose square roots are taken must be positive. For a maximum separation of these two states (noise protection), β_D/β_L must be a large quantity. This is shown in the gate design illustrated in Fig. 11.17b.

Propagation delays of about 3 nsec are typical for NMOS logic gates. Fanouts of more than 20 are possible and the power dissipation per gate is about 1 mW.

EXAMPLE 11.4

Determine the output voltage of the gate of Fig. 11.17a in the ON (conducting) state, driven by a gate input voltage of 5.0 V given $V_{TD} = 0.8$ V, $V_{TL} = -1.6$, $t_{ox} = 0.05$ μm, $\bar{\mu} = 700$ cm^2/V-sec, $Z_D/L_D = 4$, $Z_L/L_L = 0.25$.

Solution

$$C_{ox} = \frac{\epsilon_0 \epsilon_{ox}}{t_{ox}} = \frac{(8.85 \times 10^{-14} \text{ F/cm})(3.9)}{0.05 \times 10^{-4} \text{ cm}}$$

$$= 6.9 \times 10^{-8} \text{ F/cm}^2.$$

From Eqs. (11.7)

$$\beta_D = Z_D \bar{\mu} C_{ox}/L_D = (4)(700 \text{ cm}^2/\text{V-sec})(6.9 \times 10^{-8} \text{ F/cm}^2)$$

$$= 1.9 \times 10^{-4} \text{ A/V}^2,$$

$$\beta_L = Z_L \bar{\mu} C_{ox}/L_L = (0.25)(700 \text{ cm}^2/\text{V-sec})(6.9 \times 10^{-8} \text{ F/cm}^2)$$

$$= 1.2 \times 10^{-5} \text{ A/V}^2.$$

Hence from Eq. (11.8b)

$$V_{oB} = (V_i - V_{TD}) - [(V_i - V_{TD})^2 - (\beta_L/\beta_D)V_{TL}^2]^{1/2}$$

$$= (5.0 - 0.8 \text{ V}) - \{[(5.0 - 0.8)^2 \text{ V}^2]$$

$$-(1.2 \times 10^{-5}/1.9 \times 10^{-4})(-1.6)^2 \text{ V}^2\}^{1/2}$$

$$= \underline{0.02 \text{ V}}.$$

F. CMOS TRANSISTOR LOGIC

The recent emphasis on complementary metal–oxide–semiconductor digital circuits stems from the negligible power dissipation in this form of logic circuitry for CMOS gates in the static condition; power is primarily dissipated only in the process of switching or changing state. In addition this type of circuitry offers high noise margin and large fanout capability. The recent emphasis on high gate packing density on the microchip has resulted in a high power dissipation density and the need for low power consumption digital circuit design. In addition the requirement for high speed switching circuits calls for large driving currents to charge gate capacitances in a minimum time. The advantages of CMOS transistor logic are to a certain extent negated by the extra chip area needed to electrically isolate the *n*-channel transistor from the *p*-channel transistor in the basic CMOS gate schematically sketched in Fig. 11.19a. However, rapid progress in the scaling down of basic transistor dimensions has made this type of circuitry feasible. A cross section of the CMOS logic gate structure is shown in Fig. 11.19b.

The operation of the CMOS logic gate can be described as follows: with V_i at ground potential (zero) and $V_{DD} = 5$ V, enhancement NMOS transistor T_n has no gate–source drive and is OFF. However, enhancement PMOS transistor T_p has a gate voltage of 5 V below its source and so is ON, providing a low drain-to-source impedance and bringing the output voltage V_o to nearly the supply voltage of 5 V. Now assume that $V_i = 5$ V. In this case transistor T_p has no gate–source drive and is OFF but the high gate–source potential of T_n

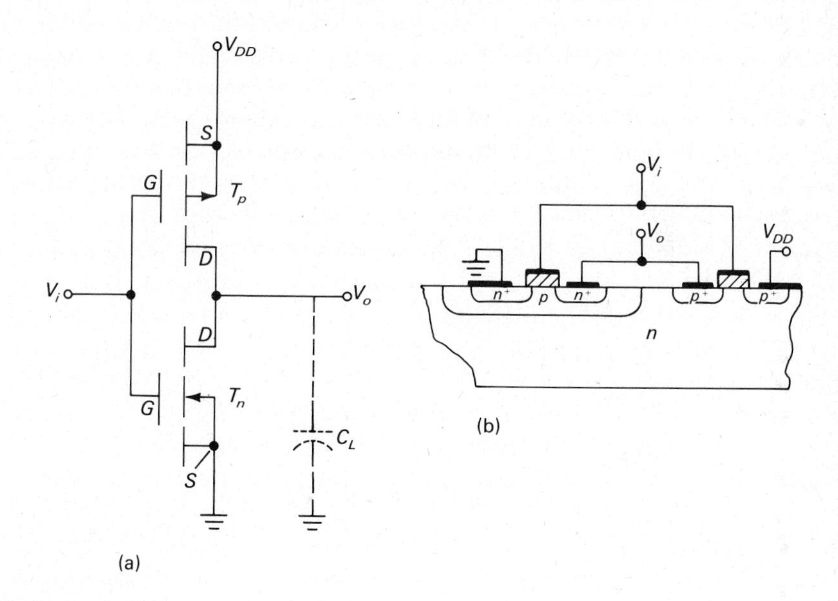

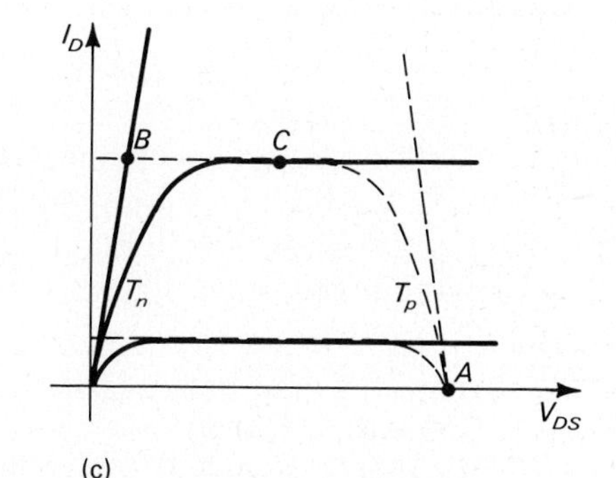

Figure 11.19 (a) Basic CMOS transistor logic gate. (b) CMOS *p*-well gate cross section. (c) Output characteristics of the *n*-channel transistor (solid curves) and *p*-channel transistor (dashed curves), plotted versus the channel drain–source voltage.

causes it to conduct, bringing the output voltage V_o to near ground potential. This assumes that the absolute values of the thresold voltages of T_n and T_p are less than 5 V, commonly about 1 V. With the input voltage halfway between the supply voltage and ground potential (2.5 V), both transistors are in saturation and the output voltage is about 2.5 V, if we assume that the gate is entirely symmetrical.[5] These operating points are represented in Fig. 11.19c by A, B, and C, respectively. In this graph the solid lines represent the output characteristics of the *n*-channel transistor and the dashed lines are those of the *p*-channel device. Because of the symmetry of this circuitry there is both active pullup and pulldown of the output voltage. Therefore the β's [see Eqs. (11.7)] of both transistors are generally designed to be about the same.

Since under high or low input conditions one of the two enhancement transistors is nonconducting, and since the output load of the logic gate is capacitive (the MOS gate capacitor of the next stage), current can only be conducted dynamically, during the charging of the load capacitance C_L. Under static conditions only the reverse bias leakage current of the drain junction of the OFF transistor is conducted, resulting in negligible static power dissipation. The energy needed to charge the load capacitance from 0 V to V_{DD} by this gate is $\frac{1}{2}C_L V_{DD}^2$. For a sinusoidal input to the logic gate the energy dissipated by the conducting *p*-channel device during a charging half-cycle is about $\frac{1}{2}C_L V_{DD}^2$, and similarly the *n*-channel device dissipates $\frac{1}{2}C_L V_{DD}^2$ during the next half-cycle in discharging the capacitive load. Hence the total energy dissipated by the logic gate in a full cycle is $C_L V_{DD}^2$ and the average power dissipated is the energy per cycle time T,

$$P = C_L V_{DD}^2/T = C_L V_{DD}^2 f, \tag{11.9}$$

where the switching frequency $f = 1/T$. Hence at low frequencies the power dissipation during dynamic switching can be very low for the CMOS logic gate. Propagation delays of about 6 nsec are typical for CMOS logic gates. Fanouts of up to 50 are possible owing to the large current driving capability of these gates. Although power dissipation in the nanowatt range is typical at low frequencies, milliwatts of dissipation can be incurred at frequencies in the megahertz range.

G. SOS TRANSISTOR LOGIC

In order to significantly increase the speed of CMOS logic gates and reduce power dissipation, the device area can be reduced. This will lower the transistor's gate and drain capacitances. Another way to reduce these capacitances is made possible by using the silicon-on-sapphire (SOS) microcircuit technology. The basic SOS logic gate is illustrated in Fig. 11.20. This is fabricated by depositing a thin (1–2 μm) epitaxial silicon layer onto a single-crystal sapphire sub-

[5]The gate is not basically symmetric in the sense that the electron mobility of the *n*-channel device is greater than the hole mobility of the *p*-channel transistor. However, the channel lengths and widths L and Z can be designed to make $\beta_p \approx \beta_n$ [see Eqs. (11.7)].

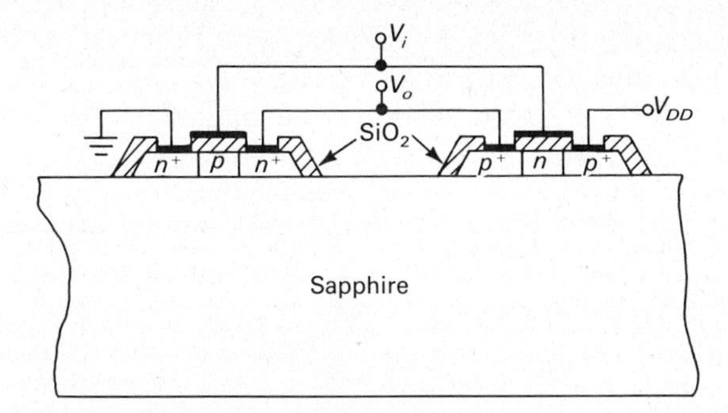

Figure 11.20 Basic SOS logic gate.

strate by pyrolysis of silane (SiH_4). Sapphire or spinel is used as a substrate since their lattice constants are not too far different from silicon and their dielectric constants are relatively low. This low dielectric constant relative to that of silicon helps reduce the isolation and junction capacitances and hence improves the switching speed of this version of CMOS circuitry. The structure shown in Fig. 11.20 is fabricated after the epitaxial deposit of silicon on sapphire by a series of impurity diffusions and etching steps. Isolation of the *n*-channel from the *p*-channel devices does not require an extra *p-n* junction well, which leads to reduced interdevice capacitance and reduced device area. The major disadvantages of SOS-CMOS technology is the high cost of the sapphire substrate and the expensive process of sapphire preparation to ensure the compatibility of the silicon and sapphire, which have a lattice mismatch.

H. DIGITAL SILICON IC SUMMARY

The fastest silicon digital logic microchips available today utilize ECL. An inexpensive chip which operates at moderately fast speeds and at moderate levels of power dissipation uses S-T^2L. I^2L technology offers the possibility of combining digital and analog functions on a single chip easily plus low speed–power product. The NMOS circuits provide the highest device packing density; CMOS circuits offer even lower power dissipation and the lowest speed–power product at low switching frequencies. Table 11.1 lists the propagation delay, power dissipation per gate, and the speed–power product, which represents the energy necessary for a switching operation.

11.6.2 Gallium Arsenide Integrated Digital Circuits

The quest for higher speed digital circuits to increase the capability of digital computers has led to the development of the faster silicon circuit technology just described. The development of smaller dimension devices generally means faster switching speeds but there is a fundamental limit to be reached here. Hence the present search for more suitable semiconductor materials for these microchips. Gallium arsenide is a compound semiconductor whose crystal

TABLE 11.1 Typical switching characteristics of silicon and GaAs integrated digital logic circuits.

Type of Circuitry	Propagation Delay per Stage (nsec)	Power Dissipation/Gate (mW)	Speed–Power Product (pJ)
T^2L	6	10	60
S-T^2L	2	10	20
ECL	1	20	20
I^2L	10	0.05	0.5
NMOS	3	1	3
CMOS	5	Frequency dependent	—
GaAs	0.06	0.5	0.03

growth technology has advanced to the point where it can be employed commercially for the production of semiconductor devices and microcircuit chips. However, this material is significantly more expensive than silicon, which is one of the most abundant elements present on Earth. But its superior electrical properties make GaAs the most likely material to compete with silicon as a substrate material for microchips at present.

The electron velocity in the channels of GaAs field-effect transistors is several times greater than in silicon MOSFETs. This means reduced carrier transit time and hence faster switching speed. In addition, semi-insulating GaAs substrates are readily available which provide for lower parasitic capacitances and hence faster logic circuit switching speeds. Other advantages of GaAs as a microchip material are the broad device operating temperature range (-200 to 200°C) possible and its radiation hardness (up to 10^8 rad).

The high electron mobility results from the small electron effective mass, as described in Section 5.2.1. The semi-insulating material is produced by introducing carrier traps which occur near the center of the GaAs energy band. The wide temperature range over which GaAs may be employed results from its large energy gap. Its radiation hardness results from the intrinsically low lifetime in this direct energy gap material which is difficult to reduce further by introducing recombination centers.

The depletion-mode metal–semiconductor field-effect transistor (D-MESFET) serves as the basic building block of today's GaAs integrated circuits. The operation of the MESFET was described in Section 10.3. The enhancement-mode MESFET (E-MESFET) and the high electron mobility transistor (HEMT, also described in Section 10.3) are less mature GaAs switching devices. The D-MESFET structure used is similar to that sketched in Fig. 10.4. However, the source and drain regions are formed by n^+-implants of 1×10^{17} atoms/cm^3 and are about 0.2 μm deep in order to reduce series source and drain resistances. The source and drain contacts are typically 3–4 μm apart and the gate metallization is 1 μm wide. A modern-day D-MESFET used in GaAs ICs is illustrated in Fig. 11.21A. The pinchoff voltage for this device is about 1 V. The transconductance of the GaAs D-MESFET is much higher than that of a comparable silicon device owing to the high electron carrier mobility and its input capacitance is lower. This leads to switching speeds as fast as 60 psec and a gain–bandwidth product of more than 15 GHz. The power dissipation per gate is about 0.5 mW, giving a speed–power product of 0.03 pJ. A comparison of the switching performance of GaAs and silicon digital logic gates is shown in Fig. 11.22.

D-MESFET GaAs integrated circuits typically require two power supplies with level shifting built into the logic gates to generate negative gate voltages required for switching from the positive drain voltages of the *n*-channel MESFETs. Enhancement-type MESFETs do not require dual power supplies and

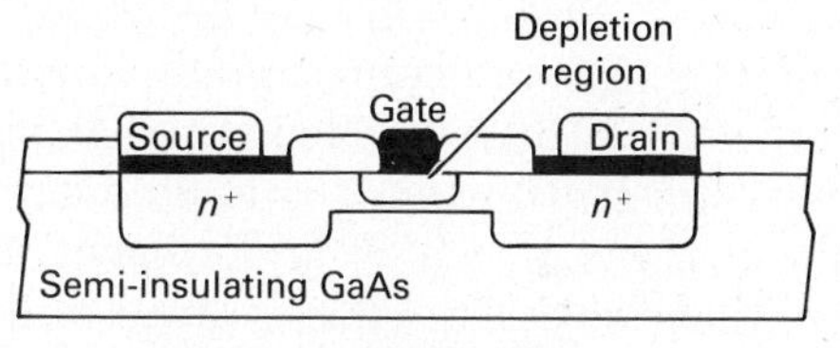

(a) Planar implanted D-MESFET

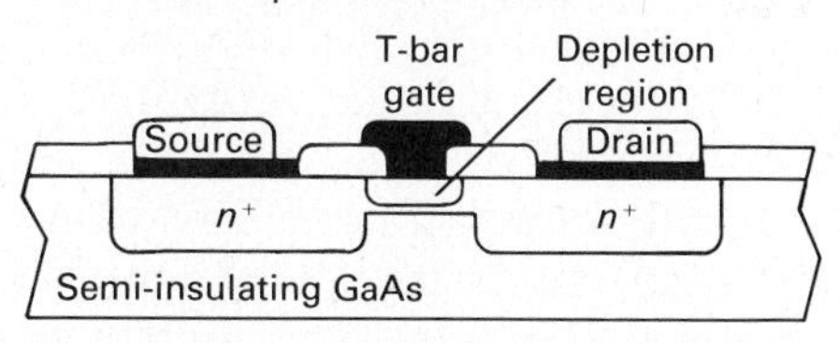

(b) Self-aligned gate E-MESFET

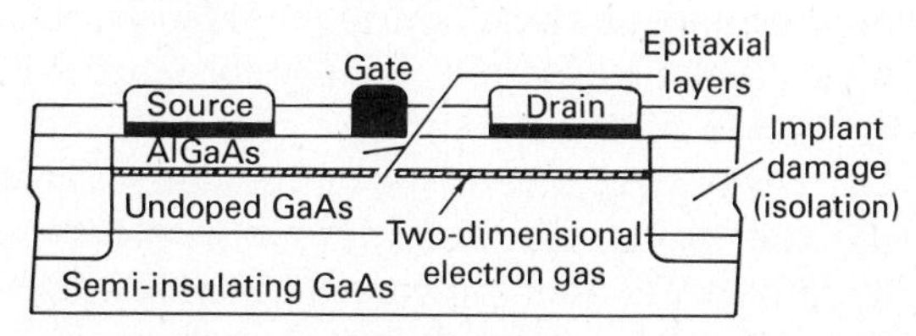

(c) Enhanced-mobility E-MESFET

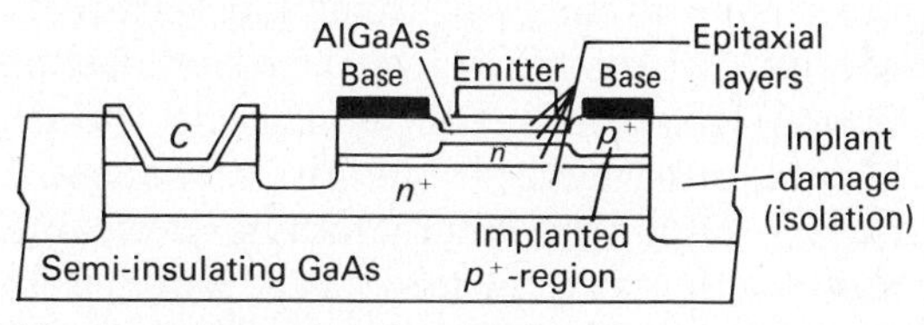

(d) Bipolar heterojunction transistor

Figure 11.21 Four types of GaAs transistor structures for use in ICs. [From R. C. Eden, A. R. Livingston, and B. M. Welch, "Integrated Circuits: The Case for Gallium Arsenide," *IEEE Spectrum,* **20,** 30–37 (1983).]

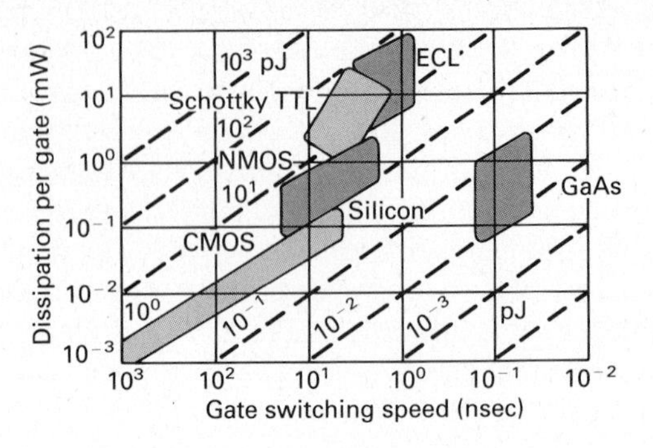

Figure 11.22 Comparison of the switching performance of silicon and GaAs digital integrated logic circuits. [Adapted from R. C. Eden, A. R. Livingston, and B. M. Welch, "Integrated Circuits: The Case for Gallium Arsenide," *IEEE Spectrum,* **20,** 30–37 (1983).]

level-shifting circuitry since a small (0.1 V), positive gate potential gives drain-to-source conduction. However, E-MESFETs are restricted to very low logic swings (0.5 V) and the degree of control of threshold voltage needed for reliable operation of large scale enhancement circuits is prohibitive. Although HEMT devices (see Fig. 11.21c) offer the potential for even faster switching circuits, this technology has not matured as yet. Similarly, bipolar heterojunction GaAs transistors (BHJT) also need additional development time. Four possible GaAs transistor structures for integrated circuits are sketched in Fig. 11.21. A comparison of the switching performance of various silicon digital ICs and GaAs integrated digital circuits is given in Fig. 11.22. The angled lines represent the power–speed product (energy) for switching in picojoules.

11.7 ANALOG INTEGRATED CIRCUITS

Analog integrated circuits form a class of microchip circuits which are used in a variety of signal processing applications. These are available as separate chips or are integrated onto digital circuit chips as auxiliary circuits. The most used analog integrated circuit is the direct-coupled high gain differential-input microchip operational amplifier, called an **op-amp.** Its function is to provide a large amplified output voltage proportional to a small difference voltage applied to its two inputs.

The design of this type of circuit uses some of the advantages provided by the monolithic technology. For example, the microchip batch processing technique permits many devices to be fabricated at once and hence at a low cost per component. A disadvantage of the monolithic technology is that only *small* resistors and capacitors can be efficiently fabricated on a single chip owing to the small resistance and capacitance available per unit area. Hence only direct-coupled circuits are practical to manufacture, which necessitates special stage biasing techniques. The inductor is seldom used as a circuit element again because of the large size of this component. Other analog circuits such as summers, integrators, differentiators, comparators, sense amplifiers, audio and video amplifiers, phase-locked loop circuits for frequency selection, and digital-to-analog and analog-to-digital converters are available.

A variety of analog circuit techniques are used to take advantage of the microchip technology as well as compensate for the unavailability of large-value capacitor and resistor circuit elements. The microchip linear operational amplifier is a basic element of many analog integrated circuits and hence its integrated circuit structure will be described here. It is not the purpose of this book to discuss the subject of analog integrated circuits in detail. Instead only those aspects of the design of the microchip op-amps will be described which illustrate the techniques dictated by the monolithic microchip technology.

A. THE INPUT STAGE

A typical microchip operational amplifier *input* circuit is shown schematically in Fig. 11.23. The design of the various elements of this circuit will be discussed next. The differential transistor circuit configuration is used for the input stage of the op-amp. The **diff-amp** uses a matched pair of transistors, Q_1 and Q_2, which are placed in close proximity to each other on the chip in order to provide for the accurate matching required. In addition, the two transistors will track accurately with temperature owing to their close proximity. The input signal applied is $V_1 - V_2$ and the output to the next stage is $V_{01} - V_{02}$.

B. CURRENT SOURCES

A constant **current source** is needed to supply a stable bias current for the diff-amp. A large resistor cannot be used for this purpose owing to the constraint of the integrated resistor technology. Instead a current source of the type shown schematically in Fig. 11.24 is generally used which provides a constant current, independent of temperature and voltage supply fluctuations. The integrated version of this current source occupies a smaller area on the chip and has a relatively low voltage drop across it to reduce the need for a high supply voltage. An expression will now be developed for the current supplied by this

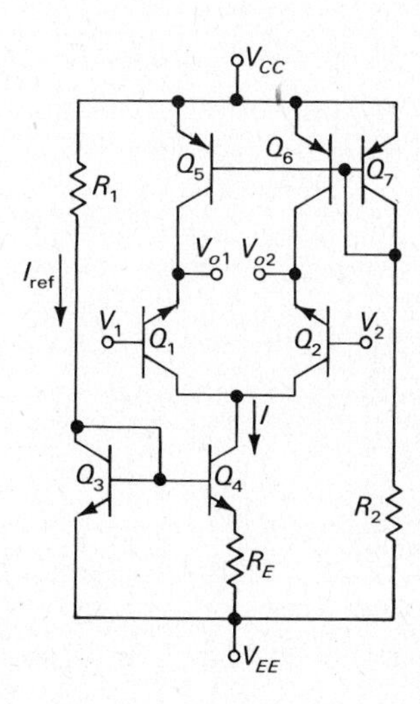

Figure 11.23 Schematic diagram for a typical input circuit of an integrated-circuit op-amp.

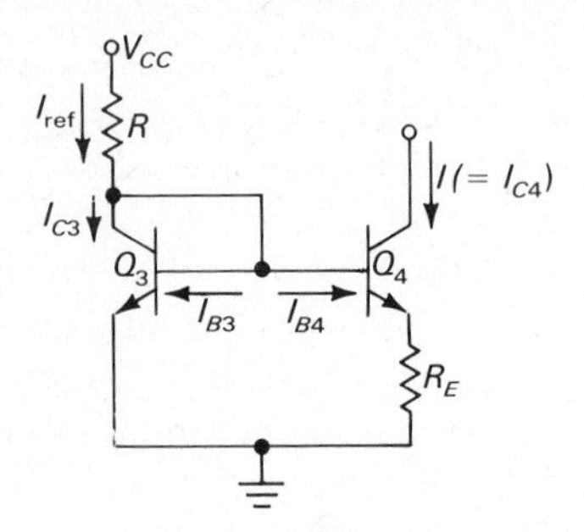

Figure 11.24 Schematic diagram of a basic integrated-circuit current source.

source in terms of design parameters. First assume $R_E = 0$. Transistors Q_3 and Q_4 are designed to be identical, so $I_{B3} = I_{B4} = I_B$, since $V_{BE3} = V_{BE4}$, and $I_{C3} = I_{C4}$, since $h_{FE3} = h_{FE4}$. Now $I_{ref} = (V_{CC} - V_{BE3})/R$ and by Kirchhoff's current law, $I_{ref} = I_{C3} + 2I_B$. Also, $I_{C3} = h_{FE}I_B$. Combining these relations we have

$$I_{C3} = \frac{h_{FE}I_{ref}}{h_{FE} + 2} = \frac{h_{FE}(V_{CC} - V_{BE3})}{R(h_{FE} + 2)}. \tag{11.10}$$

If h_{FE} is large and $V_{BE3} \ll V_{CC}$ ($V_{BE3} \approx 0.7$ V), then $I_{C3} \approx I_{ref} \approx V_{CC}/R = I_{C4}$. So the bias current source I is nearly constant, independent of the transistor parameters and temperature variations. (This would not be true for a single transistor used as a current source whose collector current is temperature dependent.) For example, if the temperature rises, I will tend to increase. However, the current through Q_3 also increases, which reduces V_{BE3} ($= V_{BE4}$), which offsets the increase in I. This circuit is generally called a **current mirror.**

An improved version of this circuit which occupies less microchip area for a given bias current level and which permits the integration of very low current sources for diff-amp input stages is the same circuit of Fig. 11.24, with $R_E > 0$. For high gain transistors, $I_E \approx I_C$ and

$$I \equiv I_{C4} \approx (V_{BE3} - V_{BE4})/R_E. \tag{11.11}$$

We may obtain a relationship between I_{C3} and I_{C4} for these identical transistors from the emitter junction diode law of Eq. (6.20) as

$$\frac{I_{C3}}{I_{C4}} = \exp\left(\frac{q(V_{BE3} - V_{BE4})}{kT}\right), \tag{11.12}$$

assuming high h_{FE}. Using Eq. (11.11) we then obtain

$$\frac{I_{C3}}{I} \approx \exp\left(\frac{IR_E}{kT/q}\right). \tag{11.13}$$

For $R_E = 0$, $I \approx I_{C3}$ as was previously found. The solution for I for nonzero values of R_E can be obtained by iterative trial and error techniques. However, if R_E is chosen so that the voltage drop IR_E is about 2.3 times the thermal voltage kT/q at 300°K, then $I_{C3}/I = 10$. Hence any value of bias current I less than I_{C3} can be obtained for a given reference current I_{C3}. This is especially important for the op-amp input circuit where the current level is low. Also it is clear that the resistor R of Fig. 11.24 can be substantially reduced in value if R_E is chosen to make I_{C3} ten times I. Since in general $R_E \ll R$ (because $V_{CC} \gg V_{BE3}$), then, for a given current source I the chip area occupied by this circuit with $R_E > 0$ will be smaller than for $R_E = 0$.

EXAMPLE 11.5

Determine the values of resistors R and R_E of Fig. 11.24 so that $I_{C3}/I = 10$. The supply voltage $V_{CC} = 30$ V and the bias current required $I = 10$ μA.

Solution

Taking the natural logarithm of Eq. (11.13) gives

$$R_E = \frac{kT}{qI} \ln\left(\frac{I_{C3}}{I}\right) = \frac{0.026\text{ V}}{10 \times 10^{-6}\text{ A}} \ln(10) \approx \underline{6000\ \Omega}.$$

$$R \approx (V_{CC} - V_{BE4} - IR_E)/I_{C3}$$

$$= [(30\text{ V}) - (0.7\text{ V}) - (10 \times 10^{-6}\text{ A})(6000\ \Omega)]/(100 \times 10^{-6}\text{ A})$$

$$= \underline{290\text{ k}\Omega}.$$

C. THE CURRENT SOURCE AS A LOAD RESISTOR

The gain of an amplifier stage generally increases as its load resistance increases. If the load resistance is increased, however, the voltage drop across it will be higher, necessitating a higher supply voltage and hence increasing the power dissipation. Also, large resistances occupy large areas on microchips. A current source used as a load provides a large resistance to current changes and hence provides a large effective resistance for the AC signals to be amplified. A diff-amp input stage utilizing a current source for bias stabilization, as well as for load devices, is shown in Fig. 11.23. There Q_1 and Q_2 serve as the input differential pair of transistors with the input voltages V_1 and V_2 applied. Transistors Q_3 and Q_4 in conjunction with resistor R form the bias current source for this differential amplifier pair. Transistors Q_5 and Q_6 along with diode Q_7 form the current source loads; Q_5, Q_6, and Q_7 are lateral *p-n-p* transistors since it is useful to tie all the emitters of current source transistors to AC ground. Also, transistors Q_5, Q_6, and Q_7 may be integrated as a single transistor with common base and emitter, but three collectors. This structure will occupy significantly less area.

D. LEVEL SHIFTING CIRCUIT

For proper circuit functioning the single-ended output voltage of an op-amp should be zero if the input voltage $V_2 - V_1$ is zero. Also, when $V_2 - V_1$ swings positive, the output voltage should rise relative to ground potential; when this quantity swings negative, the output voltage should go below ground potential. In general this will not be the case, as was discussed for the ECL circuit of Section 11.6.1C, which also has a differential pair input. DC level shifting can be obtained by using the level-shifting circuit shown in Fig. 11.25, which consists of an emitter follower-connected transistor with a current source load.

Transistors Q_2 and Q_3 along with resistors R and R_E form the current source load. The emitter follower transistor is Q_1 with a resistor R_1 chosen to ensure that with the diff-amp input signal equal to zero, the output $v_o = 0$, where $v_i = v_{DC}$ is some single-ended input voltage from the previous stage. The value of R_1 can be obtained by assuming that with no AC signal, $v_i = V_{DC}$, $i_L = 0$, and hence

$$V_{DC} - v_{BE1} - IR_1 = 0$$

or

$$R_1 = (V_{DC} - v_{BE1})/I, \tag{11.14}$$

where I is the value of the source current. When v_i is lowered such that transistor Q_1 is nonconducting, $i_L = -I$. With Q_1 made conducting by raising v_i correspondingly, the load current $i_L = I$. Hence the load current swings symmetrically about zero.

E. CLASS B OUTPUT STAGE

The schematic diagram for a typical integrated op-amp output stage is shown in Fig. 11.26. Transistors Q_1 (n-p-n) and Q_2 (p-n-p) comprise a **push–pull** output circuit. In this Class B amplifier transistor Q_1 and Q_2 will alternately conduct during each half-cycle of a sinusoidal signal; hence the term "push–pull" amplifier. Transistors Q_3 and Q_4 and resistors R_{E1} and R_{E2} form a protective

Figure 11.25 A DC level-shifting circuit.

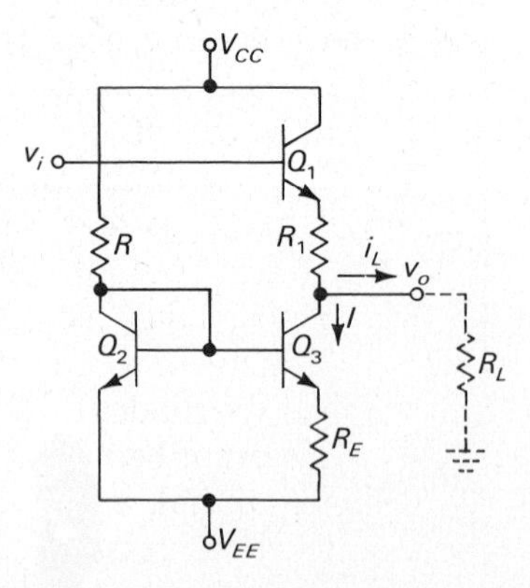

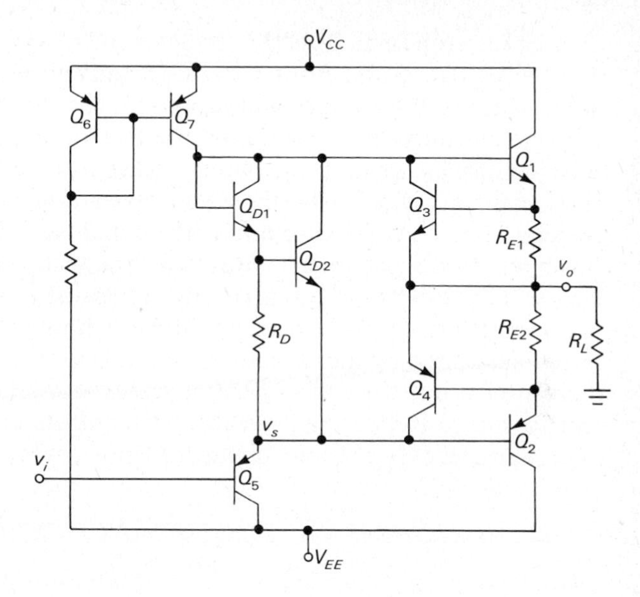

Figure 11.26 Integrated-circuit Class B amplifier output stage.

circuit should the load R_L be accidentally short circuited. If this occurs, the current through Q_1 and Q_2 could become destructively large. However, this current will provide a voltage drop across the *small-value* emitter resistors R_E, turning on transistors Q_3 and Q_4 which shunt off the smaller base currents of Q_1 and Q_2; this causes their collector current to decrease, in a negative feedback action. Since R_{E1} and R_{E2} have low values their influence will be inconsequential in normal circuit operation.

Consider a sinusoidal input voltage v_i applied to transistor Q_5 which results in a signal voltage v_s $(=v_{BE2}$, see Fig. 11.26). For the moment take transistor Q_{D2} to have a collector-to-emitter short circuit. Then v_s will drive the bases of Q_1 and Q_2 directly. Assume that the input voltage v_i swings around ground potential or zero. Then, since the two supply voltages V_{CC} and V_{EE} are equal in magnitude but opposite in sign, by symmetry the quiescent output voltage v_o will be zero. When v_s goes positive, transistor Q_1 will conduct and supply current to the load R_L and Q_2 will be cut off; if v_s goes negative Q_2 will conduct and Q_1 will be cut off.

The function of the circuit consisting of the diode Q_{D1} and the emitter–base diode of Q_{D2} is to minimize **crossover distortion.** This is caused by the lack of significant output current when the emitter–base voltages of Q_1 and Q_2 are driven by less than a diode voltage drop (0.6 V). Transistors Q_6 and Q_7 and resistor R constitute a constant source current circuit. This supplies a bias current through the diodes of Q_{D1} and Q_{D2} which maintains a potential difference of two diode drops between the bases of Q_1 and Q_2. Since the base of transistor Q_1 is above ground, even with zero input signal, Q_1 will conduct making v_o positive; this v_o increase also raises the potential of the emitter of transistor Q_2

causing it to conduct, shunting current from R_L and lowering v_o. The potential of v_o will settle to a small voltage above ground potential. Now as the signal goes positive, Q_1 will conduct more current and drive the load R_L in the emitter follower mode; when the signal goes negative Q_2 will similarly drive the load. Although the auxiliary circuitry including Q_{D1}, Q_{D2}, Q_3, and Q_4 fill space on the microchip, the former three devices are common collector connected, can be placed in one isolation box, and hence occupy correspondingly less area. The *n-p-n* transistors are of a vertical type, while the *p-n-p*'s Q_6 and Q_7 are of a lateral variety; *p-n-p*'s Q_2 and Q_5 can be vertical since their collectors are connected to the substrate, the lowest circuit potential.

In summary, a typical op-amp will contain the elements just described: a differential input stage plus perhaps a second diff-amp stage for additional gain, current sources, voltage level-shifting circuits, an output stage, and perhaps a load short-circuit protection circuit.

*11.8 LIMITATIONS ON MICROMINIATURIZATION

There have been significant advances in the information-handling capability of microchips in recent years. The device density in these integrated circuits has increased, the gate propagation delays have been reduced, and the chips have increased in area, constituting a larger portion of a computer. However, there are basic limitations to this progress. These are provided by the following factors:

1. Device size reduction and hence speed limitations.
2. Semiconductor material limitations.
3. Manufacturing chip yield and hence economic limitations.
4. Maximum power dissipation density or thermal limitations.
5. Reliability limitations.

A. SIZE LIMITATIONS

Microchip device geometries are now defined by photolithographic techniques. A fundamental law of optics states that the limit of resolution of a lithographic process is of the order of the wavelength of radiation used to define an image. Hence the recent emphasis on utilizing deep ultraviolet light ($\lambda \backsim 0.2$–$0.3\ \mu m$) to delineate small microstructures on chips. This has made possible the fabrication of integrated device structures with linewidths of $0.5\ \mu m$ and less. However, these linewidths will vary depending on the accuracy of their edge definition, according to a normal distribution. The edge variational tolerance will be of the order of the wavelength of the radiation used to establish this pattern. Hence X-ray ($\lambda \backsim 1$ Å) and electron ($\lambda < 0.1\ \mu m$) radiation will be needed to define even smaller microstructures. However, scattering by the substrate of these high energy radiations restricts the definition to considerably greater than the wavelength. Also, the small throughput of these procedures significantly increases the cost of chip manufacture.

Of course in the fabrication of an integrated circuit not only the edge definition is of importance, but also the number of device structures, which must *all* be made within a given tolerance to produce a properly functioning electronic circuit. Hence small structure definition plus the total number of logic gates per chip (area of the chip) determine the manufacturing yield of a microchip.[6]

B. SEMICONDUCTOR MATERIAL LIMITATIONS

Silicon and gallium arsenide are the most commonly used semiconductor crystals presently employed as substrate materials for the fabrication of integrated-circuit microchips. Since the silicon crystal growth process is more mature than that of gallium arsenide, the former material is by far the most utilized starting material at this time. Material parameters and characteristics which limit the performance of integrated circuits are

1. The limiting (or peak) velocity of current carriers.
2. The avalanche breakdown voltage.
3. The high and low limits of electrical conductivity.
4. The thermal conductivity.
5. The susceptibility to particle and nuclear radiation.
6. The insulation properties of chemical compounds of the material.

The high electron peak velocity of about 2×10^7 cm/sec in GaAs is about twice that in silicon. In addition, the mobility of electrons in GaAs is about five times that in silicon. This minimizes the carrier transit time and provides the possibility of improving transistor switching speed.

Submicrometer channel length MOSFETs must have heavily doped substrates to avoid voltage punchthrough. However, since the electric field in the channel depends on the drain voltage divided by the channel length, a high electric field is present which can give rise to carrier multiplication. Hence a high avalanche breakdown voltage is important for short channel MOSFETs as well as narrow-base transistors.

High electrical conductivity of semiconductors is needed to minimize electrical series resistances in devices such as the base resistance of a bipolar transistor or the source resistance of a MESFET. Low electrical conductivity is, however, necessary for example in fabricating MESFETs onto semi-insulating substrates.

High device density microchips containing many submicrometer compo-

[6]For a fuller discussion, see J. T. Wallmark and S. M. Marcus, "Minimum Size and Maximum Packing Density of Nonredundant Semiconductor Devices," *Proc. IRE,* **50,** 286–298 (1962).

nents are usually operated at high current density in order to speed up digital switching performance. This causes significant thermal effects. For efficient heat removal from microchips the semiconductor must have a high thermal conductivity. Silicon has a very high thermal conductivity for semiconductors (1.5 W/cm-°C), about three times greater than that of GaAs (and about one-third that of copper).

Susceptibility to radiation such as α-particles and other types of nuclear radiation limits the operation of microchips in such hostile environments. Even one cosmic ray burst a day can produce soft errors in digital computing devices, reducing their reliability. Irradiation can introduce deep-lying energy levels in the energy gap of a semiconductor material which can reduce minority carrier lifetime as well as modify the electrical conductivity. GaAs is more radiation resistant than silicon.

Finally, the ability of the semiconductor to chemically form insulating compounds is important. For example, thermal oxidation of Si to SiO_2 is critical in the fabrication of MOSFETs. Even in the fabrication of bipolar device circuits SiO_2 and Si_3N_4 are used for surface passivation, device isolation, and insulation between interconnecting metal layers. GaAs unfortunately does not form a stable insulating oxide or nitride. Instead SiO_2 or Si_3N_4 may be deposited onto GaAs substrates for insulation. However, an insulator is not an essential element of the GaAs MESFET.

C. MANUFACTURING CHIP YIELD

For the economical manufacture of microchips, a high percentage yield of circuits operating within specification limits must be obtained. Since most semiconductor wafers processed contain some defects, either basic crystal defects or processing-induced defects, these determine the yield of saleable chips. Naturally the larger the chip size manufactured, the greater the possibility for containing one or more chip defects and the lower the production yield. An expression which has been found useful in predicting the fractional yield Y of microchips on a semiconductor wafer having an average defect density D and an active device area A is

$$Y = (1 + \gamma DA)^{-1/\gamma}, \tag{11.15}$$

where γ is found empirically to vary from $\frac{1}{3}$ to 1.

D. POWER DISSIPATION MAXIMUM

All semiconductor devices are constrained to operate below a temperature at which the semiconductor material becomes "intrinsic" and p-n junctions can no longer exhibit a high impedance owing to excessive leakage. This maximum temperature is determined by the thermal hole–electron excitation rate across the energy gap. Hence high energy gap (1.42 eV) GaAs devices can operate at higher temperatures than Si (1.12 eV) devices. In the process of achieving high device density chips operating at high currents to obtain fast switching speed, a high power dissipation density is inevitable. Hence low power density circuitry and proper thermal heat sinking of the microchip are essential for the reliable operation of high efficiency digital circuits.

It is difficult to conceive of semiconductor bipolar transistors that operate with junction voltage swings of less than a few thermal voltages, say $2kT/q$ {p-n junctions lose their nonlinear behavior at voltages less than kT/q [see Eq. (6.20)]}. The minimum emitter current for such a transistor is $2kT/qr_e$, where r_e is the dynamic emitter resistance. The minimum bipolar switching transistor power dissipation is given by the product of the minimum voltage swing and the collector current (approximately equal to the emitter current). That is,

$$\begin{aligned}(P_{\text{diss}})_{\text{min}} &= V_{\text{min}} I_{\text{min}} \approx \frac{(2kT/q)^2}{r_e} \\ &= 2\pi (2kT/q)^2 C_e f_{\text{co}}.\end{aligned} \tag{11.16}$$

Here the cutoff switching frequency $f_{\text{co}} = 1/2\pi r_e C_e$, where the switching speed is considered to be limited by the time necessary to charge the emitter junction capacitance C_e. Finally we can write

$$(P_{\text{diss}})_{\text{min}} = 1/2\pi f_{\text{co}} \approx 2[\tfrac{1}{2} C_e (2kT/q)^2]. \tag{11.17}$$

Since $1/2\pi f_{\text{co}}$ is the minimum switching time and $\tfrac{1}{2} C_e (2kT/q)^2$ is the energy necessary to charge the emitter junction capacitance, the power–speed product for a transistor switch is a constant, proportional to the energy necessary to charge the device input capacitance. Hence devices which can reliably perform the switching function while requiring a minimum energy to do so will generally exhibit the highest power–speed product.

E. RELIABILITY LIMITATIONS

It has already been mentioned that soft errors in computers can be induced even by rare bursts of nuclear radiation on otherwise reliable chips. However, other reliability problems exist even in the solid-state devices that constitute the most essential components of computers. This is particularly of concern for the more complex chips such as microprocessors. In these complex chips many layers of interconnection metallization may be necessary, which increases the complexity of the fabrication process. Defects in the insulating layers which separate these metallization layers may cause short circuits after some hours of operation. Defective gate oxides in MOSFETs may fail when subjected to high electric field stress. Welded connections to lead wires are notorious sources of circuit unreliability.

A fundamental failure mechanism of narrow device-interconnecting metallization patterns is called **electromigration.**[7] This phenomenon is particularly important when the linewidth of these metallizations approaches the submicrometer range. In these thin, narrow metal lines the electric current den-

[7]For a study of this phenomenon, see J. R. Black, "Physics of Electromigration," *IEEE Trans. Electron Devices,* **ED-16,** 338 (1969).

sity can become exceedingly high. Then the conducting electrons in the metal will impact the metal atoms, actually physically displacing them. It is found that the metal atoms are always displaced in the direction of the most positively charged electrode,. This can leave a metal void (open circuit) near the negative polarity region. The mean-free time-to-failure by electromigration depends on the square of the current density and is an exponential function of temperature. Hence the importance of this phenomenon in high density, high speed microcircuits. The recent emphasis on the use of high atomic weight metals for interconnection patterns is due to the higher current densities needed to move these heavy atoms.

Since batch processing used in the fabrication of microchips significantly reduces the cost per device, the possible use of redundant circuits offers a possibility to increase chip reliability. Parallel circuit redundancy may be used when "open" failures are dominant; series redundant circuits are helpful when the primary failure is by short circuit.[8]

*11.9 COMPUTER-AIDED DESIGN OF INTEGRATED CIRCUITS

Recent progress in the development of complex microcircuits on a single chip of semiconductor material has been made possible not only by the significant advances in device and materials technology, but also through the extensive use of computers. The time-saving aspects of computer use makes the employment of these machines essential for the economical design and development of advanced computers-on-a-chip or large-scale semiconductor memory devices. Computer-aided design (CAD) is used in microchip device layout design, semiconductor processing modeling, device design, static logic modeling, transient circuit analysis and propagation delay simulation, and in the specification of testing procedures for the completed circuits.

The sequence of events that go into the development of a set of glass plate masks for a new digital circuit microchip are

1. Logic specification.
2. Circuit design.
3. Process design.
4. Device designs.
5. Chip layout for the specification of a set of fabrication masks.

Because of the extreme complexity of modern-day digital integrated circuits and the need of microcircuit manufacturers to develop expeditiously a wide variety of advanced ICs, computers are employed for each of these steps.

[8]For a review of redundancy techniques, see "Redundancy in Digital Systems," F. A. Inskip, *Electron. Eng.*, **38,** 244 (1967).

A. SPICE CIRCUIT ANALYSIS

Once the logic design has been sketched out, static logic level analysis is carried out to determine the potentials at the various circuit nodes and the currents in the different circuit branches. The software for this analysis is most often a circuit solving program called SPICE.[9] This is a general-purpose circuit simulation program capable of DC, nonlinear transient, and linear AC analyses. The circuits may contain resistors, capacitors, inductors, mutual inductors, independent voltage and current sources, four types of dependent sources, transmission lines, and semiconductor devices such as diodes, bipolar junction transistors, JFETs, and MOSFETs.

SPICE has built-in models for the semiconductor devices so that the user need specify only the pertinent model parameter values. For example, for MOSFETs either the electrical parameters can be specified such as the zero-bias gate threshold voltage, the transconductance parameters, and the zero-bias capacitances, or else such processing parameters as substrate doping, oxide thickness, and channel length. In the latter case the electrical parameters will be automatically calculated from the processing data. The Shickman–Hodges field-effect transistor circuit model is computed from these data. For bipolar junction transistors the Gummel–Poon integral charge-control model is used (see Section 8.4.1); however, if the Gummel–Poon parameters are not specified the model reduces to the simpler Ebers–Moll model. Device parameters such as junction leakage saturation current, ideal maximum forward current gain, current for forward current gain rolloff, and zero-bias depletion-layer capacitance must be specified.

Once the circuit configuration of the transistors, diodes, resistors, capacitances, etc., are decided upon this arrangement of circuit elements is entered into the computer. This is done by specifying the circuit values of the various components as well as their circuit locations. The information is entered into the computer by numbering the various circuit nodes and identifying the element values between two numbered nodes for a resistor or capacitor, and three nodes for a transistor. When the circuit has been completely specified and the supply and input voltages given, the DC voltage values at each node and the current in each branch of the circuit are obtained. In addition, a pulse input may be specified and the transient response of the circuit ascertained. In principle, circuits including thousands of transistors may be analyzed in this way.

B. SUPREM PROCESS MODELING

For the advanced development of new integrated devices and circuits, it is necessary to determine the appropriate impurity distribution and junction locations in the semiconductor. A commonly used computer-aided design program

[9] Simulation Program for Integrated Circuit Evaluation, developed at the University of California, Berkeley. A magnetic tape containing this software is available from this university.

for the purpose is SUPREM.[10] This computer algorithm permits the user to model the processing steps used in fabricating silicon integrated devices and circuits. The various processing steps modeled by this program are impurity deposition and drive-in cycles in diffusion, oxidation of silicon and silicon nitrides, ion implantation, epitaxial growth of silicon, and low temeprature deposition and etching of various materials. SUPREM is a one-dimensional process modeling program, but recently a two-dimensional version called SUPRA has been developed.

SUPREM simulates the changes that result in a device structure as a result of the various processing steps used in fabrication. The program calculates the thicknesses of the different layers of materials that are formed in the structure as well as the distribution of impurities within these layers. The high temperature diffusion of boron, phosphorus, arsenic, and antimony in silicon can be modeled given the temperature, time, gas concentration of these impurities, and oxidation conditions. This program calculates the spatial impurity distribution as a result of a sequence of processing steps, including segregation into the oxide. SUPREM also computes the sheet resistance of all diffused regions in the semiconductor. The dose and energy of implanted impurity atoms must be specified to determine their spatial distribution in the process of ion implantation. These simulations are carried out by the numerical solution of nonlinear partial differential equations governing the physics of these processes.

C. DEVICE MODELING

In pursuing research connected with the development of new types of device structures for use in advanced ICs, computer programs are available which calculate the current flow and potential distribution in these devices. A few details of this type of modeling were given in Section 8.4.2. This computation can be done under a variety of DC bias conditions as well as to predict transient behavior. Popular computer codes employed for this purpose are FIELDAY, MINIMOS, and MAGNUMS.[11] FIELDAY, a FInite-Element Device AnalYsis program developed by the IBM Corporation, simulates the operation of semiconductor devices under steady-state or transient conditions. Bipolar

[10]Stanford University Program for integrated-circuit PRocess Modeling, described in R. W. Dutton et al., "Correlation of Process and Electrical Device Parameter Variations," *IEEE J. Solid State Circuits,* **SC-12,** 349–355 (1977).

[11]See E. M. Burturla et al., "Finite-Element Analysis of Semiconductor Devices: the FIELDAY Program," *IBM J. Res. Dev.,* **25,** 218–231 (1981). Also see S. Selberberr et al., MINIMOS—A Two-Dimensional MOS Transistor Analyzer," *IEEE Trans. Electron Devices,* **ED-27,** 1540–1550 (1980). MAGNUMS is a bipolar and field-effect transistor modeling including thermal effects developed by the MAssachusetts Group for NUmerical Modeling of Semiconductors, University of Massachusetts, Amherst. See S. P. Gaur and D. H. Navon, "Two-Dimensional Current Flow in a Transistor Structure under Nonisothermal Conditions," *IEEE Trans. Electron Devices,* **ED-23,** 50–57 (1976).

junction transistors, field-effect transistors, and other devices can be analyzed. FIELDAY is a general purpose computer program which numerically solves the semiconductor electrical carrier transport equations at a given temperature in one, two, or three dimensions using the finite-element mathematical technique. MAGNUMS uses the finite-difference method.

The input data to this program are the semiconductor material parameters, device geometry, and the impurity doping profile throughout the device structure. The impurity distribution may be determined by using SUPREM. In addition, information must be supplied about the carrier mobilities, minority carrier recombination parameters, and avalanche multiplication coefficients. The hole and electron mobilities must be specified as a function of impurity concentration and electric field. The minority carrier lifetime in silicon depends on the number of recombination centers and their capture cross section as well as the recombination process (see Section 5.4.1).

The solution of the semiconductor equations yields values of the electric potential, and the hole and electron carrier densities at every point of the device structure for one set of electrode potentials. Now the electric field can be calculated from the gradient of the electric potential; the diffusion and drift components of the current density can be obtained by computing the gradients of the carrier densities and electric potential, respectively [see Eq. (5.22)]. Now the electrode potentials (base and collector for a BJT) can be varied and the base and collector currents recalculated. In this way a complete current–voltage characteristic can be computed for the device, be it a BJT, MOSFET, or Schottky diode. Solution of the heat flow equation in addition gives the temperature everywhere in the structure.[11] Computer CPU times of a few minutes are needed to solve this problem numerically, using relatively large digital computer systems (e.g., the CDC Cyber 175 computer).

D. COMPUTER-AIDED CHIP LAYOUT

A number of manufacturers supply systems which aid in the design and layout of circuits on a microchip. Both the hardware and software required to design the circuit topology are supplied. A user-oriented design language called XYMASK was devised for this purpose by scientists at the Bell Laboratories. The generation of the artwork is initiated by calling forth particular device configurations previously designed and stored in the computer. A CRT is used to display this pattern and it can be positioned in the layout with the aid of a light pen. When an appropriate design is completed, all the information is digitized in a form useful for generating the artwork needed for mask making. These data are stored on a magnetic tape and used as input to a coordinatograph which cuts a pattern in Rubylith, a sheet of clear Mylar laminated with a red overlay. The red material is removed in certain areas to yield a pattern which is either transparent or opaque when imaged on an appropriate high resolution photographic plate. Alternatively the tape may be used to guide a laser or electron beam, which can define the pattern in a photosensitive material on a glass

Engineer engaged in computer-aided chip-layout design. (Courtesy of Analog Devices Semiconductors, Wilmington, MA.)

plate. A computer-aided design system for MOSFET artwork was reported by Mattison.[12]

The testing of finished IC microchips is a formidible problem considering that measurements must be made at 40 or more output terminals and that the chip may be used in a variety of circuit applications. A significant amount of work on computer-aided chip testing algorithms has been reported. In fact aspects of the chip internal design are often dictated by the ability to provide simple final electrical testing procedures for the chip.

□ *SUMMARY*

The design principles and technology of integrated circuits which make microchips the primary electronic components in use today were discussed. The properties of silicon and gallium arsenide that cause these semiconductors to be the most popular materials today for the fabrication of the IC chips were analyzed. Impurity diffusion and ion implantation technologies for selectively introducing doping atoms into semiconductor materials were described. Then a discussion was presented of the manner in which the various IC device components differed from their discrete versions. The different types of silicon digital electronic gates and the basic elements of analog ICs were discussed. The forms of GaAs ICs were introduced. Next an analysis was given of the fundamental limits on microchip component density and the switching speed of digital devices used for computational purposes. Factors such as size definition, material characteristics, manufacturing yield, thermal limitations, and reliability were considered. Finally, computer-aided methods for microchip design were described. The program SPICE is used for microcircuit analysis, SUPREM for process simulation, FIELDAY for device design, and XYMASK as a chip layout aid.

[12]R. L. Mattison, "Design Automation of MOS Artwork," *Computer,* **7**, 21 (1974).

PROBLEMS

11.1 Find approximately how many 25-mm^2 integrated-circuit chips can be obtained from a slice of a silicon crystal 12 cm in diameter.

11.2 Using the data presented in Figs. 11.3, 11.4, and 11.5, determine by calculation whether successful *n*-pocket isolation is obtained by an error-function boron diffusion through a 3-μm-thick *n*-type silicon epitaxial layer containing 5×10^{15} donor impurity atoms/cm^3 grown onto a *p*-type silicon substrate, if the diffusion takes place
(a) at 1200°C for 1 h,
(b) at 1100°C for 1.5 h.
(*Hint:* For successful isolation diffusion the density of boron impurities which penetrate the 3-μm-thick epitaxial layer must be greater than the density of *n*-type impurities everywhere in the penetrated region of that layer, where conversion to *p*-type is necessary.)

11.3 Determine the depth below the surface of a *p-n* junction produced by the Gaussian diffusion of boron into the *n*-type epitaxial layer of Problem 11.2. The drive-in diffusion is carried out at 1100°C for 30 min after a boron surface deposition of 1×10^{12}/cm^2.

11.4 Boron is found to diffuse about 350 times more slowly in silicon dioxide (SiO_2) than in silicon between the temperatures 1100 and 1300°C. Using the data shown in Fig. 11.3, calculate roughly the minimum thickness of oxide required for masking against a 1-h boron diffusion at 1200°C by estimating the diffusion length $\sqrt{Dt}$ of boron in SiO_2. (*Hint:* Assume that the minimum thickness is three times the boron diffusion length.)

11.5 Determine the depth below the surface of a *p-n* junction produced by the ion implantation of boron into the epitaxial layer of Problem 11.2 at an implant energy of (a) 100, (b) 200 keV.

Boron Ion Energy (keV)	Range R_p (μm)	Straggle ΔR_p (μm)
100	0.30	0.07
200	0.50	0.09

The incident boron ion flux is 1.5×10^{15} ions/cm^2.

11.6 Compare the relative effectiveness of three device isolation methods by calculating the unit-area capacitance of
(a) a p^+-n junction,
(b) SiO_2,
(c) sapphire.

The donor doping in the p^+-n junction is 5×10^{15}/cm^3. Use an insulator thickness of 1.0 μm in parts (b) and (c).

11.7 The average resistivity of the diffused *p*-layer used to fabricate a monolithic integrated resistor is 0.1 Ω-cm and the depth of this layer is 1.5 μm. Assuming that the width of the diffused resistor is limited to no less than 2.0 μm by the ability to etch a narrow line in the oxide (caused by the basic resolution of the photoresist)

(a) Calculate the length of resistor line required to produce a 10,000-Ω resistor.

(b) If the change of this diffused resistor value with temperature is 3000 PPM/°C, by what percentage will the resistor value change for a temperature rise of 10°C?

11.8 **(a)** Determine the percentage chip surface area of a 5.0-mm^2 chip occupied by a monolithic integrated capacitor whose value is 500 pF. Assume that the active dielectric material is a silicon dioxide layer whose thickness is 0.10 μm.

(b) Calculate the breakdown voltage for this capacitor. The dielectric breakdown voltage for SiO_2 is 6×10^6 V/cm.

11.9 Refer to the standard T^2L logic NAND gate of Fig. 11.13a. Consider that the gate inputs V_1 and V_2 are high, 3.6 V relative to ground.

(a) Determine the current in the resistor R_1.

(b) Determine the current in R_{E2}.

11.10 Refer to the standard ECL OR-NOR logic gate of Fig. 11.14a. Take the reference voltage V_R as -1.17 V. Calculate the current in resistor R_E assuming both inputs V_1 and V_2 are low, at -1.325 V.

11.11 The I^2L logic gate sketched schematically in Fig. 11.15a drives a similar gate and is driven by a gate similar to that shown in Fig. 11.16a. When the logic gate is ON, the average power dissipation is 50 μW.

(a) Calculate the current drawn by this gate when it is ON.

(b) What is the total voltage swing of the output of this gate from the OFF to the ON state?

11.12 Refer to the NMOS transistor logic gate of Fig. 11.17. Assume that this gate operates with $V_{DD} = 5$ V and is driven ON by a similar gate. Using the transistor designs indicated in Fig. 11.17b and Example 11.4, calculate the average static power dissipated in this gate.

11.13 Refer to the CMOS transistor logic gate of Fig. 11.19 with $V_{DD} = 5.0$ V, driving a capacitive load $C_L = 25$ pF. Find the average power dissipated in switching this gate at a clock frequency of

(a) 10 kHz,

(b) 10 MHz.

11.14 Refer to the current source shown schematically in Fig. 11.24. Taking $V_{CC} = 15$ V, $R = 10$ kΩ, and $R_E = 1$ kΩ, compute

(a) the reference current I_{ref},

(b) the source current I.

11.15 Using the expression of Eq. (11.15) with $\gamma = \frac{1}{2}$, find the wafer yield in percent of IC chips, where the average wafer defect density is 10/cm^2, for

(a) a chip with an active device area of 5 mm^2,

(b) a chip with an active device area of 25 mm^2.

11.16 The defect density for the wafers of Problem 11.15 is reduced to 5/cm^2. Repeat the calculations of Problem 11.15 for these improved wafers.

11.17 Given that the switching speed of a bipolar transistor is approximately limited by the time necessary to charge its emitter capacitance to a voltage of $2kT/q$,

(a) Estimate the minimum power dissipation needed to operate such a transistor at an emitter current of 100 μA.

(b) Compute the power–time delay product for this transistor if the emitter capacitance including parasitics is 0.04 pF.

(c) Repeat (b) if the emitter voltage required for switching is 0.6 V.

11.18 Calculate the "intrinsic" temperature for

(a) silicon doped with 1×10^{15} donors/cm^3,

(b) gallium arsenide with 1×10^{16} donors/cm^3.

The intrinsic temperature is defined as the temperature at which the intrinsic carrier concentration equals the impurity concentration. [*Hint:* Use Eq. (4.30).]

11.19 Using SPICE, determine the electric potentials at all the nodes of the ECL circuit of Fig. 11.14a as well as all the branch currents in this gate for

(a) an input $V_1 = -1.015$ V,

(b) $V_1 = V_2 = -1.325$ V.

11.20 Using SUPREM, obtain the diffusion profile of acceptors on diffusing boron into a thick 1.0-Ω-cm n-type silicon epitaxial layer. A two-step diffusion is carried out with a predeposition at 1000°C for 10 min and a drive-in at 1100°C for 1 h.

CHAPTER 12

Other Solid-State Devices

12.1 INTRODUCTION

The discussion thus far has centered mainly on the physics of semiconductor materials and examples of a few basic semiconductor devices. In this chapter other semiconductor devices will be discussed which are of interest for high frequency microwave applications, optoelectronics, and high power switching functions. **Gunn-effect devices** are used for high frequency amplification and for generating high frequency oscillations. These two-terminal (diode) devices utilize some unique properties of certain semiconductor materials, such as GaAs and InP, to produce a negative differential conductance *I*-*V* characteristic which inherently provides the possibility of amplification. The electron mobility decreases at increasing values of electric field.

IMPATT devices are diodes which also exhibit a negative differential conductance characteristic. These components utilize a combination of avalanche multiplication and transit-time effects. The diode is DC reverse biased into the avalanche multiplication region and the AC signal voltage to be amplified is superimposed on it. The avalanche-generated carriers sweep throught the drift region to the device terminal. With appropriate device design, the resulting current will be approximately 180° out of phase with the applied voltage, giving rise to a negative conductance *I-V* characterictic.

The **semiconductor laser** is a small semiconductor diode which emits intense, coherent, unidirectional light radiation under forward current bias. Hence its light output may be easily modulated by varying the current through it. The device structure is basically a *p-n* junction diode and the fundamental light emission mechanism is similar to that of the LED described in Section 7.2.2.

The basic element of the **charge-coupled device** (CCD) is the MOS capacitor which is used to store charge. A microchip containing a large matrix of these components can be used as an optical imaging device or in a large-scale memory circuit.

The **thyristor** is the generic name for the members of a family of devices that are designed to perform low frequency electronic switching functions. For switching frequencies up to about 10 kHz and moderate power handling applications, thyristors and transistors are both employed. However, thyristors are mainly used for the turning ON and OFF of electric currents of hundreds and even thousands of amperes, with a voltage blocking capability of a thousand or more volts. On the other hand for switching rates and operation frequencies in the megahertz and gigahertz range, transistors are generally employed.

12.2 THE GUNN-EFFECT DEVICE

A semiconductor device, the Gunn oscillator, is capable of generating high frequency oscillations in the gigahertz range. This two-terminal device extends the frequency spectrum of solid-state electronic components beyond the few gigahertz limitation of silicon transistors, into the 10- and 100-GHz range. This microwave component operates by a **transferred-electron** mechanism, is a *bulk* device, and requires no *p-n* junctions in order to operate. Its fundamental mode of operation depends on the *transfer* of conduction electrons from the normal state of high mobility to low mobility as the voltage across the diode is increased beyond a critical value. This mobility decrease produces a decreasing conductivity with electric field (or voltage/length of the sample) and hence a decreasing current with voltage as shown in Fig. 12.1. This negative resistance makes possible the generation of high frequency oscillations and even high frequency, high power amplification. The device is an example of a component which requires the use of certain compound semiconductors as starting materials. For example, the semiconductors gallium arsenide, indium phosphide, and cadmium telluride make good Gunn oscillators whereas silicon and germanium do not exhibit this phenomenon. The explanation of this fact requires a quantum mechanical discussion of the band theory of solids which follows.

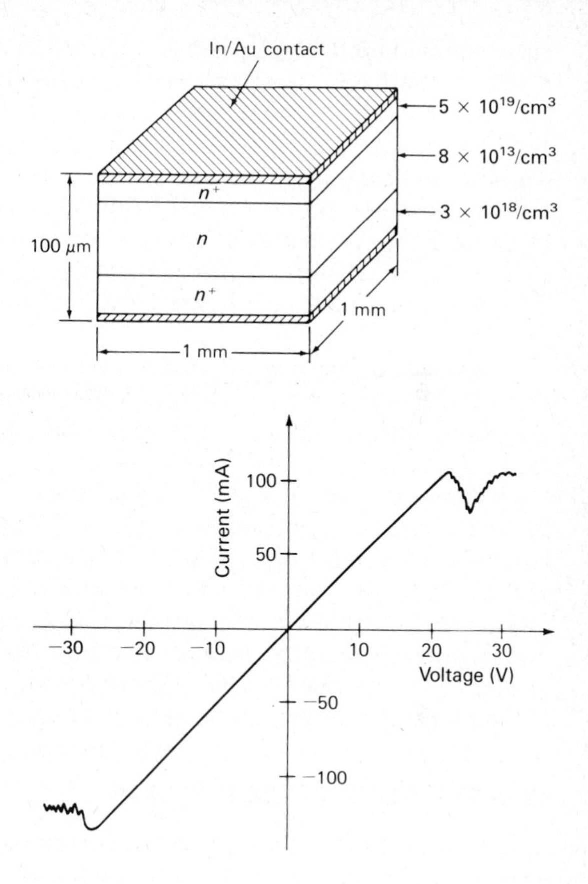

Figure 12.1 Current–voltage characteristic of a typical Gunn oscillator. The device breaks into oscillations at voltages of +25 or − 25 V, where a negative resistance occurs. Very high frequency (gigahertz) oscillations are normally observed. Also shown is a sketch of a typical design of a Gunn oscillator chip. Indicated are the dimensions of the GaAs chip as well as the impurity content of the different device regions. The purest section, the central region, is the active part of the device, while the end sections provide low resistance contacts.

12.2.1 Gunn-Effect Physics

An explanation for the occurrence of a negative resistance region in the *I-V* characteristic (Gunn effect) of the device pictured in Fig. 12.1 will now be presented using knowledge which has been derived from the band theory of solids. Figure 12.2 illustrates schematically the band structure of GaAs, a direct-gap semiconductor material, as discussed in Section 5.4.4 and shown in Fig. 5.8a. An energy gap of 1.42 eV is indicated between the top of the valence band and the bottom of the conduction band at k = 0. However, in addition to these extrema values of energy, calculations predict a set of subsidiary or **satellite** conduction band minima, two of which are shown in Fig. 12.2 at values of $k = \pm 2\pi/a$, *a* being the lattice constant. In *n*-type GaAs at room temperature the valence band is nearly filled and the conduction band contains some electrons near k = 0. Energetically the probability is extremely low for the excitation of electrons from the bottom of the conduction band to these satellite **valleys,** since an additional 0.31 eV is necessary for this transition and the ther-

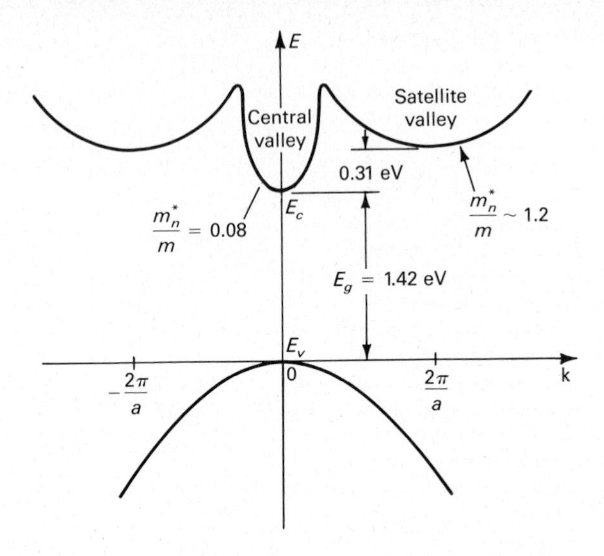

Figure 12.2 Band structure of gallium arsenide showing a direct energy band gap of 1.42 eV as well as two satellite (indirect) conduction band minima at $k = \pm 2\pi/a$ separated from the minimum of the central valley at $k = 0$ by 0.31 eV. Note the larger E versus k curvature at $k = 0$ compared with that near $k = 2\pi/a$, indicating the larger effective mass of electrons in the satellite valleys relative to the central valley; k shown here is in the [1 1 1] crystal direction. In this direction there are a total of eight satellite bands.

mal energy available, kT, is only 0.026 eV. Also, a participating phonon is needed to provide the additional k (or momentum) needed to bring a valence electron to the minimum of one of these valleys (see Section 5.4.5).

However, when a GaAs crystal is subjected to a high electric field (about 3 kV/cm), some electrons in the normal conduction band gain sufficient energy from this field to be excited into one of the satellite valleys. Now the effective mass of electrons in this subsidiary band is considerably larger than that of electrons in the normal conduction band centered about $k = 0$, as indicated by the small curvature of E versus k for $k = 2\pi/a$ [see Eq. (5.7)]. Hence these excited electrons will be less mobile than those in the central valley. As the electric field is raised beyond the threshold value for excitation, more electrons appear in the satellite conduction band and so the average electron mobility, and hence the electrical conductivity of the crystal, decreases. This results in a reduction in current as the voltage is raised above the threshold value and accounts for the negative resistance region in the Gunn diode I-V characteristic of Fig. 12.1. This of course offers the possibility of using the device to generate electrical oscillations; the physical mechanism by which this occurs will be described soon.

The tendency for the occurrence by this mechanism of a negative resistance region in the I-V characteristic of a diode fabricated from a particular semiconductor depends on the following material characteristics:

1. The material must have a satellite band with an energy minimum which is above the normal conduction band minimum.

2. The E versus k curvature for this satellite band must be less than for the normal conduction band minimum. (The effective mass must be greater in this satellite band.)

3. The satellite band must have an appreciable density of energy states.
4. The energy separation between the minima in the normal and satellite conduction bands must be greater than kT to ensure that the excitation to the low mobility band is caused by the electric field and not by thermal energy.
5. The energy separation between the minima in the normal and satellite conduction bands must be much less than the product of the avalanche breakdown voltage and the electronic charge; otherwise avalanching would preclude the occurrence of a negative resistance region.
6. Some mechanism must be available to supply the k-value (momentum) change required for transition to the satellite band.

That gallium arsenide is a material which meets all of these requirements should be clear from the previous discussion. Note that a large density of available states in the satellite band is ensured by the large effective mass there compared to the normal conduction band, according to Eq. (4.10). It is believed that a GaAs lattice phonon supplies the momentum required for the k-value change necessary for transition into the satellite band.

12.2.2 The Growth of Space-Charge Domains and Electrical Oscillations

Biasing of the Gunn diode in the negative resistance region results in an unstable situation. In a uniformly doped conducting semiconductor material any localized space-charge will tend to be eliminated rapidly with a time constant which for gallium arsenide is about 10^{-12} sec. This so-called **dielectric relaxation time** τ_d is given by the theory of electromagnetism as

$$\tau_d = \epsilon/\sigma, \tag{12.1}$$

where ϵ is the dielectric constant and σ the electrical conductivity of the material. The rate of charge dispersement is exponential in time according to a factor e^{-t/τ_d}. In the negative resistance region of GaAs the conductivity is negative, making τ_d negative according to Eq. (12.1). This predicts that the space-charge magnitude will rise exponentially in time; an **instability** will result. That is, random fluctuations in space-charge will *build up* suddenly rather than be neutralized after a short time if the Gunn device is biased in the negative conductivity region. If a small nonuniformity in the electron density occurs at some place in the device, a charge dipole will occur locally as shown in Fig. 12.3a. If the Gunn diode is biased in the negative conductivity region, this unneutralized charge will build up, and according to Poisson's equation so will the field there, as shown in Fig. 12.3b. This charged region will grow as it drifts along in the applied electric field. When the drifting space-charge region or **domain** reaches the ohmic contact at the anode, it sends a current pulse into the connected wire and hence into the external circuit. Now another space-charge

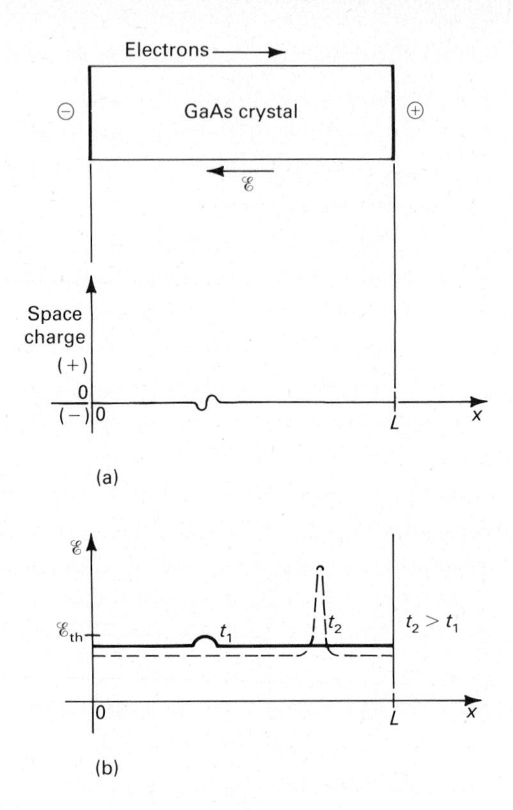

Figure 12.3 (a) Charge dipole due to local nonuniformity in the electron density in a crystal of gallium arsenide.
(b) Field buildup due to an initial localized space-charge, with the device biased in the negative mobility region, at two successive times, t_1 and a later time t_2; $\mathscr{E}_{th}$ represents the minimum value of electric field to make charge buildup possible.

domain forms at the cathode and the process repeats. Therefore the pulse frequency is inversely proportional to the drift length.

The movement of the high density electron region may be described in more detail as follows: Figure 12.4 shows a plot of the average electron drift velocity versus electric field for GaAs, indicating a drop in velocity above a critical electrical field caused by electron scattering into the high effective mass or low mobility state. Consider the device biased so that the average field is $\mathscr{E}_1$, directed to the left. Since the field in the space-charge region is somewhat higher, the electrons there will move more slowly than those in the neutral regions. Hence the electron density will increase, causing augmented accumulation of electrons on the left side and depletion on the right side of this space-charge layer. This will cause the field in the space-charge region to rise until it reaches $\mathscr{E}_2$ (see Fig. 12.4), where a steady state is achieved. Then the charged domain drifts toward the right without growth at a constant velocity v_s and with a field inside this region of $\mathscr{E}_2$ and outside this region of $\mathscr{E}_s$.

In this mode of operation, originally observed by J. B. Gunn,[1] the space-charge domain grows to its stable configuration before the ohmic contact at the

[1] J. B. Gunn. "Microwave Oscillation of Current in III-V Semiconductors," *Solid State Commun.*, **1**, 88 (1963).

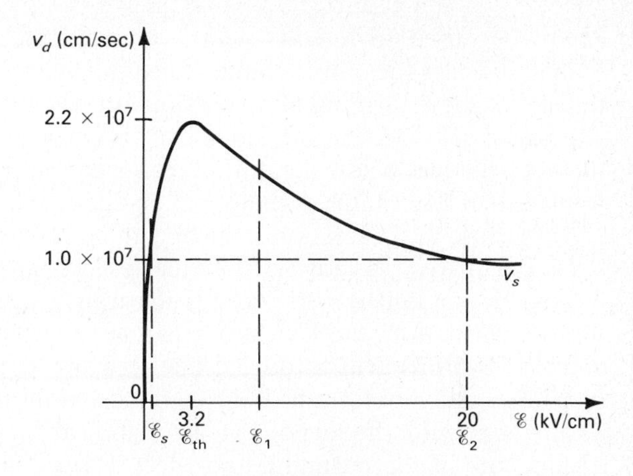

Figure 12.4 Electron drift velocity versus electric field in a gallium arsenide crystal showing a peak value of $v_d = 2.2 \times 10^7$ cm/sec at the critical field of 3.2 kV/cm and the saturation velocity $v_s = 1 \times 10^7$ cm/sec at fields of about 20 kV/cm.

end of the device is reached. In order for this to happen the time for domain formation τ_d must be less than the transit time through a device of length L. That is,

$$\frac{\epsilon}{\sigma} = \frac{\epsilon_0 \epsilon_s}{q|\mu|n} < \frac{L}{v_s}, \tag{12.2a}$$

or

$$nL > \frac{\epsilon_0 \epsilon_s v_s}{q|\mu|} \qquad (\approx 10^{12}/\text{cm}^2) \tag{12.2b}$$

for n-type GaAS containing n electrons/cm^3 where $v_s \approx 10^7$ cm/sec, $\epsilon_s \approx 13$, and the negative mobility is about -100 cm^2/V-sec. Hence when this critical product of electron density and sample length is present, short output current pulses are observed, separated in time by the space-charge domain transit time. This mode of operation is simple to implement since only a DC source is required; however, the efficiency of conversion to AC, high frequency energy is only a few percent because of the short pulses. Since for a typical device as sketched in Fig. 12.1 the pulse rate is in the tens of gigahertz range, the diode is usually operated in a microwave cavity whose resonant frequency is about that of the reciprocal of the domain transit time in the device. DC to microwave conversion efficiency approaching 10% has been achieved in this mode with a power output of about 1 W at 10 GHz. Similar performance with InP Gunn devices has also been reported.

Another more efficient mode of operation of this Gunn device does not involve charge domain formation. In fact the LSA **(limited space-charge accumulation)** mode of operation requires the suppression of these space-charge domains during operation in the negative resistance region.[2] If a microwave

[2]J. A. Copeland, "A New Mode of Operation for Bulk Negative Conductance Oscillators," *Proc. IEEE,* **54,** 1479 (1966).

cavity is caused to interact with a Gunn device so that, when the charge domain reaches the ohmic contact, the total field across the device is below the threshold, the reforming of a new domain will be delayed. In addition, if the AC field created within the diode by this external circuit is large enough that the net field goes below the minimum sustaining value while the domain is in transit, then any charge domain will be dispersed. Both of these effects can be used to cause the oscillation frequency to be controlled by the resonant microwave circuit over a range of frequencies *exceeding* the frequency capability limited by domain transit time. If the transit time is less than the time for domain formation, the LSA mode can arise. Then the domains do not form and the electrons drift through a device which exhibits negative conductance in an essentially uniform field through most of the cycle. In this mode the maximum oscillation frequency can be significantly greater than the pulse frequency rate of the normal Gunn-effect device. The useful range of LSA frequency of operation f is given by[3]

$$2 \times 10^4 > n/f > 2 \times 10^5 \text{ sec/cm}^3. \tag{12.3}$$

The lower frequency limit is determined by the need to quench any accumulation of electrons near the cathode while the field is below threshold. Gunn-effect mode-operated devices must be very thin (L small) to yield high frequency operation. Since the operation frequency of the LSA mode device is circuit determined, this device chip need not be thinned down correspondingly for increased frequency requirements. Thinning down means reduced voltage for the required threshold field, so LSA devices have potentially higher power capability than Gunn-effect-mode-operated devices. The efficiency of conversion of DC to microwave energy for the Gunn device operation in the LSA mode can approach 20%.

12.3 IMPATT DEVICES

IMPATT (or IMPact Avalanche Transit Time) devices can also be used to convert DC to microwave frequency signals, and at even higher power levels and efficiency. The device structure is basically a p-n junction diode used in the reverse-biased, avalanche mode. The device's operation is most easily explained by referring to the **Reed** diode structure sketched in Fig. 12.5a.

The important regions in the device are the p^+-n space-charge region where avalanche multiplication occurs and the intrinsic (or lightly doped) region through which the avalanche-generated electrons drift.[4] The negative differential conductance exhibited by this device in a microwave cavity when reverse biased with a DC voltage V_{DC} sufficient for avalanche carrier multiplication

[3] J. A. Copeland and R. R. Spiwak, "LSA Operation of Bulk GaAs Diodes," in *Digest of the International Solid-State Circuits Conference* (IEEE), New York, (1967) p. 26; J. A. Copeland, "LSA Oscillator Diode Theory," *J. Appl. Phys.*, **38**, 3096 (1967).

[4] The generated holes move toward the p^+-contact.

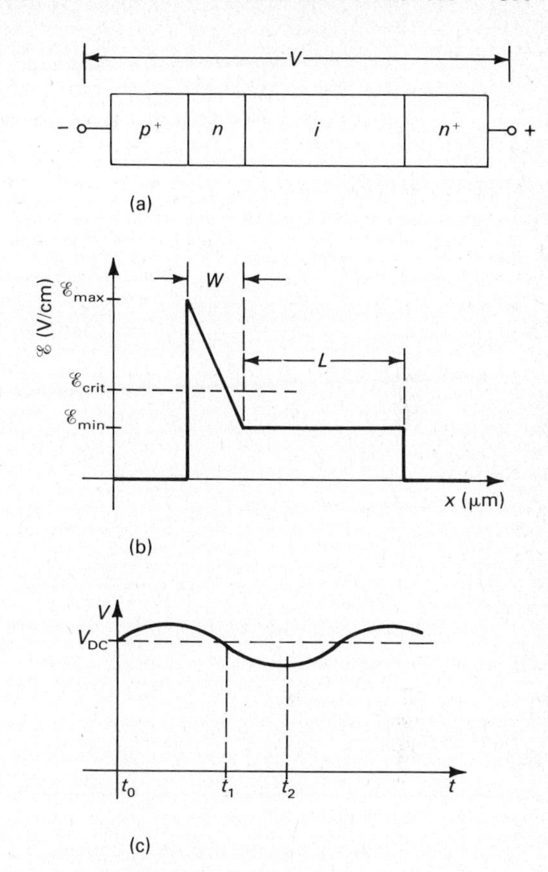

Figure 12.5 (a) The basic Reed diode structure. (b) Electric field distribution in this device, reverse biased in the avalanche region. (c) Time variation of voltage V across the device.

may be explained as follows: the DC field distribution present in the device is shown in Fig. 12.5b. Consider that the resonant circuit causes a small AC signal to be superimposed on the DC bias, starting at $t = t_0$ (Fig. 12.5c). This will cause additional electron-hole pairs to be generated in the narrow p^+-n space-charge region. There is a time constant associated with the avalanching process. This causes the number of generated charges to increase exponentially in time while the electric field is above the critical field for avalanching, $\mathcal{E}_{\text{crit}}$. If this pulse of carriers reaches its peak just before the field goes below $\mathcal{E}_{\text{crit}}$, at time $t = t_1$, a large pulse of electrons so generated will begin to drift in the intrinsic region where the field is uniform. Consider that these electrons drift during the time interval $t_1 < t < t_2$ when the AC signal voltage is negative directed and the pulse reaches the n^+-contact at a time t_2. The electrons will then give rise to a current pulse in the output circuit. Hence the generated current is 180° out of phase with the AC signal which caused it, resulting in a negative differential conductance.

The optimum conditions just described will occur if

$$L/v_s = ½(1/f), \tag{12.4a}$$

where L is the drift region length, f is the AC signal frequency, and v_s is the electron saturation velocity. It is assumed that the electric field in which the electrons drift is high enough that these carriers move at their saturation velocity. Hence in this diode structure the optimum operating frequency f from Eq. (12.4a) is given by

$$f = v_s/2L. \tag{12.4b}$$

EXAMPLE 12.1

Find the optimum operating frequency for the Reed diode of Fig. 12.5 if the length L of its drift region is 2.5 μm and it is fabricated by using a silicon crystal.

Solution

From Eq. (12.4b), $f = v_s/2L$. For silicon $v_s = 1 \times 10^7$ cm/sec. Hence

$$f = (1 \times 10^7 \text{ cm/sec})/2(2.5 \times 10^{-4} \text{ cm}) = 20 \times 10^9 \text{ Hz},$$

or

$$f = \underline{20 \text{ GHz}}.$$

The Reed diode structure has been used to explain the fundamental mode of operation of IMPATT devices. However, in practice simple p^+-n-n^+ or n^+-p-p^+ diodes are often used. In these structures it is more difficult to separate the avalanche from the drift regions. IMPATT devices have been fabricated from both Si and GaAS crystals. For optimum high frequency performance the electric field should be high enough that the carriers drift at their maximum (saturation) velocity. For Si this field is in excess of 10^4 V/cm, whereas for GaAs the field is of the order of only 10^3 V/cm. IMPATT diodes can generate higher power output at high frequencies (millimeter range) than other solid-state devices available today. Both Si and GaAs IMPATTs have been operated at a power output of 10 W at 10 GHz. The GaAs devices have exhibited a DC to microwave conversion efficiency as high as 30%; Si devices have been operated at better than 10% efficiency. This is due to the lower electric field needed to achieve the higher peak velocity of 2×10^7 cm/sec in GaAs. However, Si has a higher operating frequency potential (300 GHz) than GaAs owing to the short time constant for avalanching of Si.

12.4 THE SEMICONDUCTOR INJECTION LASER

In Section 7.2.2 the semiconductor laser was referred to as a small size, efficient transmitter in an optoelectronic communication system. A single-frequency

(1.55 μm) InGaAsP laser has been used to send information through a fiber optic glass transmission cable at 420 Mbits/sec; the error rate was only 5 per 10^{10} bits. Signal attenuation and spreading of digital pulses limit the distance and bit rate at which digital information can be transmitted. At 0.9 μm, GaAs LED or laser signals can travel only about 10 to 15 km owing to attenuation and signal dispersion at this frequency. Although the simpler light-emitting diode (LED) can be effectively used as an optical transmitter, the potentially nearly monochromatic (single frequency) emission obtained from the semiconductor laser will dictate its use as soon as its technology has matured.

The name **laser** is drawn from the words which describe the device's operation: **light amplification by stimulated emission of radiation.** The semiconductor laser is a small size, very efficient source of highly **coherent,**[5] nearly **monochromatic,** and **very directional** light. The laser is fundamentally an electronic device which provides coherent light emission when electrons in upper, excited energy states in a material fall to lower energy levels, giving off this energy difference as photons according to Planck's law. When two specific energy levels are involved, Planck's law provides for the monochromatic nature of the radiation. In a laser, stimulated emission is made possible by means of a **population inversion** which results when an upper energy level has a greater probability of being occupied by an electron than a lower energy level. (This is energetically an unstable situation and must be provided for by **pump** energy.) In the presence of a field of photons whose energy is equal to that between the two energy levels under consideration, an inverted electron population in a semiconductor will provide stimulated emission since the probability of a downward electron transition exceeds that of an upward transition. Light amplification can occur if a proper resonant structure **(Fabry-Perot**[6]**)** is provided. This is necessary since **spontaneous emission** due to **random** electron–hole recombination in the manner described in Section 5.4.2 must be overwhelmed by stimulated light emission in an enclosure which then provides a nearly coherent light source. Light coherence occurs when all the photons emitted are **in phase** with one another, that is, interfering in a constructive or reinforcing sense.

In this way 50-W pulses of light power have been produced by semiconductor lasers, and when focused by a lens to a spot of diameter of the order of the light radiation wavelength, can produce a local power density in excess of 10^{10} W/cm^2; this significantly exceeds the power density of about 10^5 W/cm^2 observed at the surface of the sun. The efficiency of production of coherent, directional light by the semiconductor laser is greatly affected by whether the downward electron transition in the semiconductor material is direct or indirect (see Sections 5.4.4 and 5.4.5). In fact all successful semiconductor lasers

[5]There exist random phase relationships between the photons emitted from an LED. When all the photons released are in phase with one another, this emitted light is said to be **coherent.** The directionality and intensity of laser-emitted light result from this property.

[6]See, for example, F. A. Jenkins and H. E. White, *Fundamentals of Physical Optics,* 3rd ed. (McGraw-Hill, New York, 1957).

are fabricated from semiconductor crystals in which the electron-hole recombination process is known to be of a direct type.

12.4.1 Population Inversion and Lasing Threshold in an Intrinsic Semiconductor

Figure 12.6 shows the usual graph of the parabolic density of states function versus energy in the conduction and valence bands of an intrinsic semiconductor. The equilibrium situation at $T = 0°K$ is indicated in Fig. 12.6a. Here all the states in the valence band are filled with electrons and the conduction band is completely empty, typical of an intrinsic semiconductor at 0°K. The Fermi energy E_F is at the center of the energy gap. If a light beam containing photons with energy $h\nu > E_g$ (the width of the forbidden energy gap) is incident on the semiconductor crystal, electrons will be excited from the valence band into the conduction band. This **nonequilibrium** state is shown in Fig. 12.6b, which indicates an inverted population of electrons due to "pumping" by the incident light beam. The conduction band at $T = 0°K$ is now filled with electrons up to E_{F_n}; the valence band is devoid of electrons (filled with holes) down to E_{F_p}. Here E_{F_n} and E_{F_p} are the quasi-Fermi levels for electrons and holes, respectively. As mentioned in Section 4.4, the Fermi energy is only defined under equilibrium conditions and is an expression then of the density of electrons and holes. It is convenient to express the steady-state density of **excess** minority carriers, as in this case, with the aid of these quasi-Fermi energies.[7] A

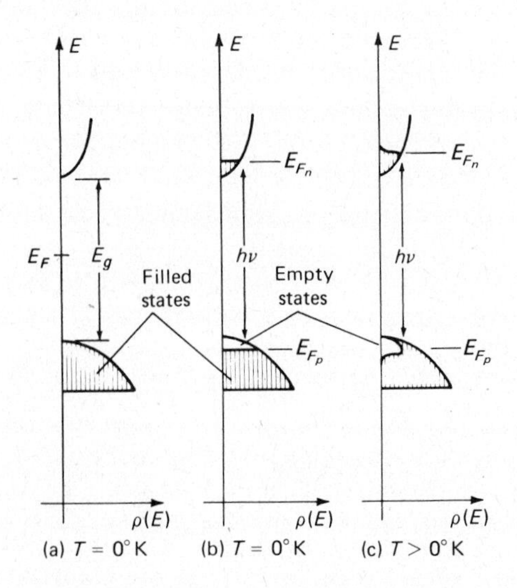

Figure 12.6 Density of states $\rho(E)$ in the conduction and valence band of an intrinsic direct-gap semiconductor versus energy E. (a) At 0°K in thermal equilibrium, with an electron-filled valence band and empty conduction band. (b) At 0°K, showing an inverted electron population caused by external excitation, with filled electron states in the conduction band at higher energies than empty states in the valence band. Radiation incident with energy $h\nu$ will cause stimulated emission.

[7]The quasi-Fermi levels for electrons and holes are defined respectively by $n = n_i \exp(E_{Fn} - E_i)/kT$ and $p = n_i \exp(E_i - E_{Fp})/kT$, where n_i and E_i are respectively the intrinsic density of carriers and the intrinsic Fermi energy for a semiconductor material.

high value of E_{F_n} represents a large nonequilibrium presence of excess electrons. If the crystal in this excited state is introduced into a cavity containing photons with energy $h\nu$ such that $E_g < h\nu < E_{F_n} - E_{F_p}$, stimulated emission corresponding to the *downward* transition of the pumped-up electrons to recombine with the holes in the valence band will take place; for if instead $h\nu > E_{F_n} - E_{F_p}$, *upward* transitions of the valence electrons into the conduction band will occur. The condition that the number of photons emitted exceed that absorbed, resulting in the possibility of lasing, is hence expressed for a block of intrinsic semiconductor material by

$$E_{F_n} - E_{F_p} > h\nu. \tag{12.5}$$

Although this discussion applies for $T = 0°K$ and a pure semiconductor, a similar description applies for a doped semiconductor at $T > 0°K$.

To ensure light amplification, stimulated emission plus a condition of sufficient gain in the cavity must exist to compensate for various loss mechanisms in the material. For a Fabry–Perot cavity the *minimum* condition for lasing is that a stream of photons make a traversal between the two reflecting end mirrors with no net loss, which is expressed by

$$R\, e^{(G_{\text{th}} - \alpha)L} = 1. \tag{12.6}$$

Here G_{th} is the gain in photons/cm, R is the reflectivity of the end mirrors, L is the cavity length, and α is the loss coefficient.

12.4.2 The Basic *p-n* Junction or Injection Laser

Semiconductor lasers have produced coherent light with wavelengths ranging from 0.32 μm in the ultraviolet to 16.5 μm in the infrared portion of the electromagnetic spectrum, depending on the energy gap of the semiconductor. The radiation output can be obtained from the injection laser of Fig. 12.7 by applying an electrical input to the *p-n* junction, causing a population inversion of electron or hole carriers which on recombination produce light emission. It has

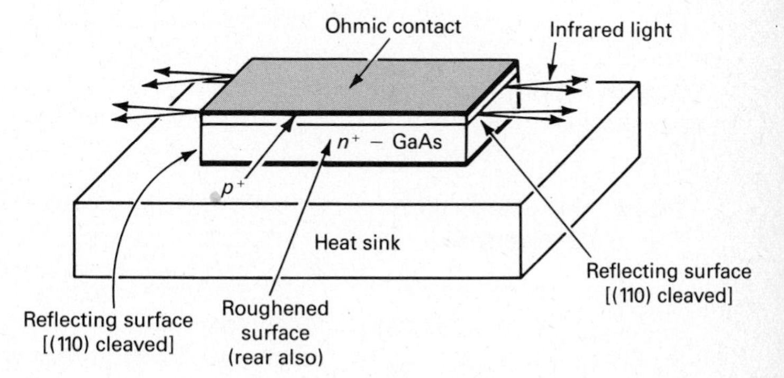

Figure 12.7 Sketch of a gallium arsenide *p-n* junction laser mounted on the heat sink showing the flat, reflecting front and back surfaces as well as the rough side surfaces. Coherent infrared light is emitted in a very narrow beam from the front and rear surfaces. Ohmic contacts are provided on the upper and lower surface of the GaAs chip for connection to a current source.

already been stressed that the semiconductor laser is used in communication systems where a vast information transmission capability was indicated as well as extreme ease of modulating the beam for signal transmission. Semiconductor injection lasers have been produced from materials such as GaAs, InSb, InP, PbTe, PbS, AlGaAs, and InGaAsP. The "pumping" of the *p-n* junction laser to produce population inversion may be provided by applying forward bias to the diode, causing carrier injection. However, other methods of pumping are possible such as the use of electron beam or optical excitation, or carrier production by avalanche multiplication. The junction laser is usually very small in size, perhaps 100 × 250 micrometers in area and 100 micrometers thick. Both the *p*- and *n*-sides of the basic laser diode must be heavily (degenerately) doped, each to at least 10^{18} impurities/cm^3, which is just short of the doping required for tunnel diodes. The light is produced in the close proximity of the *p-n* junction and this radiation must undergo many internal reflections on the front and rear faces of the semiconductor chip (which must be perfectly flat and parallel to each other) to produce gain and hence light amplification. In the case of a gallium arsenide laser chip the front and rear, plane and parallel reflecting surfaces may be produced by **cleaving** the chip on two edges. Cleaving is the process by which a crystal separates on certain specific crystal planes when the material is abruptly struck in the appropriate crystallographic direction with a sharp instrument. This eliminates the necessity of accurately cutting and polishing the two parallel faces. The resulting structure is known as a **Fabry-Perot resonant cavity.** A high percentage of reflection is obtained automatically on these faces because of the high index of refraction of most semiconductors like GaAs (n_s = 3.6), and as provided by Fresnel's law of reflection.[8] The light transmitted through one of the faces is the output of the laser. For good light amplification, the gain of photons along the cavity must exceed the transmission at the ends. With a Fabry-Perot cavity, the wavelength λ of the various possible oscillation modes is given by

$$m\lambda = 2\, n_s L, \tag{12.7}$$

where m is the mode number (an integer) and L is the distance between the front and rear reflecting faces. The value of L is not critical, since $\lambda \ll L$, and many values of integral mode numbers and wavelengths will satisfy this relationship. However, in practical situations often only one mode is favored accounting for monochromatic emission. To inhibit multimode lasing the two faces perpendicular to the reflecting faces are roughened.

12.4.3 *p-n* Junction Laser Population Inversion and Lasing Threshold

The energy band diagram of the *p-n* junction laser will now be utilized to illustrate how the various requirements necessary for successful lasing are met by the device illustrated in Fig. 12.7. Figure 12.8a shows the energy band diagram

[8] See footnote 6.

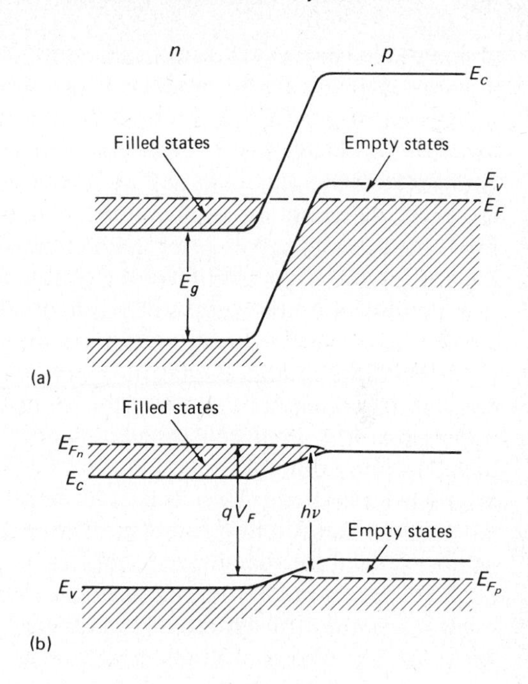

Figure 12.8 (a) A gallium arsenide *p-n* junction laser energy band diagram under thermal equilibrium conditions, showing electron-filled states in the conduction band and empty states in the valence band.
(b) Energy band diagram in the steady-state situation with a forward potential applied. An inverted population situation occurs near the junction and stimulated emission of radiation energy $h\nu$ is obtained. The quasi-Fermi levels in the conduction and valence bands are indicated by E_{Fn} and E_{Fp}, respectively, and are separated by qV_F, where V_F is the applied forward voltage.

for a semiconductor *p-n* junction in thermal equilibrium which is very heavily doped both with acceptor impurities on the *p*-side and donor impurities on the *n*-side. This drawing may be compared with the energy band diagram for a conventional *p-n* diode in equilibrium as shown in Fig. 6.4b. Note that because of the equilibrium condition, the Fermi energy is the same throughout the structure. However, whereas the Fermi level lies in the forbidden energy gap in the conventional diode, in the laser diode the Fermi energy lies in the conduction band on the *n*-side and the valance band on the *p*-side. This is consistent with the raising of the Fermi energy on the *n*-side as donor impurities are added to the semiconductor according to the description in Fig. 4.11. Also, enough electrons are supplied by the large density of donor impurities on the *n*-side to fill the conduction band up to the Fermi level there. The lowering of the Fermi level due to heavy doping of a *p*-type semiconductor was also discussed in Section 4.6 and this accounts for the Fermi level even below the valence band edge of the *p*-side of the laser diode of Fig. 12.8a. The states free of electrons, above the Fermi level in the valence band, are filled with holes.

Now if a forward bias V_F is applied to the structure, this will tend to raise the energy levels on the *n*-side relative to the *p*-side; this nonequilibrium energy situation is pictured in Fig. 12.8b. Note that in this diagram the forward bias was taken large enough that $qV_F > E_g$, the energy gap. Again the Fermi level is not defined for this nonequilibrium situation and as before the concept of the quasi-Fermi energy level is useful, the level on the *n*-side being indicated by E_{F_n} and by E_{F_p} on the *p*-side. The forward bias of course reduces the barrier height, causing holes to flow from the *p*-side to the *n*-side and electrons from the n-side to the p-side. For a sufficiently large voltage bias the electrons flowing downhill across the barrier into the p-side cause **population inversion** there and downward transitions into the many empty states available in the valence band then take place; these **electron–hole recombinations** tend to release energy approximately equal to E_g in the form of photons, efficiently, if this takes place in a direct-transition semiconductor. In direct semiconductors of GaAs, AlGaAs, and InGaAsP the electric to light energy conversion efficiency can exceed 50%. If the Fermi level is not originally in the conduction and valence bands, then the voltage required for population inversion would reduce the barrier height nearly to zero and a damaging high current would tend to flow.

Hence if the forward current is sufficient, $\backsim 5 \times 10^4$ A/cm^2 for a basic GaAs laser, a population inversion of minority electrons will appear just on the p-side of the junction and this particular requisite for lasing will be satisfied. Here pumping is provided for by minority carrier injection. In fact just as in the case of the laser action already described for an intrinsic semiconductor, the condition for lasing is given by $E_g < h\nu < E_{F_n} - E_{F_p}$. This expresses again the need for the heavy doping required so that the quasi-Fermi levels lie in the conduction band on the n-side of the diode and the valence band on the p-side. A symmetric argument would apply for the holes flowing over the junction barrier and recombining on the n-side.

As the forward voltage applied to the diode is increased from zero, initially normal injection of minority electrons into the p-region and holes into the n-region takes place with corresponding electron–hole recombination, accompanied by light output. This will be the case of spontaneous light emission (see Section 5.4.2). When the voltage is still further raised until the condition pictured in Fig. 12.8b is reached, a sudden increase in light output will be observed, if the device is in a Fabry-Perot cavity, as the condition of population inversion is achieved. Whereas the initial spontaneous radiation is of a random nature, at the **lasing threshold** the emission of coherent light is suddenly observed. At first, the emission may take place in a number of the possible resonant modes provided by the cavity. However, when sufficient bias is applied a few very intense modes tend to be observed and even a single line of monochromatic output can result as shown in Fig. 12.9. There is, however, always a small background radiation due to spontaneous emission.

The point of lasing is usually expressed by a **current density threshold** J_{th} given by

$$J_{\text{th}} = \frac{8\pi q n_s^2 w \nu^2 \,\Delta\nu}{\eta c^2}\left[\frac{1}{L}\ln\left(\frac{1}{R}\right) + \alpha\right], \tag{12.8}$$

where η is the quantum efficiency of light production, w is the width of the lasing region, c is the velocity of light, and the other quantities are already defined in relation to Eqs. (12.5)–(12.7); $\Delta\nu$ is the emitted linewidth. Some of the parameters in this expression are quite temperature sensitive and in the case of a basic GaAs laser, a current density of about 10^3 A/cm^2 is the lasing threshold at 77°K, but it rises to more than 10^5 A/cm^2 at room temperature.

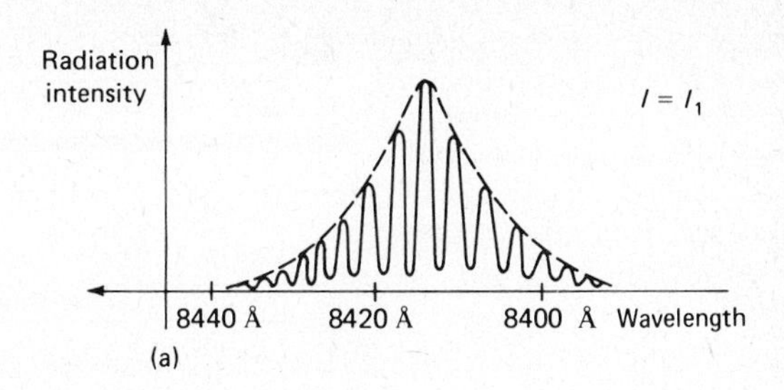

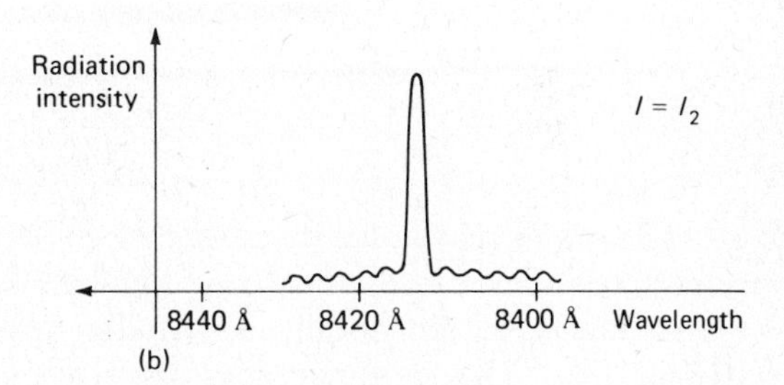

Figure 12.9 Radiation spectrum of a gallium arsenide *p-n* junction laser showing the many modes of oscillation: (a) output at current I_1, (b) output at current $I_2 > I_1$, indicating how the radiated energy can tend to be confined to a single mode as the current is increased.

This coupled with the approximately 1 V required to drive this current indicates that a power density of about 10^3 W/cm^2 must be extracted at 77°K by an appropriate heat sink if the device is to be operated in the continuous, so-called **CW**, mode. This accounts for the use of the basic GaAs *p-n* junction laser in the pulsed mode pulses (less than 1 μsec wide) and at low temperature to prevent excessive heating of the diode. This is the reason why modern semiconductor injection lasers utilize a multilayer, heterojunction structure to increase conversion efficiency and hence permit even continuous (CW) operation at room temperature. However, pulsed operation is quite adequate for the purpose of communication using digital signals. Increasing the height of the input current pulse increases the light pulse output so that pulse height modulation may be readily used to transmit information.

12.4.4 Heterojunction Lasers

Present-day semiconductor injection lasers for communication applications use a multilayer heterojunction structure grown by liquid or molecular beam epitaxy (LFE or MBE). The GaAs active layer, in which light is generated by lasing action, is sandwiched between two $Al_xGa_{1-x}As$ layers, as shown in Fig. 12.10. $Al_xGa_{1-x}As$ is a single-crystal compound semiconductor in which a fraction x of the Ga atoms in GaAs is replaced by aluminum atoms. This **ternary** (three-atom) crystal has a lattice constant approaching that of GaAs, but a larger energy gap and a smaller index of refraction. For example, for $x = 0.3$, the band gap of $Al_{0.3}Ga_{0.7}As$ is 1.798 eV or about 25% larger than that of GaAs; its index of refraction n_s is 3.385, or about 6% smaller than that of GaAs.

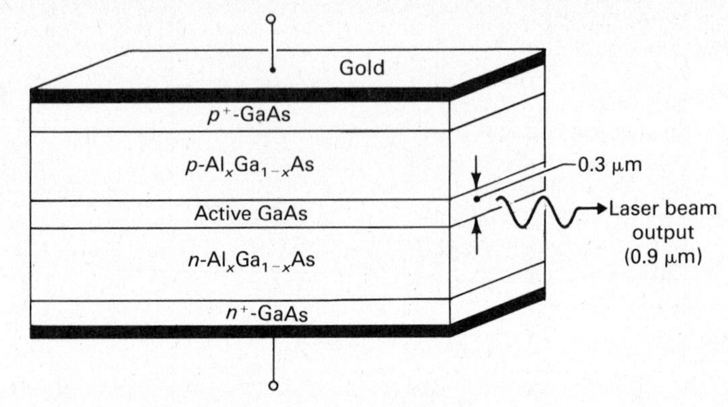

Figure 12.10 Typical GaAs/AlGaAs double-heterojunction laser structure showing the two heterojunctions, the thin, lasing GaAs layer, and the contacting, heavily doped GaAs and metallic gold layers.

The major reason for the choice of an $Al_xGa_{1-x}As$ layer is that its increased refractive index causes the lasing light waves, propagating inside the central GaAs layer, to be totally internally reflected from its interfaces with $Al_xGa_{1-x}As$.[9] This confinement significantly increases the laser light output. Also, the heterojunction provides a barrier which tends to confine the injected carriers (whose recombination results in the laser light output) to a small region ($\sim$0.3 μm) near the *p-n* junction. At room temperature the threshold current for lasing is about 10^3 A/cm^2 for this **double-heterojunction** structure compared with 5×10^4 A/cm^2 for a basic GaAs homojunction laser. This minimizes the heat removal problem.

As previously discussed in Section 7.2.2, a more efficient communication signal transmission can occur at a wavelength of 1.3 μm owing to a minimum in the absorption by the glass in the fiber optic transmission cable. The semiconductor laser can be designed to emit radiation of this wavelength by an appropriate choice of active region. A **quaternary** (four-atom) compound semiconductor InGaAsP, made of a mixture of GaAs and InP, used as the active region gives an efficient laser output at 1.3 μm for use in long distance communication systems.

EXAMPLE 12.2

(a) Derive an expression for the mode separation of light emitted from a GaAs semiconductor injection laser.

(b) Compute the mode separation for a laser with a separation of 400 μm between reflecting, cleaved surfaces, if the emitted radiation has a wavelength of 0.9 μm.

[9] The process of light confinement is equivalent to that of the confinement of microwaves by a dielectric waveguide. See also footnote 6.

Solution

(a) From Eq. (12.7), $m\lambda = 2n_sL$. Differentiating with respect to λ, we have

$$\frac{dm}{d\lambda} = -\frac{2Ln_s}{\lambda^2} + \frac{2L}{\lambda}\frac{dn_s}{d\lambda}.$$

Assuming that the refractive index change with respect to wavelength is negligibly small, we obtain

$$d\lambda = -(\lambda^2/2Ln_s)\ \mathrm{dm}.$$

(b) For the separation of adjacent modes, $dm = 1$ and

$$d\lambda = -(0.9 \times 10^{-4})^2/2(400 \times 10^{-4})(3.6)$$

$$|d\lambda| = \underline{2.8 \times 10^{-8}\ \mathrm{cm} = 2.8\ \text{Å}}.$$

12.5 THE CHARGE-COUPLED DEVICE (CCD)

As a further example of an electronic device with considerable capability whose operation may be explained by applying the principles of semiconductor physics covered in previous chapters, consider the family of charge-coupled devices (CCDs).[10] These devices have unique applications in computer circuits and optical imaging. The microminiature nature of the device structure and ease of fabrication yield the possibility of extremely high density arrays of such devices which can provide memory, logic, delay-line, and optical imaging functions, economically.

The charge-coupled device structure bears a certain relationship to the MOS structure in that the input configuration consists of a metal–insulator–semiconductor (MIS) sandwich. The charge-coupled device can produce and store minority charge carriers in potential wells created along the surface of a semiconductor just below a series of small metallic regions which are biased with respect to the underlying crystal by an applied potential of a few volts. Just as in the case of the MOS device, surface effects of the type discussed in Section 5.5 effect the operation of the CCD. The concept of space-charge or depletion-layer formation described in Section 10.4.1 will be useful in describing this device. When excess mobile charges are introduced into one of these depleted potential wells, either electrically or by optical or other radiation, they can then be moved along the semiconductor surface by altering the potential on the successive metallizations. The device operation involves the introduction of some charge into the potential well, the shifting of these mobile carriers

[10] W. S. Boyle and G. E. Smith, "Charge-Coupled Devices—A New Approach to MIS Device Structures," *IEEE Spectrum,* **8,** 18–27 (1971).

tion of some charge into the potential well, the shifting of these mobile carriers to another location along the semiconductor surface, and the detection of the presence and the magnitude of this charge at some other location in the semiconductor crystal. Since a finite time elapses between the introduction of charge into the first unit and the charge reaching the last unit, the device can serve as a delay line.

12.5.1 The MIS Structure

An MIS structure on an *n*-type semiconductor is sketched in Fig. 12.11a. A simple implementation of this CCD device uses silicon as a semiconductor, its

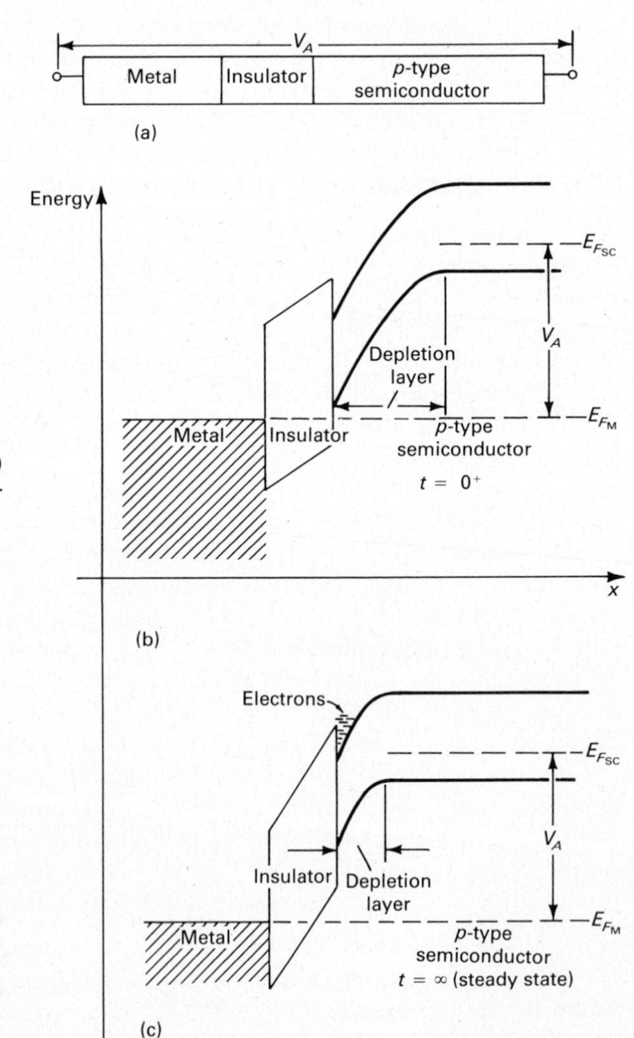

Figure 12.11 (a) The basic charge-coupled device (CCD) metal–insulator–semiconductor (MIS) structure. (b) Energy band diagram for an MIS structure at $t = 0^+$, just after the voltage V_A is applied. (c) A long time later, $t = \infty$, when the steady-state is reached and electrons from the bulk *p*-type semiconductor are collected at the oxide–semiconductor interface. Note that in the steady state, the potential drop across the insulator is greater than at $t = 0^+$, and the drop in the semiconductor is less than at $t = 0^+$, owing to the presence of the stored electron charge at the surface of the semiconductor.

grown oxide as the insulator, and aluminum as the metal. The technology for fabricating this type of device is entirely compatible with the integrated-circuit technology already described in Sections 11.1–11.4; in fact the CCD was generated from the development of that technology. Figure 12.11b shows the energy band diagram for the MIS device structure just after ($t = 0^+$) a potential V_A has been applied to the metal electrode with respect to the *p*-type semiconductor substrate taken as ground or zero potential. If this potential is only a few volts positive with respect to the semiconductor substrate, a depletion region will be set up under the oxide in a very short time (of the order of the semiconductor dielectric relaxation time). The bending of the conduction and valence energy band edges due to this applied voltage is shown in the figure and in addition the quasi-Fermi level E_{FSC} in the semiconductor is indicated. After some time delay, minority electrons in the *p*-type semiconductor tend to collect in the crystal at the oxide–semiconductor interface, where they are swept by the electric field present in the depletion layer, after diffusing into this region. As the electrons accumulate at the semiconductor surface, the potential there becomes more negative and the depletion layer tends to shrink in thickness. That is, since there are some additional negative electron charges in the depleted region now, fewer negatively charged ionized acceptors are needed there to terminate the electric field flux lines generating from the positive charges placed on the metal electrode by the applied voltage. Eventually the steady-state situation ($t = \infty$) pictured in Fig. 12.11c is established. Note that in this case more of the applied voltage appears across the insulating layer compared to the voltage drop associated with the potential barrier in the semiconductor. Elecrons may also be introduced and trapped at the surface in this potential well by optical excitation of electron–hole pairs by light or other radiation incident on the semiconductor.

The manner in which some electron charges trapped in the potential well of an MIS device like that sketched in Fig. 12.11b can be moved along the semiconductor surface to a location further along may be explained by referring to Fig. 12.12. In Fig. 12.12a, three metal electrodes are shown; the outer two are charged quickly to 2 V with respect to the *p*-type substrate, while the central electrode is charged to 3 V. The depletion region produced in the semiconductor as a result of this electrode potential distribution is indicated in the figure on the left and shows a potential well under the central electrode. The energy band diagram for this situation is shown on the right-hand side of Fig. 12.12a. This space-charge configuration will persist for the order of a *second* before holes diffuse into this depletion region and are swept to the silicon–silicon dioxide interface, forming an inversion layer there. Assume that in a time much shorter than the time necessary to establish this steady-state inversion layer (say $\backsim 1$ μsec) some electron charges are optically injected into the potential well at the central electrode. This situation is sketched in Fig. 12.12b. Note that this accounts for a small reduction of the space-charge layer width under the central metallization and hence a small increase in the capacitance of this central electrode relative to ground. (This provides one technique for detecting the quantity of electron charge stored in this potential well.) If now the third electrode on the right has its potential increased to 4 V, an electric field will be set up parallel to the semiconductor surface, drawing the electrons stored in the central region to a position under the metal electrode on the right. This is shown in Fig. 12.12c, which also indicates the deepened potential well under this third electrode. If finally the two electrodes to the left are brought to 2 V and the one at the right is maintained at 3 V, the original charge and space-charge configuration as shown in Fig. 12.12b is reestablished except that now the stored electrons have been moved one electrode distance to the right. The successful operation of this device requires that the charge transfer time be much shorter than the time for the MIS device to achieve its steady-state condition.

This type of operation is characteristic of a device known as a **shift-register.** This particular shift register has three stations, but it is obvious that devices with many more stations can be constructed by using standard integrated-circuit fabrication techniques. Each station can store a **bit** of information—this in a digital sense. That is, a capacitance measurement can detect whether charge is stored at any electrode or not. However, **analog** information can be stored as well since the magnitude of capacitance change in each MIS structure is related to the quantity of charge stored at that station.

12.5.2 A Three-Phase Charge-Coupled Shift Register

The shift register is an extremely important device used extensively in computers to store information temporarily and then to **serialize** this information. That is, the information which is shifted down to the end of the register can be "read" there, arriving in sequence in relation to the location of the data as placed in the shift register. The manner in which this shifting may be implemented is illustrated by the charge-coupled shift register driven by the three-phase clock, as shown in Fig. 12.13. Note that the electrodes are connected together in groups so that every third electrode is tied to a common line. In this way all the electrodes are contacted by using three separate lines, labeled A, B, and C. Suppose initially that the top two lines A and B shown are connected to a commmon potential V_1, and the third line, C, is at a potential V_2, where V_1 and V_2 are positive potentials relative to the p-silicon substrate and $V_2 > V_1$. This is illustrated in Fig. 12.13a, which shows deeper potential wells under electrodes 1, 4, and 7. Consider now that some negative electron charges are introduced in the depletion layer of station 1. It is now desired to transfer this hole charge along the p-type semiconductor surface, to the right. The manner in which this may be accomplished is as follows: Line B is now set at a potential V_3, where $V_3 > V_2$ and lines A and C are left at their previous values, respectively V_1 and V_2. This is pictured in Fig. 12.13b. The more positive potential of electrode 2 has now succeeded in transferring the excess electrons previously introduced at station 1 to station 2. To complete this transfer sequence line C is set back to V_1 and line B is set at V_2. Now the situation is

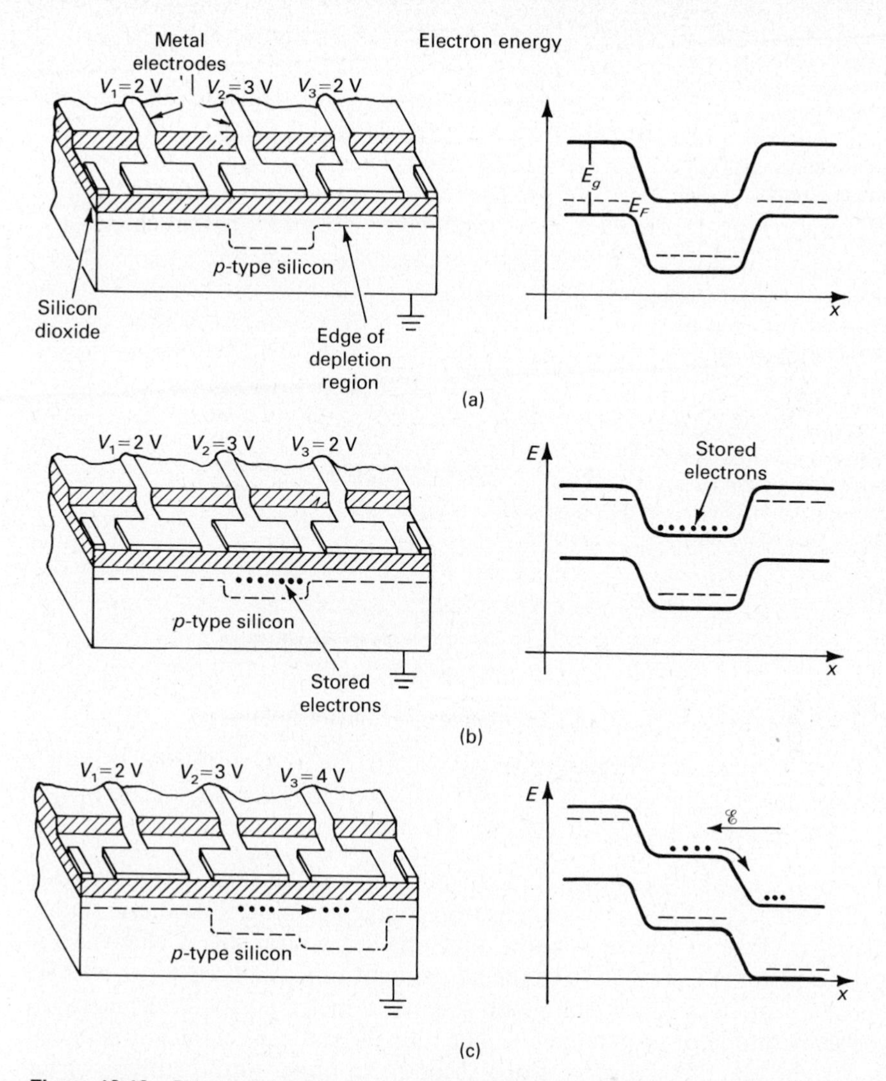

Figure 12.12 Schematic diagrams showing a string of three MIS devices capable of the movement of some electron charges along the surface of a *p*-type semiconductor, from place to place. On the left of the figure, a sketch of the physical appearance of the device is shown. To the right, the equivalent energy band diagram for the situation just to the left is indicated. The dotted lines represent the quasi-Fermi levels in the three regions.
(a) The depletion region formed in the *p*-type semiconductor just after the application of the potentials V_1, V_2, and V_3 as shown.
(b) As in (a) except some electrons are introduced in the central region and trapped there.
(c) The start of transfer of the electron charge stored in the central region to the right when the right-hand electrode is raised to a potential more positive than the central electrode, producing an electric field directed to the left.

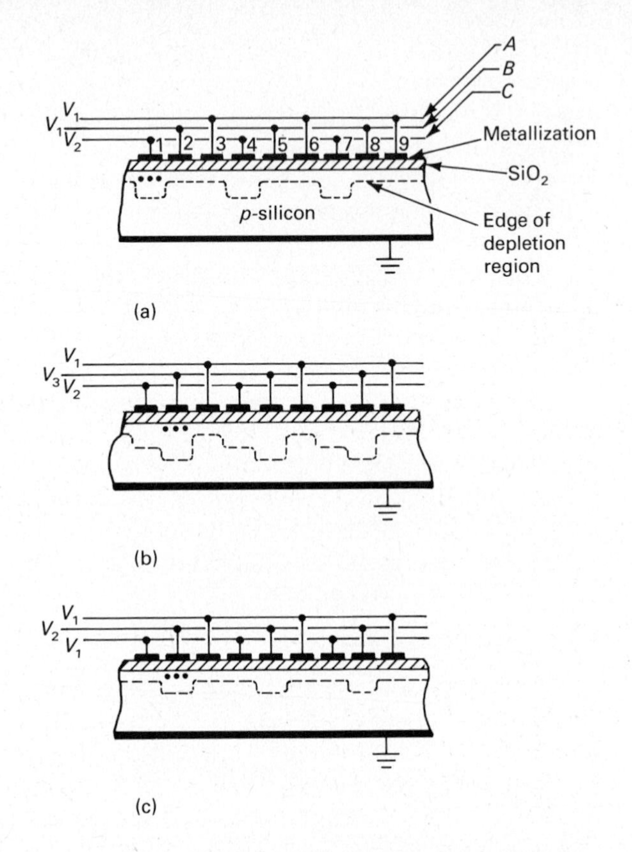

Figure 12.13 Schematic illustration of the operation of a three-phase shift register. A three-phase voltage supply is needed to drive the three lines shown (*A*,*B*,*C*) and the drive voltages are indicated by V_1, V_2, and V_3, where $V_3 > V_2 > V_1$. After electron charges are introduced into the potential well under electrode 1 in (a), (b) they are shifted sequentially to the right to the potential well under electrode 2, since $V_3 > V_2$; (c) shows the stable location of the charge under electrode 2. Repeating this sequence can move the charge to the end of the register, under electrode 9.

as shown in Fig. 12.13c, which is identical to that at the start, as indicated in Fig. 12.13a, except that the stored electron charge has been shifted one station to the right. This of course assumes that this charge transfer is efficient and that the transfer is fast relative to the adjustment time required for these MIS devices to come to a steady state with respect to the minority electrons in the *p*-type semiconductor substrate bulk. The time to establish this steady state has been found to be of the order of one second. An upper limit estimate of the time for this charge transfer to take place can be calculated by assuming that the electrons move from one station to the next by diffusion. Assuming that the distance d between electrode centers is about 3 μm and that the electron diffusion coefficient in silicon is about 20 cm^2/sec, we may approximate the transfer time by calculating the time t_{tr} it takes an electron to traverse an electron diffusion length (see Section 6.2.1). That is,

$$t_{tr} = d^2/D_n = [(3 \times 10^{-4})^2 \text{ cm}^2]/(20 \text{ cm}^2/\text{sec}) \qquad (12.9)$$
$$= 4.5 \times 10^{-9} \text{ sec} = 4.5 \text{ nsec}.$$

Of course there is some aiding field along the semiconductor surface which urges the electrons to the right, so the transfer certainly takes place in a time of the order of a few nanoseconds. Hence the transfer not only takes place in a time short enough for proper device functioning but also the shift register should be capable of operating at a few hundreds of megahertz clock rate (the rate at which the potentials on Lines A, B, and C are shifted). Measurements have also shown that charge transfer efficiencies in excess of 99% can occur. This transfer loss however tends to limit the length of the string of CCDs in a shift register application.

Shift registers can be constructed of typical switching devices such as bipolar or MOS transistors. However, the extreme simplicity of the CCD offers the possibility of implementing ultrahigh density electronic functions in a very small area on the surface of a semiconductor chip. No solid-state impurity diffusion is necessary to create the CCD, in contrast to several diffusions in the case of bipolar devices and at least one diffusion for the MOS transistor.[11] This makes the surface area of the chip occupied by a CCD about five times smaller than a comparable MOS transistor and perhaps 25 times smaller than a bipolar device, which accounts for its importance in constructing high density memory arrays and high resolution optical imaging devices.

12.5.3 Input Schemes for CCD Shift Registers

It is clear that a primary application of the CCD is in some sort of shift register application. Hence it is necessary to provide a method to introduce information at the input of the register as well as provide some method of detecting the output signal at the end of the string of charge-coupled devices.

The schemes for inputting a CCD shift register are sketched in Fig. 12.14. Figure 12.14a shows how an *n-p* junction can be utilized as an input device. A positive potential applied to the MIS device just to the right of the *n-p* junction produces an inversion layer of electrons just under its metallization; this provides a conducting channel between the *n*-region of the *n-p* junction on its left and the CCD on its right, in a manner similar to MOS transistor operation. The heavily doped *n*-region is a source of electrons which are conducted to the right along the semiconductor surface to the first CCD, which is biased to attract the electrons. The potential well of the first CCD device is sketched in the figure. Note that it is simply necessary to tie the *n*-region of the *n-p* junction to the semiconductor substrate to provide an electron source, similar to the manner in which the source of an MOS transistor is connected to substrate.

This first-described method involves the diffusion of an *n*-type impurity into the semiconductor material to form an *n-p* junction. This negates one of the advantages of the CCD in that no high temperature diffusion step is needed to form a shift register. Two techniques which require no additional diffusion step are discussed next. The first involves just using a regular MIS structure in the avalanche mode for an input device, as shown in Fig. 12.14b. Here a high

[11]This makes possible the fabrication of CCDs using semiconductor materials in which it is difficult to produce *p-n* junctions but where surface states are low in density.

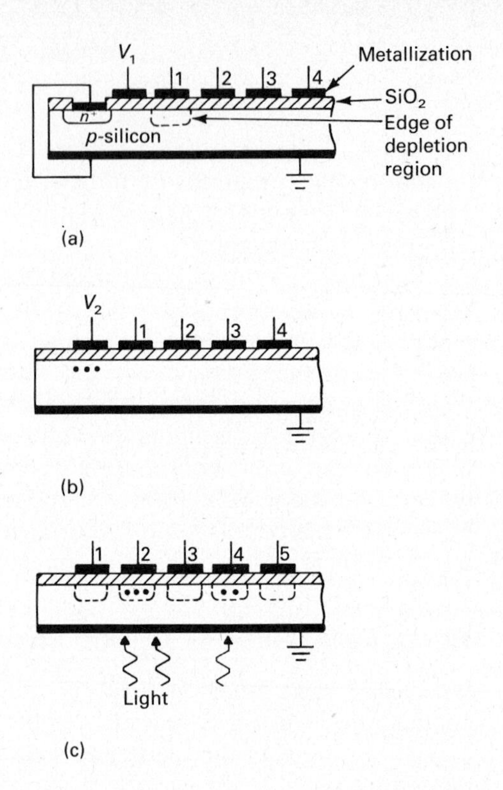

Figure 12.14 Schematic representation of three input schemes for a shift register using charge-coupled devices.
(a) Field-effect-type input using an n^+-region as a source of electrons and a positive voltage V_1 relative to substrate applied to a metal gate to produce an *n*-channel to conduct electrons to the first CCD device, which has a potential well to trap these electrons.
(b) A positive potential V_2 is applied to the device at the left, which is high enough to create the avalanche electric field for silicon which provides hole–electron pairs and hence a source of electrons.
(c) Hole–electron pairs are excited by light at the back surface of the silicon and the electrons drift to the top surface to be stored in the potential wells there.

positive voltage pulse is applied to the first device on the left, high enough that the electric field in the depletion layer reaches the avalanche value. The avalanche mechanism in the depletion layer of a *p-n* junction, described in Section 6.3.3A, is satisfactory for describing the avalanching in this space-charge region in an MIS structure. There it was seen that impact ionization caused the creation of hole–electron pairs. These minority electrons then diffuse into the first CCD of the shift register to the right. The last hole injection input method uses photoinjected carriers as illustrated in Fig. 12.14c. This may be used as an optical imaging technique, to be described later, which uses hole–electron pair production by photons whose energy $h\nu > E_g$, the semiconductor energy gap width.

12.5.4 Output Schemes for CCD Shift Registers

The schemes for detecting the output signal at the end of a CCD shift register are sketched in Fig. 12.15. Figure 12.15a shows how an *n-p* junction with a large reverse potential is used to detect the electrons in the last CCD just to its left. Because the reverse junction potential is even more positive than the CCD potentials, electrons are collected by this diode, as by the collector junction of a bipolar transistor, and output voltage is developed across the resistor R_o in proportion to the collected current and hence collected holes.

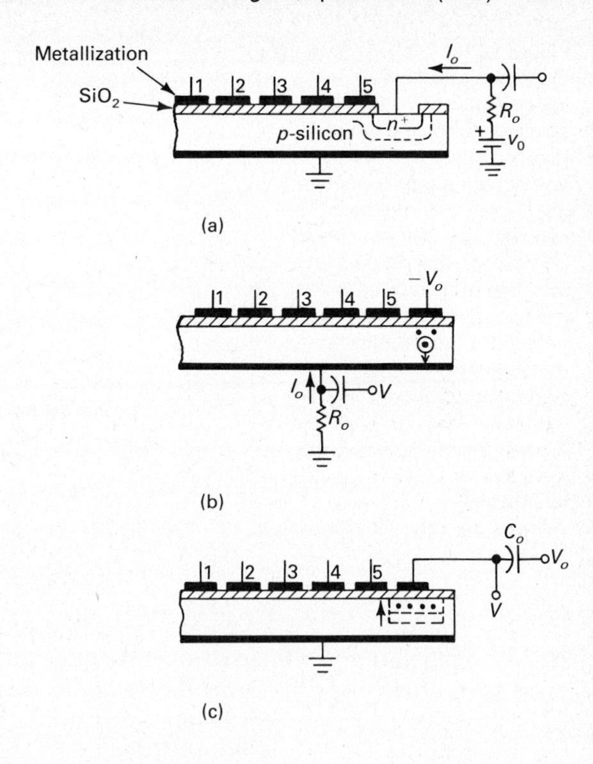

Figure 12.15 Schematic representation of three output schemes for a shift register using charge-coupled devices.
(a) An *n-p* junction diode at the end of the register, with a large reverse potential V_o, acts like a bipolar transistor collector to attract electrons, resulting in an output current I_o.
(b) A negatively pulsed MIS element yields an electric field in the silicon which drives electrons from the last register element to ground, giving rise to an output current I_o and an output voltage V.
(c) The positive voltage V_o is shared by the output MIS device and the output capacitor in series. As electrons are collected in the potential well of the MIS device, the well depth decreases and the MIS capacitance increases, causing it to take a lesser share of V_o and hence V assumes a smaller positive potential relative to ground.

Again the creation of an *n-p* junction requires additional technology; hence two methods not using *n-p* junctions will now be described. The first uses an MIS output device which is negatively biased relative to substrate, as shown in Fig. 12.15b. When electrons from the last CCD in the shift register drift toward this output electrode they are repelled into the substrate and constitute a current to ground which is detected by the voltage drop across the resistor R_o. Another detection technique uses the increase in output capacitance of a conventional positively biased MIS device when excess electrons from the previous CCD device diffuse into its depletion layer, reducing its width. This increased capacitance was described in Section 12.5.1. It will give rise to a shift in the voltage marked V, since this reflects the change in the proportion of the applied voltage V_o across the output MIS device capacitance compared with the fixed-output capacitor C_o. In this way the relative quantity of charge on the last CCD in the shift register may be detected.

12.5.5 A Solid-State Optical Imaging Device

The CCD array also offers other exciting application possibilities such as a solid-state TV camera or for optical imaging processing. A large, closely spaced two-dimensional array of CCDs is fabricated on one face of a thin slice of semiconductor. This is sketched in Fig. 12.16. All the individual devices are biased with a small positive potential relative to the *p*-type semiconductor substrate.

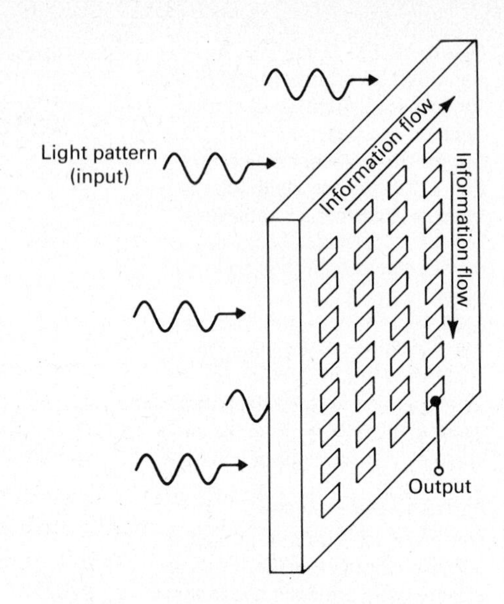

Figure 12.16 Schematic diagram of a solid-state image processor using CCD elements to store charges produced by optical excitation of hole–electron pairs in the silicon substrate by an imposed light pattern. These charges are then moved to the edge of the device line by line by means of shift register action. The charges on the edge line are then sequentially shifted to the output terminal. This raster-type arrangement serializes the optical pattern information.

Now a light pattern is focused on the underside of the semiconductor slab, creating hole–electron pairs in the material in proportion to the local light intensity. The electrons will be attracted to the seimconductor surface under the nearest positively charged metal electrode, with the charge there representing an analog of the incident light pattern. The various magnitudes of charge stored in a single line of CCD elements in the matrix can be "read" by rapidly shifting these charges to the end of the row in a serial manner, in the shift register mode, by sequentially changing the potentials on the adjacent electrodes. Then the next row can be read and serialized after the first row in a manner which is a solid-state analog of a TV raster. The envelope of this pulse train now can be transmitted and represents a video signal translation of the original focused light pattern. Of course it is necessary that the storage time of the trapped electrons at the CCD elements be longer than the video frame time for successful **vidicon** operation. Devices of this type have been fabricated with as many as one million CCD elements.

12.5.6 Polysilicon CCDs

For high speed operation of the CCD, the separation between the MOS elements must be made small. This must be accomplished in fabrication without electrically shorting the adjoining elements. An overlapping insulated electrode structure has been developed to permit very close spacing of the elements and to provide field-aided transport between them. This is illustrated in Fig. 12.17, where alternating aluminum and polysilicon electrodes (gates) are used. As shown, the overlapping aluminum electrodes are separated from the polysilicon gates by a thin layer of silicon dioxide, providing for a very close spacing

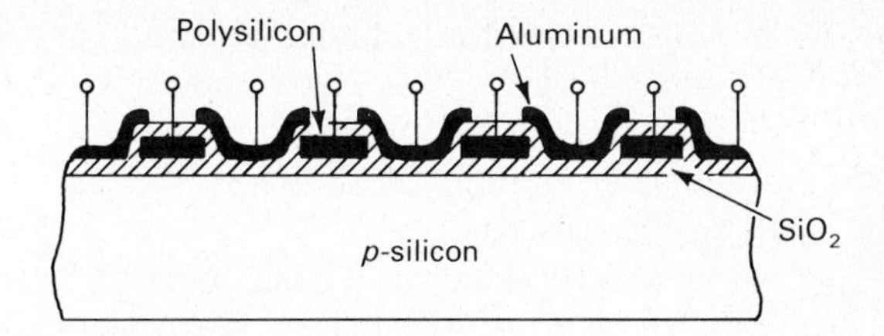

Figure 12.17 A CCD structure with polysilicon electrodes and overlapping aluminum electrodes in order to permit the adjacent elements to be spaced only an oxide thickness from each other.

of adjacent elements. The number of sequential elements that can be used in this charge transfer process depends critically on the efficiency of transfer. Although the transfer process approaches 100%, a minute loss of charge occurs through trapping of a few electrons in surface states. When this is compounded on transfer of a charge packet from element to element, the amount of charge can diminish drastically after a large number of transfers.

The transfer efficiency can be improved by forcing the electron transport to take place away from the Si/SiO_2 interface and hence avoiding any surface traps present there. A **buried-channel** CCD has an ion-implanted *n*-type layer between the gate oxide and the *p*-type substrate. This causes the electric potential under the electrodes to peak at the *n*-*p* interface rather than the Si/SiO_2 interface, enhancing the transport along the former interface.

EXAMPLE 12.3

The transfer efficiency of electrons from one electrode to the next in a CCD structure is 99.9%. A CCD shift register contains 1000 elements. Calculate the fraction of electrons left after transfer through these 1000 elements.

Solution

The transfer efficiency of the total device is the product of the transfer efficiency between any two elements. Hence the fraction of electrons left after 1000 transfers, f_{1000}, is

$$f_{1000} = (0.999)^{1000} = \underline{0.368}.$$

12.6 THE THYRISTOR

The term **thyristor** refers to a family of semiconductor devices with four or more layers which are used as electronic current switches. These devices undergo a transition from the high impedance, voltage-blocking OFF state to the low impedance, current-conductive ON state by regenerative action. They are used extensively as high power electronic switches. The structure of one such device, commonly called a **silicon controlled rectifier (SCR),** consists of two bipolar transistors, a *p-n-p* and an *n-p-n,* integrated onto a single chip of silicon; these are interconnected so that the *n-p-n* drives the *p-n-p* transistor in a regenerative manner as shown in the equivalent circuit illustrated schematically in Fig. 12.18a.

A schematic sketch of a basic thyristor integrated structure is shown in Fig. 12.18b. A small *n-p-n* transistor base current can cause a large *p-n-p* transistor current to flow. Note that the collector region of the *n-p-n* transistor structure is also the base region of the *p-n-p* transistor in this integrated device. Also note that the *p-n-p* transistor base region is generally much wider than that of the *n-p-n* transistor. This is necessary in order to provide room for the expansion of the *n-p-n* transistor collector depletion region and accounts for the high voltage blocking capability of this device. Unfortunately the wide *p-n-p* transistor base region constrains the thyristor to be a relatively low frequency power control device, operating at frequencies below 10 kHz. Its most frequent application is at 60 Hz.

12.6.1 Silicon Controlled Rectifier Operation

In order to demonstrate the regenerative action in thyristor devices consider the two transistor analogs shown in Fig. 12.18a. Using equations like (8.39) for

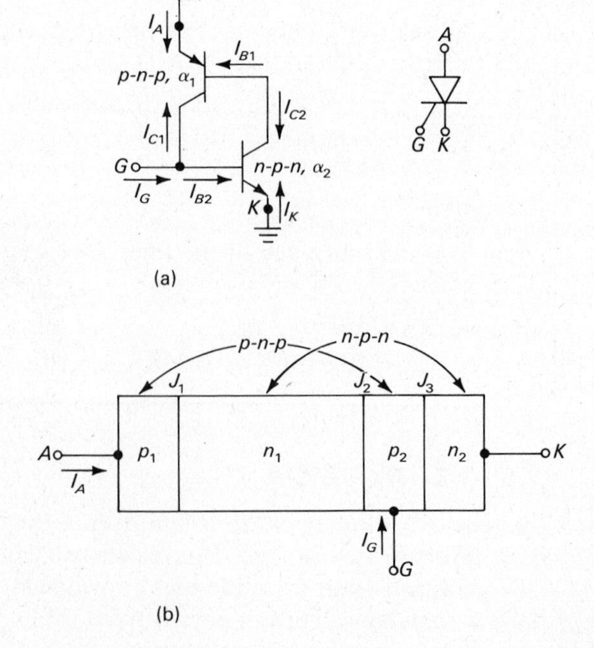

Figure 12.18 (a) Equivalent circuit of the four-layer thyristor device (SCR); also shown is the circuit symbol for this device.
(b) A schematic drawing of the basic thyristor structure. A small gate (base) current I_G can cause a large anode current I_A to flow by regenerative action.

the *p-n-p* (transistor 1) and the *n-p-n* (transistor 2) collector currents give

$$I_{C1} = -\alpha_1 I_A + I_{CBO1} \tag{12.10a}$$

and

$$I_{C2} = -\alpha_2 I_K + I_{CBO2}. \tag{12.10b}$$

Here it is assumed that the anode A is at a positive potential relative to the cathode K, so that both transistors operate in the active modes. Also, Kirchhoff's current law gives

$$I_G + I_K + I_A = 0. \tag{12.11}$$

Combining Eqs. (12.10) and (12.11) yields

$$I_A = \frac{[\alpha_2 I_G + (I_{CBO1} + I_{CBO2})]}{[1 - (\alpha_1 + \alpha_2)]} \tag{12.12}$$

Equation (12.12) demonstrates the regenerative action which can occur in this circuit configuration. In general the current gain α of a bipolar transistor will increase with current, as the current increases from small values (see Section 9.1.2). When the sum of the current gains of the *p-n-p* and *n-p-n* transistors, $\alpha_1 + \alpha_2$, approaches one, the anode current is predicted by Eq. (12.12) to increase, limited only by the external circuit. For the output of the *n-p-n* transistor drives the input of the *p-n-p,* which in turn feeds back current to the input of the *n-p-n,* raising its output, increasing the input to the *p-n-p,* etc., in a regenerative manner. The current gains can rise to a point where gate current which began this cycle is no longer needed to maintain the output current. This is predicted by Eq. (12.12), which indicates that I_A increases without limit when $\alpha_1 + \alpha_2$ approaches one, even with $I_G = 0$.

This increase of current, limited only by the external circuit, implies that the transistors both assume a low impedance state. This requires that the col-

Plastic packaged silicon controlled rectifier ready for mounting onto a heat sink. (Supplied by the Unitrode Corporation, Lexington, MA.)

lectors of both transistors become forward biased (operate in the saturation mode). Under this condition the total drop across the composite device is only one diode voltage drop (the two others cancel each other) plus some series resistance voltage drop. Hence the device, when triggered by sufficient gate current to increase the current gains appropriately, will switch from the high impedance state to a low impedance state. Regenerative action at that point does not necessitate a supply of gate current and the device is said to be **latched ON.** Hence the device can be switched ON by a pulse of sufficient height and duration.

A. PHYSICS OF SWITCHING

Thyristor device operation can also be understood by considering electron and hole flow in the structure drawn schematically in Fig. 12.18b. Assume that the anode A is placed at a moderately high positive voltage relative to the cathode K, so that junction J_2 is a high impedance (reverse biased) while junctions J_1 and J_3 are forward biased (low impedance). Since the n-p-n transistor is biased in the active region, a base (gate) current I_G causes electrons from the emitter region n_2, to be injected through base region p_2 and to be collected in region n_1. But electrons injected into n_1 correspond to majority carrier base current for the p-n-p transistor and cause holes to be injected into n_1 from the emitter p_1. These holes traverse the base region n_1 and are collected in the collector region p_2. But holes injected into p_2 correspond to majority base current for the n-p-n transistor calling for more electrons to be injected from its emitter n_2. This process can reach a steady-state situation in which both transistors are operating in the active region carrying a limited current. However, if the current gains are sufficiently high, $\alpha_1 + \alpha_2$ becoming equal or greater than one, the internal device current would tend to exceed the external current I_A. This of course is not possible and the transistor current gains adjust themselves to lower values by switching into the saturated mode. Both transistors are now in saturation, with all three junctions biased in the forward direction and the device is operating in the low impedance, latched ON state.

It should be noted that the four-layered device, illustrated in Fig. 12.18b, can also be switched from the high impedance, **blocking** mode to the low impedance, **conducting** mode, operated as a two-terminal device ($I_G = 0$). Consider again that a moderate positive voltage is applied to the anode A relative to the cathode K. Junction J_2 will be in reverse bias, with junctions J_1 and J_3 conducting current corresponding to the leakage current of the junction J_2 as modified by the current gains of the composite device. This corresponds to common emitter leakage I_{CEO} as described in Section 8.2.3. When the anode voltage is raised, the leakage current rises somewhat (see Section 6.2.4). This will cause the current gains of the two transistors to increase. When $M(\alpha_1 + \alpha_2)$ reaches one the device will switch into the low impedance state and become latched. Here the avalanche multiplication factor M is introduced since the avalanche current contributes to the total current. The switching of this two-terminal device from the high voltage blocking mode to the conducting mode where there is only a small voltage drop (∽ 0.7 V) across the device is illustrated in the device *I-V* characteristic shown in Fig. 12.19a. Indicated there is the behavior of the device with zero gate current. The forward **breakover voltage** V_{BF} occurs as a result of the voltage dependence of the current gains α_1 and α_2, and avalanche multiplication M at junction J_2. Also shown are the device characteristics with $I_G > 0$. As the gate current is increased the voltage blocking capability of junction J_2 is progressively reduced in a manner already discussed for a transistor in Section 9.1.3.

B. CIRCUIT SWITCHING

A simple DC electronic circuit which is convenient for describing the switching performance of the *p-n-p-n* device is shown schematically in Fig. 12.19b. When the *p-n-p-n* switch is ON, almost all of the positive supply voltage V_{AA} (∽100 V or more) appears across the load resistor and appreciable current is conducted (see point L on the load line of Fig. 11.19a); with the switch OFF nearly all of the voltage appears across the thyristor device (see point H). Should the supply voltage V_{AA} be made negative relative to the cathode, again all of the

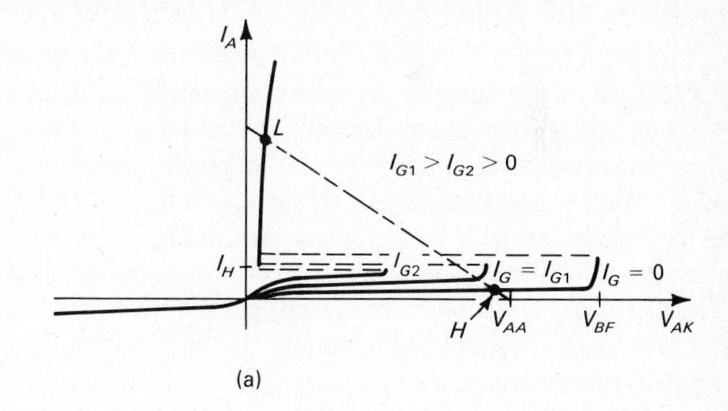

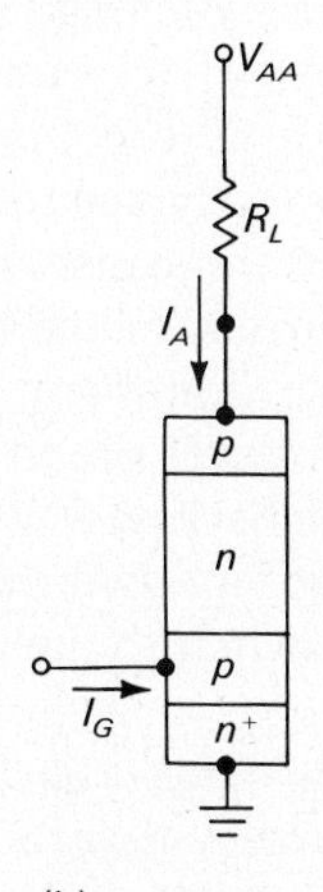

Figure 12.19 (a) The current–voltage characteristics of *p-n-p-n* switch with different gate currents. The dash-dotted line represents a resistive load line whose slope is $-1/R_L$.
(b) Schematic diagram of a simple DC switching circuit utilizing a *p-n-p-n* device.

voltage appears across the four-layer device. In this case both junctions J_1 and J_3 (see Fig. 12.18b) will be in reverse voltage bias[12] and the two-transistor action which can give rise to switching to the low impedance state is not possible. However, with a sufficiently high negative voltage the anode junction J_1 will eventually undergo avalanche breakdown or punchthrough, causing a large flow of current.

Consider the switching circuit of Fig. 12.19b. With $I_G = 0$ the device will support a positive anode-to-cathode voltage V_{AK}, indicated by the voltage corresponding to point H on the load line of Fig. 12.19a. The device may be switched ON to the low impedance state by increasing the supply voltage V_{AA}. This corresponds to the parallel displacement of the load line in the direction of increasing V_{AK}. When V_{BF} is exceeded the device switches ON. Alternatively, with V_{AA} as shown, a supply of gate current I_{G1} will reduce the voltage blocked by a small amount, as indicated by the intersection of the *p-n-p-n I-V* characteristic and the load line. However, when the gate current is increased to I_{G2} the only point of intersection of the load line and the *p-n-p-n* characteristic is point L, representing the switching to the low voltage, high current ON state.

12.6.2 The Physics of Transistor Gain Variation

The switching to the ON state of a thyristor has already been attributed to the increase of the sum of the *n-p-n* and *p-n-p* transistor gains. The mechanisms by which bipolar transistor current gain varies with collector voltage and base (or emitter) current have already been discussed in Chapter 9.

With a positive anode voltage, junction J_2 (Fig. 12.18b) supports most of the voltage with $I_G = 0$. This causes the space-charge region of this junction to widen into the *n-p-n* and *p-n-p* base regions. Because of the wide base region of the *p-n-p* transistor, its current gain is limited by recombination of minority holes in this region and hence by the base width. Increasing the anode voltage positively decreases the base width and increases the gain (see Section 9.1.2), eventually leading to the switching ON of the device. Also, a large increase of current can be caused by voltage punchthrough or avalanche multiplication.

In Section 9.1.2 it was also pointed out that the current gain of transistors at low emitter current is partly due to the dominance of emitter junction space-charge-generated current; this causes majority current flow which does not contribute to the output current of the transistor and hence reduces the current gain. This effect is less dominant at higher emitter current levels. Also, increased emitter current gives rise to an aiding field in the base region, increasing the current gain (see Section 9.1.4). Both of these effects can increase the composite current gain of the two transistors of the *p-n-p-n* device as its current rises, causing it to switch to the ON, low impedance state.

[12]Junciton J_1 will support nearly all of this reverse voltage since the breakdown voltage of J_3 is only about 10 V owing to the heavy doping of regions n_2 and p_2 (see Fig. 12.18b).

A. SHORTED-CATHODE CONFIGURATION

The variety of turnon mechanisms just described cause the gate current for switching to be a sensitive function of several device parameters, making it vary considerably even among devices fabricated at the same time. In order to uniformly manufacture thyristor devices which all switch ON within a small range of gate currents, a **shorted-cathode** device configuration is commonly used. This is illustrated in Fig. 12.20a. The metallization that interconnects the gate and cathode regions at a location remote from the gate electrode causes a reduction in current gain of the *n-p-n* transistor at low anode currents, since the current bypasses the cathode *p-n* junction and is a majority carrier current. This current is directed transversely in the gate region, parallel to the cathode junction.

As the device current increases, this creates a transverse voltage drop across the gate resistance, causing the potential at point B to rise above that at point A, the location of the gate–cathode short circuit. At sufficiently high anode current the potential at point B rises to about 0.7 V above A (essentially at the cathode potential) and electrons begin to be injected across the cathode junction at B. This current calls for more anode current, as already described,

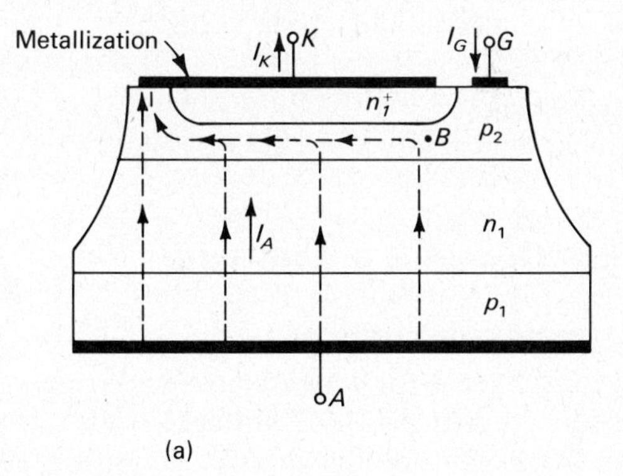

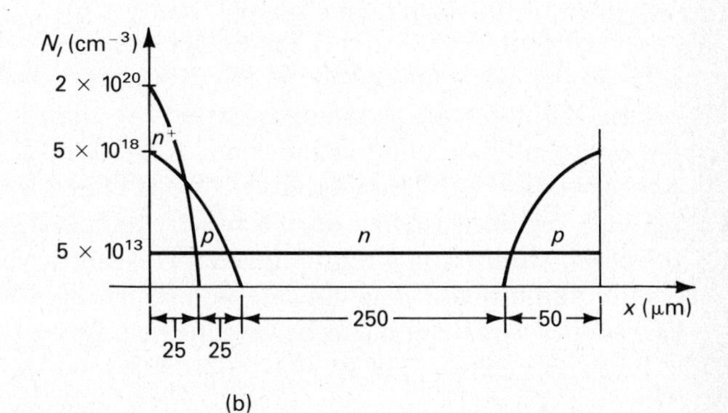

Figure 12.20 (a) Cross section of a thyristor device showing the transverse current flow to the shorted cathode. (b) Impurity distribution, anode to cathode, in a typical silicon thyristor design.

which raises the potential at B and causes more of the cathode junction to inject electrons. Eventually the device turns ON and the entire cathode junction conducts uniformly. The gate current for turn ON may be adjusted by varying the gate region doping and hence the gate resistance.

At this point the gate current which initiated this switching action may be removed and the anode current will be maintained, self-sustained.[13] Since the current flow is now one-dimensional, no transverse voltage drop is present along the cathode junction; this was the cause of nonuniform emitter current conduction or "crowding" as was described for the bipolar transistor in Section 9.1.5. The uniformity of current flow in the thyristor in the ON state produces an even distribution of heat dissipation in the device, which is an important factor in accounting for its higher power-handling capability compared with the bipolar transistor. However, the anode-to-cathode voltage drop of the thyristor in the ON state is about one volt or a little more than one *p-n* junction forward voltage drop. This is in contrast to the few tenths of a voltage drop across a turned-ON bipolar transistor.

For some circuit application the self-sustained ON state or latching of the thyristor offers an advantage. For a turned-ON bipolar transistor to remain conducting its base current must be maintained. When the base current is removed the device turns OFF after a time delay. The thyristor *must* be turned OFF by removing the anode voltage, or in some cases extracting gate current. Problems related to the turn OFF of a highly conducting thyristor will be discussed later. The high voltage capability of the thyristor of course is related to its thick, lightly doped *n*-type region (see Fig. 12.20b).

12.6.3 Triggering the Thyristor

A. THYRISTOR TURNON

The four-layer *p-n-p-n* device may be **triggered** or turned ON with $I_G = 0$ by exceeding a critical anode voltage, or else by providing sufficient gate current drive. As a result of regenerative action the device has a large current gain, so only tens of milliamperes are sufficient to turn ON thousands of amperes of anode current. Alternatively the device may be switched ON by light. If light of a proper wavelength ($h\nu > E_g$) produces hole–electron pairs in the gate region, the holes act as majority carriers in the *p*-type gate region. Even with the electrically supplied gate current $I_G = 0$, these majority holes satisfy the recombination requirements for electrons to be injected from the cathode into the gate region, just like an electrically supplied gate current. This electron injection will initiate a regenerative, switching action if the electrons are sufficiently numerous. Hence an intense enough light pulse of sufficient duration and proper wavelength can be used to turn ON a thyristor switch. This input

[13]Of course the thyristor can also be turned ON by supplying sufficient anode voltage, without gate current, as previously discussed.

light signal can cause the thyristor to trigger and yet permit complete electrical isolation of the device input circuit. A device designed to operate in this manner is called a **light-activated silicon controlled rectifier** or LASCR.

The extremely high gain of the thyristor device makes it susceptible to inadvertant switching in a noisy environment. This can be remedied at the expense of device triggering sensitivity by using a shorted-cathode configuration and adjusting the transverse gate resistance. The thyristor may also be inadvertently turned ON if the device is exposed to a high temperature. The mechanism for turn ON due to heating is as follows: Consider the thyristor device sketched in Fig. 12.19b to be voltage biased with the anode positive with respect to the cathode. The middle junction will then be reverse biased and the leakage current through this junction will increase substantially with increasing device temperature. Even with $I_G = 0$ this current may be sufficient to initiate switching even without the aid of gate current, owing to increased transistor current gain. Hence the blocking voltage of the thyristor decreases with increasing temperature. The shorted-cathode device configuration tends to provide higher temperature stability by bypassing this leakage. However, when the leakage current flowing transversely in the gate region to the short establishes a sufficient voltage drop, the cathode junction will begin to inject electrons and thyristor turn ON will follow. The maximum operating temperature of the silicon thyristor is in the range 125–150°C.

B. THYRISTOR TURNOFF

The thyristor may *not* be turned OFF by removing its gate drive; regenerative action keeps it ON. The device may be turned OFF by reducing its anode current below the **holding current** I_H, as shown in Fig. 12.19a. This is defined as the lowest anode current for which the thyristor can maintain its conducting condition. Below I_H the device reverts to the high impedance state since there is then insufficient current to make the composite current gain of the device, $\alpha_1 + \alpha_2$, equal to one. The reduction in the anode current can be accomplished by decreasing the positive anode voltage relative to that of the cathode. This can be visualized by translating the load line of Fig. 12.19a parallel to itself until the only intersection with the device's *I-V* characteristic can occur on the high impedance segment of this curve. In AC voltage operation, the anode potential is automatically lowered each half cycle.

Alternatively the thyristor should be able to be turned OFF by reversing the gate current, opposite to the direction that is needed to turn the device ON. A thyristor especially designed for gate current turnoff is called a **GTO** thyristor. However, the gate current which must be extracted for turn OFF is usually large, so a substantial voltage drop can occur transversely in the gate region (see Fig. 12.20a). This is in such a direction as to reverse-bias a portion of the cathode junction and hence shut off electron injection. However, the voltage developed then may be sufficient to cause avalanche multiplication in the heavily doped, low breakdown voltage cathode *p-n* junction. When this occurs the gate current will simply maintain this avalanche current without removing the current needed to shut down the device.

The turn OFF gain is usually less than ten, in contrast to the turn ON gain, which is often in the millions. The turn OFF gain may be predicted by referring to Eq. (12.12), this time considering that I_G will be negative. Then the turn OFF gain I_A/I_G becomes, from Eq. (12.12),

$$\frac{I_A}{I_G} = \frac{\alpha_2}{(\alpha_1 + \alpha_2) - 1}, \tag{12.13}$$

if we neglect the small leakage currents. Since the composite gain $\alpha_1 + \alpha_2$ can be much greater than one when the device is conducting high current, I_A/I_G will be small. For example, take $\alpha_1 + \alpha_2 = 1.2$ with $\alpha_2 = 0.8$. From Eq. (12.13) the turn OFF gain will be only $0.8/(1.2 - 1) = 4$. To turn OFF 1000 A would require a gate current of 250 A in contrast to about 10 mA for turn ON.

12.6.4 Limits of Thyristors in Transient Operation

A. THE *di/dt* LIMITATION

In initiating the turn ON of a thyristor by supplying a gate current pulse the *n-p-n* transistor cathode injects electrons into the gate region in a manner entirely analogous to that already discussed for a bipolar transistor. As discussed in Section 9.1.5 the emitting junction current will be nonuniformly conducted, being concentrated or crowded to the junction edge adjacent to the base (or gate) contact. This current concentration can cause local heating. In the thyristor, if the gate current pulse is large and rises rapidly, this will cause the conduction of a high cathode current density at its edge nearest the gate electrode. Since a large current is conducted in this region for a time too short for most of the heat so produced to diffuse from the small region, an excessive temperature rise can take place locally. This can lead to the device destruction mechanism known as ***di/dt* burnout.** If the gate current pulse is applied more gradually, however, the longer rise time will permit the gradual spreading of the turned ON portion of the cathode current away from its edge to yield more uniform conduction and heating. A large, short duration gate current pulse is used in an attempt to achieve fast thyristor turn ON by providing strong gate drive (see Section 9.6.1). However, care must be taken in this regard to avoid thermal device failure. Thyristors have been specially designed to permit triggering by current pulses with rise times of up to 100 A/μsec.

B. THE *dv/dt* LIMITATION

Consider that an anode-to-cathode voltage pulse is suddenly applied to a *p-n-p-n* device with $I_G = 0$. It is observed that the height of the pulse that may be blocked becomes less as the pulse rise time is reduced. That is, the thyristor can become conducting prematurely if the anode voltage is applied too suddenly. This can occur under AC operation if the sinusoidal frequency is too high or if a high voltage noise spike occurs.

Up to now current conduction in the thyristor has been described as being initiated by electrical carrier or light injection or by thermal generation of electron–hole pairs. However, if a fast-rising reverse voltage pulse is applied to a *p-n* junction, a majority carrier displacement current will flow owing to the capacitive nature of the junction. Current conduction will than take place according to

$$i = \frac{dq}{dt} = \frac{d(cv)}{dt}. \tag{12.14}$$

If a positive anode potential is suddenly applied to the thyristor of Fig. 12.18b with $I_G = 0$, this voltage will tend to be mainly supported by the central junction J_2. However, there is a finite time necessary for the charging of the capacitance of this junction during which a charging current given by Eq. (12.14) flows. This current of course must be conducted by the *n-p-n* and *p-n-p* injecting junctions J_1 and J_3. The current may be sufficiently high, causing the effective current gain of the device, $\alpha_1 + \alpha_2$, to reach unity, initiating regenerative action and switching the thyristor ON. This premature triggering is termed ***dv/dt* turn ON.** In an attempt to design a thyristor for higher current capability the cathode current injecting area is increased. Hence the area of junction J_2 will correspondingly increase as well as its capacitance. This will facilitate *dv/dt* turn ON according to Eq. (12.14). Special thyristor device designs have yielded *dv/dt* capability of up to 2000 V/μsec.

C. SILICON CONTROLLED RECTIFIER TURNOFF OR REVERSE RECOVERY

The limiting switching frequency of an SCR is determined generally by minority carrier charge storage. The manner in which charge storage effects limit the switching speed of transistors has already been described in Section 9.6.2. The thyristor is used primarily to control large currents and in its ON condition can be considered as operating as two heavily conducting transistors in saturation. The main problem in turning OFF this SCR is to remove the minority carriers stored in the wide base of its *p-n-p* transistor. The *p-n-p* transistor turn OFF time is long since it depends on the square of the base width.

The simplest technique for turning OFF a thyristor is to reduce its conducted current below the holding current. For this to occur the minority carriers must disappear by recombination or be extracted from the device through the gate lead. The problems connected with GTO thyristor turn OFF have already been described. Hence a short recombination time (lifetime) in the *p-n-p* transistor base region is necessary for rapid turn OFF. This of course is contrary to the requirement of long diffusion length for heavy conductivity modulation of the thick *n*-region necessary for a low forward voltage drop. Hence a design compromise is required which limits the frequency of operation of high voltage, high power thyristors to the 10-kHz range.

12.6.5 Bilateral Thyristor Devices (The Triac)

The *p-n-p-n* devices already described can be switched to the low-voltage, conducting state when the anode voltage is positive relative to the cathode. With the anode voltage negative (and below the avalanche voltage), these devices are always in the high impedance, nonconducting state. Hence in the control of AC power, two silicon controlled rectifiers would have to be used back-to-back in order to provide for power control over the full 360° AC cycle. A device useful for efficient AC power control over the full time cycle is called a **triac.** This device contains four or more junctions and functions as two back-to-back SCRs; one conducts during the positive part of the cycle while the other conducts during the negative part. This gives rise to an *I-V* characteristic as shown in Fig. 12.21.

A cross section of a typical triac structure is shown in Fig. 12.22a. A variety of other triac structures have also been proposed which can conduct current in both directions, and which can be turned ON with positive- or negative-directed gate voltage pulses. A full description of the physics of operation of the device shown in Fig. 12.22a is rather complex. The essence of its mode of operation, however, can be discussed by describing the carrier flow in the two-terminal bilateral device illustrated in Fig. 12.22b ($I_G = 0$).

The two-terminal triac is a laterally symmetric device, so the marking of anode A and cathode K is only for discussion purposes. Consider that the anode is placed at a postive potential relative to the cathode. Because of the shorted-gate structure (see Fig. 12.22b), leakage current will be conducted by the transistor p_1-n_1-p_2, with junction J_2 reverse biased. The current will bypass the electrically shorted junctions J_3 and J_4. This will cause a transverse flow of current in region p_2, parallel to junction J_3. As described in connection with the shorted-cathode configuration discussed in the previous section, a transverse voltage drop will now occur in region p_2, making point B attain a positive

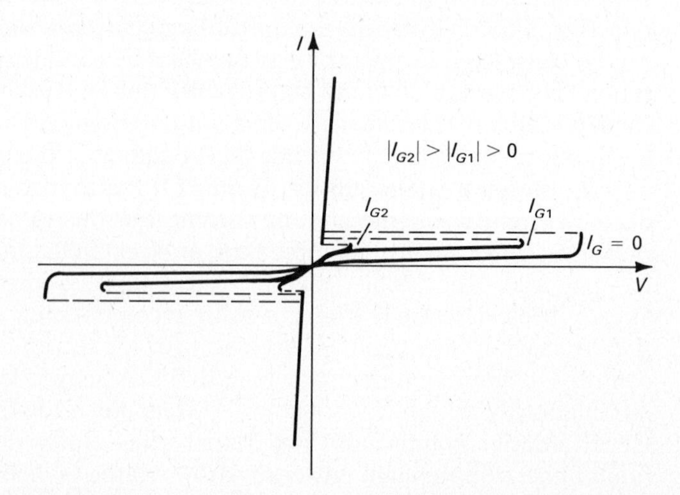

Figure 12.21 Current-voltage characteristic of a triac.

Figure 12.22 (a) Cross section of a typical triac device with circuit symbol.
(b) Cross section of a bilateral, two-terminal thyristor device showing the transverse current flow in region p_2.

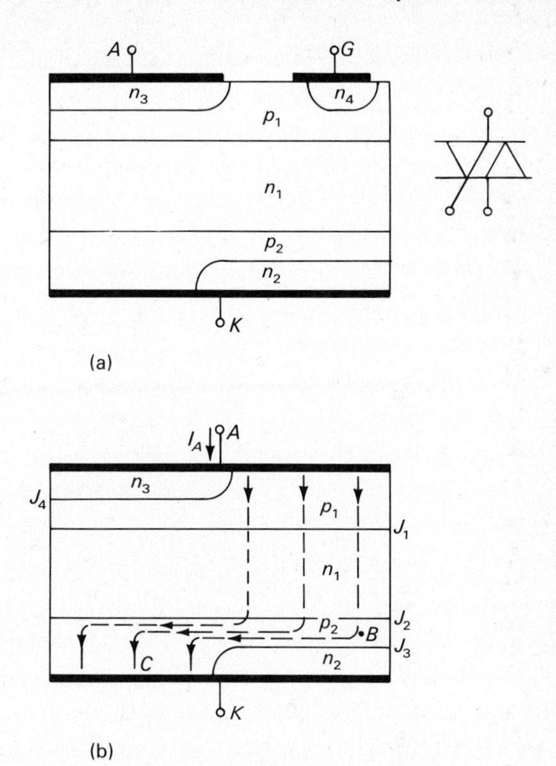

potential relative to the potential of the cathode metallization, marked C. This will cause junction J_3 to begin injecting electrons through region p_2 and into n_1. There they serve as majority electrons, playing the part of a base current for the transistor p_1-n_1-p_2, causing holes to be injected from region p_1 through n_1 into p_2. This causes more electrons to be injected by junction J_3, etc., and regenerative switching can occur. Of course if the roles of anode and cathode are interchanged the same description applies by symmetry. In the gate-triggered device of Fig. 12.22a a gate current initiates the action rather than anode voltage.

12.6.6 Phase-Control Circuitry

Since one of the most important applications of thyristors is for the regulation of AC power by **phase control,** a simple silicon controlled rectifier circuit for this purpose will now be discussed. Consider the application of AC power to an electrical load which is regulated by an SCR as shown in Fig. 12.23a. If the potential at point B is positive relative to that at point A, the SCR is forward biased for conduction. Current will be conducted through the RC network and point E will become electrically positive relative to A, raising the gate electrode above the cathode in potential. The point in the AC cycle at which the voltage

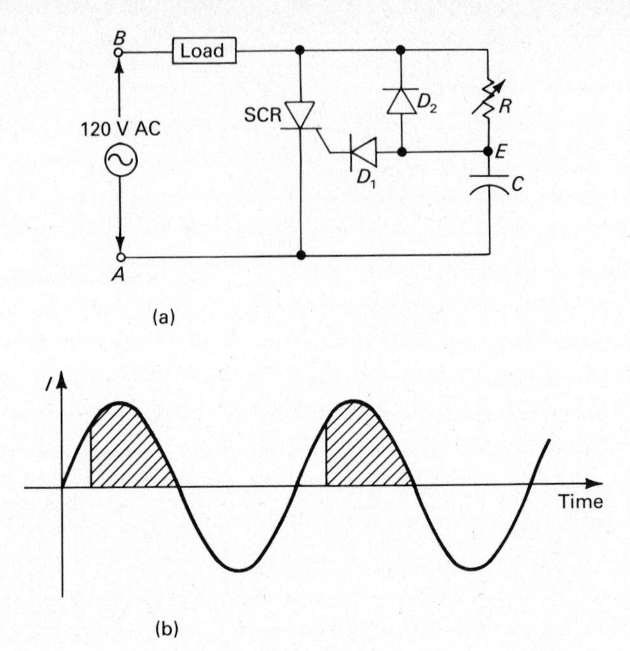

Figure 12.23 (a) A half-wave AC phase-control circuit using an SCR and an RC circuit for triggering. Diodes D_1 and D_2 prevent a high reverse voltage from being applied to the gate junction. (b) The AC current conducted (crosshatched) by the load. If a triac were used instead of an SCR control could be achieved over the full AC cycle.

at E is sufficient to trigger the SCR will depend on the ratio of the resistance R to the capacitive impedance $1/2\pi fC$. Hence the variable resistor can be used to adjust the point on the AC cycle at which the SCR will turn ON, causing current to be conducted through the load. This is illustrated in Fig.12.23b where the portion of the AC cycle during which load current conduction takes place is shown crosshatched.

When the potential at point B is negative relative to A, the SCR will be reverse biased and will not conduct current. Hence this circuit will only be useful in controlling the power through the load over 180°, or one-half the AC cycle. A similar circuit using a triac will offer power regulation over the full 360° of the AC cycle. This type of phase-control circuitry can be used for light dimming, controlling the speed of motors, electric heater temperature control, etc.

Some advantages of using thyristors rather than power transistors in power control are

1. The thyristor requires little gate drive current to turn ON very high currents.
2. The thyristor can block voltage for both polarities of AC power.
3. The thyristor has a very high blocking voltage capability coupled with a low voltage drop at high currents.
4. Current is conducted uniformly when the thyristor in ON.

5. No second breakdown destructive burnout mechanism occurs in the thyristor.

Some disadvantages of thyristors compared to power transistors are

1. Thyristors cannot be turned OFF by turning OFF the gate current.
2. Thyristors cannot operate at high frequencies.
3. Thyristors are subject to turn ON by noise voltage spikes.
4. Thyristors have a limited temperature range of operation.

☐ *SUMMARY*

Two devices exhibiting a negative differential conductance *I-V* characteristic were analyzed. The GaAs Gunn, high frequency oscillator operates on the principle that at high electric fields electrons are transferred from a central valley in the conduction band where its mass is low to a satellite valley where its mass is high. This results in a decrease in average electron mobility, and hence a negative conductance. The IMPATT diode develops a negative differential conductance when a current pulse generated by avalanche multiplication is collected 180° out of phase with the voltage signal that created it. Si and GaAs IMPATT devices can generate higher power output at high frequencies (greater than 10 GHz) than other solid-state devices available today. The mechanism by which the semiconductor laser efficiently emits intense, coherent, unidirectional light radiation is described in terms of the theory of solids. CW lasers which operate reliably at room temperature and radiate light of a wavelength appropriate to optical communication can be constructed by using heterojunctions of compound semiconductors. The CCD whose basic element is the MOS capacitor is described. The shift register mode of operation of this easily fabricated device can be used to construct large-scale electronic memories or high resolution optical imaging devices. The thyristor contains four or more *p-n* junctions and can operate as an electronic switch handling hundreds of amperes of current and blocking thousands of applied volts. However, this device can only switch at frequencies in the 10-kHz range and tends to be temperature sensitive. An application of a type of thyristor, called an SCR, in phase control of power is described.

PROBLEMS

12.1 (a) Show that for space-charge domain formation in the Gunn oscillator mode the nL product (electron density times active sample length) is of the order of $10^{12}/\text{cm}^2$ for gallium arsenide, using the data given following Eq. (12.2).

(b) For $nL \approx 10^{12}/\text{cm}^2$, is the inequality of Eq. (12.3) satisfied for the LSA mode? Take the oscillation period to be about one-third the transit time.

12.2 Calculate the ratio of the density of states at some energy E just above the band edge in the satellite valley compared with the central valley of GaAs at 300°K using the expression of Eq. (4.10) and the information given in Fig. 12.2.

12.3 **(a)** Estimate the DC power dissipated in a GaAs Gunn diode oscillator chip 1.0 mm^2 in area, with active region of 10.0 Ω-cm resistivity and 70 μm thickness, just at the threshold of oscillation.
(b) Calculate the time between pulses generated by this device operating in the Gunn mode.

12.4 The Reed type of IMPATT diode shown in Fig. 12.5 can produce high frequency oscillations when used in the avalanche mode. Determine the minimum DC bias voltage for this device when operated as a microwave oscillator, if the n-region thickness is 0.5 μm and its doping is $2.0 \times 10^{16}/cm^3$; the i-region thickness is 3.0 μm. The device is fabricated using crystals of (a) GaAs, (b) Si. Assume that the breakdown field is 4.0×10^5 V/cm for both GaAs and Si. Also assume that the electrons drift at their saturation velocity in the intrinsic region.

12.5 Find the optimum operating frequency for the Reed diodes fabricated of (a) GaAs, (b) Si, for the design given in Problem 12.4.

12.6 Lasers have been produced yielding radiation from 0.32 to 16.5 μm in wavelength. Estimate the range of energy gap values for the semiconductor materials used to fabricate these devices.

12.7 For a homojunction GaAs laser chip 100×250 μm in area and 100 μm thick,
(a) Estimate the power dissipated in the device at 300°K if the current density threshold is 5×10^4 A/cm^2 and the diode requires about 1 V drop across it for lasing.
(b) For pulsed operation using 1-μsec-wide pulses at a repetition rate of 1kHz, calculate the time-averaged power dissipated in the device.

12.8 **(a)** Determine the energy $h\nu$ corresponding to the peak radiation indicated in Fig. 12.9b. What does this tell about the value of the energy gap of the material from which this semiconductor laser is fabricated?
(b) Calculate the spacing in angstrom units of the oscillation modes in a GaAs injection laser, according to Eq. (12.7), if the distance between the front and rear reflecting surfaces is 250 μm.

12.9 **(a)** For a GaAs junction laser whose length $L = 250$ μm, determine the time for the radiation to traverse the sample length. (*Note:* the velocity of propagation of electromagnetic radiation in GaAs equals the velocity in vacuum divided by the index of refraction in GaAs.)
(b) Calculate the reflectivity R of a GaAs-air interface using the Fresnel law $R = [(n_s - 1)/(n_s + 1)]^2$, where n_s is the index of refraction of GaAs.
(c) If α, the loss factor, is 120 cm^{-1} at 300°K, find the gain G_{th} necessary for lasing using Eq. (12.6).

12.10 Using Eq. (12.8) and the results of Problem 12.9, estimate the threshold current density J_{th} at 300°K for the GaAs junction laser whose output is shown in Fig. 12.9b. Take $w = 0.2$ μm and $\eta = 0.9$.

12.11 **(a)** Estimate the capacitance in the steady state of a silicon CCD of the type illustrated in Fig. 12.12, with electrode area 3×20 μm, if the silicon dioxide thickness is 0.1 μm and the depletion-layer width in the silicon under the electrode is about 1.0 μm. (*Hint:* this capacitance may be calculated by determining separately the capacitances due to the oxide and the depletion layer and treating the structure as two capacitors in series.)

(b) If these electrodes are spaced 4.0 μm apart center-to-center, calculate the maximum number of such devices it is possible to construct on a silicon chip 5.0 mm^2 in area.

12.12 The capacitance of a silicon CCD of the type shown in Fig. 12.12 is 0.01 pF in the steady state, when 3.0 V is applied to the metal electrode relative to the *p*-type silicon substrate. The electrode dimensions are 3 $\times$ 20 μm and the silicon dioxide thickness is 0.1 μm.

(a) Calculate the charge on the metal electrode.

(b) Estimate the depletion-layer width in the silicon. Assume that it extends into the silicon only in a direction normal to the silicon surface. (*Hint:* See hint to Problem 12.11a.)

(c) Speculate on the capacitance value of this device instantly, at $t = 0^+$, on the application of this voltage, compared with the steady-state $t = \infty$ capacitance value.

(d) Estimate the density of electron charges which, if trapped in the depletion region at $t = 0^+$, would significantly affect the capacitance value at that time.

12.13 Redraw the energy band diagram of Fig. 12.11 for the case of an *n*-type silicon substrate. In this case the potential applied to the metal electrodes must be negative relative to the substrate.

12.14 Find the anode current I_A at which the SCR of Fig. 12.18b switches from the forward blocking state to the conducting state. Assume that the current gain of the *n-p-n* structure is $\alpha_2 = 0.6 + 0.1I_A$, and that of the *p-n-p* transistor is $\alpha_1 = 0.2 + 0.1I_A$.

12.15 **(a)** Determine the maximum reverse blocking voltage V_{KA} of the SCR of Fig. 12.18b, if the impurity concentration in the n_1 region is 5×10^{13}/cm donors and the width of that region is 200 μm. Take the avalanche breakdown field for this doping to be 2.5×10^5 V/cm; assume that the doping everywhere is uniform and that all junctions are abrupt.

(b) Repeat (a) if the width of the n_1 region is reduced to 100 μm.

12.16 **(a)** Determine the effective base width of the p_1-n_1-p_2 transistor structure of Fig. 12.18b and Problem 12.15a when the SCR is forward biased with 500 V.

(b) Calculate the ratio of the current gain α of the p_1-n_1-p_2 transistor when the applied forward bias is 1000 V compared with α when this bias is reduced to 500 V. Assume that the diffusion length of holes in n_1 is 100 μm. Take the injection efficiency of the p_1-n_1 junction to be unity.

12.17 Compute approximately the anode turn ON current for the shorted-cathode silicon thyristor structure sketched in Fig. 12.20a. Assume that $I_G = 0$ and that the transverse gate region resistance parallel to the cathode junction is 50 Ω.

12.18 **(a)** Determine the longest wavelength of light (in μm) which can be used to trigger a silicon light-activated SCR.

(b) Prove that a GaAs LED or laser may be used as a light source to trigger a silicon LASCR.

12.19 Calculate the maximum dv/dt capability (in V/μsec) of an SCR device with $I_G = 0$ which has a middle junction (J_2) capacitance of 200 pF and which switches ON with an anode current of 10 mA. Take the junction capacitance to be approximately voltage independent.

12.20 The turnoff or recovery time of a switched ON SCR is 20 μsec. Determine the maximum operation frequency of a phase-control circuit using this SCR. Assume that the turnoff time dominates the total switching time for this device.

Appendixes

APPENDIX A

FUNDAMENTAL PHYSICAL CONSTANTS

Quantity	Symbol	Value	Unit
Electronic charge (magnitude)	q	1.60×10^{-19}	coulomb (C)
Electronic mass (magnitude)	m	9.11×10^{-28}	gram (g)
Unit nucleonic mass	m_{nucl}	1.66×10^{-24}	gram (g)
Avogadro's number	N_A	6.02×10^{23}	atoms/g-atom (mole)
Boltzmann's constant	k	1.38×10^{-23}	joule/kelvin (J/°K)
Thermal energy (300°K)	kT	0.026	electron-volt (eV)
Thermal voltage (300°K)	kT/q	0.026	volt (V)
Speed of light (vacuum)	c	3.0×10^{10}	centimeter/second (cm/sec)
Planck's constant	h	6.63×10^{-34}	joule-second (J-sec)
Frequency (cycles/sec)	Hz	1.0	hertz (Hz)
Permittivity of free space	ϵ_o	8.85×10^{-14}	farad/centimeter (F/cm)
Relative dielectric constants:			
Germanium	$(\epsilon_r)_{\text{Ge}}$	16.0	—
Silicon	$(\epsilon_r)_{\text{Si}}$	11.9	—
Gallium arsenide	$(\epsilon_r)_{\text{GaAs}}$	13.1	—
Silicon dioxide	$(\epsilon_r)_{\text{SiO2}}$	3.9	—
Silicon nitride	$(\epsilon_r)_{\text{Si3N4}}$	7.5	—
Sapphire	$(\epsilon_r)_{\text{Sapp}}$	9.2	—
Thermal conductivities (300°K)	κ_{Ge}	0.6	watts/cm-°C (W/cm-°C)
	κ_{Si}	1.5	watts/cm-°C (W/cm-°C)
	κ_{GaAs}	0.46	watts/cm-°C (W/cm-°C)
Lattice constants (300°K)	a_{Ge}	5.646	angstrom (Å)
	a_{Si}	5.430	angstrom (Å)
	a_{GaAs}	5.653	angstrom (Å)
Energy gaps (300°K)	$E_{g\text{Ge}}$	0.66	electron-volt (eV)
	$E_{g\text{Si}}$	1.12	electron-volt (eV)
	$E_{g\text{GaAs}}$	1.42	electron-volt (eV)
Intrinsic concentrations (300°K)	$n_{i\text{Ge}}$	2.4×10^{13}	centimeter^{-3} (cm^{-3})
	$n_{i\text{Si}}$	1.5×10^{10}	centimeter^{-3} (cm^{-3})
	$n_{i\text{GaAs}}$	1.8×10^{6}	centimeter^{-3} (cm^{-3})
Micrometer (micron)	μm	1.0×10^{-4}	centimeter (cm)
Angstrom unit	Å	1.0×10^{-8}	centimeter (cm)
Electron-volt	eV	1.6×10^{-19}	joules (J)

(*continues*)

Prefixes:

- milli (m-) = 10^{-3}
- micro (μ-) = 10^{-6}
- nano (n-) = 10^{-9}
- pico (p-) = 10^{-12}
- kilo (k-) = 10^{3}
- mega (M-) = 10^{6}
- giga (G-) = 10^{9}

APPENDIX B

ELECTRICAL RESISTIVITY VERSUS IMPURITY CONCENTRATION AT 300°K FOR SILICON AND GALLIUM ARSENIDE

Adapted from S. M. Sze, *Physics of Semiconductor Devices,* 2nd ed. (Wiley, New York, 1981), p. 32.

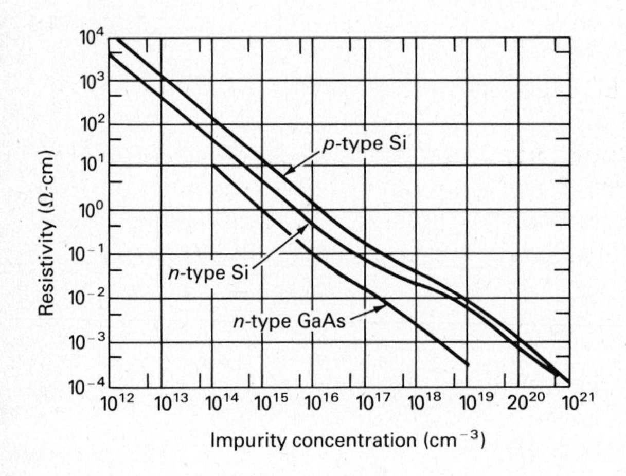

APPENDIX C

SOLUTION OF THE SCHRÖDINGER EQUATION FOR A PERIODIC (KRÖNIG–PENNEY) POTENTIAL

If the Bloch function of Eq. (3.16) is a solution of the Schrödinger equation (3.14), it can be substituted back into the latter equation. This substitution leads to the requirement that $u(x)$ satisfy

$$\frac{d^2u}{dx^2} + 2j\mathrm{k}\frac{du}{dx} - \left(\mathrm{k}^2 - \alpha^2 + \frac{8\pi^2 m_0 V(x)}{h^2}\right) u = 0, \tag{C.1}$$

where $\alpha^2 = 8\pi^2 m_0 E/h^2$; i.e., the Bloch function is a solution of the Schrödinger equation if $u(x)$ satisfies Eq. (C.1).

It is now necessary to introduce the proper potential energy values into this equation. Since the potential function repeats itself with period $(a + b)$ as shown in Fig. 3.1, it will only be necessary to solve the equation in the interval $-b < x < a$. The solution in any other similar interval will be identical with this one. Substituting the appropriate potential energy values from Eq. (3.15), we get

$$\frac{d^2u_1}{dx^2} + 2j\mathrm{k}\frac{du_1}{dx} - (\mathrm{k}^2 - \alpha^2)\, u_1(x) = 0 \qquad \text{(Region I, } 0 < x < a\text{)} \tag{C.2a}$$

and

$$\frac{d^2u_2}{dx^2} + 2j\mathrm{k}\frac{du_2}{dx} - (\mathrm{k}^2 - \beta^2)\, u_2(x) = 0 \qquad \text{(Region II, } -b < x < 0\text{).} \tag{C.2b}$$

Hence $u_1(x)$ and $u_2(x)$ are the values of $u(x)$ in Regions I and II, respectively, and $\beta^2 = 8\pi^2 m_0(E - V_0)/h^2$. Equations (C.2a) and (C.2b) are recognizable as of the form of the differential equation for an electric circuit containing resistance, inductance, and capacitance and can be solved by the method of the Laplace transform. One can show that the solutions of these two equations are of the form

$$u_1(x) = Ae^{j(\alpha-\mathrm{k})x} + Be^{-j(\alpha+\mathrm{k})x}, \qquad 0 < x < a, \tag{C.3a}$$

and

$$u_2(x) = Ce^{j(\beta-\mathrm{k})x} + De^{-j(\beta+\mathrm{k})x}, \qquad -b < x < 0. \tag{C.3b}$$

Boundary Conditions

In order to determine the integration constants A, B, C, and D, it is convenient to apply the conditions of finiteness and continuity of ψ and $d\psi/dx$, at $x = a$,

$x = -b$, and $x = 0$. It therefore follows that the functions $u(x)$ obey these requirements since e^{jkx} is known to be a bounded and continuous function for all x. Also, because of the periodicity of $u(x)$ in the distance $(a + b)$ [see Eq. (3.16)], it follows that $u_1(a) = u_2(-b)$ and $u_1'(a) = u_2'(-b)$. Introducing these boundary conditions in a straightforward but somewhat laborious manner gives

$$A + B = C + D, \tag{C.4a}$$

since $u_1(0) = u_2(0)$;

$$j(\alpha - \mathrm{k})A - j(\alpha + \mathrm{k})B = j(\beta - \mathrm{k})C - j(\beta + \mathrm{k})D, \tag{C.4b}$$

since $u_1'(0) = u_2'(0)$;

$$e^{j(\alpha-\mathrm{k})a}A + e^{-j(\alpha+\mathrm{k})a}B = e^{-j(\beta-\mathrm{k})b}C + e^{j(\beta+\mathrm{k})b}D, \tag{C.4c}$$

since $u_1(a) = u_2(-b)$;

$$j(\alpha-\mathrm{k})e^{j(\alpha-\mathrm{k})a}A - j(\alpha+\mathrm{k})e^{-j(\alpha+\mathrm{k})a}B = j(\beta-\mathrm{k})e^{-j(\beta-\mathrm{k})b}C - j(\beta+\mathrm{k})e^{j(\beta+\mathrm{k})b}D, \tag{C.4d}$$

since $u_1'(a) = u_2'(-b)$.

The simultaneous solution of these equations is very complicated and, in fact, is not necessary. It is presently more interesting physically to obtain the energy values for the electron in a periodic crystal lattice for which valid solutions of Eqs. (C.4) are possible. A theorem of algebra states that in a set of simultaneous linear homogeneous equations [like (C.4)], there is only a nontrivial solution if the determinant of the coefficients of A, B, C, and D equals zero. The evaluation of this determinant again is straightforward but laborious and leads to

$$-[(\alpha^2 + \beta^2)/2\alpha\beta] \sin \alpha a \sin \beta b + \cos \alpha a \cos \beta b = \cos[\mathrm{k}(a + b)]. \tag{C.5}$$

Now the electron energy E can, in general, be less than or greater than V_0. If $E > V_0$ then β is previously defined as a real quantity. In cases in which $E < V_0$, it is convenient to introduce $\beta = j\gamma$ and then we have for Eq. (C.5), noting that $\cos jx = \cosh x$ and $\sin jx = j \sinh x$,

$$[(\gamma^2 - \alpha^2)/2\alpha\gamma] \sinh \gamma b \sin \alpha a + \cosh \gamma b \cos \alpha a = \cos[\mathrm{k}(a + b)]. \tag{C.6}$$

Here γ^2 is real and positive for electron energies such that $0 < E < V_0$; β^2 is real and positive for energies such that $E > V_0$, where Eq. (C.5) is most conveniently applied.

Allowed Energy Bands

Equations (C.5) and (C.6) are complicated transcendental equations involving the electron energy E through the quantity α. To solve for α, the method of graphical solution is suitable. However, to obtain an equation more susceptible to solution, Krönig and Penney suggest a special case. Let the potential barrier width b shrink to zero, and the barrier height V_0 at the same time grow to infinity, with the product of the two, V_0b, remaining finite. Then Eq. (C.6) is most appropriate and as $\gamma \rightarrow \infty$, Eq. (C.6) reduces to

$$[P \sin(\alpha a)/\alpha a] + \cos \alpha a = \cos ka, \tag{C.7}$$

where $P = 4\pi^2 m V_0 ba/h^2$ is sometimes referred to as the **scattering power** of the potential barrier. It is a measure of the strength with which electrons in a crystal are attracted to the ions on the crystal lattice sites.

There are only two variables in this equation, namely α and k. The right-hand side of Eq. (C.7) is bounded since it can only assume values between +1 and −1. If we plot the left-hand side of this equation against αa, it will be possible to determine those values of α (and hence energy) which are permissible; that is, permit $[P \sin(\alpha a)/\alpha a + \cos \alpha a]$ to take values between +1 and −1. When each of these same values is set equal to cos ka, k is determined; α can then be found from Eq. (C.7).

Figure C.1 shows this plot for a value of P assumed arbitrarily as $3\pi/2$. Indicated in the figure are the permitted values of this function, shown as a solid line. This then gives rise to the concept of ranges of permitted values of α for a given ion lattice spacing a, and since $E = \alpha^2 h^2/8\pi^2 m$, allowed **bands of energy** are predicted. Note that the width of each allowed band decreases as P increases; i.e., *tighter binding* results in narrower energy bands. In fact, as $P \rightarrow \infty$, since the left-hand side of Eq. (C.7) must remain bounded and hence finite, $\sin(\alpha a) = 0$. The solutions of this latter equation are $\alpha a = \pm n\pi$, where n is any integer; then the permitted energies reduce to unique values $E_n = n^2h^2/$

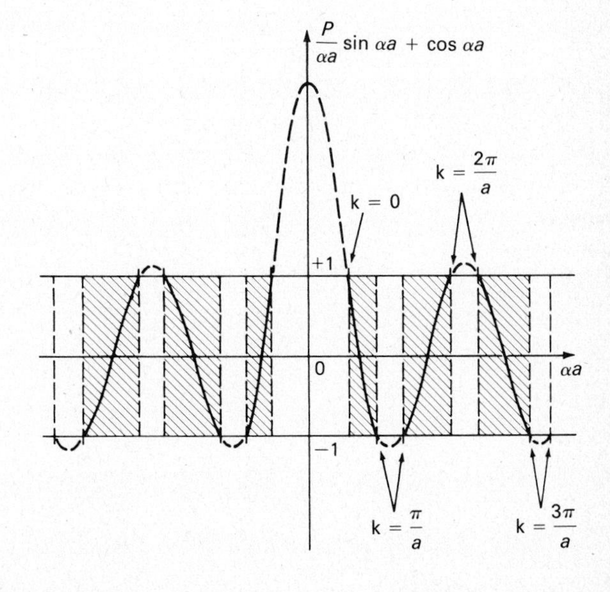

Figure C.1 A graph of $(P/\alpha a)(\sin \alpha a + \cos \alpha a)$ versus αa with $P = 3\pi/2$, for the purpose of solving transcendental equation (C.7). The allowed values of αa are indicated by the crosshatched regions. Since α is related to electron energy the allowed energy values are determined.

$8ma^2$. These are recognized as the eigenvalues for the electron in a box with impenetrable walls. This seems reasonable since in this case the electron is prohibited from leaving its ion attachment. Finite values of P lead to a smearing out of these discrete levels into energy bands.

At the other extreme, i.e., $P \rightarrow 0$, then $\cos \alpha a = \cos \mathrm{k}a$ or $\mathrm{k} = \alpha$. When we introduce the latter value of α, this result leads to $E = h^2\mathrm{k}^2/8\pi^2 m$ and corresponds physically to the "free" electron with all energies permitted. Since for the free electron, $E = p^2/2m$, k can be identified with the electron momentum as $p = h\mathrm{k}/2\pi$. The general relationship (for any P) between the electron energy E and the wave vector k is of course contained in Eq. (C.7).

APPENDIX D

RATIO OF DRIFT TO DIFFUSION CURRENT IN BIPOLAR TRANSISTORS

To Prove:

At low injection levels the minority electron current in the *p*-region near the junction of an n^+-p diode is almost exclusively by diffusion, with the electron drift current being small in comparison.

Proof:

Calling R the numerical ratio of the drift to diffusion current, then from Eqs. (5.20) and (5.22) we can write

$$R = \frac{q\mu_n n\mathscr{E}}{qD_n(dn/dx)}. \tag{D.1}$$

Hence the subscript n on n_p is left off since the derivation refers only to the *p*-region. The hole current portion of Eq. (5.22) may be rewritten as

$$\frac{dp}{dx} = \frac{J_p - q\mu_p p\mathscr{E}}{qD_p}. \tag{D.2}$$

Now the charge neutrality assumption requires equality of injected holes and electrons. Hence

$$p - p_0 = n - n_0, \tag{D.3}$$

where again the subscript 0 refers to equilibrium conditions. Differentiation of Eq. (D.3) shows the equality required of the hole and electron gradients, that is,

$$\frac{dp}{dx} = \frac{dn}{dx}. \tag{D.4}$$

Introducing Eq. (D.4) into Eq. (D.2), we can write

$$\frac{dn}{dx} = \frac{J_p}{qD_p} - \frac{\mu_p}{D_p}\mathscr{E}. \tag{D.5}$$

Substituting this value for dn/dx into Eq. (D.1) yields

$$R = \frac{(\mu_n/D_n)n\mathscr{E}}{(J_p/qD_p) - (\mu_p/D_p)p\mathscr{E}}. \tag{D.6}$$

Dividing numerator and denominator by $p\mathscr{E}$ then gives

$$R = \frac{n/p}{(J_p/q\mu_p p\mathscr{E}) - 1}. \tag{D.7}$$

Now the total hole current J_p consists in general of a drift plus a diffusion component $J_{\text{drift}} + J_{\text{diffusion}}$. Introducing this into Eq. (D.7) using Eq. (D.2) we can write

$$R = \left| \frac{(J_{\text{drift}})_p}{(J_{\text{diffusion}})_p} \right| \frac{n}{p}. \tag{D.8}$$

Now for an n^+-p diode the current in the p-region near the junction is primarily electron current and so the hole current there $J_p = (J_{\text{drift}})_p + (J_{\text{diffusion}})_p \approx 0$. Hence

$$|(J_{\text{drift}})_p| / |(J_{\text{diffusion}})_p| \approx 1, \tag{D.9}$$

and since by definition low level injection means $n/p \ll 1$, Eq. (D.8) states that near the junction $R \ll 1$ and the proposition is proved. Note that far from the junction nearly all injected electrons have recombined, there is little diffusion current left, and so the drift current now dominates.

APPENDIX E

p-n DIODE SPECIFICATION SHEET

1N4001 thru 1N4007

Designers▲Data Sheet

LEAD MOUNTED
SILICON RECTIFIERS

50-1000 VOLTS
DIFFUSED JUNCTION

"SURMETIC"▲ RECTIFIERS

. . . subminiature size, axial lead mounted rectifiers for general-purpose low-power applications.

Designers Data for "Worst Case" Conditions

The Designers▲ Data Sheets permit the design of most circuits entirely from the information presented. Limit curves – representing boundaries on device characteristics – are given to facilitate "worst case" design.

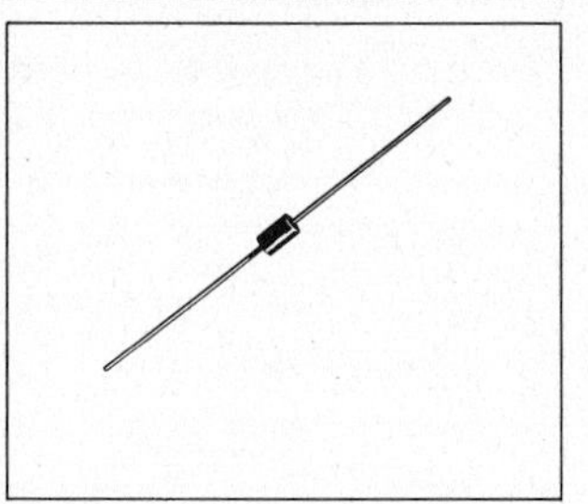

*MAXIMUM RATINGS

Rating	Symbol	1N4001	1N4002	1N4003	1N4004	1N4005	1N4006	1N4007	Unit
Peak Repetitive Reverse Voltage Working Peak Reverse Voltage DC Blocking Voltage	V_{RRM} V_{RWM} V_R	50	100	200	400	600	800	1000	Volts
Non-Repetitive Peak Reverse Voltage (halfwave, single phase, 60 Hz)	V_{RSM}	60	120	240	480	720	1000	1200	Volts
RMS Reverse Voltage	$V_{R(RMS)}$	35	70	140	280	420	560	700	Volts
Average Rectified Forward Current (single phase, resistive load, 60 Hz, see Figure 8, $T_A = 75^oC$)	I_O	1.0							Amp
Non-Repetitive Peak Surge Current (surge applied at rated load conditions, see Figure 2)	I_{FSM}	30 (for 1 cycle)							Amp
Operating and Storage Junction Temperature Range	T_J, T_{stg}	-65 to +175							^oC

*ELECTRICAL CHARACTERISTICS

Characteristic and Conditions	Symbol	Typ	Max	Unit
Maximum Instantaneous Forward Voltage Drop (i_F = 1.0 Amp, $T_J = 25^oC$) Figure 1	v_F	0.93	1.1	Volts
Maximum Full-Cycle Average Forward Voltage Drop (I_O = 1.0 Amp, $T_L = 75^oC$, 1 inch leads)	$V_{F(AV)}$	–	0.8	Volts
Maximum Reverse Current (rated dc voltage) $T_J = 25^oC$ $T_J = 100^oC$	I_R	0.05 1.0	10 50	μA
Maximum Full-Cycle Average Reverse Current (I_O = 1.0 Amp, $T_L = 75^oC$, 1 inch leads	$I_{R(AV)}$	–	30	μA

*Indicates JEDEC Registered Data.

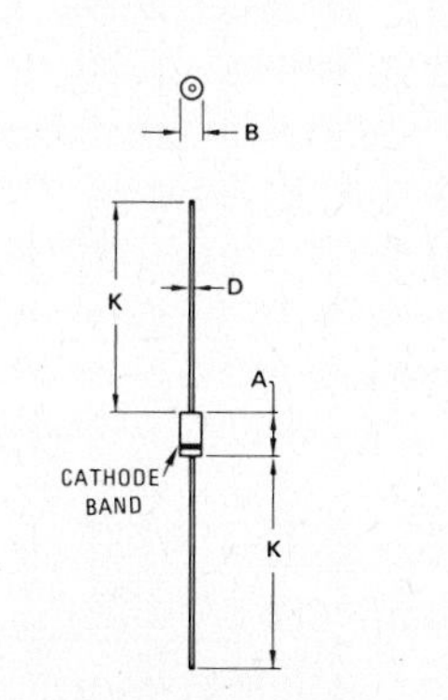

DIM	MILLIMETERS MIN	MILLIMETERS MAX	INCHES MIN	INCHES MAX
A	5.97	6.60	0.235	0.260
B	2.79	3.05	0.110	0.120
D	0.76	0.86	0.030	0.034
K	27.94	–	1.100	–

CASE 59-04
Does Not Conform to DO-41 Outline.

MECHANICAL CHARACTERISTICS

CASE: Void free, Transfer Molded
MAXIMUM LEAD TEMPERATURE FOR SOLDERING PURPOSES: 350^oC, 3/8" from case for 10 seconds at 5 lbs. tension
FINISH: All external surfaces are corrosion-resistant, leads are readily solderable
POLARITY: Cathode indicated by color band
WEIGHT: 0.40 Grams (approximately)

▲Trademark of Motorola Inc.

DS 6015 R3

APPENDIX F1

n-p-n BIPOLAR TRANSISTOR SPECIFICATION SHEETS

2N2218,A 2N2219,A
2N2221,A 2N2222,A
2N5581 2N5582

NPN SILICON ANNULAR♦ HERMETIC TRANSISTORS

. . . widely used "Industry Standard" transistors for applications as medium-speed switches and as amplifiers from audio to VHF frequencies.

- DC Current Gain Specified – 1.0 to 500 mAdc
- Low Collector-Emitter Saturation Voltage –
  $V_{CE(sat)}$ @ I_C = 500 mAdc
  = 1.6 Vdc (Max) – Non-A Suffix
  = 1.0 Vdc (Max) – A-Suffix
- High Current-Gain–Bandwidth Product –
  f_T = 250 MHz (Min) @ I_C = 20 mAdc – All Types Except
  = 300 MHz (Min) @ I_C = 20 mAdc – 2N2219A, 2N2222A, 2N5582
- Complements to PNP 2N2904,A thru 2N2907,A
- JAN/JANTX Available for 2N2218,A Series
- 2N2218 and 2N2219 available in TO-39 Package With 1/2" Leads (1)

NPN SILICON SWITCHING AND AMPLIFIER TRANSISTORS

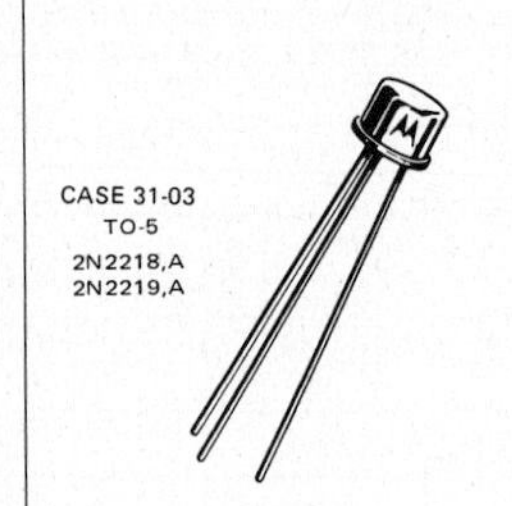

CASE 31-03
TO-5
2N2218,A
2N2219,A

CASE 22-03
TO-18
2N2221,A
2N2222,A

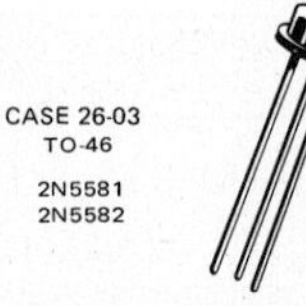

CASE 26-03
TO-46
2N5581
2N5582

SELECTION GUIDE

Device Type	Characteristic			Package
	BV_{CEO} I_C = 10 mAdc Volts	h_{FE} I_C = 150 mAdc Min/Max	I_C = 500 mAdc Min	
2N2218 2N2219	30	40/120 100/300	20 30	TO-5
2N2221 2N2222	30	40/120 100/300	20 30	TO-18
2N5581 2N5582	40	40/120 100/300	25 40	TO-46
2N2218A 2N2219A	40	40/120 100/300	25 40	TO-5
2N2221A 2N2222A	40	40/120 100/300	25 40	TO-18

*MAXIMUM RATINGS

Rating	Symbol	2N2218 2N2219 2N2221 2N2222	2N2218A 2N2219A 2N2221A 2N2222A	2N5581 2N5582	Unit
Collector-Emitter Voltage	V_{CEO}	30	40	40	Vdc
Collector-Base Voltage	V_{CB}	60	75	75	Vdc
Emitter-Base Voltage	V_{EB}	5.0	6.0	6.0	Vdc
Collector Current – Continuous	I_C	800	800	800**	mAdc
		2N2218,A 2N2219,A	2N2221,A 2N2222,A	2N5581 2N5582	
Total Device Dissipation @ T_A = 25°C Derate above 25°C	P_D	0.8 5.33	0.5 3.33	0.5 3.33	Watt mW/°C
Total Device Dissipation @ T_C = 25°C Derate above 25°C	P_D	3.0 20	1.8 12	2.0 11.43	Watts mW/°C
Operating and Storage Junction Temperature Range	T_J, T_{stg}	← -65 to +200 →			°C

*Indicates JEDEC Registered Data.
**Motorola Guarantees this Data in Addition to JEDEC Registered Data.

♦Annular Semiconductors Patented by Motorola Inc.
(1) To order 1/2" leads, TO-39 Package, specify 2N2218(s) and 2N2219(s).

DS 5120 R1

2N2218,A • 2N2219,A • 2N2221,A
2N2222,A • 2N5581 • 2N5582

*ELECTRICAL CHARACTERISTICS (T_A = 25°C unless otherwise noted)

Characteristic		Symbol	Min	Max	Unit
OFF CHARACTERISTICS					
Collector-Emitter Breakdown Voltage		BV_{CEO}			Vdc
(I_C = 10 mAdc, I_B = 0)	Non-A Suffix		30	–	
	A-Suffix, 2N5581,2N5582		40	–	
Collector-Base Breakdown Voltage		BV_{CBO}			Vdc
(I_C = 10 μAdc, I_E = 0)	Non-A Suffix		60	–	
	A-Suffix, 2N5581,2N5582		75	–	
Emitter-Base Breakdown Voltage		BV_{EBO}			Vdc
(I_E = 10 μAdc, I_C = 0)	Non-A Suffix		5.0	–	
	A-Suffix, 2N5581,2N5582		6.0	–	
Collector Cutoff Current		I_{CEX}			nAdc
(V_{CE} = 60 Vdc, $V_{EB(off)}$ = 3.0 Vdc)	A-Suffix, 2N5581,2N5582		–	10	
Collector Cutoff Current		I_{CBO}			μAdc
(V_{CB} = 50 Vdc, I_E = 0)	Non-A Suffix		–	0.01	
(V_{CB} = 60 Vdc, I_E = 0)	A-Suffix, 2N5581,2N5582		–	0.01	
(V_{CB} = 50 Vdc, I_E = 0, T_A = 150°C)	Non-A Suffix		–	10	
(V_{CB} = 60 Vdc, I_E = 0, T_A = 150°C)	A-Suffix, 2N5581,2N5582		–	10	
Emitter Cutoff Current		I_{EBO}			nAdc
(V_{EB} = 3.0 Vdc, I_C = 0)	A-Suffix, 2N5581,2N5582		–	10	
Base Cutoff Current		I_{BL}			nAdc
(V_{CE} = 60 Vdc, $V_{EB(off)}$ = 3.0 Vdc)	A-Suffix		–	20	
ON CHARACTERISTICS					
DC Current Gain		h_{FE}			–
(I_C = 0.1 mAdc, V_{CE} = 10 Vdc)	2N2218,A,2N2221,A,2N5581(1)		20	–	
	2N2219,A,2N2222,A,2N5582(1)		35	–	
(I_C = 1.0 mAdc, V_{CE} = 10 Vdc)	2N2218,A,2N2221,A,2N5581		25	–	
	2N2219,A,2N2222,A,2N5582		50	–	
(I_C = 10 mAdc, V_{CE} = 10 Vdc)	2N2218,A,2N2221,A,2N5581(1)		35	–	
	2N2219,A,2N2222,A,2N5582(1)		75	–	
(I_C = 10 mAdc, V_{CE} = 10 Vdc, T_A = -55°C)	2N2218A,2N2221A,2N5581		15	–	
	2N2219A,2N2222A,2N5582		35	–	
(I_C = 150 mAdc, V_{CE} = 10 Vdc)(1)	2N2218,A,2N2221,A,2N5581		40	120	
	2N2219,A,2N2222,A,2N5582		100	300	
(I_C = 150 mAdc, V_{CE} = 1.0 Vdc)(1)	2N2218A,2N2221A,2N5581		20	–	
	2N2219A,2N2222A,2N5582		50	–	
(I_C = 500 mAdc, V_{CE} = 10 Vdc)(1)	2N2218,2N2221		20	–	
	2N2219,2N2222		30	–	
	2N2218A,2N2221A,2N5581		25	–	
	2N2219A,2N2222A,2N5582		40	–	
Collector-Emitter Saturation Voltage(1)		$V_{CE(sat)}$			Vdc
(I_C = 150 mAdc, I_B = 15 mAdc)	Non-A Suffix		–	0.4	
	A-Suffix, 2N5581,2N5582		–	0.3	
(I_C = 500 mAdc, I_B = 50 mAdc)	Non-A Suffix		–	1.6	
	A-Suffix, 2N5581,2N5582		–	1.0	
Base-Emitter Saturation Voltage(1)		$V_{BE(sat)}$			Vdc
(I_C = 150 mAdc, I_B = 15 mAdc)	Non-A Suffix		0.6	2.0	
	A-Suffix, 2N5581,2N5582		0.6	1.2	
(I_C = 500 mAdc, I_B = 50 mAdc)	Non-A Suffix		–	2.6	
	A-Suffix, 2N5581,2N5582		–	2.0	

***MOTOROLA** Semiconductor Products Inc.*

*ELECTRICAL CHARACTERISTICS (Continued)

Characteristic		Symbol	Min	Max	Unit
SMALL-SIGNAL CHARACTERISTICS					
Current-Gain—Bandwidth Product(2) (I_C = 20 mAdc, V_{CE} = 20 Vdc, f = 100 MHz)	All Types, Except	f_T	250	–	MHz
	2N2219A,2N2222A,2N5582		300	–	
Output Capacitance(3) (V_{CB} = 10 Vdc, I_E = 0, f = 100 kHz)		C_{ob}	–	8.0	pF
Input Capacitance(3) (V_{EB} = 0.5 Vdc, I_C = 0, f = 100 kHz)	Non-A Suffix	C_{ib}	–	30	pF
	A-Suffix, 2N5581,2N5582		–	25	
Input Impedance (I_C = 1.0 mAdc, V_{CE} = 10 Vdc, f = 1.0 kHz)	2N2218A,2N2221A,2N5581	h_{ie}	1.0	3.5	k ohms
	2N2219A,2N2222A,2N5582		2.0	8.0	
(I_C = 10 mAdc, V_{CE} = 10 Vdc, f = 1.0 kHz)	2N2218A,2N2221A,2N5581		0.2	1.0	
	2N2219A,2N2222A,2N5582		0.25	1.25	
Voltage Feedback Ratio (I_C = 1.0 mAdc, V_{CE} = 10 Vdc, f = 1.0 kHz)	2N2218A,2N2221A,2N5581	h_{re}	–	5.0	X 10^{-4}
	2N2219A,2N2222A,2N5582		–	8.0	
(I_C = 10 mAdc, V_{CE} = 10 Vdc, f = 1.0 kHz)	2N2218A,2N2221A,2N5581		–	2.5	
	2N2219A,2N2222A,2N5582		–	4.0	
Small-Signal Current Gain (I_C = 1.0 mAdc, V_{CE} = 10 Vdc, f = 1.0 kHz)	2N2218A,2N2221A,2N5581	h_{fe}	30	150	–
	2N2219A,2N2222A,2N5582		50	300	
(I_C = 10 mAdc, V_{CE} = 10 Vdc, f = 1.0 kHz)	2N2218A,2N2221A,2N5581		50	300	
	2N2219A,2N2222A,2N5582		75	375	
Output Admittance (I_C = 1.0 mAdc, V_{CE} = 10 Vdc, f = 1.0 kHz)	2N2218A,2N2221A,2N5581	h_{oe}	3.0	15	μmhos
	2N2219A,2N2222A,2N5582		5.0	35	
(I_C = 10 mAdc, V_{CE} = 10 Vdc, f = 1.0 kHz)	2N2218A,2N2221A,2N5581		10	100	
	2N2219A,2N2222A,2N5582		25	200	
Collector-Base Time Constant (I_E = 20 mAdc, V_{CB} = 20 Vdc, f = 31.8 MHz)	A-Suffix, 2N5581,2N5582	$r_b'C_c$	–	150	ps
Noise Figure (I_C = 100 μAdc, V_{CE} = 10 Vdc, R_S = 1.0 k ohm, f = 1.0 kHz)	2N2219A,2N2222A	NF	–	4.0	dB
SWITCHING CHARACTERISTICS (A-Suffix, 2N5581 and 2N5582)					
Delay Time	(V_{CC} = 30 Vdc, $V_{BE(off)}$ = 0.5 Vdc, I_C = 150 mAdc, I_{B1} = 15 mAdc) (Figure 14)	t_d	–	10	ns
Rise Time		t_r	–	25	ns
Storage Time	(V_{CC} = 30 Vdc, I_C = 150 mAdc, I_{B1} = I_{B2} = 15 mAdc) (Figure 15)	t_s	–	225	ns
Fall Time		t_f	–	60	ns
Active Region Time Constant** (I_C = 150 mAdc, V_{CE} = 30 Vdc)		T_A	–	2.5	ns

*Indicates JEDEC Registered Data.
**Motorola Guarantees this Data in Addition to JEDEC Registered Data.
(1)Pulse Test: Pulse Width ≤ 300 μs, Duty Cycle ≤ 2.0%.
(2)f_T is defined as the frequency at which $|h_{fe}|$ extrapolates to unity.
(3)2N5581 and 2N5582 are Listed C_{cb} and C_{eb} for these conditions and values.

MOTOROLA *Semiconductor Products Inc.*

APPENDIX F2

p-n-p BIPOLAR TRANSISTOR SPECIFICATION SHEETS

2N2944
2N2945
2N2946

PNP SILICON ANNULAR♦ TRANSISTORS

... designed for low-level, high-speed chopper applications.

- Low Offset Voltage – $V_{EC(ofs)}$ = 0.18 mVdc (Typ) @ I_B = 200 μAdc
- High Emitter-Base Voltage – V_{EB} = 40 Vdc (2N2946)
- Low Dynamic "On" Series Resistance – $r_{ec(on)}$ = 4.0 Ohms (Typ) @ I_B = 1.0 mAdc (2N2944)

PNP SILICON CHOPPER TRANSISTORS

MAXIMUM RATINGS

Rating	Symbol	2N2944	2N2945	2N2946	Unit
*Emitter-Collector Voltage	V_{CEO}	10	20	35	Vdc
*Collector-Base Voltage	V_{CB}	15	25	40	Vdc
*Emitter-Base Voltage	V_{EB}	15	25	40	Vdc
*Collector Current	I_C		100		mAdc
*Total Power Dissipation @ $T_A = 25^oC$	P_D		400		mW
Derate above 25°C			2.3		mW/°C
*Total Power Dissipation @ $T_C = 25^oC$	P_D		2.0		Watts
Derate above 25°C			11.43		mW/°C
*Operating and Storage Junction Temperature Range	T_J, T_{stg}		-65 to +200		°C

THERMAL CHARACTERISTICS

Characteristic	Symbol	Max	Unit
*Thermal Resistance, Junction to Ambient	$R_{\theta JA}$	435	°C/W
Thermal Resistance, Junction to Case	$R_{\theta JC}$	87.5	°C/W

*Indicates JEDEC Registered Data.
♦Annular Semiconductors Patented by Motorola Inc.

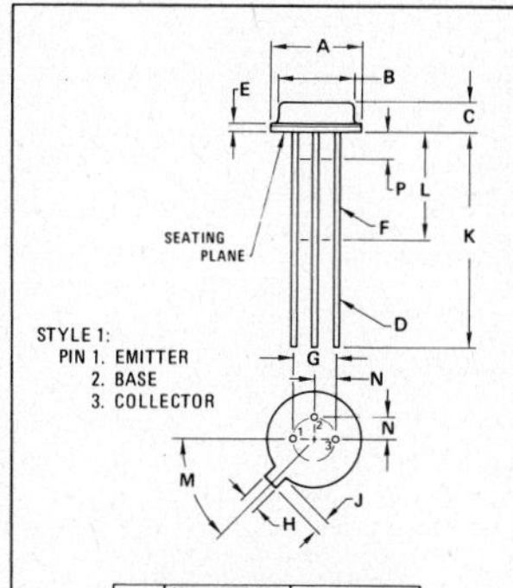

DIM	MILLIMETERS MIN	MILLIMETERS MAX	INCHES MIN	INCHES MAX
A	5.31	5.84	0.209	0.230
B	4.52	4.95	0.178	0.195
C	1.65	2.16	0.065	0.085
D	0.406	0.533	0.016	0.021
E	–	1.02	–	0.040
F	0.305	0.483	0.012	0.019
G	2.54 BSC		0.100 BSC	
H	0.914	1.17	0.036	0.046
J	0.711	1.22	0.028	0.048
K	12.70	–	0.500	–
L	6.35	–	0.250	–
M	45° BSC		45° BSC	
N	1.27 BSC		0.050 BSC	
P	–	1.27	–	0.050

All JEDEC dimensions and notes apply

CASE 26-03
TO-46

DS 5286 R1

2N2944 • 2N2945 • 2N2946

ELECTRICAL CHARACTERISTICS (T_A = 25°C unless otherwise noted.)

Characteristic		Symbol	Min	Typ	Max	Unit
***OFF CHARACTERISTICS**						
Collector Cutoff Current		I_{CBO}				nAdc
(V_{CB} = 15 Vdc, I_E = 0)	2N2944		–	–	0.1	
(V_{CB} = 25 Vdc, I_E = 0)	2N2945		–	–	0.2	
(V_{CB} = 40 Vdc, I_E = 0)	2N2946		–	–	0.5	
Emitter Cutoff Current		I_{EBO}				nAdc
(V_{EB} = 15 Vdc, I_C = 0)	2N2944		–	–	0.1	
(V_{EB} = 25 Vdc, I_C = 0)	2N2945		–	–	0.2	
(V_{EB} = 40 Vdc, I_C = 0)	2N2946		–	–	0.5	
ON CHARACTERISTICS						
*DC Current Gain		h_{FE}				–
(I_C = 1.0 mAdc, V_{CE} = 0.5 Vdc)	2N2944		80	180	–	
	2N2945		40	160	–	
	2N2946		30	130	–	
DC Current Gain (inverted connection)		$h_{FE(inv)}$				–
(I_E = 200 μAdc, V_{EC} = 0.5 Vdc)	2N2944		6.0	20	–	
	2N2945		4.0	17	–	
	2N2946		3.0	15	–	
Emitter-Collector Offset Voltage		$V_{EC(ofs)}$				mVdc
(I_B = 200 μAdc, I_E = 0)	2N2944		–	0.18	0.3	
	2N2945		–	0.23	0.5	
	2N2946		–	0.27	0.8	
(I_B = 1.0 mAdc, I_E = 0)*	2N2944		–	0.4	0.6	
	2N2945		–	0.5	1.0	
	2N2946		–	0.6	2.0	
(I_B = 2.0 mAdc, I_E = 0)	2N2944		–	0.8	1.0	
	2N2945		–	0.9	1.6	
	2N2946		–	1.0	2.5	
***DYNAMIC CHARACTERISTICS**						
Current-Gain—Bandwidth Product		f_T				MHz
(I_C = 1.0 mAdc, V_{CE} = 6.0 Vdc, f = 1.0 MHz)	2N2944		10	15	–	
	2N2945		5.0	13	–	
	2N2946		3.0	12	–	
Output Capacitance (V_{CB} = 6.0 Vdc, I_E = 0, f = 500 kHz)		C_{ob}	–	3.2	10	pF
Input Capacitance (V_{EB} = 6.0 Vdc, I_C = 0, f = 500 kHz)		C_{ib}	–	1.9	6.0	pF
Dynamic "On" Series Resistance		$r_{ec(on)}$				Ohms
(I_B = 1.0 mAdc, I_E = 0, I_c = 100 μArms, f = 1.0 kHz)	2N2944		–	4.0	20	
	2N2945		–	4.5	35	
	2N2946		–	5.0	45	

*Indicates JEDEC Registered Data.

MOTOROLA Semiconductor Products Inc.

BOX 20912 • PHOENIX, ARIZONA 85036 • A SUBSIDIARY OF MOTOROLA INC.

3984-1 PRINTED IN USA 2-74 IMPERIAL LITHO 842053 10M DS-5286-R1

APPENDIX G1

n-CHANNEL MOSFET SPECIFICATION SHEETS

3N128

SILICON N-CHANNEL MOS FIELD-EFFECT TRANSISTOR

. . . designed for VHF amplifier and oscillator applications in communications equipment.

- High Forward Transadmittance – $|y_{fs}|$ = 5000 μmhos (Min) @ f = 1.0 kHz
- Low Input Capacitance – C_{iss} = 7.0 pF (Max) @ f = 1.0 MHz
- Low Noise Figure – NF = 5.0 dB (Max) @ f = 200 MHz
- High Power Gain – P_G = 13.5 dB (Min) @ f = 200 MHz
- Complete "y" Parameter Curves
- Third Order Intermodulation Distortion Performance Curve Provided

N-CHANNEL MOS FIELD-EFFECT TRANSISTOR

*MAXIMUM RATINGS

Rating	Symbol	Value	Unit
Drain-Source Voltage	V_{DS}	+20	Vdc
Drain-Gate Voltage	V_{DG}	+20	Vdc
Gate-Source Voltage	V_{GS}	±10	Vdc
Drain Current	I_D	50	mAdc
Power Dissipation @ T_A = 25°C Derate above 25°C	P_D	330 2.2	mW mW/°C
Operating and Storage Junction Temperature Range	T_J, T_{stg}	–65 to +175	°C

*Indicates JEDEC Registered Data.

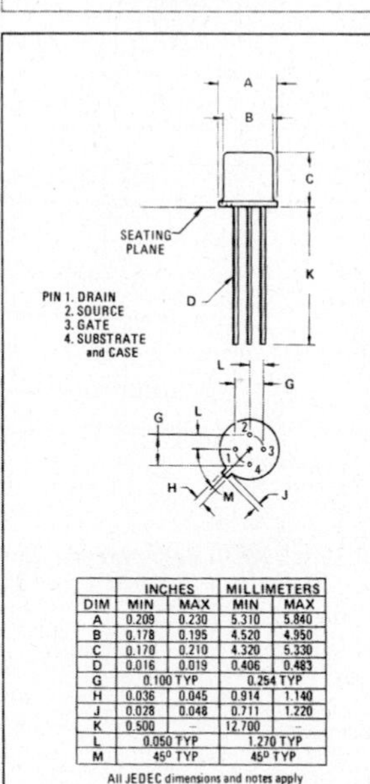

	INCHES		MILLIMETERS	
DIM	MIN	MAX	MIN	MAX
A	0.209	0.230	5.310	5.840
B	0.178	0.195	4.520	4.950
C	0.170	0.210	4.320	5.330
D	0.016	0.019	0.406	0.483
G	0.100 TYP		0.254 TYP	
H	0.036	0.045	0.914	1.140
J	0.028	0.048	0.711	1.220
K	0.500	–	12.700	–
L	0.050 TYP		1.270 TYP	
M	45° TYP		45° TYP	

All JEDEC dimensions and notes apply

CASE 20(7)
TO-72

HANDLING PRECAUTIONS

MOS field-effect transistors have extremely high input resistance. They can be damaged by the accumulation of excess static charge. Avoid possible damage to the devices while handling, testing, or in actual operation, by following the procedures outlined below:

1. To avoid the build-up of static charge, the leads of the devices should remain shorted together with a metal ring except when being tested or used.
2. Avoid unnecessary handling. Pick up devices by the case instead of the leads.
3. Do not insert or remove devices from circuits with the power on because transient voltages may cause permanent damage to the devices.

DS 5506

3N128

***ELECTRICAL CHARACTERISTICS** (T_A = 25°C unless otherwise noted)

Characteristic	Symbol	Min	Max	Unit
OFF CHARACTERISTICS				
Gate-Source Breakdown Voltage (I_G = −10 μAdc, V_{DS} = 0)	$V_{(BR)GSS}$	−50	–	Vdc
Gate-Source Cutoff Voltage (V_{DS} = 15 Vdc, I_D = 50 μAdc)	$V_{GS(off)}$	−0.5	−8.0	Vdc
Gate Reverse Current (V_{GS} = −8.0 Vdc, V_{DS} = 0) (V_{GS} = −8.0 Vdc, V_{DS} = 0, T_A = 125°C)	I_{GSS}	– –	0.05 5.0	nAdc
ON CHARACTERISTICS				
Zero-Gate-Voltage Drain Current (1) (V_{DS} = 15 Vdc, V_{GS} = 0)	I_{DSS}	5.0	25	mAdc
SMALL-SIGNAL CHARACTERISTICS				
Forward Transadmittance (V_{DS} = 15 Vdc, I_D = 5.0 mAdc, f = 1.0 kHz)	$\lvert y_{fs}\rvert$	5000	12,000	μmhos
Forward Transconductance (V_{DS} = 15 Vdc, I_D = 5.0 mAdc, f = 200 MHz)	$Re(y_{fs})$	5000	–	μmhos
Output Conductance (V_{DS} = 15 Vdc, I_D = 5.0 mAdc, f = 200 MHz)	$Re(y_{os})$	–	500	μmhos
Input Conductance (V_{DS} = 15 Vdc, I_D = 5.0 mAdc, f = 200 MHz)	$Re(y_{is})$	–	800	μmhos
Input Capacitance (V_{DS} = 15 Vdc, I_D = 5.0 mAdc, f = 1.0 MHz)	C_{iss}	–	7.0	pF
Reverse Transfer Capacitance (V_{DS} = 15 Vdc, I_D = 5.0 mAdc, f = 1.0 MHz)	C_{rss}	–	0.28	pF
Noise Figure (V_{DS} = 15 Vdc, I_D = 5.0 mAdc, f = 200 MHz)	NF	–	5.0	dB
Power Gain (V_{DS} = 15 Vdc, I_D = 5.0 mAdc, f = 200 MHz)	P_G	13.5	–	dB

*Indicates JEDEC Registered Data.
(1) Pulse Test: Pulse Width = 300 μs, Duty Cycle = 2.0%.

TYPICAL CHARACTERISTICS
(T_A = 25°C)

FIGURE 1 – DRAIN CHARACTERISTICS

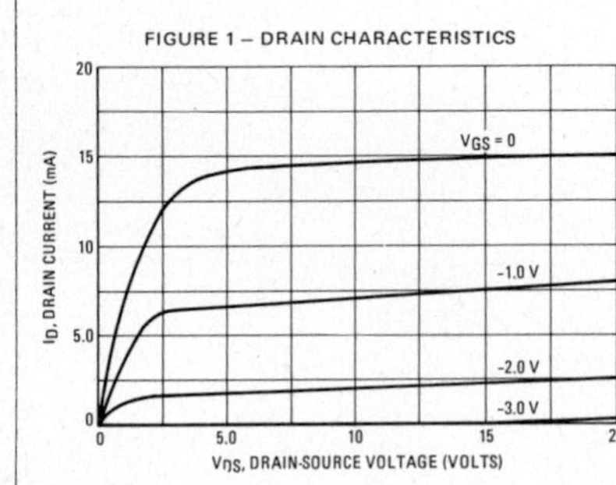

FIGURE 2 – TRANSFER CHARACTERISTICS

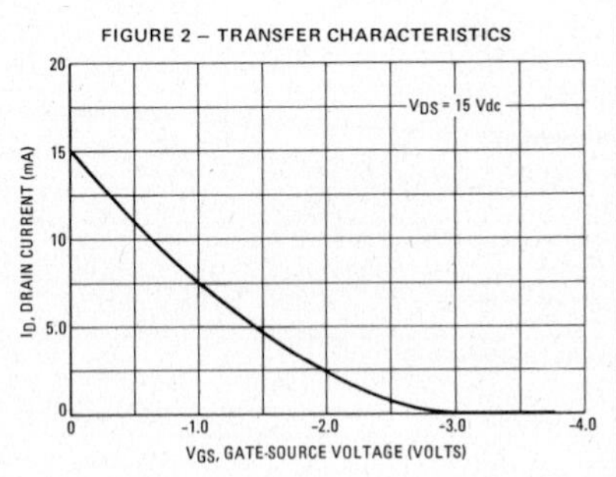

APPENDIX G2

p-CHANNEL MOSFET SPECIFICATION SHEETS

PRELIMINARY SPECIFICATION

MFE3003

P-CHANNEL

MOS FIELD-EFFECT TRANSISTOR

(Type C)

AUGUST 1968 – PS 69 R3
(Replaces PS 69 R2)

P-CHANNEL SILICON NITRIDE PASSIVATED MOS FIELD-EFFECT TRANSISTOR

Enhancement Mode (Type C) MOS Field-Effect transistor designed for chopper applications.

- Low Drain-Source Resistance – $r_{ds(on)}$ less than 200 ohms
- Low Reverse Transfer Capacitance – C_{rss} less than 1.0 pF
- High DC Input Resistance – I_{GSS} less than 100 pAdc

MAXIMUM RATINGS

Rating	Symbol	Value	Unit
Drain-Source Voltage	V_{DS}	-15	Vdc
Drain-Gate Voltage	V_{DG}	±20	Vdc
Gate-Source Voltage	V_{GS}	±20	Vdc
Drain Current	I_D	30	mAdc
Total Device Dissipation @ T_C = 25°C Derate above 25°C	P_D	200 1.33	mW mW/°C
Operating & Storage Junction Temperature Range	T_J, T_{stg}	-65 to +175	°C

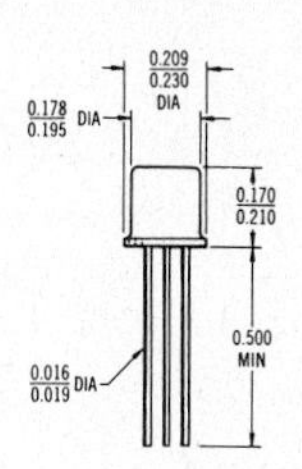

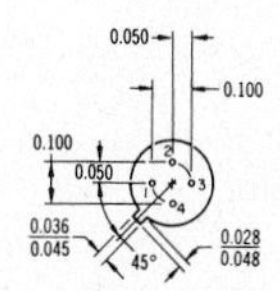

Pin
1. Drain
2. Source
3. Gate
4. Case and Substrate

TO-72
CASE 20 (7)

HANDLING PRECAUTIONS:

MOS field-effect transistors have extremely high input resistance. They can be damaged by the accumulation of excess static charge. Avoid possible damage to the devices while handling, testing, or in actual operation, by following the procedures outlined below:

1. To avoid the build-up of static charge, the leads of the devices should remain shorted together with a metal ring except when being tested or used.
2. Avoid unnecessary handling. Pick up devices by the case instead of the leads.
3. Do not insert or remove devices from circuits with the power on because transient voltages may cause permanent damage to the devices.

ELECTRICAL CHARACTERISTICS (T_A = 25°C unless otherwise noted)

Characteristic	Symbol	Min	Max	Unit
OFF CHARACTERISTICS				
Drain-Source Breakdown Voltage (V_{GS} = 0, I_D = -10 μAdc)	$V_{(BR)DSS}$	-15	-	Vdc
Gate Leakage Current (V_{GS} = ±10 Vdc, V_{DS} = 0)	I_{GSS}	-	±100	pAdc
ON CHARACTERISTICS				
Zero-Gate Voltage Drain Current (V_{DS} = -10 Vdc, V_{GS} = 0)	I_{DSS}	-	-10	nAdc
(V_{DS} = -10 Vdc, V_{GS} = 0, T_C = 125°C)		-	-100	
Gate-Source Threshold Voltage (V_{DS} = -10 Vdc, I_D = -10 μAdc)	$V_{GS(TH)}$	-	-4.0	Vdc
SMALL-SIGNAL CHARACTERISTICS				
Drain-Source Resistance (V_{GS} = -10 Vdc, I_D = 0, f = 1.0 kHz)	$r_{ds(on)}$	-	200	Ohms
Drain-Substrate Capacitance ($V_{D(SUB)}$ = -10 Vdc, f = 1.0 MHz)	$C_{d(sub)}$	-	4.0	pF
Input Capacitance (V_{DS} = -10 Vdc, V_{GS} = 0, f = 1.0 MHz)	C_{iss}	-	5.0	pF
Reverse Transfer Capacitance (V_{DS} = 0, V_{GS} = 0, f = 1.0 MHz)	C_{rss}	-	1.0	pF

2666-6 PRINTED IN USA 11-68 IMPERIAL LITHO B8630 5M
CODE 6.1.4-4 PS 69 R3

Index

Note: Page numbers in italic indicate tabular or graphical data.